AN INTRODUCTION TO INVARIANTS AND MODULI

Incorporated in this volume are the first two books in Mukai's series on moduli theory. The notion of a moduli space is central to geometry. However, its influence is not confined there; for example, the theory of moduli spaces is a crucial ingredient in the proof of Fermat's last theorem. Researchers and graduate students working in areas ranging from Donaldson or Seiberg-Witten invariants to more concrete problems such as vector bundles on curves will find this to be a valuable resource. Among other things, this volume includes an improved presentation of the classical foundations of invariant theory that, in addition to geometers, will be useful to those studying representation theory. This translation gives an accurate account of Mukai's influential Japanese texts.

Cambridge Studies in Advanced Mathematics

Editorial Board:

B. Bollobas, W. Fulton, A. Katok, F. Kirwin, P. Sarnak

AN INTRODUCTION TO INVARIANTS AND MODULI

SHIGERU MUKAI

Translated by W. M. Oxbury

CAMBRIDGE UNIVERSITY PRESS
Cambridge, New York, Melbourne, Madrid, Cape Town,
Singapore, São Paulo, Delhi, Mexico City

Cambridge University Press
The Edinburgh Building, Cambridge CB2 8RU, UK

Published in the United States of America by Cambridge University Press, New York

www.cambridge.org
Information on this title: www.cambridge.org/9781107406360

Originally published in Japanese by Iwanami Shoten, Publishers, Tokyo, 1998, 2000.
First published by Cambridge University Press 2003
Reprinted 2006
First paperback edition 2012

A catalogue record for this publication is available from the British Library

Library of Congress Cataloguing in Publication Data
Mukai, Shigeru, 1953–
[Mojurai riron. English]
An introduction to invariants and moduli / S. Mukai; translated by W.M. Oxbury.
p.   cm. – (Cambridge studies in advanced mathematics)
Includes bibliographical references and index.
ISBN 0-521-80906-1
1. Invariants.   2. Moduli theory.   I. Title.   II. Series.
QA244 .M8413   2003
512.5–dc21   2002023422

ISBN 978-1-107-40636-0 Hardback
ISBN 978-0-521-80906-1 Paperback

Contents

Preface

The aim of this book is to provide a concise introduction to algebraic geometry and to algebraic moduli theory. In so doing, I have tried to explain some of the fundamental contributions of Cayley, Hilbert, Nagata, Grothendieck and Mumford, as well as some important recent developments in moduli theory, keeping the proofs as elementary as possible. For this purpose we work throughout in the category of algebraic varieties and elementary sheaves (which are simply order-reversing maps) instead of schemes and sheaves (which are functors). Instead of taking GIT (Geometric Invariant Theory) quotients of projective varieties by $PGL(N)$, we take, by way of a shortcut, Proj quotients of affine algebraic varieties by the general linear group $GL(N)$. In constructing the moduli of vector bundles on an algebraic curve, Grothendieck's Quot scheme is replaced by a certain explicit affine variety consisting of matrices with polynomial entries. In this book we do not treat the very important analytic viewpoint represented by the Kodaira-Spencer and Hodge theories, although it is treated, for example, in Ueno [113], which was in fact a companion volume to this book when published in Japanese.

The plan of the first half of this book (Chapters 1–5 and 7) originated from notes taken by T. Hayakawa in a graduate lecture course given by the author in Nagoya University in 1985, which in turn were based on the works of Hilbert [20] and Mumford et al. [30]. Some additions and modifications have been made to those lectures, as follows.

(1) I have included chapters on ring theory and algebraic varieties accessible also to undergraduate students. A strong motivation for doing this, in fact, was the desire to collect in one place the early series of fundamental results of Hilbert that includes the Basis Theorem and the Nullstellensatz.

(2) For the proof of linear reductivity (or complete reductivity), Cayley's Ω-process used by Hilbert is quite concrete and requires little background

knowledge. However, in view of the importance of algebraic group representations I have used instead a proof using Casimir operators. The key to
the proof is an invariant bilinear form on the Lie space. The uniqueness
property used in the Japanese edition was replaced by the positive definite-
ness in this edition.

(3) I have included the Cayley-Sylvester formula in order to compute explicitly the Hilbert series of the classical binary invariant ring since I believe
both tradition and computation are important. I should add that this and
Section 4.5 are directly influenced by Springer [8].

Both (2) and (3) took shape in a lecture course given by the author at Warwick
University in the winter of 1998.

(4) I have included the result of Nagata [11], [12] that, even for an algebraic
group acting on a polynomial ring, the ring of invariants need not be finitely
generated.

(5) Chapter 1 contains various introductory topics adapted from lectures given
in the spring of 1998 at Nagoya and Kobe Universities.

The second half of the book was newly written in 1998–2000 with two main
purposes: first, an elementary invariant-theoretic construction of moduli spaces
including Jacobians and, second, a self-contained proof of the Verlinde formula
for $SL(2)$. For the first I make use of Gieseker matrices. Originally this idea was
invented by Gieseker [72] to measure the stability of the action of $PGL(N)$ on
the Quot scheme. But in this book moduli spaces of bundles are constructed
by taking quotients of a variety of Gieseker matrices themselves by the general linear group. This construction turns out to be useful even in the case of
Jacobians. For the Verlinde formula, I have chosen Zagier's proof [115] among
three known algebraic geometric proofs. However, Thaddeus's proof [112] uses
some interesting birational geometry, and I give a very brief explanation of this
for the case of rank 2 parabolic bundles on a pointed projective line.

Acknowledgements

This book is a translation of *Moduli Theory I, II* published in 1998 and 2000 in Japanese. My warmest gratitude goes to the editorial board of Iwanami Publishers, who read my Japanese very carefully and made several useful comments, and to Bill Oxbury, who was not only the translator but also a very kind referee. He corrected many misprints and made many useful suggestions, and following these I was able to improve the presentation of the material and to make several proofs more complete.

I am also very grateful to K. Nishiyama, K. Fujiwara, K. Yanagawa and H. Nasu, from whom I also received useful comments on the Japanese edition, and to H. Saito for many discussions.

Finally, I would like to thank on this occasion the Graduate School of Mathematics (including its predecessor, the Department of Mathematics in the Faculty of Sciences) of Nagoya University, where I worked for 23 years from my first position and was able to complete the writing of this book.

Partial financial support was made by Grant-in-Aid for Scientific Research (B)(2)0854004 of the Japanese Ministry of Education and the JSPS's (A)(2)10304001.

Shigeru Mukai
Kyoto, June 2002

Introduction

(a) What is a moduli space?

A *moduli space* is a manifold, or variety, which parametrises some class of geometric objects. The j-invariant classifying elliptic curves up to isomorphism and the Jacobian variety of an algebraic curve are typical examples. In a broader sense, one could include as another classical example the classifying space of a Lie group. In modern mathematics the idea of moduli is in a state of continual evolution and has an ever-widening sphere of influence. For example:

- By defining a suitable height function on the moduli space of principally polarised abelian varieties it was possible to resolve the Shafarevich conjectures on the finiteness of abelian varieties (Faltings 1983).
- The moduli space of Mazur classes of 2-dimensional representations of an absolute Galois group is the spectrum of a Hecke algebra.

The application of these results to resolve such number-theoretic questions as Mordell's Conjecture and Fermat's Last Theorem are memorable achievements of recent years. Turning to geometry:

- Via Donaldson invariants, defined as the intersection numbers in the moduli space of instanton connections, one can show that there exist homeomorphic smooth 4-manifolds that are not diffeomorphic.

Indeed, Donaldson's work became a prototype for subsequent research in this area.

Here's an anology. When natural light passes through a prism it separates into various colours. In a similar way, one can try to elucidate the hidden properties of an algebraic variety. One can think of the moduli spaces naturally associated to the variety (the Jacobian of a complex curve, the space of instantons on a complex surface) as playing just such a role of 'nature's hidden colours'.

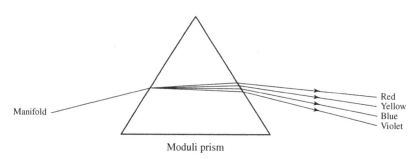

Moduli prism

   The aim of this book is to explain, with the help of some concrete examples, the basic ideas of moduli theory as they have developed alongside algebraic geometry – in fact, from long before the modern viewpoint sketched above. In particular, I want to give a succinct introduction to the widely applicable methods for constructing moduli spaces known as *geometric invariant theory*.

   If a moduli problem can be expressed in terms of algebraic geometry then in many cases it can be reduced to the problem of constructing a quotient of a suitable algebraic variety by an action of a group such as the general linear group $GL(m)$. From the viewpoint of moduli theory this variety will typically be a Hilbert scheme parametrising subschemes of a variety or a Quot scheme parametrising coherent sheaves. From a group-theoretic point of view it may be a finite-dimensional linear representation regarded as an affine variety or a subvariety of such. To decide what a solution to the quotient problem should mean, however, forces one to rethink some rather basic questions: What *is* an algebraic variety? What does it mean to take a quotient of a variety? In this sense the quotient problem, present from the birth and throughout the development of algebraic geometry, is even today sadly lacking an ideal formulation. And as one sees in the above examples, the 'moduli problem' is not determined in itself but depends on the methods and goals of the mathematical area in which it arises. In some cases elementary considerations are sufficient to address the problem, while in others much more care is required. Maybe one cannot do without a projective variety as quotient; maybe a stack or algebraic space is enough. In this book we will construct moduli spaces as projective algebraic varieties.

(b) Algebraic varieties and quotients of algebraic varieties

An algebraic curve is a rather sophisticated geometric object which, viewed on the one hand as a Riemann surface, or on the other as an algebraic function field in one variable, combines analysis and algebra. The theory of meromorphic functions and abelian differentials on compact Riemann surfaces, developed by Abel, Riemann and others in the nineteenth century, was, through the efforts of many later mathematicians, deepened and sublimated to an 'algebraic function theory'. The higher dimensional development of this theory has exerted a profound influence on the mathematics of the twentieth century. It goes by the general name of 'the study of algebraic varieties'. The data of an algebraic variety incorporate in a natural way that of real differentiable manifolds, of complex manifolds, or again of an algebraic function field in several variables. (A field K is called an algebraic function field in n variables over a base field k if it is a finitely generated extension of k of transcendence degree n.) Indeed, any algebraic variety may be defined by patching together (the spectra of) some finitely generated subrings $R_1, \ldots, R_N$ of a function field K. This will be explained in Chapter 3.

This ring-theoretic approach, from the viewpoint of varieties as given by systems of algebraic equations, is very natural; however, the moduli problem, that is, the problem of constructing quotients of varieties by group actions, becomes rather hard. When an algebraic group G acts on an affine variety, how does one construct a quotient variety? (An *algebraic group* is an algebraic variety with a group structure, just as a Lie group is a smooth manifold which has a compatible group structure.) It turns out that the usual quotient topology, and the differentiable structure on the quotient space of a Lie group by a Lie subgroup, fail to work well in this setting. Clearly they are not sufficient if they fail to capture the function field, together with its appropriate class of subrings, of the desired quotient variety. The correct candidates for *these* are surprisingly simple, namely, the subfield of G-invariants in the original function field K, and the subrings of G-invariants in the integral domains $R \subset K$ (see Chapter 5). However, in proceeding one is hindered by the following questions.

(1) Is the subring of invariants R^G of a finitely generated ring R again finitely generated?
(2) Is the subfield of invariants $K^G \subset K$ equal to the field of fractions of R^G?
(3) Is $K^G \subset K$ even an algebraic function field? – that is, is K^G finitely generated over the base field k?
(4) Even if the previous questions can be answered positively and an algebraic variety constructed accordingly, does it follow that the points of this variety can be identified with the G-orbits of the original space?

In fact one can prove property (3) quite easily; the others, however, are not true in general. We shall see in Section 2.5 that there exist counterexamples to (1) even in the case of an algebraic group acting linearly on a polynomial ring. Question (2) will be discussed in Chapter 6.

So how should one approach this subject? Our aim in this book is to give a concrete construction of some basic moduli spaces as quotients of group actions, and in fact we will restrict ourselves exclusively to the general linear group $GL(m)$. For this case property (1) does indeed hold (Chapter 4), and also property (4) if we modify the question slightly. (See the introduction to Chapter 5.) A correspondence between G-orbits and points of the quotient is achieved provided we restrict, in the original variety, to the open set of *stable points* for the group action. Both of these facts depend on a representation-theoretic property of $GL(m)$ called linear reductivity.

After paving the way in Chapter 5 with the introduction of affine quotient varieties, we 'globalise' the construction in Chapter 6. Conceptually, this may be less transparent than the affine construction, but essentially it just replaces the affine spectrum of the invariant ring with the projective spectrum (Proj) of the semiinvariant ring. This 'global' quotient, which is a projective variety, we refer to as the *Proj quotient*, rather than 'projective quotient', in order to distinguish it from other constructions of the projective quotient variety that exist in the literature.

An excellent example of a Proj quotient (and indeed of a moduli space) is the Grassmannian. In fact, the Grassmannian is seldom considered in the context of moduli theory, and we discuss it here in Chapter 8. This variety is usually built by gluing together affine spaces, but here we construct it globally as the projective spectrum of a semiinvariant ring and observe that this is equivalent to the usual construction. For the Grassmannian $\mathbb{G}(2, n)$ we compute the Hilbert series of the homogeous coordinate ring. We use this to show that it is generated by the Plücker coordinates, and that the relations among these are generated by the Plücker relations.

In general, for a given moduli problem, one can only give an honest construction of a moduli space if one is able to determine explicitly the stable points of the group action. This requirement of the theory is met in Chapter 7 with the numerical criterion for stability and semistability of Hilbert and Mumford, which we apply to some geometrical examples from Chapter 5. Later in the book we construct moduli spaces for line bundles and vector bundles on an algebraic curve, which requires the notion of stability of a vector bundle. Historically, this was discovered by Mumford as an application of the numerical criterion, but in this book we do not make use of this, as we are able to work directly with the semiinvariants of our group actions. Another important application, which we do not touch on here, is to the construction of a compactification of the moduli space of curves as a projective variety.

(c) Moduli of bundles on a curve

In Chapter 9 algebraic curves make their entry. We first explain:

(1) what is the genus of a curve?
(2) Riemann's inequality and the vanishing of cohomology (or index of speciality); and
(3) the duality theorem.

In the second half of Chapter 9 we construct, as the projective spectrum of the semiinvariant ring of a suitable group action on an affine variety, an algebraic variety whose underlying set of points is the Picard group of a given curve, and we show that over the complex numbers this is nothing other than the classical Jacobian.

In Chapter 10 we extend some essential parts of the line bundle theory of the preceding chapter to higher rank vector bundles on a curve, and we then construct the moduli space of rank 2 vector bundles. This resembles the line bundle case, but with the difference that the notion of stability arises in a natural way. The moduli space of vector bundles, in fact, can be viewed as a Grassmannian over the function field of the curve, and one can roughly paraphrase Chapter 10 by saying that a moduli space is constructed as a projective variety by explicitly defining the Plücker coordinates of a semistable vector bundle. (See also Seshadri [77].) One advantage of this construction – although it has not been possible to say much about this in this book – is the consequence that, if the curve is defined over a field k, then the same is true, a priori, of the moduli space.

In Chapter 11 the results of Chapters 9 and 10 are reconsidered, in the following sense. Algebraic varieties have been found whose sets of points can be identified with the sets of equivalence classes of line bundles, or vector bundles, on the curve. However, to conclude that 'these varieties are the moduli spaces for line bundles, or vector bundles' is not a very rigorous statement. More mathematical would be, first, to give some clean definition of 'moduli' and 'moduli space', and then to prove that the varieties we have obtained are moduli spaces in the sense of this definition. One answer to this problem is furnished by the notions of representability of a functor and of coarse moduli. These are explained in Chapter 11, and the quotient varieties previously constructed are shown to be moduli spaces in this sense. Again, this point of view becomes especially important when one is interested in the field over which the moduli space is defined. This is not a topic which it has been possible to treat in this book, although we do give one concrete example at the end of the chapter, namely, the Jacobian of an elliptic curve.

In the final chapter we give a treatment of the Verlinde formulae for rank 2 vector bundles. Originally, these arose as a general-dimension formula for objects that are somewhat unfamiliar in geometry, the spaces of conformal blocks from 2-dimensional quantum field theory. (See Ueno [113].) In our context, however, they appear as elegant and precise formulae for the Hilbert polynomials for the semiinvariant rings used to construct the moduli of vector bundles. Various proofs are known, but the one presented here (for odd degree bundles) is that of Zagier [115], making use of the formulae for the intersection numbers in the moduli space of Thaddeus [111]. On the way, we observe a curious formal similarity between the cohomology ring of the moduli space and that of the Grassmannian $\mathbb{G}(2, n)$.

Convention: Although it will often be unnecessary, we shall assume throughout the book that the field k is algebraically closed and of characteristic zero.

1

Invariants and moduli

This chapter explores some examples of parameter spaces which can be constructed by elementary means and with little previous knowledge as an introduction to the general theory developed from Chapter 3 onwards. To begin, we consider equivalence classes of plane conics under Euclidean transformations and use invariants to construct a parameter space which essentially corresponds to the eccentricity of a conic.

This example already illustrates several essential features of the construction of moduli spaces. In addition we shall look carefully at some cases of finite group actions, and in particular at the question of how to determine the ring of invariants, the fundamental tool of the theory. We prove Molien's Formula, which gives the Hilbert series for the ring of invariants when a finite group acts linearly on a polynomial ring.

In Section 1.3, as an example of an action of an algebraic group, we use classical invariants to construct a parameter space for $GL(2)$-orbits of binary quartics.

In Section 1.4 we review plane curves as examples of algebraic varieties. A plane curve without singularities is a Riemann surface, and in the particular case of a plane cubic this can be seen explicitly by means of doubly periodic complex functions. This leads to another example of a quotient by a discrete group action, in this case parametrising lattices in the complex plane. The group here is the modular group $SL(2, \mathbb{Z})$ (neither finite nor connected), and the Eisenstein series are invariants. Among them one can use two, g_2 and g_3, to decide when two lattices are isomorphic.

1.1 A parameter space for plane conics

Consider the curve of degree 2 in the (real or complex) (x, y) plane

$$ax^2 + 2bxy + cy^2 + 2dx + 2ey + f = 0. \tag{1.1}$$

1

If the left-hand side factorises as a product of linear forms, then the curve is a union of two lines; otherwise we say that it is *nondegenerate* (Figure 1.1).

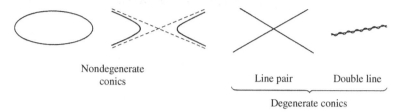

Nondegenerate
conics

Line pair   Double line

Degenerate conics

Figure 1.1

Let us consider the classification of such curves of degree 2, up to Euclidean transformations, from the point of view of their invariants. The Euclidean transformation group G contains the set of translations

$$x \mapsto x + l, \qquad y \mapsto y + m$$

as a normal subgroup and is generated by these and the rotations. Alternatively, G can be viewed as the group of matrices

$$X = \begin{pmatrix} p & q & l \\ -q & p & m \\ 0 & 0 & 1 \end{pmatrix}, \qquad p^2 + q^2 = 1. \tag{1.2}$$

Curves of degree 2 correspond to symmetric 3×3 matrices by writing the equation (1.1) as

$$(x, y, 1) \begin{pmatrix} a & b & d \\ b & c & e \\ d & e & f \end{pmatrix} \begin{pmatrix} x \\ y \\ 1 \end{pmatrix} = 0,$$

and then under the Euclidean transformation (1.2) the symmetric matrix of the curve transforms by

$$\begin{pmatrix} a & b & d \\ b & c & e \\ d & e & f \end{pmatrix} \mapsto X^{\mathrm{t}} \begin{pmatrix} a & b & d \\ b & c & e \\ d & e & f \end{pmatrix} X.$$

In other words, the 6-dimensional vector space V of symmetric 3×3 matrices is a representation of the Euclidean transformation group G (see Section 1.21.10). Now, geometry studies properties which are invariant under groups of transformations, so let us look for invariants under this group action, in the form of polynomials $F(a, b, \ldots, f)$.

The transformation matrix (1.2) has determinant 1, and so the first invariant polynomial we encounter is

$$D = \det \begin{pmatrix} a & b & d \\ b & c & e \\ d & e & f \end{pmatrix}.$$

Here $D \neq 0$ exactly when the degree 2 curve is nondegenerate, and for this reason D is called the *discriminant* of the curve. Next we observe that the trace and determinant of the 2×2 submatrix $\begin{pmatrix} a & b \\ b & c \end{pmatrix}$ are also invariant; we will denote these by $T = a + c$ and $E = ac - b^2$. Moreover, any invariant polynomial can be (uniquely) expressed as a polynomial in D, T, E. In other words, the following is true.

Proposition 1.1. *The set of polynomials on V invariant under the action of G is a subring of $\mathbb{C}[a, b, c, d, e, f]$ and is generated by D, T, E. Moreover, these elements are algebraically independent; that is, the subring is $\mathbb{C}[D, T, E]$.* □

Proof. Let $G_0 \subset G$ be the translation subgroup, with quotient $G/G_0 \cong O(2)$, the rotation group of the plane. We claim that it is enough to show that the subring of polynomials invariant under G_0 is

$$\mathbb{C}[a, b, c, d, e, f]^{G_0} = \mathbb{C}[a, b, c, D]. \tag{1.3}$$

This is because the polynomials in $\mathbb{C}[a, b, c]$ invariant under the rotation group $O(2)$ are generated by the trace T and discriminant E.

We also claim that if we consider polymonials in a, b, c, d, e, f and $1/E$, then

$$\mathbb{C}\left[a, b, c, d, e, f, \frac{1}{E}\right]^{G_0} = \mathbb{C}\left[a, b, c, D, \frac{1}{E}\right]. \tag{1.4}$$

It is clear that this implies (1.3), and so we are reduced to proving (1.4). The point here is that the determinant D can be written

$$D = Ef + (2bde - ae^2 - cd^2),$$

so that

$$f = \frac{D + ae^2 + cd^2 - 2bde}{E},$$

and hence

$$\mathbb{C}\left[a, b, c, d, e, f, \frac{1}{E}\right] = \mathbb{C}\left[a, b, c, d, e, D, \frac{1}{E}\right].$$

So a polynomial F in this ring (that is, a polynomial in a, b, c, d, e, f with coefficients which may involve powers of $1/E$) which is invariant under G_0 has to satisfy

$$F(a, b, c, d + al + bm, e + bl + cm, D) = F(a, b, c, d, e, D)$$

for arbitrary translations (l, m). Taking $(l, m) = (-bt, at)$ shows that F cannot have terms involving e, while taking $(l, m) = (-ct, bt)$ shows that it cannot have terms involving d; so we have shown (1.4).   □

Remark 1.2. One can see that Proposition 1.1 is consistent with a dimension count as follows. First, V has dimension 6. The Euclidean group G has dimension 3 (that is, Euclidean motions have 3 degrees of freedom). A general curve of degree 2 is preserved only by the finitely many elements of G (namely, 180° rotation about the centre and the trivial element), and hence we expect that 'the quotient V/G has dimension 3'. Thus we may think of the three invariants D, T, E as three 'coordinate functions on the quotient space'.   □

The space of all curves of degree 2 is $V \cong \mathbb{C}^6$, but here we are only concerned with polynomials, viewed as functions, on this space. Viewed in this sense the space is called an affine space and denoted $\mathbb{A}^6$. (See Chapter 3.) We shall denote the subset corresponding to nondegenerate curves by $U \subset V$. This is an open set defined by the condition $D \neq 0$. The set of 'regular functions' on this open set is the set of rational functions on V whose denominator is a power of D, that is,

$$\mathbb{C}\left[a, b, c, d, e, f, \frac{1}{D}\right].$$

Up to now we have been thinking not in terms of curves but rather in terms of their defining equations of degree 2. In the following we shall want to think in terms of the curves themselves. Since two equations that differ only by a scalar multiple define the same curve, we need to consider functions that are invariant under the larger group $\widetilde{G}$ generated by G and the scalar matrices $X = rI$. The scalar matrix rI multiplies the three invariants D, E, T by r^6, r^4, r^2, respectively. It follows that the set

$$\mathbb{C}\left[a, b, c, d, e, f, \frac{1}{D}\right]^{\widetilde{G}}$$

of $\widetilde{G}$-invariant polynomial functions on U is generated by

$$A = \frac{E^3}{D^2}, \quad B = \frac{T^3}{D}, \quad C = \frac{ET}{D}.$$

Among these three expressions there is a relation

$$AB - C^3 = 0,$$

so that:

a moduli space for nondegenerate curves of degree 2 in the Euclidean plane is the affine surface in $\mathbb{A}^3$ defined by the equation $xz - y^3 = 0$.

(The origin is a singular point of this surface called a rational double point of type A_2.)

One can also see this easily in the following way. By acting on the defining equation (1.1) of a nondegenerate degree 2 curve with a scalar matrix rI for a suitable $r \in \mathbb{C}$ we can assume that $D(a, b, \ldots, f) = 1$. The set of curves normalised in this way is then an affine plane with coordinates T, E. Now, the ambiguity in choosing such a normalisation is just the action of ωI, where $\omega \in \mathbb{C}$ is an imaginary cube root of unity, and so the parameter space for nondegenerate degree 2 curves is the surface obtained by dividing out the (T, E) plane by the action of the cyclic group of order 3,

$$(T, E) \mapsto (\omega T, \omega^2 E).$$

The origin is a fixed point of this action, and so it becomes a quotient singularity in the parameter space.

Next, let us look at the situation over the real numbers $\mathbb{R}$. We note that here cube roots are uniquely determined, and so by taking that of the discriminant D of equation (1.1) we see that for real curves of degree 2 we can take as coordinates the numbers

$$\alpha = \frac{E}{\sqrt[3]{D^2}}, \quad \beta = \frac{T}{\sqrt[3]{D}}.$$

In this way the curves are parametrised simply by the real (α, β) plane:

(i) Points in the (open) right-hand parabolic region $\beta^2 < 4\alpha$ and the (closed) 4th quadrant $\alpha \geq 0$, $\beta \leq 0$ do not correspond to any curves over the real numbers. (It is natural to refer to the union of these two sets as the 'imaginary region' of the (α, β) plane. See Figure 1.2.) The points of the parameter space are real, but the coefficients of the defining equation (1.1) always require imaginary complex numbers. For example, the origin $(0, 0)$ corresponds to the curve

$$\sqrt{-1}(x^2 - y^2) + 2xy = 2x.$$

(ii) Points of the parabola $\beta^2 = 4\alpha$ in the 1st quadrant correspond to circles of radius $\sqrt{2/\beta}$.

(iii) Points of the open region $\beta^2 > 4\alpha > 0$ between the parabola and the β-axis parametrise ellipses.

(iv) Points of the positive β-axis $\alpha = 0$, $\beta > 0$ parametrise parabolas.

(v) Points in the left half-plane $\alpha < 0$ parametrise hyperbolas. Within this region, points along the negative α-axis parametrise rectangular hyperbolas (the graph of the reciprocal function), while points in the 2nd and 3rd quadrants correspond respectively to acute angled and obtuse angled hyperbolas.

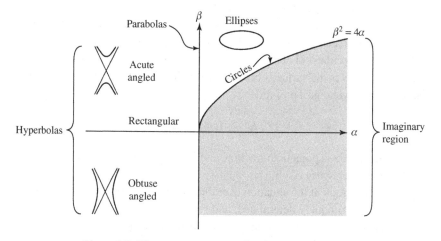

Figure 1.2: The parameter space of real curves of degree 2

Let us now follow a rotation of this figure in the positive direction about the origin.

Beginning with a circle (eccentricity $e = 0$), our curve grows into an ellipse through a parabolic phase ($e = 1$) before making a transition to a hyperbola. The angle between the asymptotes of this hyperbola is initially close to zero and gradually grows to $180°$, at which point ($e = \infty$) the curve enters the imaginary region. After passing through this region it turns once again into a circle. (This is Kepler's Principle.)

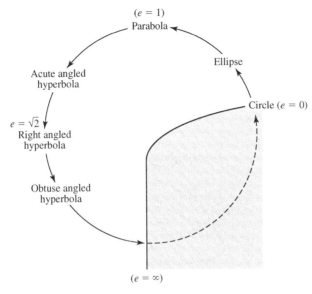

Figure 1.3: Transmigration of a conic

Remark 1.3. In the case of an ellipse, our curve has a (Euclidean invariant) area which is equal to $\pi/\sqrt{\alpha}$. In particular, this area increases as the curve approaches the β-axis, and one may think of a parabola, corresponding to a point on the axis, as having infinite area. Taking this point of view a step further, one may think of a hyperbola as having imaginary area. □

We have thus established a correspondence between real curves of degree 2 up to Euclidean transformations and points of the (α, β) plane. The group G does not have the best properties (it is not linearly reductive – this will be explained in Chapter 4), but nevertheless in this example we are lucky and every point of the (α, β) plane corresponds to some curve.

Plane curves of degree 2 are also called *conics*, as they are the curves obtained by taking plane cross sections of a circular cone (an observation which goes back to Apollonius and Pappus). From this point of view, the eccentricity e of the curve is determined by the angle of the plane (Figure 1.4).

To be precise, let ϕ be the angle between the axis of the cone and the circular base, and let ψ be the angle between the axis and the plane of the conic. If we now let

$$e = \frac{\sin \psi}{\sin \phi},$$

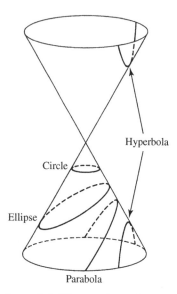

Figure 1.4: Plane sections of a cone

then for $e < 1$, $e = 1$ and $e > 1$, respectively, the conic section is an ellipse, a parabola or a hyperbola. As is well known, the eccentricity can also be expressed as

$$e = \frac{\text{distance from the focus}}{\text{distance to the directrix}}.$$

(For a curve with equation $(x/a)^2 \pm (y/b)^2 = 1$, where $a \leq b$, we find that $e = \sqrt{1 \mp (a/b)^2}$.) This is not an invariant polynomial function, but it satisfies an algebraic equation whose coefficients are invariants. Namely, it is the invariant multivalued function satisfying the quartic equation

$$(e^2 - 1) + \frac{1}{e^2 - 1} = 2 - \frac{T^2}{2E}.$$

Although e is properly speaking multivalued, we can take advantage of the fact that we are considering conics over the real numbers. In this case it is possible to choose a branch so that the function is single-valued for conics with real coefficients.

Suppose we extend the Euclidean transformation group to include also similarities (dilations and contractions). Transforming a conic by a scale factor k multiplies α by $\sqrt[3]{k^2}$ and multiplies β by $\sqrt[3]{k}$. So the 'moduli space' is now the

(α, β) plane, minus the origin, divided out by the action of scalars

$$(\alpha, \beta) \mapsto (\sqrt[3]{k^2}\alpha, \sqrt[3]{k}\beta).$$

In other words, it is a projective line (more precisely, the weighted projective line $\mathbb{P}(1 : 2)$; see Example 3.46 in Chapter 3). The one dimensional parameter that we obtain in this way is essentially the eccentricity e.

The aim of the first part of this book is to generalise the construction of this sort of parameter space to equivalence classes of polynomials in several variables under the action of the general linear group. In geometric language, our aim is to construct parameter spaces for equivalence classes of general-dimensional projective hypersurfaces with respect to projective transformations.

1.2 Invariants of groups

To say that a polynomial $f(x_1, \ldots, x_n)$ in n variables is an *invariant* with respect to an $n \times n$ matrix $A = (a_{ij})$ can have one of two meanings:

(i) f is invariant under the coordinate transformation determined by A. That is, it satisfies

$$f(Ax) := f\left(\sum_i a_{1i}x_i, \ldots, \sum_i a_{ni}x_i\right) = f(x). \tag{1.5}$$

(ii) f is invariant under the derivation

$$\mathcal{D}_A = \sum_{i,j} a_{ij}x_i \frac{\partial}{\partial x_j}$$

determined by A. In other words, it satisfies

$$\mathcal{D}_A f = \sum_{i,j} a_{ij}x_i \frac{\partial f}{\partial x_j} = 0. \tag{1.6}$$

In both cases, the invariant polynomials under some fixed set of matrices form a subring of $\mathbb{C}[x_1, \ldots, x_n]$. The idea of a Lie group and of a Lie algebra, respectively, arises in a natural way out of these two notions of invariants.

(a) Hilbert series

To begin, we review the first notion 1.2(i) of invariance. (The second will reappear in Chapter 4.) Given a set of nonsingular matrices $T \subset GL(n)$, we consider the set of all invariant polynomials

$$\{f \in \mathbb{C}[x_1, \ldots, x_n] \mid f(Ax) = f(x) \text{ for all } A \in T\}.$$

Clearly this is a subring of $\mathbb{C}[x_1, \ldots, x_n]$, called the *ring of invariants* of T. Notice that if $f(x)$ is an invariant under matrices A and B, then it is an invariant under the inverse A^{-1} and the product AB. It follows that in the definition of the ring of invariants we may assume without loss of generality that T is closed under taking products and inverses. This is just the definition of a group; moreover, in essence we have here the definition of a group representation.

Definition 1.4. Let $G \subset GL(n)$ be a subgroup. A polynomial $f \in \mathbb{C}[x_1, \ldots, x_n]$ satisfying

$$f(Ax) = f(x) \text{ for all } A \in G$$

is called a *G-invariant*. $\qquad\qquad\qquad\qquad\qquad\qquad\qquad\qquad\qquad\square$

We shall write $S = \mathbb{C}[x_1, \ldots, x_n]$ for the polynomial ring and S^G for the ring of invariants of G. Let us examine some cases in which G is a finite group.

Example 1.5. Let G be the symmetric group consisting of all $n \times n$ permutation matrices – that is, having a single 1 in each row and column, and 0 elsewhere. The invariants of G in $\mathbb{C}[x_1, \ldots, x_n]$ are just the symmetric polynomials. These form a subring which includes the n elementary symmetric polynomials

$$\begin{aligned}
\sigma_1(x) &= \sum_i x_i \\
\sigma_2(x) &= \sum_{i<j} x_i x_j \\
&\cdots \\
\sigma_n(x) &= x_1 \ldots x_n,
\end{aligned}$$

and it is well known that these generate the subring of all symmetric polynomials. $\qquad\qquad\qquad\qquad\qquad\qquad\qquad\qquad\qquad\qquad\qquad\qquad\square$

Example 1.6. Suppose G is the alternating group consisting of all even permutation matrices (matrices as in the previous example, that is, with determinant $+1$). In this case a G-invariant polynomial can be uniquely expressed as the sum of a symmetric and an alternating polynomial:

$$\left\{\begin{matrix} \text{invariant} \\ \text{polynomials} \end{matrix}\right\} \cong \left\{\begin{matrix} \text{symmetric} \\ \text{polynomials} \end{matrix}\right\} \oplus \left\{\begin{matrix} \text{alternating} \\ \text{polynomials} \end{matrix}\right\}.$$

Moreover, the set of alternating polynomials is a free module over the ring of symmetric polynomials with the single generator

$$\Delta(x) = \prod_{1 \le i < j \le n} (x_i - x_j).$$

□

Example 1.7. Let $G = \{\pm I_n\} \subset GL(n)$, the subgroup of order 2, where I_n is the identity matrix. This time the set of invariant polynomials is a vector space with basis consisting of all monomials of even degree. As a ring it is generated by the monomials of degree 2; in the case $n = 2$, for example, it is generated by $x_1^2, x_1 x_2, x_2^2$.

□

Let $S = \mathbb{C}[x_1, \ldots, x_n]$. Any polynomial $f(x) = f(x_1, \ldots, x_n)$ can written as a sum of homogeneous polynomials:

$$f(x) = f_0 + f_1(x) + f_2(x) + \cdots + f_{\text{top}}(x) \quad \text{with deg } f_i(x) = i.$$

Invariance of $f(x)$ is then equivalent to invariance of all the summands $f_i(x)$. Denoting by $S_d \subset S$ the subspace of homogeneous polynomials of degree d, it follows that there are direct sum decompositions

$$S = \bigoplus_{d \ge 0} S_d, \quad S^G = \bigoplus_{d \ge 0} S^G \cap S_d.$$

(S and S^G are *graded rings*. See Section 2.5(a).) We can introduce a generating function for the dimensions of the homogeneous components of S^G. This is the formal power series in an indeterminate t, called the *Hilbert series* (also called the *Poincaré series*, or the *Molien series*) of the graded ring S^G:

$$P(t) := \sum_{d \ge 0} (\dim S^G \cap S_d) t^d \in \mathbb{Z}[[t]].$$

Example 1.8. The Hilbert series of the matrix groups in Examples 1.5 and 1.6 are given, respectively, by the generating functions:

(i)
$$\frac{1}{(1 - t)(1 - t^2) \cdots (1 - t^n)},$$

(ii)
$$\frac{1 + t^{n(n-1)/2}}{(1 - t)(1 - t^2) \cdots (1 - t^n)}.$$

One sees this in the following way. First, if we expand the expression

$$\frac{1}{(1 - \sigma_1)(1 - \sigma_2) \cdots (1 - \sigma_n)}$$

as a formal series, the terms form a basis of the infinite-dimensional vector space of symmetric polynomials. So, substituting t^i for σ_i we obtain (i) for the Hilbert series of Example 1.5. For Example 1.6, a similar argument gives (ii) after noting that

$$S^G = \mathbb{C}[\sigma_1, \ldots, \sigma_n] \oplus \mathbb{C}[\sigma_1, \ldots, \sigma_n]\Delta,$$

where $\deg \Delta = n(n-1)/2$. $\qquad\square$

Note that by a similar argument the full polynomial ring $S = \bigoplus S_d$ has Hilbert series $P(t) = (1 - t)^{-n}$. In particular, this gives the familiar fact that $\dim S_d = \binom{n-1+d}{n-1}$. We will make more systematic use of this idea in the proof of Molien's Theorem below.

The Hilbert series is a very important invariant of the ring S^G which, as these examples illustrate, measures its 'size and shape':

Proposition 1.9. *If S^G is generated by homogeneous polynomials $f_1, \ldots, f_r$ of degrees $d_1, \ldots, d_r$, then the Hilbert series of S^G is the power series expansion at $t = 0$ of a rational function*

$$P(t) = \frac{F(t)}{(1 - t^{d_1}) \cdots (1 - t^{d_r})}$$

for some $F(t) \in \mathbb{Z}[t]$.

Proof. We use induction on r, observing that when $r = 1$ the ring S^G is just $\mathbb{C}[f_1]$ with the Hilbert series

$$P(t) = 1 + t^{d_1} + t^{2d_1} + \cdots = \frac{1}{1 - t^{d-1}}.$$

For $r > 1$ we consider the (injective complex linear) map $S^G \to S^G$ defined by $h \mapsto f - rh$. We denote the image by $R \subset S^G$ and consider the Hilbert series for the graded rings R and S^G/R. These satisfy

$$P_{S^G}(t) = P_R(t) + P_{S^G/R}(t).$$

On the other hand, $\dim(S^G \cap S_d) = \dim(R \cap S_{d+d_r})$, so that $P_R(t) = t^{d_r} P_{S^G}(t)$, and hence

$$P_{S^G}(t) = \frac{P_{S^G/R}(t)}{1 - t^{d_r}}.$$

But S^G/R is isomorphic to the subring of S generated by the polynomials $f_1, \ldots, f_{r-1}$, and hence by the inductive hypothesis $P_{S^G/R}(t) = F(t)/(1 - t^{d_1}) \cdots (1 - t^{d_{r-1}})$ for some $F(t) \in \mathbb{Z}[t]$. □

(b) Molien's formula

There is a formula which gives the Hilbert series explicitly for the ring of invariants of any finite group. Given an $n \times n$ matrix A, we call the polynomial

$$\det(I_n - tA) \in \mathbb{C}[t]$$

the *reverse characteristic polynomial* of A. Its degree is equal to n minus the multiplicity of 0 as an eigenvalue of A. Note that since its constant term is always 1, it is invertible in the formal power series ring $\mathbb{C}[[t]]$.

Molien's Theorem 1.10. *The ring of invariants $S^G \subset S = \mathbb{C}[x_1, \ldots, x_n]$ of any finite group $G \subset GL(n)$ has a Hilbert series given by:*

$$P(t) = \frac{1}{|G|} \sum_{A \in G} \frac{1}{\det(I_n - tA)} \in \mathbb{C}[[t]].$$

Before proving this we recall some facts from the representation theory of finite groups. First, a linear representation of a group G is a homomorphism

$$\rho : G \to GL(V)$$

from G to the automorphism group $GL(V)$ of a vector space V. (One could also allow the case where ρ is an antihomomorphism. Then the composition with the antiautomorphism $G \to G, g \mapsto g^{-1}$ is a homomorphism. However, in this book we are concerned mainly with invariant elements, so we will not worry about the distinction between left- and right-actions.) In our situation, where G is a subgroup of $GL(n)$, each homogeneous summand S_d of the polynomial ring S becomes such a finite-dimensional representation of G. If V is a finite-dimensional representation, its *character* is the map

$$\chi : G \to \mathbb{C}, \qquad g \mapsto \operatorname{tr} \rho(g).$$

One may consider the invariant subspace:

$$V^G := \{v \in V \mid \rho(g)v = v \text{ for all } g \in G\}.$$

The dimension of this subspace is precisely the average value of the character:

Dimension Formula 1.11.

$$\dim V^G = \frac{1}{|G|} \sum_{g \in G} \chi(g).$$

Proof. Consider the averaging map

$$E : V \to V, \quad v \mapsto \frac{1}{|G|} \sum_{g \in G} \rho(g)v.$$

This is a linear map which restricts to the identity map on $V^G \subset V$ and whose image is V^G. It follows from this that $\dim V^G = \operatorname{tr} E = \frac{1}{|G|} \sum_{g \in G} \chi(g)$. (See Exercise 1.1 at the end of the chapter.) $\qquad\Box$

Proof of Molien's Theorem 1.10. Pick an element $A \in G$, and let $\{x_1, \ldots, x_n\}$ be a basis of eigenvectors of A in S_1, belonging to eigenvalues $a_1, \ldots, a_n$. Note that A is diagonalisable since it has finite order (G is finite!). (In fact, for the present proof it would be enough for A to be upper triangular.) The reverse characteristic polynomial of A is then

$$\det(I_n - tA) = (1 - a_1 t)(1 - a_2 t) \cdots (1 - a_n t).$$

Now consider the formal power series expansion of

$$\frac{1}{(1 - x_1)(1 - x_2) \cdots (1 - x_n)}$$

whose terms are precisely the monomials of the ring S, without multiplicity. The action of A on this series gives

$$\frac{1}{(1 - a_1 x_1)(1 - a_2 x_2) \cdots (1 - a_n x_n)},$$

from which the trace of A acting on S_d may be read off as the sum of the coefficients in degree d. So if we make the substitution $x_1 = \cdots = x_n = t$, we see that the character $\chi_d(A)$ of A on S_d is precisely the coefficient of t^d in the expansion of

$$\frac{1}{(1 - a_1 t)(1 - a_2 t) \cdots (1 - a_n t)}.$$

In other words, we obtain

$$\sum_d \chi_d(A) t^d = \frac{1}{\det(I_n - tA)}.$$

If we now take the average over G and apply the Dimension Formula 1.11, we obtain Molien's formula. □

Example 1.12. The Hilbert series of the invariant ring of Example 1.7 is

$$P(t) = \frac{1}{2} \left\{ \frac{1}{(1-t)^n} + \frac{1}{(1+t)^n} \right\}.$$

When $n = 2$ this reduces to

$$P(t) = \frac{1+t^2}{(1-t^2)^2} = \frac{1-t^4}{(1-t^2)^3}.$$

The last expression can also be deduced (see Proposition 1.9) from the fact that the invariant ring S^G has three generators, $A = x_1^2$, $B = x_1 x_2$, $C = x_2^2$, and one relation in degree 4, $AC - B^2 = 0$.

(c) Polyhedral groups

We will consider some examples where G is the symmetry group of a regular polyhedron.

Example 1.13. The quaternion group. The two matrices of order 4

$$\begin{pmatrix} i & 0 \\ 0 & -i \end{pmatrix}, \quad \begin{pmatrix} 0 & 1 \\ -1 & 0 \end{pmatrix}$$

generate a subgroup $G \subset SL(2, \mathbb{C})$ of order 8, consisting of $\pm I_2$ and six elements of order 4. (These are isomorphic to ± 1, $\pm i$, $\pm j$, $\pm k$ obeying the rules of quaternion multiplication – hence the terminology.) By Molien's Theorem, therefore, the invariant ring of G has a Hilbert series equal to

$$P(t) = \frac{1}{8} \left\{ \frac{1}{(1-t)^2} + \frac{1}{(1+t)^2} + \frac{6}{1+t^2} \right\} = \frac{1+t^6}{(1-t^4)^2} = \frac{1-t^{12}}{(1-t^4)^2(1-t^6)}.$$

We can use this fact to determine the structure of the invariant ring. First (suggested by the denominator on the right-hand side) we observe two invariants of degree 4:

$$A = x^4 + y^4, \qquad B = x^2 y^2.$$

These elements generate a subring $\mathbb{C}[A, B] \subset S^G$ with the Hilbert series

$$\frac{1}{(1 - t^4)^2}.$$

Next we observe an invariant of degree 6:

$$C = xy(x^4 - y^4).$$

Although C is not in $\mathbb{C}[A, B]$, its square C^2 is, since

$$C^2 = A^2B - 4B^3. \tag{1.7}$$

In fact, by Proposition 1.9, this relation in degree 12 shows that the subring $\mathbb{C}[A, B] \oplus C\mathbb{C}[A, B] \subset S^G$ has the same Hilbert series as S^G – so the two rings coincide, and we have shown:

Proposition 1.14. *The quaternion group* $G \subset SL(2, \mathbb{C})$ *has an invariant ring*

$$S^G = \mathbb{C}[A, B, C]/(C^2 - A^2B + 4B^3),$$

where $A = x^4 + y^4$, $B = x^2y^2$, $C = xy(x^4 - y^4) \in \mathbb{C}[x, y]$. $\square$

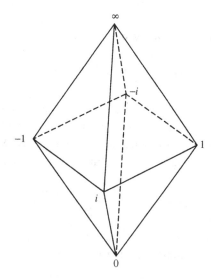

Figure 1.5: The degree 6 invariant and the octahedron

This example, the quaternion group of order 8, is also the binary dihedral group of the 2-gon. The zeros of the degree 6 invariant C, viewed as points of the Riemann sphere, are the vertices of a regular octahedron.

One may consider, in the space with coordinates A, B, C, the surface with equation (1.7): the origin is a singular point of this surface, called a rational double point of type D_4.

Let us examine these ideas for the case of the binary group of a regular icosahedron.

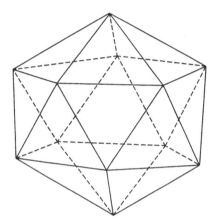

Figure 1.6: The icosahedron

Remark 1.15. One can show that every finite subgroup of $SL(2, \mathbb{C})$ is conjugate to a subgroup of the special unitary group $SU(2)$. On the other hand, there is a natural double cover $SU(2) \to SO(3)$, and it is well known that every finite subgroup of $SO(3)$ is cyclic, dihedral or the symmetry group of a Platonic solid. We therefore have a nice classification of finite subgroups of $SL(2, \mathbb{C})$, of which Examples 1.13 and 1.16 are examples. $\qquad\square$

Example 1.16. The binary icosahedral group. This is the group $G_{120} \subset SU(2)$ containing $\pm I_2$, whose quotient

$$G_{60} = G_{120}/\{\pm I_2\} \subset SU(2)/\{\pm I_2\} = SO(3, \mathbb{R})$$

is the rotation group of a regular icosahedron (recall that $G_{60} \cong A_5$, the alternating group which permutes five embedded octahedra). The orders of elements in the group G_{120} are distributed as follows:

order	1	2	3	4	5	6	10
no. of elements	1	1	20	30	24	20	24

Using Molien's Theorem, the Hilbert series is

$$P(t) = \frac{1}{120} \left\{ \frac{1}{(1-t)^2} + \frac{1}{(1+t)^2} + \sum \frac{N}{t^2 - 2t\cos\theta + 1} \right\},$$

where the summation is over $N = 30$ edge rotations with $\theta = \pi/2$, $N = 20$ face rotations with $\theta = \pi/3$ and another $N = 20$ with $2\pi/3$, and $N = 12$ vertices rotations each, with $\theta = \pi/5, 2\pi/5, 3\pi/5$ and $4\pi/5$, respectively. This gives an expression

$$P(t) = \frac{1}{120} \left\{ \frac{1}{(1-t)^2} + \frac{1}{(1+t)^2} + \frac{20}{1+t+t^2} + \frac{30}{1+t^2} \right.$$

$$\left. + \frac{24(t^2 + t/2 + 1)}{1+t+t^2+t^3+t^4} + \frac{20}{1-t+t^2} + \frac{24(t^2 - t/2 + 1)}{1-t+t^2-t^3+t^4} \right\},$$

which simplifies to

$$P(t) = \frac{1+t^{30}}{(1-t^{12})(1-t^{20})} = \frac{1-t^{60}}{(1-t^{12})(1-t^{20})(1-t^{30})}.$$

Remark 1.17. As in the previous example, the significance of the right-hand side is that it suggests the existence of generators of the invariant ring of degrees 12, 20 and 30 satisfying a single algebraic relation in degree 60. Geometrically, we do not have far to look: if we inscribe the icosahedron in a sphere S^2, viewed as the Riemann number sphere $\mathbb{CP}^1 = \mathbb{C} \cup \{\infty\}$, then its vertices will determine 12 points of $\mathbb{C} \cup \{\infty\}$, which are the roots of a polynomial f_{12} of degree 12. Similarly, the midpoints of the edges and the faces are the zero-sets of polynomials J_{30} and H_{20}, respectively. This is a general recipe for the binary polyhedral groups; for the icosahedron we construct these polynomials algebraically next. (See also Klein [4], [5], Schur [26] Chapter II §5 or Popov and Vinberg [6].)   □

Let $\alpha_1, \ldots, \alpha_{12}$ be the coordinates of the 12 vertices, then, in $S^2 = \mathbb{C} \cup \{\infty\}$. Then the homogeneous polynomial

$$f_{12}(x, y) = \prod_{i=1}^{12} (x - \alpha_i y)$$

is an invariant of G_{120}, its linear factors permuted by the rotations of the icosahedron. By choosing coordinates suitably, in fact, we can write

$$f_{12}(x, y) = xy(x^{10} + 11x^5y^5 - y^{10}).$$

The Hessian determinant of $f = f_{12}$,

$$H_{20}(x, y) = \frac{1}{121} \begin{vmatrix} f_{xx} & f_{xy} \\ f_{yx} & f_{yy} \end{vmatrix}$$

$$= -x^{20} + 228x^{15}y^5 - 494x^{10}y^{10} - 228x^5y^{15} - y^{20},$$

is then an invariant of degree 20. Moreover, the Jacobian determinant of $f = f_{12}$ and $H = H_{20}$,

$$J_{30}(x, y) = \frac{1}{20} \begin{vmatrix} f_x & f_y \\ H_x & H_y \end{vmatrix}$$

$$= x^{30} + 522x^{25}y^5 - 10005x^{20}y^{10} - 10005x^{10}y^{20} - 522x^5y^{25} + y^{30},$$

is an invariant of degree 30. The polynomials f_{12}, H_{20}, J_{30} are pairwise algebraically independent but together satisfy the relation

$$J^2 + H^3 = 1728 f^5.$$

It follows from these computations (i.e. from comparison with our expression for the Hilbert series) that the invariant ring $\mathbb{C}[x, y]^{G_{120}}$ is generated by f, H, J. To say this in other language, the quotient space $\mathbb{A}^2/G_{120}$ is isomorphic to the surface in $\mathbb{A}^3$ with equation

$$w^2 = 1728u^5 - v^3.$$

(This is made precise in the discussion of Section 5.1, and in particular in Corollary 5.17.) The origin is a singular point of this surface called a rational double point of type E_8. In the minimal resolution of this singular point the exceptional set is a configuration of eight intersecting $\mathbb{P}^1$s whose dual graph is the Dynkin diagram E_8. (See Figure 1.7.)

1.3 Classical binary invariants

(a) Resultants and discriminants

Given polynomials of degrees d, e,

$$\begin{aligned} f(x) &= a_0x^d + a_1x^{d-1} + \cdots + a_{d-1}x + a_d = a_0 \prod_{i=1}^d (x - \lambda_i) \\ g(x) &= b_0x^e + b_1x^{e-1} + \cdots + b_{e-1}x + b_e = b_0 \prod_{j=1}^e (x - \mu_j), \end{aligned} \tag{1.8}$$

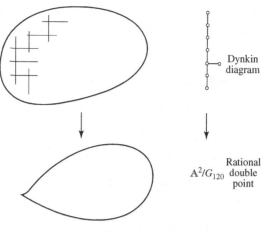

Figure 1.7

with $a_0 b_0 \neq 0$, we set

$$R(f, g) = a_0^e b_0^d \prod_{i,j} (\lambda_i - \mu_j).$$

The vanishing of $R(f, g)$ is the condition for the equations $f(x) = 0$, $g(x) = 0$ to have a common root. It follows from Example 1.5 (applied to permutations of the variables in the polynomial rings $\mathbb{C}[\lambda_1, \ldots, \lambda_d]$ and $\mathbb{C}[\mu_1, \ldots, \mu_e]$) that $R(f, g)$ can be expressed as a polynomial in the coefficients of f and g:

Lemma-Definition 1.18. *The $R(f, g)$ is equal to the $(d + e) \times (d + e)$ determinant*

$$R(f, g) = \begin{vmatrix} a_0 & a_1 & \cdots & \cdots & a_d & & & \\ & a_0 & a_1 & \cdots & \cdots & a_d & & \\ & & \ddots & \ddots & \ddots & \ddots & \ddots & \\ & & & a_0 & a_1 & \cdots & \cdots & a_d \\ b_0 & b_1 & \cdots & \cdots & b_e & & & \\ & b_0 & b_1 & \cdots & \cdots & b_e & & \\ & & \ddots & \ddots & \ddots & \ddots & \ddots & \\ & & & b_0 & b_1 & \cdots & \cdots & b_e \end{vmatrix},$$

and is called the resultant *of f and g.*

Proof. First we observe:

Claim: $f, g \in \mathbb{C}[x]$ possess a nonconstant common factor if and only if there exist nonzero polynomials $u, v \in \mathbb{C}[x]$, with $\deg u < \deg f$ and $\deg v < \deg g$, such that $vf + ug \equiv 0$.

One direction is trivial: if f, g have a nonconstant common factor, then we can write $f = uh, g = vh$ for some $u, v, h \in \mathbb{C}[x]$, and these u, v have the required properties. In the other direction, let $f = p_1^{r_1} \ldots p_s^{r_s}$ be the unique factorisation of f into irreducibles. Then each $p_i^{r_i}$ divides ug, so p_i divides u or g. Since $\deg u < \deg f$, some p_i must divide g, and this proves the claim.

Let $\mathbb{C}[x]_r$ denote the subset of polynomials of degree at most r; this is a finite-dimensional vector space with basis $1, x, \ldots, x^r$. The preceding claim can be interpreted in terms of a $\mathbb{C}$-linear map:

$$\rho_{f,g} : \mathbb{C}[x]_{n-1} \oplus \mathbb{C}[x]_{m-1} \to \mathbb{C}[x]_{n+m-1}, \qquad (u, v) \mapsto vf + ug,$$

and says that f, g have a nonconstant common factor if and only if this linear map has nonzero kernel. Since it maps between spaces of equal dimension $n + m$, this is equivalent to the vanishing of $\det \rho_{f,g}$. Now use the standard basis of each space to write $\rho_{f,g}$ as a matrix, to deduce that the determinant of $\rho_{f,g}$ is that given in the lemma. This shows that f and g have a common root if and only the resultant vanishes, and the lemma follows easily from this. $\square$

Given a single polynomial $f(x)$ as in (1.8), the expression

$$D(f) = a_0^{2d-1} \prod_{1 \le i \ne j \le d} (\lambda_i - \lambda_j)$$

is called its *discriminant*.

Lemma 1.19. *The discriminant of $f(x)$ is equal to the resultant of $f(x)$ and its derivative $f'(x)$,*

$$D(f) = R(f, f').$$

Proof. Consider the polynomial

$$g(x) := f(x + \varepsilon) = a_0' x^d + a_1' x^{d-1} + \cdots + a_{d-1}' x + a_d', \qquad a_0' = a_0.$$

By Lemma 1.18 we have

$$a_0^{2d} \prod_{i,j}(\lambda_i + \varepsilon - \lambda_j) = \begin{vmatrix} a_0 & a_1 & \cdots & \cdots & a_d & & & \\ & a_0 & a_1 & \cdots & \cdots & a_d & & \\ & & \ddots & \ddots & \ddots & \ddots & \ddots & \\ & & & a_0 & a_1 & \cdots & \cdots & a_d \\ a_0 & a_1' & \cdots & \cdots & a_d' & & & \\ & a_0 & a_1' & \cdots & \cdots & a_d' & & \\ & & \ddots & \ddots & \ddots & \ddots & \ddots & \\ & & & a_0 & a_1' & \cdots & \cdots & a_d' \end{vmatrix}$$

$$= \varepsilon^d \times \begin{vmatrix} a_0 & a_1 & \cdots & \cdots & a_d & & & \\ & a_0 & a_1 & \cdots & \cdots & a_d & & \\ & & \ddots & \ddots & \ddots & \ddots & \ddots & \\ & & & a_0 & a_1 & \cdots & \cdots & a_d \\ 0 & \frac{a_1'-a_1}{\varepsilon} & \cdots & \cdots & \frac{a_d'-a_d}{\varepsilon} & & & \\ & 0 & \frac{a_1'-a_1}{\varepsilon} & \cdots & \cdots & \frac{a_d'-a_d}{\varepsilon} & & \\ & & \ddots & \ddots & \ddots & \ddots & \ddots & \\ & & & 0 & \frac{a_1'-a_1}{\varepsilon} & \cdots & \cdots & \frac{a_d'-a_d}{\varepsilon} \end{vmatrix}.$$

Cancelling $a_0 \varepsilon^d$ and letting $\varepsilon \to 0$ now shows that

$$a_0^{2d-1} \prod_{i \neq j}(\lambda_i - \lambda_j) = \begin{vmatrix} a_0 & a_1 & \cdots & \cdots & a_d & & & \\ & a_0 & a_1 & \cdots & \cdots & a_d & & \\ & & \ddots & \ddots & \ddots & \ddots & \ddots & \\ & & & a_0 & a_1 & \cdots & \cdots & a_d \\ da_0 & (d-1)a_1 & \cdots & \cdots & a_{d-1} & & & \\ & da_0 & (d-1)a_1 & \cdots & \cdots & a_{d-1} & & \\ & & \ddots & \ddots & \ddots & \ddots & \ddots & \\ & & & da_0 & (d-1)a_1 & \cdots & \cdots & a_{d-1} \end{vmatrix}$$

$$= R(f, f').$$

$\square$

It will be convenient to consider projective coordinates $(x : y) \in \mathbb{P}^1$ and homogeneous polynomials

$$f(x, y) = \sum_{i=0}^{d} a_i \binom{d}{i} x^{d-i} y^i = a_0 x^d + d a_1 x^{d-1} y + \binom{d}{2} a_2 x^{d-2} y^2 + \cdots + a_d y^d.$$

(It is traditional, and useful, to include the binomial coefficients in our forms.) Note that $a_0 = 0$ is now allowed: in this case, $\infty = (1 : 0)$ is a zero of $f(x, y)$. A multiple root of the equation $f(x, y) = 0$ is now a common root of the equations

$$\frac{\partial f}{\partial x}(x, y) = \frac{\partial f}{\partial y}(x, y) = 0,$$

and so the necessary and sufficient condition for the existence of a multiple root is the vanishing of the resultant of the partials:

$$R\left(\frac{\partial f}{\partial x}, \frac{\partial f}{\partial y}\right) = 0.$$

We introduce $d + 1$ independent variables $\xi_0, \ldots, \xi_d$ for the coefficients of a general form and write, in the classical notation of Cayley and Sylvester,

$$f(x, y) = (\xi_0, \ldots, \xi_d \,)\! x, y) := \sum_{i=0}^{d} \xi_i \binom{d}{i} x^{d-i} y^i. \tag{1.9}$$

Definition 1.20. The resultant

$$D(\xi) := R\left(\frac{1}{d}\frac{\partial f}{\partial x}, \frac{1}{d}\frac{\partial f}{\partial y}\right)$$

$$= \begin{vmatrix}
\xi_0 & (d-1)\xi_1 & \cdots & \cdots & \xi_{d-1} & & & \\
& \xi_0 & (d-1)\xi_1 & \ddots & & \ddots & \xi_{d-1} & \\
& & \ddots & \ddots & & & \ddots & \ddots & \ddots \\
& & & \xi_0 & (d-1)\xi_1 & \cdots & \cdots & \xi_{d-1} \\
\xi_1 & (d-1)\xi_2 & \cdots & \cdots & \xi_d & & & \\
& \xi_1 & (d-1)\xi_2 & \ddots & & \ddots & \xi_d & \\
& & \ddots & \ddots & & & \ddots & \ddots & \ddots \\
& & & \xi_1 & (d-1)\xi_2 & \cdots & \cdots & \xi_d
\end{vmatrix}$$

is called the *discriminant* of the form $(\xi \,)\! x, y)$. $\square$

Example 1.21. For the quadratic form

$$f(x, y) = \xi_0 x^2 + 2\xi_1 xy + \xi_2 y^2$$

we get the familiar discriminant

$$D(\xi) = \begin{vmatrix} \xi_0 & \xi_1 \\ \xi_1 & \xi_2 \end{vmatrix} = \xi_0 \xi_2 - \xi_1^2.$$

Example 1.22. The discriminant of the cubic form

$$f(x, y) = \xi_0 x^3 + 3\xi_1 x^2 y + 3\xi_2 xy^2 + \xi_3 y^3$$

is

$$\begin{vmatrix} \xi_0 & 2\xi_1 & \xi_2 & 0 \\ 0 & \xi_0 & 2\xi_1 & \xi_2 \\ \xi_1 & 2\xi_2 & \xi_3 & 0 \\ 0 & \xi_1 & 2\xi_2 & \xi_3 \end{vmatrix} = \xi_0^2 \xi_3^2 - 3\xi_1^2 \xi_2^2 - 3\xi_0 \xi_1 \xi_2 \xi_3 + 4\xi_1^3 \xi_3 + 4\xi_0 \xi_2^3.$$

This is equal, in fact, to the discriminant of the quadratic Hessian form

$$H(x, y) = \frac{1}{6^2} \begin{vmatrix} f_{xx} & f_{xy} \\ f_{yx} & f_{yy} \end{vmatrix} = (\xi_1^2 - \xi_0 \xi_2)x^2 + (\xi_1 \xi_2 - \xi_0 \xi_3)xy + (\xi_2^2 - \xi_1 \xi_3)y^2.$$

$\square$

We consider next the action of matrices

$$g = \begin{pmatrix} \alpha & \beta \\ \gamma & \delta \end{pmatrix} \in GL(2)$$

on forms $(\xi \, \langle \, x, y)$ in (1.9). That is, under the coordinate transformation $(x, y) \mapsto (\alpha x + \beta y, \gamma x + \delta y)$ we obtain a new form,

$$(\xi \, \langle \, gx) = (\xi \, \langle \, \alpha x + \beta y, \gamma x + \delta y).$$

One can expand this and rewrite it as

$$(\xi g \, \langle \, x) = \sum_{i=0}^{d} \xi_i(g) \binom{d}{i} x^{d-i} y^i.$$

The coefficients $\xi_i(g)$ can be written

$$\xi_i(g) = \sum_j \tilde{g}_i^j(\alpha, \beta, \gamma, \delta)\xi_j,$$

where $\tilde{g}_i^j$ is homogeneous of degree d in $\alpha, \beta, \gamma, \delta$.

Proposition 1.23. *For all* $g \in GL(2)$ *the discriminant satisfies*

$$D(\xi g) = (\det g)^{d(d-1)} \times D(\xi).$$

Proof. Viewing

$$(\xi \mathbin{\lozenge} x, 1) = \sum_{i=0}^{d} \xi_i \binom{d}{i} x^{d-i} = 0$$

as an equation of degree d over the rational function field $\mathbb{C}(\xi_0, \ldots, \xi_d)$, we can denote its roots (in a splitting field) by $\lambda_1, \ldots, \lambda_d$. Thus

$$(\xi \mathbin{\lozenge} x, 1) = \xi_0 \prod_{i=1}^{d} (x - \lambda_i y)$$

and the discriminant is

$$D(\xi) = \xi_0^{2d-2} \prod_{1 \le i, j \le d} (\lambda_i - \lambda_j).$$

Transforming by the matrix $g = \begin{pmatrix} \alpha & \beta \\ \gamma & \delta \end{pmatrix} \in GL(2)$ gives

$$(\xi \mathbin{\lozenge} \alpha x + \beta y, \gamma x + \delta y) = \xi_0 \prod_{i=1}^{d} (\alpha x + \beta y - \lambda_i (\gamma x + \delta y))$$

$$= (\xi \mathbin{\lozenge} \alpha, \gamma) \prod_{i=1}^{d} \left(x - \frac{\delta \lambda_i - \beta}{-\gamma \lambda_i + \alpha} y \right).$$

It follows that g transforms the differences $(\lambda_i - \lambda_j)$ to

$$\frac{\delta \lambda_i - \beta}{-\gamma \lambda_i + \alpha} - \frac{\delta \lambda_j - \beta}{-\gamma \lambda_j + \alpha} = \frac{(\alpha \delta - \beta \gamma)(\lambda_i - \lambda_j)}{(\gamma \lambda_i - \alpha)(\gamma \lambda_j - \alpha)},$$

and hence

$$D(\xi g) = (\xi \mathbin{\lozenge} \alpha, \gamma)^{2d-2} \prod_{1 \le i, j \le d} \frac{(\alpha \delta - \beta \gamma)(\lambda_i - \lambda_j)}{(\gamma \lambda_i - \alpha)(\gamma \lambda_j - \alpha)} = D(\xi)(\det g)^{d(d-1)}.$$

$\square$

More generally, we can consider arbitrary homogeneous polynomials in the coefficients $\xi_0, \ldots, \xi_d$:

Definition 1.24. If a homogeneous polynomial $F(\xi_0, \ldots, \xi_d)$ satisfies

$$F(\xi g) = F(\xi), \qquad \text{for all } g \in SL(2),$$

then F is called a *classical binary invariant*.

$\square$

The set of binary forms of degree d is a vector space V_d of dimension $d + 1$ on which the general linear group $GL(2)$ acts by $\xi \mapsto \xi g$. (The spaces $V_0, V_1, V_2, \ldots$ are all the irreducible linear representations of $GL(2)$ and will reappear in Chapter 4; see Section 4.4.) This induces an action of $GL(2)$ on the polynomial ring $S = \mathbb{C}[\xi_0, \ldots, \xi_d]$, and a classical binary invariant (of degree e) is an invariant homogeneous polynomial for the restriction of this action to $SL(2)$; that is, it is an element of $S_e^{SL(2)}$. Proposition 1.23 says that the discriminant is a classical binary invariant of degree $2d - 2$.

(b) Binary quartics

The general binary quartic form, with variable coefficients $\xi_0, \xi_1, \xi_2, \xi_3, \xi_4$, looks like:

$$(\xi \, \lozenge x, y) = \xi_0 x^4 + 4\xi_1 x^3 y + 6\xi_2 x^2 y^2 + 4\xi_3 x y^3 + \xi_4 y^4. \tag{1.10}$$

As we shall see in Chapter 4 (see Proposition 4.69), the classical invariant ring $\mathbb{C}[\xi_0, \xi_1, \xi_2, \xi_3, \xi_4]^{SL(2)}$ has the Hilbert series

$$P(t) = \frac{1}{(1 - t^2)(1 - t^3)}.$$

In particular, this indicates (see Proposition 1.9) the existence of invariants of degree 2 and 3; we can verify this as follows. We make a change of variables $U = x^2, V = 2xy, W = y^2$ and note that

$$x^4 = U^2, \quad 2x^3 y = UV, \quad 4x^2 y^2 = V^2 = 4UW, \quad 2xy^3 = VW, \quad y^4 = W^2.$$

It follows that the quartic equation $(\xi \, \lozenge x, y) = 0$ transforms to a pair of simultaneous quadratic equations in U, V, W:

$$\xi_0 U^2 + 2\xi_1 UV + \xi_2(V^2 + 2UW) + 2\xi_3 VW + \xi_4 W^2 = 0,$$
$$4UW - V^2 = 0.$$

These two quadratic forms are represented, respectively, by the symmetric matrices

$$\begin{pmatrix} \xi_0 & \xi_1 & \xi_2 \\ \xi_1 & \xi_2 & \xi_3 \\ \xi_2 & \xi_3 & \xi_4 \end{pmatrix}, \quad \begin{pmatrix} & & 2 \\ & -1 & \\ 2 & & \end{pmatrix},$$

which have relative characteristic polynomial

$$\det\left(\begin{pmatrix} \xi_0 & \xi_1 & \xi_2 \\ \xi_1 & \xi_2 & \xi_3 \\ \xi_2 & \xi_3 & \xi_4 \end{pmatrix} + \lambda \begin{pmatrix} & & 2 \\ & -1 & \\ 2 & & \end{pmatrix}\right) = \begin{vmatrix} \xi_0 & \xi_1 & \xi_2 + 2\lambda \\ \xi_1 & \xi_2 - \lambda & \xi_3 \\ \xi_2 + 2\lambda & \xi_3 & \xi_4 \end{vmatrix}$$

$$= 4\lambda^3 - g_2(\xi)\lambda - g_3(\xi).$$

Proposition 1.25. *The coefficients*

$$g_2(\xi) = \begin{vmatrix} \xi_0 & \xi_2 \\ \xi_2 & \xi_4 \end{vmatrix} - 4\begin{vmatrix} \xi_1 & \xi_2 \\ \xi_2 & \xi_3 \end{vmatrix} = \xi_0\xi_4 - 4\xi_1\xi_3 + 3\xi_2^2,$$

$$g_3(\xi) = \begin{vmatrix} \xi_0 & \xi_1 & \xi_2 \\ \xi_1 & \xi_2 & \xi_3 \\ \xi_2 & \xi_3 & \xi_4 \end{vmatrix} = \xi_0\xi_2\xi_4 - \xi_0\xi_3^2 - \xi_1^2\xi_4 + 2\xi_1\xi_2\xi_3 - \xi_2^3,$$

are classical invariants of binary quartics.

(For an interpretation of g_2 and g_3, see Remark 1.29 at the end of this section.)

Proof. A matrix

$$g = \begin{pmatrix} \alpha & \beta \\ \gamma & \delta \end{pmatrix} \in SL(2)$$

transforms U, V, W into

$$\alpha^2 U + \alpha\beta V + \beta^2 W, \quad 2\alpha\gamma U + (\alpha\delta + \beta\gamma)V + 2\beta\delta W, \quad \gamma^2 U + \gamma\delta V + \delta^2 W,$$

leaving the quadratic $4UW - V^2$ invariant (it gets multiplied by $(\det g)^2 = 1$). In other words, $SL(2)$ acts by orthogonal transformations of (U, V, W)-space – that is, $SL(2) \to SO(3)$ – with respect to the inner product

$$\begin{pmatrix} & & 2 \\ & -1 & \\ 2 & & \end{pmatrix}.$$

In particular, it acts with determinant 1. (Not -1, since $SL(2)$ is connected.) The matrix

$$T(\lambda) = \begin{pmatrix} \xi_0 & \xi_1 & \xi_2 + 2\lambda \\ \xi_1 & \xi_2 - \lambda & \xi_3 \\ \xi_2 + 2\lambda & \xi_3 & \xi_4 \end{pmatrix}$$

transforms to

$$\begin{pmatrix} \alpha^2 & 2\alpha\gamma & \gamma^2 \\ \alpha\beta & \alpha\delta+\beta\gamma & \gamma\delta \\ \beta^2 & 2\beta\delta & \delta^2 \end{pmatrix} T(\lambda) \begin{pmatrix} \alpha^2 & \alpha\beta & \beta^2 \\ 2\alpha\gamma & \alpha\delta+\beta\gamma & 2\beta\delta \\ \gamma^2 & \gamma\delta & \delta^2 \end{pmatrix}.$$

It follows that the transformation leaves the relative characteristic polynomial, the determinant of $T(\lambda)$, invariant. $\qquad\Box$

Remarks 1.26.

(i) Geometrically, the characteristic equation $4\lambda^3 - g_2(\xi)\lambda - g_3(\xi) = 0$ determines the three reducible elements (line pairs) in the pencil of plane conics

$$\xi_0 U^2 + 2\xi_1 UV + (\xi_2 - \lambda)V^2 + 2(\xi + 2\lambda)UW + 2\xi_3 VW + \xi_4 W^2 = 0$$

– corresponding, in other words, to the linear combinations $T(\lambda)$ of the two quadratic forms.

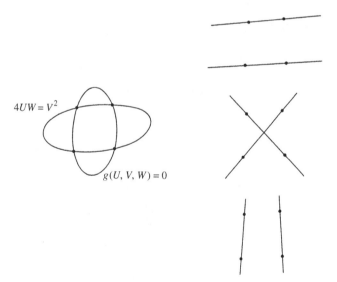

$4UW = V^2$

$g(U, V, W) = 0$

Figure 1.8: Reducible elements in the pencil of conics

(ii) Taking a root λ of the cubic characteristic polynomial, the corresponding quadratic form factorises as a product of linear forms. The simultaneous quadratic equations for U, V, W therefore reduce to a pair of independent quadratic equations, which can be solved. This gives a method of solution of the general quartic equation.

(iii) In Chapter 11 we will show that the Jacobian of the elliptic curve

$$\tau^2 = f(x, y)$$

determined by the binary quartic $f(x, y) = (\xi \lozenge x, y)$ has the equation

$$\tau^2 = 4\lambda^3 - g_2(\xi)\lambda - g_3(\xi).$$

(See Section 11.3(c). Another good reference is Cassels [1].)

The discriminant of the quartic form (1.10) can be expressed in terms of the invariants $g_2(\xi)$ and $g_3(\xi)$:

$$D(\xi) = \begin{vmatrix} \xi_0 & 3\xi_1 & 3\xi_2 & \xi_3 & & \\ & \xi_0 & 3\xi_1 & 3\xi_2 & \xi_3 & \\ & & \xi_0 & 3\xi_1 & 3\xi_2 & \xi_3 \\ \xi_1 & 3\xi_2 & 3\xi_3 & \xi_4 & & \\ & \xi_1 & 3\xi_2 & 3\xi_3 & \xi_4 & \\ & & \xi_1 & 3\xi_2 & 3\xi_3 & \xi_4 \end{vmatrix}$$

$$= g_2(\xi)^3 - 27g_3(\xi)^2.$$

Note that under the action of a scalar matrix $\begin{pmatrix} \alpha & 0 \\ 0 & \alpha \end{pmatrix} \in GL(2)$ the expressions $g_2(\xi)$, $g_3(\xi)$ are multiplied by α^8, α^{12}, respectively. Consequently, the expression

$$J(\xi) = \frac{g_2(\xi)^3}{D(\xi)}$$

is invariant under the action of $GL(2)$ and is finite-valued on binary quartics with no multiple factors. $GL(2)$-invariance means that the function

$$J : \begin{Bmatrix} \text{quartics without} \\ \text{multiple factors} \end{Bmatrix} \to \mathbb{C}, \qquad (a \lozenge x, y) \mapsto g_2(a)^3/D(a),$$

is constant on the orbits of $GL(2)$, that is, it maps each orbit to a single point of the line. Conversely (and as we would expect, since the space of quartics is 5-dimensional and $GL(2)$ is 4-dimensional):

Proposition 1.27. *The fibres of J are precisely the orbits of $GL(2)$.*

Proof. We observe that every binary quartic without multiple factors is equivalent under $GL(2)$ to the form

$$(x^2 + y^2)^2 - \lambda(x^2 - y^2)^2 \tag{1.11}$$

for suitable choice of $\lambda \in \mathbb{C}$. To see this, one can first change coordinates so that the four zeros of the quartic are 0, ∞ and some b, $b^{-1} \in \mathbb{C}$. Next, applying the Cayley transformation, $z \mapsto (z-1)/(z+1)$, these four zeros become $-1, 1, c, -c \in \mathbb{C}$. The quartic has therefore been transformed to

$$(x^2 - y^2)(x^2 - c^2 y^2).$$

Finally, this can be brought to the form (1.11) by rescaling y suitably.

For the quartic form (1.11), the function J takes the value

$$\frac{4(\lambda^2 - \lambda + 1)^3}{27\lambda^2(\lambda - 1)^2},$$

and this expression has the property that

$$\frac{4(\mu^2 - \mu + 1)^3}{27\mu^2(\mu - 1)^2} = \frac{4(\lambda^2 - \lambda + 1)^3}{27\lambda^2(\lambda - 1)^2}$$

for precisely the six values

$$\mu = \lambda, \quad \frac{1}{1-\lambda}, \quad 1 - \frac{1}{\lambda}, \quad \frac{1}{\lambda}, \quad 1 - \lambda, \quad \frac{\lambda}{1-\lambda}. \qquad (1.12)$$

To complete the proof it therefore suffices to show that for each of these μ the form $(x^2 + y^2)^2 - \mu(x^2 - y^2)^2$ is equivalent under $GL(2)$ to (1.11). For example, the transformation

$$x \mapsto \frac{1}{\sqrt[4]{i\lambda}}x, \qquad y \mapsto \frac{1}{\sqrt[4]{-i\lambda}}y$$

takes (1.11) to

$$(x^2 + y^2)^2 - \frac{1}{\lambda}(x^2 - y^2)^2.$$

Likewise,

$$x \mapsto \frac{1}{\sqrt{2}}(x+y), \qquad y \mapsto \frac{1}{\sqrt{2}}(x-y)$$

transforms (1.11) to

$$(\lambda - 1)(x^2 + y^2)^2 - \lambda(x^2 - y^2)^2.$$

The remaining cases are similar. (See Exercise 1.3). $\qquad \square$

What we have shown here is that the affine line $\mathbb{A}^1 \cong \mathbb{C}$ (see Section 3.1(a)) is a parameter space for $GL(2)$-equivalence classes of binary quartics without repeated factors. However, this was possible because we had a particularly concrete description of the invariant ring, and such cases are rare. Indeed, the above proof is also very special to the case of binary quartics. Nevertheless, it is still possible to construct moduli spaces in much more general situations, and it will be our aim to show this in the rest of this book.

It is also important to understand the orbits under $SL(2)$. Consider the map

$$\{\text{binary quartics}\} \to \mathbb{A}^2, \qquad (a \lozenge x, y) \mapsto (g_2(a), g_3(a)).$$

Since g_2 and g_3 are classical invariants, Ψ maps $SL(2)$-equivalence classes to points of the plane, and conversely each fibre is a single orbit (as before, this is suggested by a dimension count: dim(quartics) $-$ dim $SL(2) = 5 - 3 = 2$):

Proposition 1.28. *Two binary quartics $(a \lozenge x, y)$, $(b \lozenge x, y)$ without multiple factors are equivalent under the action of $SL(2)$ if and only if $(g_2(a), g_3(a)) = (g_2(b), g_3(b))$.*   □

This follows from Proposition 1.27 together with the general theory of Chapter 5 (see Example 5.25). However, if both $(a \lozenge x, y)$ and $(b \lozenge x, y)$ have repeated factors, then Proposition 1.28 no longer holds. For example, the quartics $x^2 y^2$ and $x^2(y^2 - x^2)$ both map to the point $(3, 1)$, although they are inequivalent under $SL(2)$. Similarly the three forms, $x^3 y$, x^4 and 0 all map to the origin $g_2 = g_3 = 0$ but belong to distinct orbits.

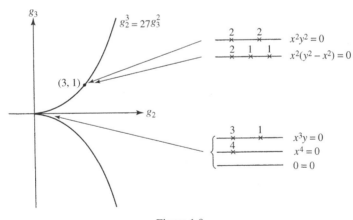

Figure 1.9

This phenomenon too will be explained in a more general setting in Chapter 5.

Remark 1.29. Here is one interpretation of the invariants g_2 and g_3. Identify a binary quartic, up to scalar, with its (unordered) zero set of four points z_1, z_2, z_3, z_4 in the projective line $\mathbb{P}^1 = \mathbb{C} \cup \{\infty\}$. Then the number λ appearing in the proof of Proposition 1.27 is – up to the ordering of the points – just the classical *cross-ratio* function

$$\lambda = \frac{(z_1 - z_3)(z_2 - z_4)}{(z_1 - z_4)(z_2 - z_3)}.$$

When the points are reordered, λ is invariant under the Klein subgroup of the permutation group S_4, and its orbit under the quotient $S_4/\text{Klein} \cong S_3$ is the set (1.12). We can express this by saying that the function J on binary quartics factorises as

$$J = j \circ \binom{\text{cross ratio}}{\text{function}},$$

where

$$j : \mathbb{P}^1 \to \mathbb{P}^1, \qquad \lambda \mapsto \frac{4(\lambda^2 - \lambda + 1)^3}{27\lambda^2(\lambda - 1)^2}$$

is a Galois extension with Galois group S_3. This map j has three branch points $0, 1, \infty \in \mathbb{P}^1$. We see that $j = \infty$ if and only if two of z_1, z_2, z_3, z_4 coincide, while:

$$j = 1 \quad \Longleftrightarrow \quad g_2 = 0$$

(that is, the cross ratio is $-1, 2$ or $\frac{1}{2}$ and z_1, z_2, z_3, z_4 are said to be *anharmonic*), and

$$j = 0 \quad \Longleftrightarrow \quad g_3 = 0$$

(the cross ratio is $-\omega$ or $-\omega^2$ where $\omega^3 = 1$, and z_1, z_2, z_3, z_4 are said to be *equianharmonic*).   □

1.4 Plane curves

While polynomials are algebraic objects, they acquire geometrical shape from their interpretation as plane curves, surfaces and hypersurfaces.

(a) Affine plane curves

For example, the zero-set of $f(x, y) = y^2 - x^2 - x^3$ (Figure 1.10) highlights at once certain features. First, one observes that there is a singular point. Recall the following well-known fact.

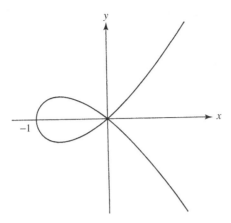

Figure 1.10: The affine plane curve $y^2 - x^2 - x^3 = 0$

Implicit Function Theorem 1.30. *If $f(a, b) = 0$ and $\frac{\partial f}{\partial y}(a, b) \neq 0$, then the condition $f(x, y) = 0$ expresses y locally as a function of x. That is, there exist positive real numbers $\epsilon, \delta > 0$ and an analytic function*

$$Y : \mathbb{D}_{a,\epsilon} \to \mathbb{D}_{b,\delta},$$

where $\mathbb{D}_{a,\epsilon} := \{z \in \mathbb{C} \mid |z - a| < \epsilon\}$, satisfying $f(z, Y(z)) = 0$ and such that the map

$$\mathbb{D}_{a,\epsilon} \to \mathbb{D}_{a,\epsilon} \times \mathbb{D}_{b,\delta}, \qquad z \mapsto (z, Y(z))$$

is an isomorphism to a neighbourhood of the point (a, b) in the curve C: $f(x, y) = 0$. ☐

If the other partial derivative $\frac{\partial f}{\partial x}(a, b)$ is nonzero, then similarly x can be expressed locally as a function of y. If in this way one of x or y can be taken as a local coordinate on the curve, then the point (a, b) is said to be a nonsingular point; otherwise it is said to be a singular point of the curve.

Definition 1.31. A point (a, b) on the plane curve $f(x, y) = 0$ is called a *singular point* if

$$\frac{\partial f}{\partial x}(a, b) = \frac{\partial f}{\partial y}(a, b) = 0.$$

☐

By definition, the set of singular points of the curve is the set of common zeros of the three polynomials $f(x, y)$, $\partial f/\partial x$, $\partial f/\partial y$.

Example 1.32. In the case $f(x, y) = y^2 - x^2 - x^3$ of Figure 1.10, the simultaneous equations

$$y^2 - x^2 - x^3 = -2x - 3x^2 = 2y = 0$$

have only one solution $(0, 0)$, which is the unique singular point of the curve. $\square$

Consider the Taylor expansion at (a, b) of the polynomial in Definition 1.31:

$$
\begin{aligned}
f(x, y) = {}& f(a, b) \\
& + (x - a)\frac{\partial f}{\partial x}(a, b) + (y - b)\frac{\partial f}{\partial y}(a, b) \\
& + \frac{1}{2}\left((x - a)^2 \frac{\partial^2 f}{\partial x^2}(a, b) + 2(x - a)(y - b)\frac{\partial^2 f}{\partial x \partial y}(a, b) \right. \\
& \left. + (y - b)^2 \frac{\partial^2 f}{\partial y^2}(a, b) \right) \\
& + \cdots .
\end{aligned}
$$

The point (a, b) is a singular point when the terms up to degree 1 in this expansion vanish. This has the following generalisation.

Definition 1.33. A point (a, b) on the curve $C : f(x, y) = 0$ is said to have *multiplicity m* on C if the partial derivatives of $f(x, y)$ all vanish at (a, b) up to degree $m - 1$,

$$\frac{\partial^{i+j} f}{\partial x^i \partial y^j}(a, b) = 0, \qquad 0 \le i + j \le m - 1,$$

but there exists a partial derivative of order m which is nonzero at (a, b). $\square$

The simplest kind of singular point is a point of multiplicity 2. In this case we can consider the quadratic equation

$$\xi^2 \frac{\partial^2 f}{\partial x^2}(a, b) + 2\xi\eta \frac{\partial^2 f}{\partial x \partial y}(a, b) + \eta^2 \frac{\partial^2 f}{\partial y^2}(a, b) = 0$$

with coefficients as in the (nonvanishing) degree 2 term in the Taylor expansion.

Definition 1.34. A point (a, b) at which the curve C has multiplicity 2 and the quadratic equation above does not have a repeated root is called an *ordinary double point*. □

Example 1.32 has an ordinary double point at the origin. An example of a curve with a double point (that is, a point of multiplicity 2) which is not ordinary is

$$y^2 = x^{n+1}, \qquad n \geq 2.$$

This is called a *simple singularity of type A_n*. When $n = 2$ it is called a *cusp*.

(b) Projective plane curves

Next, instead of a polynomial in two variables $f(x, y)$, we shall consider the geometry of a homogeneous polynomial in three variables

$$f(x, y, z) = \sum_{i+j+k=d} a_{ijk} x^i x^j x^k, \qquad a_{ijk} \in \mathbb{C}.$$

If $f(a, b, c) = 0$, then $f(\alpha a, \alpha b, \alpha c) = 0$ for any nonzero scalar $\alpha \in \mathbb{C}$. It is therefore natural, given a homogeneous polynomial, to consider the zero-set

$$C : \quad f(x, y, z) = 0$$

as defining a subset of the projective plane

$$\mathbb{P}^2 = \{(a : b : c) \mid (a, b, c) \neq (0, 0, 0)\}.$$

The projective plane $\mathbb{P}^2$ is covered by three affine planes

$$U_1 = \{(1 : b : c)\}, \quad U_2 = \{(a : 1 : c)\}, \quad U_3 = \{(a : b : 1)\}.$$

Consequently, the projective plane curve $C \subset \mathbb{P}^2$ is obtained by gluing the three affine plane curves

$$\begin{aligned}
\mathbb{A}^2 \cong U_1 \supset \quad C_1 : & \quad f(1, y, z) = 0 \\
\mathbb{A}^2 \cong U_2 \supset \quad C_2 : & \quad f(x, 1, z) = 0 \\
\mathbb{A}^2 \cong U_3 \supset \quad C_3 : & \quad f(x, y, 1) = 0.
\end{aligned}$$

(Gluing, in general, will be explained in Chapter 3.)

Example 1.35. Let $f(x, y, z) = y^2 z - x^2 z - x^3$. The projective plane curve

$$C : \quad y^2 z - x^2 z - x^3 = 0$$

does not pass through the point $(1 : 0 : 0)$ and is therefore obtained by gluing
the two affine curves

$$U_2 \supset C_2 : \quad z - x^2 z - x^3 = 0$$
$$U_3 \supset C_3 : \quad y^2 - x^2 - x^3 = 0$$

via the isomorphism

$$C_2 - \{R\} \to C_3 - \{P, Q\}, \qquad (x, z) \mapsto (x/z, 1/z),$$

where $P = \{x = y = 0\}$, $Q = \{x = -1, \ y = 0\}$, $R = \{x = z = 0\}$. (See
Figure 1.11.)   □

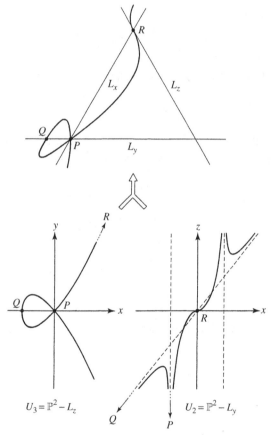

Figure 1.11: The projective plane curve $y^2 z - x^2 z - x^3 = 0$

The singular points of the projective curve $C \subset \mathbb{P}^2$ are just the singular points
of the affine curves C_i. As we have seen in Example 1.32, C_3 has the origin as its

only singular point. The curve C_2 is nonsingular. So in this example, C is a cubic curve with an ordinary double point at $(0 : 0 : 1)$ and nonsingular elsewhere.

The following fact is very convenient for locating the singular points of a projective plane curve directly from the homogeneous polynomial $f(x, y, z)$.

Proposition 1.36. *The singular points $(a : b : c)$ of the plane curve C : $f(x, y, z) = 0$ are the common zeros of the three partial derivatives:*

$$\frac{\partial f}{\partial x}(a, b, c) = \frac{\partial f}{\partial y}(a, b, c) = \frac{\partial f}{\partial z}(a, b, c) = 0.$$

Proof. Suppose that the polynomial $f(x, y, z)$ has degree d. Then f satisfies Euler's identity:

$$x\frac{\partial f}{\partial x} + y\frac{\partial f}{\partial y} + z\frac{\partial f}{\partial z} = d \cdot f(x, y, z).$$

It follows that the zero-set of the partials in the proposition coincides with the zero-set of the polynomials

$$f(x, y, z), \quad \frac{\partial f}{\partial x}(x, y, z), \quad \frac{\partial f}{\partial y}(x, y, z).$$

In the affine plane U_3, this is just the set of singular points of the curve $C_3 = C \cap U_3$. Similarly, the zeros in U_1 and U_2 are the singular points of C_1 and C_2. □

In the example $f(x, y, z) = y^2z - x^2z - x^3$, the singular points are found by solving the simultaneous equations

$$-2xz - 3x^2 = 2yz = y^2 - x^2 = 0,$$

and the only solution is $(0 : 0 : 1)$.

The following can be proved in much the same way.

Proposition 1.37. *A point $(a : b : c) \in \mathbb{P}^2$ has multiplicity $\geq m$ on the curve $C : f(x, y, z) = 0$ if and only if all partials of order $m - 1$ vanish at $(a : b : c)$,*

$$\frac{\partial^{m-1} f}{\partial x^i \partial y^j \partial z^k}(a, b, c) = 0, \qquad i + j + k = m - 1.$$

□

Just as in Euclidean geometry one investigates properties of figures which are invariant under rotations and translations, so in projective geometry one studies properties that are invariant under projective transformations, or, to say the same thing, properties which do not depend on a choice of projective coordinates.

In the case of a projective plane curve $C : f(x, y, z) = 0$, performing a projective transformation – or, equivalently, changing to a different system of homogeneous coordinates – gives a curve with defining equation

$$f^M(x, y, z) = f(ax + by + cz, a'x + b'y + c'z, a''x + b''y + c''z) \qquad (1.13)$$

for some invertible matrix

$$M = \begin{pmatrix} a & b & c \\ a' & b' & c' \\ a'' & b'' & c'' \end{pmatrix} \in GL(3, \mathbb{C}).$$

Thus, the projective geometry of plane curves amounts to the study of properties of homogeneous polynomials which are invariant under transformations coming from invertible matrices in this manner. A typical example of such a property is singularity. Another example is irreducibility: a curve is said to be irreducible if its defining equation does not factorise into polynomials of lower degree.

Definition 1.38. Two plane curves are said to be *projectively equivalent* if their defining equations are transformed into each other by some invertible matrix $M \in GL(3, \mathbb{C})$. $\qquad \square$

Clearly projective equivalence is an equivalence relation, and in fact the classification of plane curves of degree 2 up to this equivalence is rather simpler than the problem of Section 1.1. A projective plane curve of degree 2 can be written

$$C : \quad a_{11}x^2 + a_{22}y^2 + a_{33}z^2 + 2a_{12}xy + 2a_{13}xz + 2a_{23}yz = 0.$$

If we take the coefficients as entries of a symmetric matrix

$$A = \begin{pmatrix} a_{11} & a_{12} & a_{13} \\ a_{21} & a_{22} & a_{23} \\ a_{31} & a_{32} & a_{33} \end{pmatrix}, \qquad a_{ij} = a_{ji},$$

then the defining equation of the curve can be expressed in matrix form:

$$(x, y, z)A \begin{pmatrix} x \\ y \\ z \end{pmatrix} = 0.$$

The change of coordinates (1.13) transforms the matrix A to

$$M^t A M = \begin{pmatrix} a & a' & a'' \\ b & b' & b'' \\ c & c' & c'' \end{pmatrix} A \begin{pmatrix} a & b & c \\ a' & b' & c' \\ a'' & b'' & c'' \end{pmatrix}.$$

Consequently, we obtain the following from well-known facts of linear algebra.

Proposition 1.39. *Over the complex field $\mathbb{C}$ the projective equivalence class of a plane conic is determined by the rank of its defining symmetric matrix.* $\square$

This means that in projective geometry there are only three equivalence classes of conic, of ranks 3, 2 and 1. A rank 3 conic is projectively equivalent to $xz - y^2 = 0$. Rank 2 is equivalent to $xz = 0$ and is therefore a union of two distinct lines. The rank 1 case is equivalent to $y^2 = 0$ and is therefore a double line.

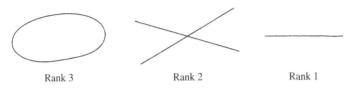

Rank 3   Rank 2   Rank 1

Figure 1.12: Conics over $\mathbb{C}$

In particular we see that, for degree 2 curves, being reducible is equivalent to being singular – however, this is a feature special to conics. Let us classify singular irreducible plane curves of degree 3. One has already been seen in Example 1.35, and there exists one other type.

Proposition 1.40. *An irreducible plane cubic curve which has a singular point is projectively equivalent to one of the following:*

(i) $y^2 z = x^3$;
(ii) $y^2 z = x^3 - x^2 z$.

Proof. Choose homogeneous coordinates so that the singular point is $(0 : 0 : 1)$. Then the defining equation $f(x, y, z)$ of the curve C cannot include the monomials z^3, xz^2, yz^2 and so is of the form

$$f(x, y, z) = zq(x, y) + d(x, y)$$

for some forms q of degree 2 and d of degree 3. By irreducibility, the quadratic form $q(x, y)$ is nonzero, and hence by making a linear transformation of the coordinates x, y it can be assumed to be one of

$$q(x, y) = xy, \ y^2.$$

In the first case the cubic form $d(x, y)$ must contain both monomials x^3, y^3 (otherwise C is reducible). Making a coordinate change $z \mapsto z + ax + by$ gives

$$f(x, y, z) = xyz + (d(x, y) + ax^2 y + bxy^2).$$

By choosing the coefficients a, b suitably we can bring the bracket to the cube of a linear form, and hence the curve is projectively equivalent to type (ii).

In the second case, $d(x, y)$ must include the monomial x^3. Changing coordinates by $x \mapsto x + ky$ for a suitable choice of coefficient k, one can kill the term $x^2 y$ in $d(x, y)$ and so obtain

$$f(x, y, z) = zy^2 + (ax^3 + bxy^2 + cy^3) = y^2(z + bx + cy) + ax^3.$$

If we now take as new coordinates $\sqrt[3]{a}x$, y and $z + bx + cy$, then the equation takes the form (i).   □

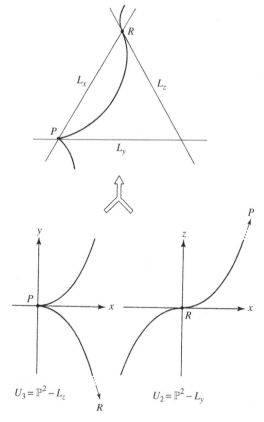

Figure 1.13: The projective curve $y^2 z - x^3 = 0$

Case (i) is the gluing of two affine cubics

$$U_2 \supset C_2 : \quad z = x^3$$
$$U_3 \supset C_3 : \quad y^2 = x^3$$

via the isomorphism

$$C_2 - \{R\} \to C_3 - \{P\}, \qquad (x, z) \mapsto (x/z, 1/z),$$

where $P = \{x = y = 0\}$ and $R = \{x = z = 0\}$.

1.5 Period parallelograms and cubic curves

The first plane curves one encounters which have moduli in a meaningful sense are the cubics. Changing our point of view somewhat, we shall approach these in this section from the direction of doubly periodic complex functions.

(a) Invariants of a lattice

Consider a parallelogram in the complex plane $\mathbb{C}$ with one vertex at the origin. The four vertices are then $0, \omega_1, \omega_2, \omega_1 + \omega_2 \in \mathbb{C}$.

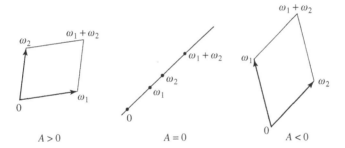

Figure 1.14: Oriented area

The area of this parallelogram is equal to the absolute value of the number

$$A = \operatorname{Im}(\overline{\omega}_1 \omega_2) = \frac{\overline{\omega}_1 \omega_2 - \omega_1 \overline{\omega}_2}{2\sqrt{-1}},$$

called the *oriented area*. The set of parallelograms with positive oriented area is parametrised by

$$\widetilde{\mathfrak{H}} = \{(\omega_1, \omega_2) \mid A(\omega_1, \omega_2) > 0\} \subset \mathbb{C}^2.$$

When $A \neq 0$ the complex plane is tesselated by the parallelogram and its translates by integer multiples of ω_1 and ω_2. The set of vertices of all the translated parallelograms then form a rank 2 free abelian subgroup of $\mathbb{C}$ called a *lattice*. This lattice is uniquely determined by the parallelogram, but the converse is not true. For a given lattice, giving a tesselating parallelogram is equivalent to specifying a $\mathbb{Z}$-basis.

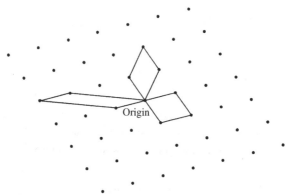

Figure 1.15: Fundamental domains of a lattice.

To say this another way, lattices in $\mathbb{C}$ are parametrised by the quotient space of the action of $GL(2, \mathbb{Z})$ on $\mathbb{C}^2$ by

$$(\omega_1, \omega_2) \mapsto (a\omega_1 + b\omega_2, c\omega_1 + d\omega_2), \qquad \begin{pmatrix} a & b \\ c & d \end{pmatrix} \in GL(2, \mathbb{Z}).$$

If we restrict to parallelograms with positive oriented area, that is, to the open set $\widetilde{\mathfrak{H}} \subset \mathbb{C}^2$, then we have an action of the modular group $SL(2, \mathbb{Z})$, and the quotient $\widetilde{\mathfrak{H}}/SL(2, \mathbb{Z})$ is a parameter space for lattices.

$SL(2, \mathbb{Z})$ is an infinite discrete group, neither finite nor connected. Nevertheless, the quotient $\widetilde{\mathfrak{H}}/SL(2, \mathbb{Z})$ can be constructed using invariant forms – that is, invariant analytic functions – on $\widetilde{\mathfrak{H}}$. The most basic of these are the *Eisenstein series*, for each even number $2k \geq 4$,

$$G_{2k}(\omega_1, \omega_2) = \sum_{(m,n) \neq (0,0)} \frac{1}{(m\omega_1 + n\omega_2)^{2k}}.$$

One can write this as a function of the lattice $\Gamma \subset \mathbb{C}$ as

$$G_{2k}(\Gamma) = \sum_{0 \neq \gamma \in \Gamma} \frac{1}{\gamma^{2k}}.$$

(This series is absolutely convergent as soon as $2k \geq 3$. Note, moreover, that if $2k$ is replaced by an odd number, then G_{2k} vanishes identically.)

Definition 1.41. An $SL(2, \mathbb{Z})$-invariant holomorphic function $F(\omega_1, \omega_2)$ on $\widetilde{\mathfrak{H}}$ which satisfies, for all $\alpha \in \mathbb{C}^*$,

$$F(\alpha\omega_1, \alpha\omega_2) = \alpha^{-w} F(\omega_1, \omega_2)$$

is called an *automorphic function of weight w* (in homogeneous form). □

Remarks 1.42.

(i) Multiplication of a lattice Γ by a nonzero complex number $\alpha \in \mathbb{C}$ corresponds to making a rotation together with a dilation (or contraction). The action of $\mathbb{C}^*$ by $\Gamma \mapsto \alpha\Gamma$ commutes with that of the modular group $SL(2, \mathbb{Z})$.

(ii) By restricting to $\omega_1 = 1$, $\omega_2 = \tau$, an automorphic function of weight w determines a holomorphic function $f(\tau)$ on the upper half-plane

$$\mathfrak{H} = \{\tau \mid \operatorname{Im} \tau > 0\}$$

which satisfies

$$f(\tau) = (c + d\tau)^{-w} f\left(\frac{a + b\tau}{c + d\tau}\right).$$

Conversely, given a function $f(\tau)$ satisfying this relation, the function

$$F(\omega_1, \omega_2) = \omega_1^{-w} f\left(\frac{\omega_2}{\omega_1}\right)$$

is an automorphic function of weight w. The definition above is therefore equivalent to the notion of a (nonhomogeneous) automorphic form in one variable.

(iii) The first examples of automorphic forms are the Eisenstein series G_{2k}, of weight $2k$. One can show (see, for example, Serre [7], chapter 7) that the ring of all automorphic forms (that is, the ring of all invariant holomorphic functions on $\widetilde{\mathfrak{H}}$) has a Hilbert series equal to

$$\frac{1}{(1 - t^4)(1 - t^6)}.$$

It follows from Proposition 1.9 that the ring is generated by G_4 and G_6. □

Consider now the holomorphic map

$$\widetilde{\mathfrak{H}} \to \mathbb{C}^2, \qquad (\omega_1, \omega_2) \mapsto (60 G_4(\omega_1, \omega_2), 140 G_6(\omega_1, \omega_2)).$$

Clearly this map factors through the quotient space (that is, the quotient complex manifold) $\widetilde{\mathfrak{H}}/SL(2, \mathbb{Z})$.

Theorem 1.43. *The holomorphic map*

$$\widetilde{\mathfrak{H}}/SL(2,\mathbb{Z}) \to \mathbb{C}^2, \qquad [\Gamma] \mapsto (u,v) = (60G_4(\Gamma), 140G_6(\Gamma))$$

is a bijection to the open set $u^3 - 27v^2 \neq 0$.

(b) The Weierstrass $\wp$ function

To prove Theorem 1.43 we shall make use of the Weierstrass $\wp$ function

$$\wp(z) = \wp(z;\omega_1,\omega_2) = \frac{1}{z^2} + \sum_{(m,n)\neq(0,0)} \left\{ \frac{1}{(z - m\omega_1 - n\omega_2)^2} - \frac{1}{(m\omega_1 + n\omega_2)^2} \right\},$$

or, alternatively,

$$\wp(z) = \wp(z;\Gamma) = \frac{1}{z^2} + \sum_{0\neq\gamma\in\Gamma} \left\{ \frac{1}{(z-\gamma)^2} - \frac{1}{\gamma^2} \right\}.$$

$\wp(z)$ is a doubly periodic meromorphic function, that is,

$$\wp(z+\omega_1) = \wp(z+\omega_2) = \wp(z),$$

with a double pole at each lattice point and regular elsewhere. Moreover, its Laurent expansion at the origin is given by

$$\wp(z) = \frac{1}{z^2} + \sum_{n=1}^{\infty}(2n+1)G_{2n+2}(\Gamma)z^{2n}.$$

(Thus, as a generating function for the Eisenstein automorphic forms, $\wp(z;\omega_1,\omega_2)$ ties together the moduli τ plane and the doubly periodic z plane.)

The following properties of doubly periodic functions are fundamental.

Liouville's Theorem 1.44. *Let $f(z)$ be a doubly periodic meromorphic function on the complex plane.*

(i) If $f(z)$ is holomorphic everywhere, then it is constant.
(ii) The sum of the residues of $f(z)$ over any period parallelogram is zero.
(iii) Over any period parallelogram $f(z)$ has the same number of poles as zeros.
(iv) In a given period parallelogram, suppose that $f(z)$ has zeros $u_1,\ldots,u_n$ and poles $v_1,\ldots,v_n$. Then

$$u_1 + \cdots + u_n + v_1 + \cdots + v_n \in \Gamma.$$

Note that in parts (iii) and (iv), poles and zeros are to be counted with their multiplicities.

Proof. For (i), observe that an entire doubly periodic function is bounded, and therefore constant. Statements (ii), (iii) and (iv) are obtained by integrating

$$f(z), \quad \frac{f'(z)}{f(z)}, \quad \frac{zf'(z)}{f(z)},$$

respectively, around the boundary of a period parallelogram. □

The derivative of $\wp(z)$ is

$$\wp'(z) = -2 \left(\frac{1}{z^3} + \sum_{0 \neq \gamma \in \Gamma} \frac{1}{(z-\gamma)^3} \right)$$
$$= -\frac{2}{z^3} - \sum_{n=1}^{\infty} 2n(2n+1)G_{2n+2}(\Gamma)z^{2n-1}.$$

This is a doubly periodic meromorphic function with a triple pole at each lattice point and regular elsewhere. Let $g_2(\Gamma) = 60G_4(\Gamma)$, $g_3(\Gamma) = 140G_6(\Gamma)$, and let $f(z) = \wp'(z)^2 - 4\wp(z)^3 + g_2(\Gamma)\wp(z) + g_3(\Gamma)$. Then, if one computes the Laurent expansion of $f(z)$ at the origin, one finds that $f(z)$ is holomorphic at the origin and vanishes there. But $f(z)$ is doubly periodic and holomorphic away from the lattice points, so by Liouville's Theorem 1.44(i) we obtain the identity

$$\wp'(z)^2 = 4\wp(z)^3 - g_2(\Gamma)\wp(z) - g_3(\Gamma). \tag{1.14}$$

Lemma 1.45. *If $f(X, Y)$ is a polynomial in two variables which vanishes when $X = \wp(z)$, $Y = \wp'(z)$, then $f(X, Y)$ is divisible by $Y^2 - 4X^3 - g_2(\Gamma)X - g_3(\Gamma)$.*

Proof. Viewing $f(X, Y)$ as a polynomial in Y, let $r(X, Y)$ be the remainder on division by $Y^2 - 4X^3 - g_2(\Gamma)X - g_3(\Gamma)$. This has degree at most 1 in Y and has the property that $r(\wp(z), \wp'(z)) \equiv 0$. On the other hand, $\wp'(z)$ is an odd function and so cannot be expressed as a rational function of the even function $\wp(z)$. Hence $r(X, Y)$ is zero. □

Lemma 1.46. *The cubic equation $4X^3 - g_2X - g_3 = 0$ corresponding to the right-hand side of (1.14) has no repeated root. In particular, its discriminant $g_2^3 - 27g_3^2$ is nonzero.*

Proof. If $\alpha \in \mathbb{C}$ were a repeated root, then we could write (1.14) as

$$\wp'(z)^2 = 4(\wp(z) - \alpha)^2(\wp(z) + 2\alpha),$$

and hence

$$\left(\frac{\wp'(z)}{\wp(z) - \alpha}\right)^2 = 4(\wp(z) + 2\alpha).$$

This implies that the function $\wp'(z)/(\wp(z)-\alpha)$ is doubly periodic with a simple pole at lattice points and regular elsewhere; but this contradicts part (ii) of Liouville's Theorem.   □

Proof of Theorem 1.43. Differentiating both sides of (1.14) gives

$$\wp''(z) = 6\wp(z)^2 - \tfrac{1}{2}g_2(\Gamma). \qquad (1.15)$$

Computing the Laurent expansion at the origin of each side of this equation, we obtain

$$\frac{1}{z^4} + \sum_{k=2}^{\infty} \binom{2k-1}{3} G_{2k} z^{2k-4} = \left(\frac{1}{z^2} + \sum_{k=2}^{\infty}(2k-1)G_{2k}z^{2k-2}\right)^2 - 5G_4,$$

and comparing coefficients yields a recurrence relation:

$$G_{2k} = \frac{3}{(2k-1)(2k+1)(k-3)} \sum_{\substack{i+j=k \\ i,j \geq 2}} (2i-1)(2j-1)G_{2i}G_{2j}, \quad k \geq 4.$$

$$(1.16)$$

It follows from this that all G_{2k} for $k \geq 4$ can be expressed as polynomials in G_4 and G_6. For example,

$$G_8 = \tfrac{3}{7}G_4^2, \qquad G_{10} = \tfrac{5}{11}G_4 G_6.$$

Now, given two lattices Γ, Γ', if

$$G_4(\Gamma) = G_4(\Gamma'), \qquad G_6(\Gamma) = G_6(\Gamma'),$$

then by the identity theorem it follows that $\wp(z; \Gamma) = \wp(z; \Gamma')$ on the whole complex plane. The Weierstrass function determines the lattice as the set of its poles, and so $\Gamma = \Gamma'$. This shows that the map in Theorem 1.43 is injective. That its image is in the complement of the curve $u^3 - 27v^2 = 0$ follows from Lemma 1.46. Surjectivity on this open set follows from results of Chapter 9 (see Section 9.6).   □

(c) The ℘ function and cubic curves

Let us consider the holomorphic map

$$\mathbb{C} \to \mathbb{P}^2, \qquad z \mapsto (\wp(z) : \wp'(z) : 1). \tag{1.17}$$

Double periodicity means that this map factors down to the Riemann surface $\mathbb{C}/\Gamma$. Moreover, (1.14) implies that the image is contained in the cubic curve

$$Y^2 Z = 4X^3 - g_2 X Z^2 - g_3 Z^3. \tag{1.18}$$

The origin $z = 0$ is a double pole of $\wp(z)$ and a triple pole of $\wp'(z)$, and so it maps to the point $(0 : 1 : 0) \in \mathbb{P}^2$. Moreover, this point is an *inflection point* of the curve; that is, its tangent line $Z = 0$ meets the curve with multiplicity 3 here and has no further intersection points.

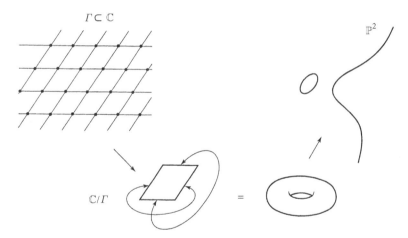

Figure 1.16: A complex cubic curve

The cubic curve (1.18) is nonsingular by Proposition 1.36 and Lemma 1.46. Both the quotient $\mathbb{C}/\Gamma$ and the cubic are compact Riemann surfaces, and so the map is an isomorphism.

Example 1.47. Let $\omega_1 = 1$. If $\omega_2 = \sqrt{-1}$, then $G_6(\Gamma) = 0$; so the Riemann surface $\mathbb{C}/\Gamma$ is isomorphic to the curve $Y^2 Z = 4X^3 - XZ^2$. If $\omega_2 = (-1 + \sqrt{-3})/2$, then $G_4(\Gamma) = 0$, and it is isomorphic to the curve $Y^2 Z = 4X^3 - Z^3$. □

The curve (1.18) can also be viewed as the Riemann surface of the multivalued function $\sqrt{4z^3 - g_2 z - g_3}$. By considering (elliptic) integrals

$$\oint_\alpha \frac{dz}{\sqrt{4z^3 - g_2 z - g_3}}$$

along closed paths α on the surface we recover our lattice in the complex plane.

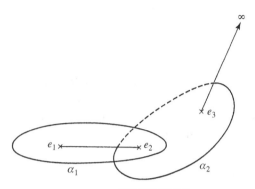

Figure 1.17: The Riemann surface of $\sqrt{4z^3 - g_2 z - g_3}$. e_1, e_2, e_3 are the three roots of $4z^3 - g_2 z - g_3 = 0$.

This is the subgroup $\Gamma \subset \mathbb{C}$ consisting of values of the integral taken along all (homology classes of) loops α and defines a mapping $(g_2, g_3) \mapsto \Gamma$ which is the inverse map of the map of Theorem 1.43. Indeed, this is a strengthening of the theorem, linking in a very pretty way automorphic functions with periodic maps. Extensions of this sort of discussion to more general algebraic varieties are one of the themes at the heart of moduli theory.

Finally, let us look at the relationship of the $\wp$ function with functions on degenerations of the cubic curve. First, consider a complex multiple $\alpha\Gamma$ of the lattice Γ and its limit as $|\alpha| \to \infty$. The Eisenstein constants $g_k(\Gamma)$ tend to zero, and the $\wp$ function and its derivative approach

$$\wp(z) \to \frac{1}{z^2}, \qquad \wp'(z) \to -\frac{2}{z^3}.$$

Thus, in the limit, the holomorphic map (1.17) becomes

$$\mathbb{C} \to \mathbb{P}^2, \qquad z \mapsto (z : -2 : z^3).$$

The image of this map is the singular cubic curve

$$Y^2 Z = 4X^3$$

with its singular point $(0 : 0 : 1)$ removed.

Next, we consider the effect of fixing one period, say, $\omega_1 = \pi$, and letting the other go to infinity, that is, $\omega_2 \mapsto k\omega_2$, and we take the limit as $k \to \infty$. The behaviour of the $\wp$ function and its derivative is then:

$$\wp(z) \to \frac{1}{z^2} + \sum_{n \in \mathbb{Z}} \left\{ \frac{1}{(z - n\pi)^2} - \frac{1}{(n\pi)^2} \right\},$$

$$= \frac{1}{\sin^2 z} - \frac{1}{3}$$

$$\wp'(z) \to \frac{-2 \cos z}{\sin^3 z},$$

and the map (1.17) becomes

$$\mathbb{C} \to \mathbb{P}^2, \qquad z \mapsto \left(\frac{1}{\sin^2 z} - \frac{1}{3} : \frac{-2 \cos z}{\sin^3 z} : 1 \right).$$

This map descends to the quotient $\mathbb{C}/\mathbb{Z}\pi \cong \mathbb{C}^*$, and its image is the singular cubic curve

$$Y^2 Z = 4 \left(X + \tfrac{1}{3}Z \right) \left(X - \tfrac{2}{3}Z \right)$$

with its singular point $(-1/3 : 0 : 1)$ removed.

In Proposition 1.40 the curve of part (i) corresponds to the origin of the (g_2, g_3) plane, while the curve of part (ii) corresponds to the curve $g_2^3 - 27g_3^2 = 0$. The preceding discussion shows that in addition the $\wp$ function behaves well under degeneration to these cases.

Remark 1.48. A nonsingular cubic curve always has exactly nine points of inflection. If we choose projective coordinates so that one of these is $(0 : 1 : 0)$ with tangent line $Z = 0$, then the defining equation of the curve takes the form (1.18). On the other hand, it is also possible to choose coordinates in such a way that the nine inflection points are

$$(-1 : \omega^i : 0), \quad (0 : -1 : \omega^i), \quad (\omega^i : 0 : -1),$$

for $i = 0, 1, 2$, where $\omega = e^{2\pi \sqrt{-1}/3}$. For these coordinates the equation of the curve takes the form

$$X^3 + Y^3 + Z^3 - 3\lambda XYZ = 0, \qquad \lambda \neq 1, \omega, \omega^2, \infty.$$

This is called the *Hessian cubic*. $\qquad\qquad\qquad\qquad\qquad\qquad\qquad\square$

Exercises

1. A linear map $E : V \rightarrow V$ from a vector space to itself which satisfies $E^2 = E$ is called a projection. Show that the dimension of the image is equal to the trace of E.

2. Show that the determinant

$$\begin{vmatrix} \xi_0 & \xi_1 & \xi_2 & \xi_3 \\ \xi_1 & \xi_2 & \xi_3 & \xi_4 \\ \xi_2 & \xi_3 & \xi_4 & \xi_5 \\ \xi_3 & \xi_4 & \xi_5 & \xi_6 \end{vmatrix}$$

is a classical invariant for sextic binary forms

$$(\xi \, \emptyset x, y) = \xi_0 x^6 + 6\xi_1 x^5 y + 15\xi_2 x^4 y^2 + 20\xi_3 x^3 y^3$$
$$+ 15\xi_4 x^2 y^4 + 6\xi_5 x y^5 + \xi_6 y^6.$$

3. (i) Show that the six linear fractional transformations defined by (1.12) form a group, and that this group is isomorphic to the symmetric group S_3.

   (ii) Show that, under the action of this group on the field of rational functions $\mathbb{C}(\lambda)$, the field of invariants is generated by

$$\frac{(\lambda^2 - \lambda + 1)^3}{\lambda^2 (\lambda - 1)^2}.$$

2

Rings and polynomials

The aim of this chapter is to give a very brief review of the basic algebraic techniques which form the foundation of invariant theory and of algebraic geometry generally. Beginning in Section 2.1 we introduce Noetherian rings, taking as our point of departure Hilbert's Basis Theorem, which was discovered in the search for a proof of finite generation of rings of invariants. (This result will appear in Chapter 4). In Section 2.2 we prove unique factorisation in polynomial rings, by induction on the number of variables using Gauss's lemma. In Section 2.3 we prove the important fact that in a finitely generated algebra over a field an element contained in all maximal ideals is nilpotent. As we will see in Chapter 3, this observation is really nothing other than Hilbert's Nullstellensatz.

A power series ring in one variable is an example of a valuation ring, and we discuss these in Section 2.4. A valuation ring (together with its maximal ideal) is characterised among subrings of its field of fractions as a maximal element with respect to the dominance relation. This will be used in Chapter 3 for proving the Valuative Criterion for completeness of an algebraic variety.

In the final section we discuss Nagata's example of a group action under which the ring of invariants which is not finitely generated – that is, his counterexample to Hilbert's 14th problem. This is constructed by taking nine points in general position in the projective plane and considering the existence and non-existence of curves of degree d with assigned multiplicity m at each of the points, and making use of Liouville's Theorem on elliptic functions.

2.1 Hilbert's Basis Theorem

We begin with a discussion of the Basis Theorem, which is the key to Hilbert's theorem of finite generatedness that we will meet in Chapter 4. In Hilbert's original paper [19] the word *ideal* is not used; and we would like to state the Basis Theorem in a form close to that expressed by Hilbert. Today we learn that an ideal of a ring S is a subset $J \subset S$ satisfying:

$$x \in S, \; y \in J \implies \quad xy \in J$$
$$y, z \in J \implies y \pm z \in J.$$

The following definition, however, is closer to the spirit of the original notion:

Definition 2.1. Given a subset $Y = \{y_\lambda \mid \lambda \in \Lambda\}$ of a ring S, the set of all linear combinations

$$\sum_{\lambda \in \Lambda} x_\lambda y_\lambda$$

(where the sum is finite – that is, $x_\lambda = 0$ for all but finitely many $\lambda \in \Lambda$) with coefficients $x_\lambda \in S$ is an *ideal* of S, called the *ideal generated by Y*.   □

This is completely analogous to the idea of a subspace spanned by some set of vectors in a vector space. If Y is a finite set $y_1, \ldots, y_m$, then we denote the ideal generated by Y by $(y_1, \ldots, y_m)$. Any ideal that can be expressed in this way is said to be *finitely generated*.

Theorem 2.2. *Let $S = k[x_1, \ldots, x_n]$ be a polynomial ring over an arbitrary (commutative) field k, and let J be an ideal generated by a subset $Y \subset S$. Then there exists a finite subset $y_1, \ldots, y_m \in Y$ which generates J.*

(This is similar to the definition of a compact topology: that for an arbitrary open cover there can be found a finite subcover.) We shall prove the theorem in the following well-known (and equivalent) form:

Theorem 2.3. *In the polynomial ring $k[x_1, \ldots, x_n]$ every ideal is finitely generated.*

In the case of a single variable the following stronger result is true, which we shall prove as preparation for the theorems above.

Theorem 2.4. *In the polynomial ring in one variable $k[x]$, every ideal is generated by a single element.*

Proof. We may assume that the ideal $I \subset k[x]$ is nonzero, and we let $f(x)$ be a nonzero polynomial in I with minimal degree. It is then enough to show the following:

Claim: Every $g(x) \in I$ is divisible by $f(x)$.

We assume that $g(x) \neq 0$, and we shall prove the claim by induction on the degree $d = \deg g(x)$. (The case $d = 0$ is trivial since $g \in I$ implies $g = 0$ if I is a proper ideal.) By the way in which $f(x)$ was chosen we have $m := d - \deg f(x) \geq 0$. Thus, for a suitable choice of constant $a \in k$, the polynomial $g_1(x) = g(x) - ax^m f(x)$ has degree strictly less than d. On the other hand, $g_1(x) \in I$, so by the inductive hypothesis it is divisible by $f(x)$. Hence so is $g(x)$. □

We can extract from this proof the 'principle of the leading term'. For any commutative ring R we denote by $R[x]$ the ring of polynomials in one variable x with coefficients in R.

Definition 2.5.

(i) Given a polynomial

$$f(x) = a_0 + a_1 x + \cdots + a_n x^n \in R[x],$$

we denote by Lt $f(x)$ the leading term $a_n x^n$ where $a_n \neq 0$. If $f(x)$ is identically zero, then we define Lt $f(x) = 0$.

(ii) If I is an ideal in $R[x]$, then we define Lt I to be the ideal generated by the set

$$\{\text{Lt } f(x) \mid f(x) \in I\}$$

of all leading terms of polynomials in I. □

For an ideal $I \subset R[x]$ let $I_{\leq k} \subset I$ be the subset consisting of polynomials in I with degree at most k, and let $\mathfrak{a}_k \subset R$ be the set of coefficients of x^k in all polynomials $f(x) \in I_{\leq k}$. Then Lt I can be expressed as

$$\sum_k \mathfrak{a}_k x^k = \mathfrak{a}_0 + \mathfrak{a}_1 x + \mathfrak{a}_2 x^2 + \cdots \subset R[x].$$

That is, Lt I consists of all polynomials for which the coefficient of each x^k belongs to $\mathfrak{a}_k$.

Lemma 2.6 (Principle of the leading term). *Let I be an ideal in $R[x]$, and suppose that $f_1(x), \ldots, f_N(x) \in I$. If the leading terms Lt $f_1(x), \ldots,$ Lt $f_N(x) \in$ Lt I generate Lt I, then the polynomials $f_1(x), \ldots, f_N(x)$ generate I.*

Proof. We shall show by induction on $\deg g(x)$ that, if $g(x) \in I$, then it is contained in the ideal $J = (f_1(x), \ldots, f_N(x))$. (As before, if $d = 0$, then $g = 0$.) By hypothesis, the leading term of $g(x)$ can be expressed in terms of $\mathrm{Lt}\, f_1(x), \ldots, \mathrm{Lt}\, f_N(x)$:

$$\mathrm{Lt}\, g(x) = \sum_{i=1}^{N} a_i x^{m_i} \mathrm{Lt}\, f_i(x)$$

for suitable $a_i \in R$, and where we have set $m_i = \deg g(x) - \deg f_i(x)$. This means that the polynomial

$$g_1(x) = g(x) - \sum_{i=1}^{N} a_i x^{m_i} f_i(x)$$

has strictly lower degree than $g(x)$ and so by the inductive hypothesis belongs to J. This implies that $g(x)$ it belongs to J. ☐

A ring R in which every ideal is finitely generated is called a *Noetherian ring*.

Theorem 2.7. *If R is a Noetherian ring, then the polynomial ring in one variable $R[x]$ over R is also Noetherian.*

Proof. Let I be an ideal in $R[x]$. The ideals $\mathfrak{a}_n \subset R$ defined above form an increasing sequence $\mathfrak{a}_n \subset \mathfrak{a}_{n+1}$, and we set

$$\mathfrak{a} = \bigcup_{n \geq 0} \mathfrak{a}_n.$$

It is clear that $\mathfrak{a}$ is an ideal in R. By hypothesis, then, $\mathfrak{a}$ is finitely generated. This means that we can find a finite number of polynomials $f_1(x), f_2(x), \ldots, f_M(x) \in I$ whose leading coefficients generate $\mathfrak{a}$. Denote by e the maximum degree of the polynomials $f_i(x)$. This implies that $\mathfrak{a}_e = \mathfrak{a}_{e+1} = \cdots = \mathfrak{a}$ (the *ascending chain condition*).

Using the Noetherian property of R again, we can choose some more polynomials $f_{M+1}(x), \ldots, f_N(x) \in I$ whose leading terms $\mathrm{Lt}\, f_{M+1}(x), \ldots,$ $\mathrm{Lt}\, f_N(x)$ provide generators for $\mathfrak{a}_0, \mathfrak{a}_1 x, \ldots, \mathfrak{a}_{e-1} x^{e-1}$. Then $\mathrm{Lt}\, f_1(x), \ldots,$ $\mathrm{Lt}\, f_M(x), \mathrm{Lt}\, f_{M+1}(x), \ldots, \mathrm{Lt}\, f_N(x)$ is a finite set of generators for $\mathrm{Lt}\, I$, and hence by Lemma 2.6 the polynomials $f_1(x), \ldots, f_N(x)$ generate I. ☐

If we repeat n times the operation of passing from the ring R to the polynomial ring $R[x]$, then we obtain the polynomial ring in n variables $R[x_1, \ldots, x_n]$. It follows that if we apply Theorem 2.7 n times, starting from the field k, then we obtain Theorem 2.3. ☐

2.2 Unique factorisation rings

It is well known that every integer m has a prime factorisation

$$m = \pm p_1^{n_1} p_2^{n_2} \dots p_l^{n_l} \quad \text{where } p_1, p_2, \dots, p_l \text{ are distinct primes,}$$

and that this factorisation is unique. Polynomials in the polynomial ring $k[x_1, \dots, x_n]$ have a similar property; first we shall collect together some ideas that we need.

Definition 2.8.

(i) A ring R in which $uv = 0$ (for $u, v, \in R$) only if one of $u, v = 0$ is called an *integral domain*.

(ii) An element $u \in R$ for which there exists some $v \in R$ satisfying $uv = 1$ is called an *invertible element*.

(iii) An element $u \in R$ with the property that $u = vw$ only if one of v, w is an invertible element is called an *irreducible element*.

(iv) An element $p \in R$ is called a *prime element* if the ideal $(p) \subset R$ that it generates is a prime ideal. In other words, if a product vw is divisible by p, then one of v, w is divisible by p.   □

(Recall that a prime ideal is an ideal $\mathfrak{p} \subset R$ with the property that, if $ab \in \mathfrak{p}$ for some $a, b \in R$, then at least one of a, b is contained in $\mathfrak{p}$. Equivalently, an ideal $\mathfrak{p} \subset R$ is prime if and only if the residue ring $R/\mathfrak{p}$ is an integral domain.)

The ring of rational integers $\mathbb{Z}$ is an integral domain, and its invertible elements are 1 and -1. Up to sign, therefore, the irreducible elements in $\mathbb{Z}$ are exactly the prime numbers. When R is the polynomial ring $k[x]$ in one variable over a field k, the invertible elements are the nonzero constant polynomials and the irreducible elements are the irreducible polynomials in the usual sense. In each of these cases, it follows from the Euclidean algorithm that every element is prime. In general, every prime element is irreducible, but the converse is not true. (See Exercise 2.1.)

Definition 2.9. An integral domain R is called a *unique factorisation domain* if the following two conditions are satisfied.

(i) Every irreducible element is prime.

(ii) An arbitrary element of R can be expressed as a product of (a finite number of) irreducible elements.   □

Theorem 2.10. *The polynomial ring $k[x_1, \dots, x_n]$ over a field k is a unique factorisation domain.*

As in the previous section, this will follow inductively from:

Theorem 2.11. *If R is a unique factorisation domain, then the polynomial ring R[x] is also a unique factorisation domain.*

In the polynomial ring $R[x]$ there are two kinds of prime elements. Let us begin with the simpler kind:

Lemma 2.12. *A prime element in the ring R is also prime in the ring R[x].*

The proof of this follows easily from Exercise 2.2.

Definition 2.13. A polynomial in $R[x]$ is said to be *primitive* if it is not divisible by any prime element of R.   □

Lemma 2.12 implies the following.

Gauss's Lemma 2.14. *A product of primitive polynomials is again primitive.*

From now on, R will be a unique factorisation domain and K will be its field of fractions. We shall view R and $R[x]$ as subrings of K and $K[x]$:

$$
\begin{array}{ccc}
K & \hookrightarrow & K[x] \\
\cup & & \cup \\
R & \hookrightarrow & R[x]
\end{array}
$$

Note that a polynomial $q(x)$ in $K[x]$ can always be expressed as a product of a primitive polynomial in $R[x]$ with an element of K; moreover, such a representation is unique up to multiplication by invertible elements of R.

Lemma 2.15. *Suppose that $q(x) \in K[x]$ and that $f(x) \in R[x]$ is primitive. Then $f(x)q(x) \in R[x]$ implies $q(x) \in R[x]$.*

Proof. As just noted, we can write $q(x) = cq'(x)$, where $q'(x)$ is a primitive polynomial and $c \in K$. By hypothesis, $cq'(x)f(x) \in R[x]$, and by the Gauss Lemma 2.14 the product $q'(x)f(x)$ is a primitive polynomial. It follows that $c \in R$ and so $q(x) \in R[x]$.   □

In other words, the property of a polynomial $g(x) \in R[x]$ being divisible by a primitive polynomial $f(x)$ is the same in the ring $K[x]$ as in the ring $R[x]$. From this observation we deduce the following.

Proposition 2.16. *For a primitive polynomial $f(x) \in R[x]$ the following three conditions are equivalent.*

(i) $f(x)$ is an irreducible element in $R[x]$.
(ii) $f(x)$ is an irreducible polynomial in $K[x]$.
(iii) $f(x)$ is a prime element in $R[x]$.

Together with Lemma 2.12, this exhausts all the prime elements of the ring $R[x]$.

Proof of Theorem 2.11. That an irreducible element in $R[x]$ is prime we have seen in Proposition 2.16. What remains is to show that an arbitrary $f(x) \in R[x]$ can be expressed as a product of irreducible polynomials in $R[x]$. First of all, we can do this in $K[x]$ and write

$$f(x) = g_1(x) \cdots g_N(x)$$

for $g_1(x), \ldots, g_N(x) \in K[x]$. Now, if we take a primitive polynomial $h_i(x) \in R[x]$ which equals $g_i(x)$ up to multiplication by an element of K, then $f(x)$ is divisible by each $h_i(x)$ and we obtain

$$f(x) = c h_1(x) \cdots h_N(x)$$

for some $c \in R$. Now, by decomposing c into primes in R, we get a prime decomposition of $f(x)$ in $R[x]$. □

Later on we shall make use of the following property of unique factorisation domains.

Proposition 2.17. *Let R be a unique factorisation domain and $\mathfrak{p} \subset R$ be a nonzero prime ideal containing no other prime ideals $\mathfrak{q} \subset \mathfrak{p}$ of R except $\mathfrak{q} = 0, \mathfrak{p}$. Then $\mathfrak{p}$ is generated by a single element.*

(A prime ideal containing no other prime ideals $\mathfrak{q} \subset \mathfrak{p}$ except $\mathfrak{q} = 0, \mathfrak{p}$ is said to be of *height* 1.)

Proof. Pick any nonzero element $u \in \mathfrak{p}$ and decompose it into primes in R. Since $\mathfrak{p}$ is a prime ideal, it must contain one of the factors in this decomposition. Call this element v. Then v generates a prime ideal (v), but by hypothesis this coincides with $\mathfrak{p}$. □

2.3 Finitely generated rings

We begin by noting the following fact.

Lemma 2.18. *The polynomial ring* $k[x_1, \ldots, x_n]$ *contains infinitely many irreducible polynomials.* □

Indeed, if the field is infinite, it is enough to take the linear polynomials $x_1 - a$ for $a \in k$. The case of a finite field is left to Exercise 2.6.

Definition 2.19. Let $S \subset R$ be a subring. An element $b \in R$ is said to be *integral* over S if it satisfies an equation $f(b) = 0$ for some monic polynomial with coefficients in S:

$$f(x) = x^n + a_1 x^{n-1} + \cdots + a_{n-1}x + a_n \in S[x].$$

R is *integral over* S if every element of R is integral over S. □

Lemma 2.20. *Suppose that R is integral over a subring $S \subset R$ and that $\mathfrak{a} \subset S$ is an ideal. If $\mathfrak{a}$ generates R, that is $\mathfrak{a}R = R$, then $\mathfrak{a} = S$.*

Proof. Since $1 \in \mathfrak{a}R$, we can write $1 = a_1 r_1 + \cdots a_m r_m$ for some $a_i \in \mathfrak{a}$ and $r_i \in R$. It is now enough to prove the result in the subring $R' = S[r_1, \ldots, r_m] \subset R$, since $\mathfrak{a}R = R$ implies $\mathfrak{a}R' = R'$. Since R' is integral over S, it is finitely generated as an S-module; so let $b_1, \ldots, b_N \in R'$ be generators. By hypothesis there exist coefficients $a_{ij} \in \mathfrak{a}$ such that

$$\begin{aligned}
b_1 &= a_{11}b_1 + a_{12}b_2 + \cdots + a_{1N}b_N \\
b_2 &= a_{21}b_1 + a_{22}b_2 + \cdots + a_{2N}b_N \\
&\cdots \\
b_N &= a_{N1}b_1 + a_{N2}b_2 + \cdots + a_{NN}b_N.
\end{aligned}$$

Let A be the determinant

$$A = \det \left| I_N - \begin{pmatrix} a_{11} & a_{12} & \cdots & a_{1N} \\ a_{21} & a_{22} & \cdots & a_{2N} \\ & & \cdots & \\ a_{N1} & a_{N2} & \cdots & a_{NN} \end{pmatrix} \right|.$$

Then $A - 1 \in \mathfrak{a}$ while $Ab_1 = Ab_2 = \cdots = Ab_N = 0$ (multiply both sides of the matrix equation by the adjugate matrix, noting that adj $M \times M = \det M \times I$). Hence $A = 0$ and so $1 \in \mathfrak{a}$. □

Lemma 2.21. *If R is integral over S and R is a field, then S is also a field.*

Proof. Let $a \in S$ be a nonzero element. Then $a^{-1} \in R$ is integral over S, so there exists some monic polynomial $f(x) = x^n + a_1 x^{n-1} + \cdots + a_{n-1} x + a_n \in S[x]$ satisfying

$$f(\frac{1}{a}) = \frac{1}{a^n} + \frac{a_1}{a^{n-1}} + \frac{a_2}{a^{n-2}} + \cdots + \frac{a_{n-1}}{a} + a_n = 0.$$

Multiplying through by a^{n-1} gives

$$-\frac{1}{a} = a_1 + a_2 a + \cdots + a_{n-1} a^{n-2} + a_n a^{n-1},$$

from which we see that $a^{-1} \in S$.   □

Lemma 2.22. *Suppose that an integral domain B is algebraic and finitely generated over a subring $A \subset B$. Then there exists a nonzero element $a \in A$ such that $B[a^{-1}]$ is integral over $A[a^{-1}]$.*

Proof. Any element $b \in B$ is algebraic over A; this means that it satisfies an equation $f(b) = 0$ for some nonzero polynomial $f(x)$ with coefficients in A. Denote by $0 \neq L(b) \in A$ the coefficient of the leading term, so we can write

$$f(x) = L(b) x^n + a_1 x^{n-1} + \cdots + a_{n-1} x + a_n \in A[x].$$

Consequently b is integral over $A[L(b)^{-1}]$. Given generators $b^{(1)}, \dots, b^{(N)}$ of B/A, the product $a = L(b^{(1)}) \cdots L(b^{(N)})$ now has the required property.   □

Proposition 2.23. *Let K be a field which is finitely generated as an algebra over a field k. Then the degree of the extension K/k is finite.*

Proof. It is enough to show that the extension K/k is algebraic. Let $y_1, \dots, y_N$ be generators of K as a k-algebra, ordered in such a way that:

(i) $y_1, \dots, y_M$ are algebraically independent; and
(ii) $y_{M+1}, \dots, y_N$ are algebraic over $k(y_1, \dots, y_M)$.

Our aim is to show that $M = 0$. In the previous lemma, take $B = K$ and $A = k[y_1, \dots, y_M]$: this gives a nonzero polynomial $f(y_1, \dots, y_M)$ such that K is integral over $k[y_1, \dots, y_M, f(y_1, \dots, y_M)^{-1}]$. By Lemma 2.21 this implies that $k[y_1, \dots, y_M, f(y_1, \dots, y_M)^{-1}]$ is a field. So for any polynomial

$g(y_1, \ldots, y_M)$ we have

$$\frac{1}{g(y_1, \ldots, y_M)} = \frac{h(y_1, \ldots, y_M)}{f(y_1, \ldots, y_M)^n}$$

for some natural number n and polynomial $h(y_1, \ldots, y_M)$. In other words, g divides a power of f. But by Lemma 2.18 there are infinitely many choices for irreducible g, if $M > 0$, and this forces $M = 0$.   □

By a *k-algebra* we will mean a commutative ring containing the field k as a subring.

Corollary 2.24. *If R is a finitely generated k-algebra and $\mathfrak{m} \subset R$ is a maximal ideal, then the composition*

$$k \to R \to R/\mathfrak{m}$$

is a finite (algebraic) field extension.

(Recall that an ideal $\mathfrak{m} \subset R$ is *maximal* if there are no ideals between $\mathfrak{m}$ and R, or, equivalently, if the residue ring $R/\mathfrak{m}$ is a field.)

Corollary 2.25. *Let R be a finitely generated k-algebra and $S \subset R$ a subring containing k. If $\mathfrak{m} \subset R$ is a maximal ideal, then $\mathfrak{m} \cap S$ is a maximal ideal in S.*

Proof. By the previous corollary $k \hookrightarrow R/\mathfrak{m}$ is a finite extension of fields; hence by Lemma 2.21 the intermediate residue ring $S/\mathfrak{m} \cap S$ is a field.   □

Proposition 2.26. *Let R be an integral domain finitely generated over a subring S. Then there exists an ideal $I \subset R$ such that $S \cap I = 0$ and the ring extension $S \hookrightarrow R/I$ is algebraic.*

Proof. Let K be the field of fractions of R, let k be the field of fractions of S and let $\widetilde{R} \subset K$ be the subring generated by R and k. This is a finitely generated k-algebra, and, if we choose any maximal ideal $\mathfrak{m} \subset \widetilde{R}$, then $k \hookrightarrow \widetilde{R}/\mathfrak{m}$ is an algebraic extension of fields. It follows that $I = \mathfrak{m} \cap R$ has the required properties.   □

Theorem 2.27. *Let R be a finitely generated k-algebra and $a \in R$ an element contained in all maximal ideals of R. Then a is nilpotent.*

Proof. We consider the linear polynomial $1 - ax$ in $R[x]$. Let $\mathfrak{m} \subset R[x]$ be an arbitrary maximal ideal; by Corollary 2.25, the intersection $\mathfrak{m} \cap R$ is a maximal ideal in R. Therefore $\mathfrak{m}$ contains a, and this implies that it cannot contain $1 - ax$. It follows (since $\mathfrak{m}$ is arbitrary) that $1 - ax$ is an invertible element: that is, there exists a polynomial $c_0 + c_1 x + \cdots + c_n x^n$ such that

$$(1 - ax)(c_0 + c_1 x + \cdots + c_n x^n) = 1.$$

From this it follows easily that $a^{n+1} = 0$ (Exercise 2.3). $\qquad\square$

This last result will be the key to Hilbert's Nullstellensatz in the next chapter.

2.4 Valuation rings

(a) Power series rings

A complex function $f(z)$, regular in a neighbourhood of the origin, has a Taylor expansion of the form

$$f(z) = \sum_{n=0}^{\infty} a_n z^n, \qquad \varlimsup_{n \to \infty} \sqrt[n]{|a_n|} < +\infty.$$

The set of all power series $\sum_{n=0}^{\infty} a_n z^n$ of this form, equipped with the usual rules of addition and multiplication, forms a ring, called the *convergent power series ring* and denoted by $\mathbb{C}\{z\}$.

A meromorphic function on a neighbourhood of the origin has a Laurent expansion of the form

$$f(z) = \sum_{n=-N}^{\infty} a_n z^n, \qquad \varlimsup_{n \to \infty} \sqrt[n]{|a_n|} < +\infty.$$

The set of these Laurent series, again with the usual algebraic operations, is a field; moreover, this field is exactly the field of fractions of $\mathbb{C}\{z\}$.

These two examples are the prototype for the valuation rings and valuation fields that we will discuss in the following. In our discussion, however, the topology of the complex number field $\mathbb{C}$ will play no part, and accordingly we can view the convergence conditions in the definitions above as dispensable; these definitions then make sense over an arbitrary field k. To emphasise this change of viewpoint we shall replace the analytic coordinate z by the formal symbol t. We then consider the set of formal power series

$$f(t) = \sum_{n=0}^{\infty} a_n t^n, \qquad a_n \in k.$$

With the usual rules of addition and multiplication this set becomes a ring, called the *formal power series ring* over k and denoted by $k[[t]]$. In contrast to the case of a convergent power series, $f(t)$ is not to be interpreted as a function. Only at $t = 0$ is evaluation of f allowed, and this determines a surjective ring homomorphism

$$\mathrm{sp} : k[[t]] \to k, \qquad f(t) \mapsto f(0) = a_0.$$

This is called the *specialisation map*, or *reduction* of the ring. Its kernel, ker sp, is the maximal ideal generated by t. If $f(0) \neq 0$, then $f(t)$ is an invertible element of $k[[t]]$. Moreover, although $f(t)$ is not a function, one can nonetheless define the multiplicity of $t = 0$ as a zero: this is the unique integer $n \geq 0$ for which there exists a nonzero element $u(t) \in k[[t]]$ such that

$$f(t) = t^n u(t), \qquad u(0) \neq 0.$$

Laurent series can also be considered formally. We consider the set of all formal series, allowing only finitely many nonzero negative powers:

$$f(t) = \sum_{n \in \mathbb{Z}} a_n z^n, \qquad a_n \in k, \quad \sharp\{n < 0 \mid a_n \neq 0\} < \infty.$$

This set becomes a field under the usual algebraic operations, called the *field of formal Laurent series* and denoted by $k((t))$. This is the field of fractions of the formal power series ring $k[[t]]$ and can be expressed as:

$$k((t)) = \lim_{N \to \infty} t^{-N} k[[t]] = \bigcup_N t^{-N} k[[t]].$$

In fact, a nonzero element $f(t)$ in $k((t))$ can be written uniquely as

$$f(t) = t^n u(t), \qquad u(t) \in k[[t]], \quad u(0) \neq 0.$$

This integer $n \in \mathbb{Z}$ is called the *valuation* of $f(t)$ (with respect to the variable t) and is written $v(f)$. Thus we have a map

$$v : k((t)) - 0 \to \mathbb{Z}, \qquad f \mapsto v(f),$$

also called the valuation. At $0 \in k((t))$ the valuation $v(0)$ is not defined, but this will cause no problems; when necessary it is convenient to adopt the convention that $v(0) = +\infty$.

It is easy to verify the following properties of the valuation.

- $v(fg) = v(f) + v(g)$;
- $v(f + g) \geq \min\{v(f), v(g)\}$;
- if $v(f) \neq v(g)$, then equality holds.

(b) Valuation rings

In analysis one has notions of limits, such as the limit $\lim_{n\to\infty} a_n$ of a sequence $\{a_n\}$, and limits of functions, such as

$$\lim_{t\to 0} \frac{\sin t}{t} = 1.$$

In algebra the notions that correspond to such limiting processes are valuation rings and their specialisations.

Definition 2.28. Let R be an integral domain with field of fractions K. If, for every $x \in K$, either $x \in R$ or $1/x \in R$, then R is called a *valuation ring*. □

If R is an integral domain with field of fractions K, then the set R^* of invertible elements in R forms a subgroup of the multiplicative group K^* of nonzero elements of K. We denote by

$$\Lambda = K^*/R^*$$

the (abelian) quotient group. Following the usual custom we shall write the group operation in Λ additively. Note in particular that the residue class of $1 \in K^*$ is denoted $0 \in \Lambda$.

Definition 2.29. For $x, y \in K^*$ with residue classes $\overline{x}, \overline{y} \in \Lambda$, we define

$$\overline{x} \geq \overline{y} \quad \text{if} \quad x/y \in R.$$

□

This defines a partial ordering on the group Λ and is a total ordering if R is a valuation ring. In this case Λ, together with the ordering $\geq$, is called the *valuation group*, and the natural homomorphism

$$v : K^* \to \Lambda, \qquad x \mapsto \overline{x}.$$

is called the *valuation* of K with respect to the ring R.

Depending on the case being considered, it may also be convenient to introduce a maximal element $+\infty$ to Λ and to define $v(0) = +\infty$. Then the valuation extends to the whole field K by

$$v : K \to \Lambda \cup \{+\infty\}, \quad v(x) = \begin{cases} \overline{x} & \text{if } x \neq 0 \\ +\infty & \text{if } x = 0. \end{cases}$$

Example 2.30. The ring of formal power series $R = k[[t]]$ is a valuation ring with valuation group isomorphic to the infinite cyclic group $\mathbb{Z}$. The valuation $v : K^* = k((t)) - 0 \to \mathbb{Z}$ then coincides with that described in the previous section.   □

Example 2.31. The collection of rational functions

$$\left\{ \frac{g(t)}{f(t)} \mid f(t), g(t) \in k[t], \ f(0) \neq 0 \right\} \subset k(t),$$

allowing only denominators which are nonzero at $t = 0$, is a valuation ring, again with valuation group $\mathbb{Z}$.   □

A valuation ring whose valuation group is infinite cyclic as in these examples is called a *discrete valuation ring*. Another typical example is the ring $\mathbb{Z}_p$ of p-adic integers.

Valuations have the following properties (already seen for the formal power series ring).

Proposition 2.32. *Let $x, y \in K$.*

(i) $v(xy) = v(x) + v(y)$.
(ii) $v(x + y) \geq \min\{v(x), v(y)\}$.

Proof. We only need to prove (ii). It is sufficient to assume that both x and y are nonzero. By definition of a valuation ring, one of x/y or y/x belongs to R; since the statement is symmetric in x, y, we may assume that $y/x \in R$. Then by definition $v(y) \geq v(x)$, while (since $1 + (y/x) \in R$) $v(1 + y/x) \geq v(1) = 0$. Hence

$$v(x + y) = v\left(x(1 + \tfrac{y}{x})\right) = v(x) + v\left(1 + \tfrac{y}{x}\right) \geq v(x) = \min\{v(x), v(y)\}.$$

□

The next fact is an easy exercise.

Proposition 2.33. *Let R be a valuation ring with field of fractions K and define*

$$\mathfrak{m} = \{x \in R \mid \frac{1}{x} \notin R\} \cup \{0\} \subset K.$$

Then $\mathfrak{m}$ is the unique maximal ideal in R. In particular, every valuation ring is a local ring.   □

Example 2.34. The formal power series ring $R = k[[t]]$ is a local ring with maximal ideal $\mathfrak{m}$ generated by t, and the quotient $R/\mathfrak{m}$ is isomorphic to k. $\square$

We next clarify the position occupied by the valuation rings among all local integral domains (Theorem 2.36 below).

Definition 2.35. Let A, B be rings and $\mathfrak{m} \subset A$, $\mathfrak{n} \subset B$ be maximal ideals. We say that $(A, \mathfrak{m})$ *dominates* $(B, \mathfrak{n})$ if $B \hookrightarrow A$ is a subring such that $\mathfrak{m} \cap B = \mathfrak{n}$. The relation of dominance is written

$$(A, \mathfrak{m}) \geq (B, \mathfrak{n}).$$

$\square$

Note that a ring/maximal ideal pair $(A, \mathfrak{m})$ is always dominated by its localisation

$$A_{\mathfrak{m}} = \left\{ \frac{x}{y} \mid x, y \in A, \ y \notin \mathfrak{m} \right\}.$$

This shows that a maximal element with respect to dominance is always a local ring. Moreover, the following is true.

Theorem 2.36. *Let K be a field and $R \subset K$ be a subring with maximal ideal $\mathfrak{m} \subset R$. If the pair $(R, \mathfrak{m})$ is maximal among subrings of K (and maximal ideals in them) with respect to dominance, then R is a valuation ring with field of fractions equal to K.*

(The converse is also true – see Exercise 2.9.)

Proof. Pick an element $x \in K$. We will show that either x or $1/x \in R$. These inclusions will follow, respectively, from the following two possibilities for x.

(1) x is integral over R.

In this case let $\tilde{R} = R[x] \subset K$ be the subring generated by R and x, and let $\tilde{\mathfrak{m}} \subset \tilde{R}$ be an arbitrary maximal ideal. Then the intersection $\mathfrak{p} = R \cap \tilde{\mathfrak{m}}$ is a prime ideal in R and there exists a natural inclusion

$$R/\mathfrak{p} \hookrightarrow \tilde{R}/\tilde{\mathfrak{m}}.$$

Here $\tilde{R}/\tilde{\mathfrak{m}}$ is a field and is integral over $R/\mathfrak{p}$, so by Lemma 2.21 $\mathfrak{p}$ is a maximal ideal. Since R is a local ring, this implies $\mathfrak{p} = \mathfrak{m}$. But this says that $(\widetilde{R}, \widetilde{\mathfrak{m}})$ dominates $(R, \mathfrak{m})$, so, by the maximality hypothesis with respect to dominance, equality holds. Thus $x \in R$.

(2) x is not integral over R.

In this case we let $\widetilde{R} = R[1/x] \subset K$ be the subring generated by R and $1/x$.

Claim: $1/x$ is not an invertible element in $\widetilde{R}$.

For otherwise we would have an element

$$a_1 + a_2 x^{-1} + \cdots + a_d x^{-d+1} + a_{d+1} x^{-d} \in R[1/x]$$

equal to x. Multiplying this equality through by the denominator we obtain an equation

$$x^{d+1} - a_1 x^d - a_2 x^{d-1} - \cdots - a_d x - a_{d+1} = 0, \quad a_i \in R,$$

contrary to hypothesis.

It follows from the claim that there exists a maximal ideal $\widetilde{\mathfrak{m}} \subset \widetilde{R}$ containing $1/x$. As before, we consider the prime ideal $\mathfrak{p} = R \cap \widetilde{\mathfrak{m}}$ and the natural inclusion

$$R/\mathfrak{p} \hookrightarrow \widetilde{R}/\widetilde{\mathfrak{m}}.$$

But this map is surjective since $1/x \in \widetilde{\mathfrak{m}}$; so in this case also $\mathfrak{p}$ is a maximal ideal and hence equal to $\mathfrak{m}$. As in case (1), we now argue from the maximality hypothesis that $1/x \in R$.   □

From this theorem one can deduce the existence, for any field K, of valuation rings with field of fractions K. (More precisely, K is any field that contains a subring which is not a field. A finite field, for example, does not satisfy this requirement.) For in the set of all pairs $(B, \mathfrak{n})$, where $B \subset K$ is a subring and $\mathfrak{n} \subset B$ is a maximal ideal, ordered by dominance, one notes that every chain has an upper bound (for this it is enough to take unions); then by Zorn's Lemma and Theorem 2.36, we see that every $(B, \mathfrak{n})$ is dominated by some valuation ring $(A, \mathfrak{m})$ in K.

This is already a strong result. However, by thinking more carefully about the residue fields involved we can give a more precise formulation. Note that whenever $(A, \mathfrak{m})$ dominates $(B, \mathfrak{n})$ there is an induced inclusion of residue fields $B/\mathfrak{n} \hookrightarrow A/\mathfrak{m}$.

Theorem 2.37. *Let B be a subring of a field K and $\mathfrak{n} \subset B$ be a maximal ideal. Then there exists a valuation ring $(R, \mathfrak{m})$ dominating $(B, \mathfrak{n})$ whose field* *of fractions is K and such that the field extension*

$$B/\mathfrak{n} \hookrightarrow R/\mathfrak{m}$$

is algebraic.

Proof. Let k be the algebraic closure of the residue field $B/\mathfrak{n}$ and fix an embedding $B/\mathfrak{n} \hookrightarrow k$. Then denote by

$$h : B \to k$$

the composition $B \to B/\mathfrak{n} \hookrightarrow k$. We now define an order relation on the set of pairs (A, g), where A is a subring of K and $g : A \to k$ is a ring homomorphism, as follows. We set $(A_1, g_1) \geq (A_2, g_2)$ if and only if $(A_1, \ker g_1)$ dominates $(A_2, \ker g_2)$ in the sense of Definition 2.35 and $g_1 : A_1 \to k$ restricts on A_2 to $g_2 : A_2 \to k$. On this partially ordered set every chain has an upper bound. Consequently, there exists a maximal element (R, g) which dominates (B, h). It is now enough to show that the pair $(R, \mathfrak{m})$, where $\mathfrak{m} = \ker g$, is a valuation ring.

For this, we look again at the proof of Theorem 2.36. Although the property of being a valuation ring followed from maximality, there were two cases to be considered. In case (1) the field extension

$$R/\mathfrak{m} \hookrightarrow \widetilde{R}/\widetilde{\mathfrak{m}}$$

was algebraic since the element $x \in K$ was integral. This means that the embedding of $R/\mathfrak{m}$ in k induced by g extends to an embedding of $\widetilde{R}/\widetilde{\mathfrak{m}}$ in k. Hence the homomorphism $g : R \to k$ extends to a homomorphism $\widetilde{g} : \widetilde{R} \to k$. In case (2) we no longer have the extension of residue fields, but we nevertheless construct an extension. For this, one can see that the maximal element $(R, \ker g)$ for the new order relation is a valuation ring. $\qquad\square$

Here is an example of a case in which the extension of residue fields is transcendental and at the same time of a valuation ring which is not discrete.

Example 2.38. Let $K = k(x, y)$ be the field of fractions of the polynomial ring in two variables $k[x, y]$. The subring

$$B = \left\{ \frac{f(x)}{g(x)} \mid f(x), g(x) \in k[x], \ g(0) \neq 0 \right\}$$

is a discrete valuation ring with residue field k and field of fractions $k(x)$. The subring

$$A = \left\{ \frac{f(x, y)}{g(x, y)} \mid f(x, y), g(x, y) \in k[x, y], \ g(0, y) \neq 0 \right\}$$

is a discrete valuation ring, with field of fractions $k(x, y)$, which dominates B. Its residue field is $k(y)$ and is a transcendental extension of the residue field k of B.

Take a valuation ring C whose field of fractions is equal to the residue field of A: for example,

$$C = \left\{ \frac{d(y)}{e(y)} \mid d(y), e(y) \in k[y], \ e(0) \neq 0 \right\},$$

and define

$$R = \left\{ \frac{f(x, y)}{g(x, y)} \mid f(x, y), g(x, y) \in k[x, y], \ g(0, y) \in C0 \right\}.$$

(This is called the *composition* of the valuation rings A and C.) This is (an example of) the valuation ring whose existence is guaranteed by Theorem 2.37. R has the same residue field as B, but note that it is no longer discrete. In fact, the valuation group of R is $\mathbb{Z} \oplus \mathbb{Z}$ equipped with the lexicographic ordering. □

2.5 A diversion: rings of invariants which are not finitely generated

At the International Congress of Mathematicians at Paris in 1900, David Hilbert, who had proved the finite generation of classical rings of invariants, posed the following question.

Hilbert's 14th Problem: *If an algebraic group acts linearly on a polynomial ring in finitely many variables, is the ring of invariants always finitely generated?*

Although it inverts the historical order of events, we will explain in the remainder of this chapter the counterexample to this question due to Nagata [11].

Remark 2.39. The fundamental result which we will prove in Chapter 4 (see Theorem 4.53) is that the answer to Hilbert's problem is yes if the group G is linearly reductive. In fact, the answer is also yes if G is the additive (non-reductive) group $\mathbb{C}$ (or, more generally, of the field k. See Weitzenböck [16] or Seshadri [15]). What about the additive group $\mathbb{C}^s$ for $s \geq 2$? Nagata found a counterexample for $s = 13$, and below we improve this to $s = 6$ (Corollary 2.46 below). (Since writing this book, the author has found counterexamples for $s = 3$ (Mukai [10]), but the case $s = 2$ remains open!) $\qquad\square$

(a) Graded rings

Definition 2.40. A ring R with direct sum decomposition $R = \bigoplus_{e \in \mathbb{Z}} R_{(e)}$ satisfying $R_{(e)} R_{(e')} \subset R_{(e+e')}$ is called a *graded ring*. $\qquad\square$

The following fact is obvious, but is of fundamental importance because of its role in the proof of Hilbert's Theorem 4.51 later on (see Exercise 2.10).

Proposition 2.41. *Suppose that R is a graded ring for which $R_{(e)} = 0$ for all $e < 0$. If the ideal $R_+ = \bigoplus_{e>0} R_{(e)}$ is finitely generated, then R is finitely generated as an algebra over $R_{(0)}$.* $\qquad\square$

More generally one can replace $\mathbb{Z}$, indexing the summands in the definition of a graded ring, by any group or semigroup. But note that in the above proposition, conversely, if the semigroup (or that part of it supporting the grading) is not finitely generated, then the ring R will also not be finitely generated.

First, two examples.

Example 2.42. Consider the set of all polynomials $f(x, y)$ in two variables, whose restriction to the x-axis is constant. This defines a subring $R \subset k[x, y]$. A polynomial in R can be written

$$f(x, y) = \text{constant} + y g(x, y).$$

Thus, as a vector space over k, the ring R has a basis consisting of monomials $x^m y^n$ such that $n \geq 1$ or $(m, n) = (0, 0)$ (see Figure 2.1). In other words, R is a semigroup ring, graded by the semigroup $G \subset \mathbb{Z}^2$ consisting of such pairs (m, n):

$$R = \bigoplus_{(m,n) \in G} k\{x^m y^n\}.$$

It is clear that G is not finitely generated, and hence R also fails to be finitely generated. $\qquad\square$

Example 2.43. The vector subspace of $k[x, y, y^{-1}]$ spanned by monomials

$$\{x^m y^n \mid -\sqrt{2}m \leq n \leq \sqrt{2}m\}$$

is also a subring. Again, this is a semigroup ring graded by a semigroup which is not finitely generated.   □

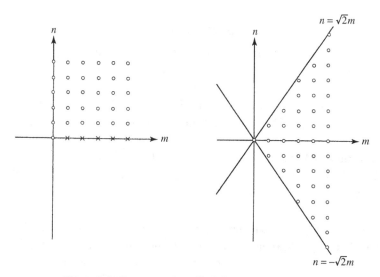

Figure 2.1: Two examples of infinitely generated rings

These examples can be generalised in the following way. Suppose that a ring R is bigraded, that is,

$$R = \bigoplus_{(m,n)\in\mathbb{Z}^2} R_{(m,n)}$$

such that the product of $R_{(m,n)}$ and $R_{(m',n')}$ is contained in $R_{(m+m',n+n')}$. Define the *support* of R to be

$$\text{Supp } R := \{(m, n) \in \mathbb{Z}^2 \mid R_{(m,n)} \neq 0\}.$$

If R is an integral domain, then Supp R is a subsemigroup of $\mathbb{Z}^2$.

Proposition 2.44. *If R is a bigraded integral domain and* Supp R *is not finitely generated as a semigroup, then R is not finitely generated as a ring.*   □

(b) Nagata's trick

Let $\mathbb{A}^{2N} \cong \mathbb{C}^{2N}$ be a $2N$-dimensional complex affine space with coordinates $(p_1, \ldots, p_N, q_1, \ldots, q_N)$, where we will assume that $N \geq 3$. (For the definition of affine space, see Section 3.1(a) in the next chapter.) We define an action of $\mathbb{C}^N$ on $\mathbb{A}^{2N}$ by

$$
\begin{aligned}
p_i &\mapsto p_i, \\
q_i &\mapsto s_i p_i + q_i,
\end{aligned}
$$

for $(s_1, \ldots, s_N) \in \mathbb{C}^N$, and an action of $(\mathbb{C}^*)^N$ on $\mathbb{A}^{2N}$ by

$$
\begin{aligned}
p_i &\mapsto t_i p_i, \\
q_i &\mapsto t_i q_i,
\end{aligned}
$$

for $(t_1, \ldots, t_N) \in \mathbb{C}^* \times \cdots \times \mathbb{C}^* = (\mathbb{C}^*)^N$. Note that these two actions commute. The ring of invariants for the action of $\mathbb{C}^N$ is $\mathbb{C}[p_1, \ldots, p_N]$, while the ring of invariants for the action of $(\mathbb{C}^*)^N$ is $\mathbb{C}$, the constant functions only.

We now pick N points $w_1 = (a_1, b_1), \ldots, w_N = (a_N, b_N)$ in the affine plane $\mathbb{A}^2$. With respect to these points, we consider the subset $G \subset \mathbb{C}^N$ of transformations $s \in \mathbb{C}^N$ which leave invariant the three rational forms

$$
A = a_1 \frac{q_1}{p_1} + \cdots + a_N \frac{q_N}{p_N},
$$

$$
B = b_1 \frac{q_1}{p_1} + \cdots + b_N \frac{q_N}{p_N},
$$

$$
C = \frac{q_1}{p_1} + \cdots + \frac{q_N}{p_N}.
$$

Thus $G \subset \mathbb{C}^N$ is a vector subspace of dimension $N - 3$.

Similarly, the set of transformations $t \in (\mathbb{C}^*)^N$ which leave invariant the product $D = p_1 \ldots p_N$ is a subgroup of codimension 1, which we shall denote by $T \subset (\mathbb{C}^*)^N$.

Theorem 2.45. *Let $N = 9$. If the points $w_1, \ldots, w_9 \in \mathbb{A}^2$ are sufficiently general, then the ring of invariants for the action of $G \cdot T \cong \mathbb{C}^6 \cdot (\mathbb{C}^*)^8 \subset GL(18, \mathbb{C})$ on the polynomial ring $\mathbb{C}[p_1, \ldots, p_N, q_1, \ldots, q_N]$ is not finitely generated.*

The following corollary follows from Hilbert's Theorem 4.53 in Chapter 4, since the invariant ring is S^T, where S is the ring of G-invariants, acted on by the (linearly reductive) group T.

Corollary 2.46. *If $N = 9$ and the points $w_1, \ldots, w_9 \in \mathbb{A}^2$ are sufficiently general, then the ring of invariants $\mathbb{C}[p_1, \ldots, p_N, q_1, \ldots, q_N]^G$ under the action of $G \cong \mathbb{C}^6$ fails to be finitely generated.* □

First of all, it is clear that the field of rational functions invariant under the action of the group $G \cdot T$ is generated by A, B, C and D. Consequently, the ring of invariants consists of those rational functions in A, B, C, D which are polynomials in $p_1, \ldots, p_N, q_1, \ldots, q_n$:

$$R := \mathbb{C}[p_1, \ldots, p_N, q_1, \ldots, q_N]^{G \cdot T}$$
$$= \mathbb{C}(A, B, C, D) \cap \mathbb{C}[p_1, \ldots, p_N, q_1, \ldots, q_n]$$
$$\subset \mathbb{C}(p_1, \ldots, p_N, q_1, \ldots, q_n).$$

It is not hard to show that an element of R is necessarily a polynomial in A, B, C, that is, $R \subset \mathbb{C}[A, B, C, D, D^{-1}]$. The invariant ring R is therefore bigraded by the degree in D and the homogeneous degree in A, B, C. We will write

$$R = \bigoplus_{\substack{d \geq 0 \\ m \in \mathbb{Z}}} R_{(d,m)},$$

where

$$R_{(d,m)} = \{D^{d-m} f(A, B, C) \mid f \text{ homogeneous of degree } d\}.$$

What is the support of this bigraded ring? Clearly, when $m \leq 0$, the homogeneous polynomial f is completely arbitrary, while for large $m > 0$ the condition that $D^{d-m} f(A, B, C)$ be a polynomial in p_i, q_j becomes nontrivial. Nagata's trick is the next lemma, which determines the support of this bigraded invariant ring in terms of the geometry of plane curves.

Lemma 2.47. *For a homogeneous polynomial $f(x, y, z)$ of degree d and for a positive integer $m > 0$, the following are equivalent:*

(i) $D^{d-m} f(A, B, C) \in R_{(d,m)}$;
(ii) $f(x, y, z)$ has a zero of multiplicity m at each of the points $w_1 = (a_1 : b_1 : 1), \ldots, w_N = (a_N : b_N : 1) \in \mathbb{P}^2$.

Proof. Condition (i) means that the expansion of

$$D^{d-m} f(A, B, C) = (p_1 \cdots p_N)^{d-m} f\left(a_1 \frac{q_1}{p_1} + \cdots + a_N \frac{q_N}{p_N}, b_1 \frac{q_1}{p_1} + \cdots \right.$$
$$\left. + b_N \frac{q_N}{p_N}, \frac{q_1}{p_1} + \cdots + \frac{q_N}{p_N}\right)$$

has no denominator. The coefficient of $(q_1/p_1)^d$ in the expansion of $f(A, B, C)$ is $f(a_1, b_1, 1)$, and so p_1^m fails to appear in the denominator if and only if

$f(a_1, b_1, 1) = 0$. The coefficient of $(q_1/p_1)^{d-1}$ is

$$f_x(a_1, b_1, 1) \left(a_2 \frac{q_2}{p_2} + \cdots + a_N \frac{q_N}{p_N} \right) + f_y(a_1, b_1, 1) \left(b_2 \frac{q_2}{p_2} + \cdots + b_N \frac{q_N}{p_N} \right)$$

$$+ f_z(a_1, b_1, 1) \left(\frac{q_2}{p_2} + \cdots + \frac{q_N}{p_N} \right),$$

where f_x, f_y, f_z are the partial derivatives of f. Thus p_1^{m-1} fails to appear in the denominator if and only if $f_x(a_1, b_1, 1) = f_y(a_1, b_1, 1) = f_z(a_1, b_1, 1) = 0$. This proves the lemma for $m = 1, 2$; the cases $m \geq 3$ are similar. □

Remark 2.48. Nagata's strategy (Nagata [11], [12]) is now to show that the set of pairs (d, m) for which there exists a projective plane curve $f(x, y, z) = 0$ as in Lemma 2.47(ii) is an infinitely generated subsemigroup of $\mathbb{Z}^2$, and hence by Proposition 2.44 that the ring of invariants is infinitely generated. In fact, this works if $N = s^2$ is the square of a natural number $s \geq 4$: that is, there exist such plane curves only if $d/m > s$ (so the ring is supported on a semigroup similar to that of Example 2.43). □

In the case $N = 9$, on the other hand, the supporting semigroup is actually finitely generated, but nevertheless the bigraded ring fails to be finitely generated. This is what we will show next, by exploiting the relationship between plane cubics and doubly periodic complex functions.

(c) An application of Liouville's Theorem

In order to prove Theorem 2.45 we are going to use the holomorphic map (1.17) (see Section 1.5(c)) to build a set of $N = 9$ points $w_1, \ldots, w_9 \in \mathbb{P}^2$ for which the ring of invariants

$$\mathbb{C}[p_1, \ldots, p_N, q_1, \ldots, q_N]^{G \cdot T} = \bigoplus_{d, m \in \mathbb{Z}} R_{(d, m)}$$

fails to be finitely generated. The key to this is the last part of Liouville's Theorem 1.44. Fix a lattice in the complex plane $\Gamma \subset \mathbb{C}$ and a period parallelogram. Inside this parallelogram we pick nine distinct points

$$\widetilde{w}_1, \ldots, \widetilde{w}_9 \in \mathbb{C} \bmod \Gamma,$$

and we let

$$w_1 = (a_1 : b_1 : 1), \ldots, w_9 = (a_9 : b_9 : 1) \in \mathbb{A}^2 \subset \mathbb{P}^2$$

be their images under the map (1.17). Theorem 2.45 will follow from the following:

Proposition 2.49. *If, for all natural numbers $n \in \mathbb{N}$,*

$$n(\widetilde{w}_1 + \cdots + \widetilde{w}_9) \notin \Gamma,$$

then the ring $\bigoplus_{d,m} R_{(d,m)}$ fails to be finitely generated.

The set of homogeneous polynomials $f(x, y, z)$ of degree d is a vector space of dimension $(d + 1)(d + 2)/2$, and the requirement that f vanish with multiplicity m at a given point imposes at most $m(m + 1)/2$ linear conditions on this space. Hence:

$$\dim R_{(d,m)} \geq \tfrac{1}{2}(d + 1)(d + 2) - \tfrac{9}{2}m(m + 1)$$

$$= \tfrac{1}{2}(d - 3m)(d + 3m + 3) + 1. \tag{2.1}$$

In particular, $R_{(d,m)} \neq 0$ whenever $d \geq 3m$. It turns out, from the way the nine points have been chosen, that the converse is also true; and moreover, that when $d = 3m$ equality holds in the above estimate.

Lemma 2.50. *Assume that $m(\widetilde{w}_1 + \cdots + \widetilde{w}_9) \notin \Gamma$. Then:*

(i) if $d < 3m$, then $R_{(d,m)} = 0$;
(ii) if $d = 3m$, then $\dim R_{(d,m)} = 1$.

Proof. Given $f(A, B, C) \in R_{(d,m)}$, we consider the function $f(\wp(z), \wp'(z), 1)$. This is holomorphic away from lattice points, with a pole of order at most $3d$ at the origin. On the other hand, note that it has a zero of order at least m at each of the points $\widetilde{w}_1, \ldots, \widetilde{w}_9$.

(i) If $d < 3m$, then it follows from Theorem 1.44(iii) that $f(\wp(z), \wp'(z), 1)$ is identically zero. By Lemma 1.45, it can therefore be expressed as

$$f(x, y, z) = (y^2 z - 4x^3 - g_2 x z^2 - g_3 z^3) h(x, y, z).$$

But then $h(x, y, z) \in R_{(d-3, m-1)}$, and we can apply the same reasoning. By induction, then, $f(x, y, z) = 0$.

(ii) If $d = 3m$, it is enough to show, again, that $f(\wp(z), \wp'(z), 1) = 0$. If not, then by Theorem 1.44 (iii) and (iv) we see that $m(\widetilde{w}_1 + \cdots + \widetilde{w}_9) \in \Gamma$, contrary to hypothesis. $\qquad\square$

By Lemma 2.50(i) together with (2.1), the bigraded ring $\bigoplus_{d,m} R_{(d,m)}$ has support

$$\{(d, m) \mid d \geq 3m, \ d \geq 0\} \subset \mathbb{Z}^2,$$

and this is finitely generated. However, we can show that the ring itself is not finitely generated by using (2.1) and Lemma 2.50(ii).

Proof of Proposition 2.49. Let $f_0(x, y, z) = y^2 z - 4x^3 - g_2 x z^2 - g_3 z^3 \in R_{(3,1)}$. By Lemma 2.50(ii), f_0 generates all $R_{(d,m)}$ along the line $d = 3m$ (see Figure 2.2); so if R is finitely generated, then the remaining generators can be chosen from the subring

$$R^\varepsilon = \bigoplus_{d \geq (3+\varepsilon)m} R_{(d,m)}$$

for some sufficiently small rational number $\varepsilon > 0$. For each such (d, m), multiplication by f_0 defines a sequence of maps

$$R_{(d,m)} \xrightarrow{\times f_0} R_{(d+3,m+1)} \xrightarrow{\times f_0} \cdots \xrightarrow{\times f_0} R_{(d+3a,m+a)},$$

and these maps must all be surjective if R is generated by f_0 and R^ε. However, this is impossible, as (2.1) gives for the dimensions an estimate

$$\dim R_{(d+3a,m+a)} \geq \tfrac{1}{2} m \varepsilon (d + 3m + 6a + 3) + 1,$$

which tends to infinity as $a \to \infty$. Hence $R = \bigoplus_{d,m} R_{(d,m)}$ cannot be finitely generated. $\qquad\square$

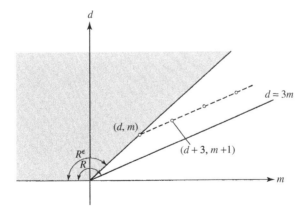

Figure 2.2

Remark 2.51. In fact, after writing the original version of this book the author realised that an alternative argument could be given using the fact that the support of the bigraded quotient ring $R/(f_0)$ is not finitely generated. This is closely related to an example of Rees (Rees [14]).

Exercises

1. Show that in the integral domain $\mathbb{Z}[\sqrt{-5}]$ the number 3 is irreducible but not prime. *Hint:* Use, for example, $6 = (1 + \sqrt{-5})(1 - \sqrt{-5})$.
2. Show that if R is an integral domain, then the polynomial ring $R[x]$ is also an integral domain.
3. For an element a in a ring R, suppose that the linear polynomial $1 - ax$ is an invertible element in the ring $R[x]$. Show that a is nilpotent.
4. Show that an integral domain with only finitely many elements is a field.
5. Prove that an integral domain which contains a field k and is finite dimensional as a vector space over k is a field.
6. Prove Lemma 2.18 in the case when k is a finite field.
7. Show that the formal power series ring $k[[t]]$ over a field k is an integral domain.
8. Show that the only ideals in $k[[t]]$ are the powers of the maximal ideal (t^n). (In particular, every ideal is principal.)
9. Prove the converse of Theorem 2.36: every valuation ring is maximal in its field of fractions with respect to dominance.
10. By studying the proof of Theorem 4.51, give a proof of Proposition 2.41.

3
Algebraic varieties

In the broadest terms, a *manifold* means a topological space equipped with a sheaf of rings which is locally isomorphic to a given ringed space or spaces, its 'local models'. For differentiable manifolds and complex manifolds, respectively, the local models are open sets in $\mathbb{R}^n$ and $\mathbb{C}^n$, together with the sheaves of differentiable and holomorphic functions on these spaces. Algebraic manifolds, or *varieties*, are defined analogously. We first fix a finitely generated field extension K of the ground field k, and then take as our local models the spectra Spm R of rings having K as their field of fractions. In other words, an algebraic variety is obtained by gluing together ringed affine varieties possessing the same algebraic function field. This chapter explains these notions of affine varieties, their sheaves of rings and their gluings.

We begin by defining the n-dimensional affine space $\mathbb{A}^n$ over the complex numbers $\mathbb{C}$ as the set $\mathbb{C}^n$ equipped with the Zariski topology and elementary sheaf of rings $\mathcal{O}$ assigning to a basic open set $D(f) \subset \mathbb{C}^n$ the ring of rational functions $\mathbb{C}[x_1, \ldots, x_n, 1/f(x)]$. These constructions are easily generalised from $\mathbb{A}^n$ to the set Spm R of maximal ideals in any finitely generated algebra R over any algebraically closed field k. One calls Spm R an affine variety, and a morphism Spm $R \to$ Spm S is the same thing as a k-algebra homomorphism $S \to R$. An *algebraic variety* is then a ringed topological space obtained by gluing together affine varieties with a common function field, and in good cases are separated: the most important examples are projective varieties (Section 3.2). Many properties of affine varieties can be defined for general algebraic varieties using a covering by affine charts.

Section 3.3 explains categories and functors in elementary terms, and how an algebraic variety X determines in a natural way a functor $\underline{X}$ from the category of algebras over the ground field k to the category of sets. From this point of view the idea of an algebraic group enters in a natural way: an *algebraic group* is simply an algebraic variety G for which the functor $\underline{G}$ takes values in the category of groups. In the affine case, $G = \mathrm{Spm}\ A$, this property is equivalent to the existence of a k-algebra homomorphism $A \to A \otimes A$ (the coproduct) satisfying various conditions.

A projective variety is a particular case of a complete algebraic variety. An variety X is called *complete* if every projection with X as a fibre is a closed map. In practice, however, completeness is usually verified by means of the Valuative Criterion. We prove this in the final section and apply it to toric varieties: completeness of a toric variety is equivalent to the property that its defining fan covers $\mathbb{R}^n$.

3.1 Affine varieties

(a) Affine space

To begin we will work over the field $k = \mathbb{C}$ of complex numbers; this is familiar and convenient, though in fact the only property of $\mathbb{C}$ that we need is its algebraic closure. By *affine space* $\mathbb{A}^n$ we shall mean a *ringed space* consisting of an n-dimensional complex space $\mathbb{C}^n$ as its underlying set, equipped with Zariski topology and structure sheaf $\mathcal{O}$, both of which we shall explain in this section.

Given an open set $U \subset \mathbb{C}^n$ in the usual Euclidean topology, we denote by $\mathcal{O}^{\mathrm{an}}(U)$ the ring of holomorphic complex-valued functions on U. This defines a sheaf

$$\{\text{open subsets of } \mathbb{C}^n\} \to \{\text{rings}\}, \qquad U \mapsto \mathcal{O}^{\mathrm{an}}(U). \tag{3.1}$$

By the *complex analytic space* $\mathbb{C}^n_{\mathrm{an}}$ we shall mean the topological space $\mathbb{C}^n$ equipped with the sheaf $\mathcal{O}^{\mathrm{an}}$. The affine space $\mathbb{A}^n$ will be a polynomial version of this object.

First of all, given an ideal $\mathfrak{a} \subset \mathbb{C}[x_1, \ldots, x_n]$, we define a subset $V(\mathfrak{a}) \subset \mathbb{C}^n$ consisting of the common zeros of all polynomials in $\mathfrak{a}$ (or, equivalently, of a set of polynomials generating $\mathfrak{a}$):

$$V(\mathfrak{a}) = \{a \in \mathbb{C}^n \mid f(a) = 0 \text{ for all } f \in \mathfrak{a}\}. \tag{3.2}$$

This has the properties that for arbitrary families of ideals $\{\mathfrak{a}_i \mid i \in I\}$ we have

$$\bigcap_{i \in I} V(\mathfrak{a}_i) = V\left(\sum_{i \in I} \mathfrak{a}_i\right),$$

while for finite families $\{\mathfrak{a}_1, \ldots, \mathfrak{a}_r\}$ we have

$$\bigcup_{1 \le i \le r} V(\mathfrak{a}_i) = V\left(\prod_{1 \le i \le r} \mathfrak{a}_i\right).$$

In other words, the collection of all subsets $V(\mathfrak{a}) \subset \mathbb{C}^n$ is closed under the operations of taking arbitrary intersections or finite unions and contains $\emptyset = V(\mathbb{C}[x_1, \ldots, x_n])$ and $\mathbb{C}^n = V(0)$.

Definition 3.1. The topology on $\mathbb{C}^n$ with the subsets $V(\mathfrak{a}) \subset \mathbb{C}^n$, over all ideals $\mathfrak{a} \subset \mathbb{C}[x_1, \ldots, x_n]$, as closed sets is called the *Zariski topology*. We shall denote $\mathbb{C}^n$ equipped with this topology by $\mathbb{C}^n_{\text{alg}}$. $\square$

Given a polynomial $f \in \mathbb{C}[x_1, \ldots, x_n]$, we consider the complement of $V((f)) \subset \mathbb{C}^n$,

$$D(f) = \{a \in \mathbb{C}^n \mid f(a) \neq 0\}. \tag{3.3}$$

This is an open set, which we shall call a *basic open set*. Given two polynomials $f, g \in \mathbb{C}[x_1, \ldots, x_n]$, we have

$$D(fg) = D(f) \cap D(g),$$

so that the collection of basic open sets is closed under finite intersections. Moreover, for an ideal $\mathfrak{a} \subset \mathbb{C}[x_1, \ldots, x_n]$,

$$\mathbb{C}^n - V(\mathfrak{a}) = \bigcup_{f \in \mathfrak{a}} D(f),$$

so the open sets $D(f)$ form a basis for the Zariski topology.

One can now define a sheaf (3.1) for the Zariski topology on $\mathbb{C}^n_{\text{alg}}$. This is in fact an example of an elementary sheaf:

Definition 3.2. Let X be a topological space, and let $\mathcal{U}_X$ be the set of its nonempty open subsets. Let K be a set and $\mathcal{P}_K$ its power set. A mapping

$$F : \mathcal{U}_X \to \mathcal{P}_K$$

with the property that, for any collection of nonempty open sets $\{U_i \mid i \in I\} \subset \mathcal{U}_X$,

$$F\left(\bigcup_{i \in I} U_i\right) = \bigcap_{i \in I} F(U_i),$$

is called an *elementary sheaf* of subsets of K. $\square$

Note that since $F(W) = F(U \cup W) = F(U) \cap F(W)$, an elementary sheaf always has the property that, for open sets $U, W \in \mathcal{U}_X$,

$$U \subset W \quad \Longrightarrow \quad F(U) \supset F(W).$$

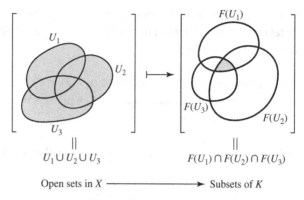

Open sets in X ⟶ Subsets of K

Figure 3.1: An elementary sheaf

The set K may have more structure: if K is a module and every $F(U)$ is a submodule, then F is called an elementary sheaf of modules; if K is a ring and every $F(U)$ is a subring, then F is called an elementary sheaf of rings, and so on.

In the situation we are considering K will be the field $\mathbb{C}(x_1, \ldots, x_n)$ of rational functions on $X = \mathbb{C}_{\text{alg}}^n$. This is the field of fractions of $\mathbb{C}[x_1, \ldots, x_n]$, consisting of rational expressions

$$\frac{g(x_1, \ldots, x_n)}{h(x_1, \ldots, x_n)}, \quad \text{where } g, h \in \mathbb{C}[x_1, \ldots, x_n] \text{ and } h \neq 0.$$

The polynomial $g(x)$ is finite-valued at all points of $\mathbb{C}^n$, but the same is not true of the function $g(x)/h(x)$. However, if we allow for $h(x)$ only powers $f(x)^m$, then we obtain rational functions which are finite-valued at all points of the basic open set $D(f)$. We shall write

$$\mathbb{C}[x_1, \ldots, x_n, 1/f(x)] = \left\{ \frac{g(x)}{f(x)^m} \mid g \in \mathbb{C}[x_1, \ldots, x_n], \ m \geq 0 \right\}$$

for this set of functions, which is a subring of $\mathbb{C}(x_1, \ldots, x_n)$.

Now, in the spirit of (3.1), we define

$$\mathcal{O}(D(f)) = \mathbb{C}[x_1, \ldots, x_n, 1/f(x)]; \qquad (3.4)$$

and, more generally, since any Zariski-open set $U \subset \mathbb{C}_{\text{alg}}^n$ is a union

$$U = \mathbb{C}^n - V(\mathfrak{a}) = \bigcup_{f \in \mathfrak{a}} D(f)$$

of such sets, we define

$$\mathcal{O}(U) = \bigcap_{f \in \mathfrak{a}} \mathcal{O}(D(f)). \tag{3.5}$$

Proposition 3.3. $\mathcal{O} : U \mapsto \mathcal{O}(U)$ *as above defines an elementary sheaf on* $\mathbb{C}^n_{\mathrm{alg}}$ *of subrings of the rational function field* $\mathbb{C}(x_1, \ldots, x_n)$.

The proof of this will appear in a more general situation in the next section (Proposition 3.16).

Definition 3.4. The Zariski topological space $\mathbb{C}^n_{\mathrm{alg}}$ equipped with the sheaf $\mathcal{O}$ is called an n-dimensional affine space, denoted by $\mathbb{A}^n$. □

Let $\mathfrak{p} \subset \mathbb{C}[x_1, \ldots, x_n]$ be a prime ideal. The closed set $V(\mathfrak{p}) \subset \mathbb{C}^n_{\mathrm{alg}}$, equipped with the induced (Zariski) topology and elementary sheaf $\mathcal{O}$ (defined by $U \cap V(\mathfrak{p}) \mapsto \mathcal{O}(U \cap V(\mathfrak{p})) := \mathcal{O}(U)/\mathfrak{p} \cap \mathcal{O}(U))$, is called an *affine variety*.

Example 3.5. Proposition 2.17, applied to the polynomial ring $R = \mathbb{C}[x_1, \ldots, x_n]$, can now be expressed in more geometrical language: an affine variety $X \subset \mathbb{A}^n$ of codimension 1 is defined by the vanishing of a single polynomial equation, $X = V((f))$, $f \in \mathbb{C}[x_1, \ldots, x_n]$. An affine variety $X \subset \mathbb{A}^n$ of codimension 1 is called an *affine hypersurface*. When $n = 2$, this is a plane curve. □

(b) The spectrum

The affine space $\mathbb{A}^n$ is a triple consisting of an underlying space $\mathbb{C}^n$, its Zariski topology, and structure sheaf $\mathcal{O}$. We shall see next that all three of these elements are determined by the polynomial ring $\mathbb{C}[x_1, \ldots, x_n]$ alone. Once this is understood it is simple to define affine algebraic varieties more generally.

Remark 3.6. We could retain the ground field $\mathbb{C}$ as in the previous section, but since the Euclidean topology is not needed (and in the algebraic setting can even be misleading) it will be clearer to return to an arbitrary algebraically closed ground field k. □

We begin by observing that the ring homomorphism which assigns $x_i \mapsto a_i \in k$ for each $i = 1, \ldots, n$, that is,

$$k[x_1, \ldots, x_n] \to k, \qquad f \mapsto f(a_1, \ldots, a_n),$$

has kernel equal to the maximal ideal generated by the n linear polynomials

$$x_1 - a_1, \quad x_2 - a_2, \quad \ldots, \quad x_n - a_n.$$

Theorem 3.7. *Every maximal ideal in the polynomial ring $k[x_1, \ldots, x_n]$ is of the form $(x_1 - a_1, \ldots, x_n - a_n)$ for some $a_1, \ldots, a_n \in k$.*

Proof. Let $\mathfrak{m} \subset k[x_1, \ldots, x_n]$ be an arbitrary maximal ideal. Then the residue field $K = k[x_1, \ldots, x_n]/\mathfrak{m}$ can be viewed as a k-algebra and is finitely generated. We consider the composition of ring homomorphisms

$$k \to k[x_1, \ldots, x_n] \to k[x_1, \ldots, x_n]/\mathfrak{m} = K.$$

This expresses K as a field extension of k, which by Proposition 2.23 is finite and therefore algebraic. But since k is algebraically closed, this shows that $k \to K$ is an isomorphism. Hence there exist elements $a_1, \ldots, a_n \in k$ such that

$$x_i \equiv a_i \bmod \mathfrak{m} \quad \text{for } i = 1, \ldots, n.$$

Therefore each $x_i - a_i \in \mathfrak{m}$, and since $(x_1 - a_1, \ldots, x_n - a_n)$ is a maximal ideal it follows that $(x_1 - a_1, \ldots, x_n - a_n) = \mathfrak{m}$. $\qquad\square$

Applying Theorem 3.7 and Theorem 2.27 to the quotient ring $k[x_1, \ldots, x_n]/\mathfrak{a}$ we obtain:

Hilbert's Nullstellensatz 3.8. *Let $\mathfrak{a} \subset k[x_1, \ldots, x_n]$ be any ideal. If a polynomial $f \in k[x_1, \ldots, x_n]$ vanishes on $V(\mathfrak{a})$, then $f^m \in \mathfrak{a}$ for some $m \in \mathbb{N}$.* $\qquad\square$

Given a ring R and an ideal $\mathfrak{a} \subset R$ we shall write

$$\sqrt{\mathfrak{a}} = \{f \in R \mid f^m \in \mathfrak{a} \text{ for some } m \in \mathbb{N}\}.$$

This is an ideal in R, called the *radical* of $\mathfrak{a}$. In this language, Hilbert's Nullstellensatz says that, for an ideal $\mathfrak{a} \subset k[x_1, \ldots, x_n]$, the ideal of all polynomials vanishing on $V(\mathfrak{a})$ is precisely $\sqrt{\mathfrak{a}}$.

Corollary 3.9. *For ideals $\mathfrak{a}, \mathfrak{b} \subset k[x_1, \ldots, x_n]$, $V(\mathfrak{a}) \subset V(\mathfrak{b}) \iff \sqrt{\mathfrak{b}} \subset \sqrt{\mathfrak{a}}$.* $\qquad\square$

Definition 3.10. Given a ring R, the set $\mathrm{Spm}\, R$ of its maximal ideals is called the (maximal) *spectrum* of R. $\qquad\square$

The full spectrum of R is the larger space Spec R consisting of all prime ideals in R, not just the maximal ideals; but in this book we shall need only the maximal spectrum.

When R is the polynomial ring $k[x_1, \ldots, x_n]$, Theorem 3.7 says that there is a bijection between the maximal spectrum Spm $R = \text{Spm}\, k[x_1, \ldots, x_n]$ and the underlying space k^n of the affine space $\mathbb{A}^n$ over k. Generalising this example, we now define the Zariski topology and sheaf of rings $\mathcal{O}$ on Spm R. First of all, given an ideal $\mathfrak{a} \subset R$ and an element $f \in R$, we define subsets:

$$\begin{aligned} V(\mathfrak{a}) &= \{\mathfrak{m} \mid \mathfrak{m} \supset \mathfrak{a}\} \subset \text{Spm}\, R, \\ D(f) &= \{\mathfrak{m} \mid f \notin \mathfrak{m}\} \subset \text{Spm}\, R. \end{aligned} \tag{3.6}$$

When $R = k[x_1, \ldots, x_n]$ these definitions agree with (3.2) and (3.3). The following agrees with Definition 3.1:

Definition 3.11. The subsets $V(\mathfrak{a})$, for arbitrary ideals $\mathfrak{a} \subset R$, are the closed sets of a topology on Spm R called the *Zariski topology*. □

Example 3.12. The nonempty closed sets in the affine line $\mathbb{A}^1$ are $\mathbb{A}^1$ itself and all *finite* sets. In fact, for dimension 1 the Zariski topology is the weakest topology for which single points are closed. □

The Zariski topology is characterised by the following property.

Proposition 3.13. *If R is a Noetherian ring, then the topological space* Spm R *is Noetherian. In other words, every descending chain of closed sets*

$$\text{Spm}\, R \supset Z_1 \supset Z_2 \supset \cdots \supset Z_m \supset \cdots$$

terminates after finitely many terms.

Proof. Let $Z_i = V(\mathfrak{a}_i)$. By the Nullstellensatz we can suppose (since $\sqrt{\sqrt{\mathfrak{a}}} = \sqrt{\mathfrak{a}}$) that $\sqrt{\mathfrak{a}_i} = \mathfrak{a}_i$. Then by Corollary 3.9 these ideals form an ascending chain

$$\mathfrak{a}_1 \subset \mathfrak{a}_2 \subset \cdots \mathfrak{a}_m \subset \cdots \subset R.$$

Since R is Noetherian, the ideal $\bigcup_i \mathfrak{a}_i \subset R$ is finitely generated, so that the chain terminates after finitely many terms. □

Note that the Euclidean topology on $\mathbb{C}^n$, for example, certainly does not have this property.

Definition 3.14. A topological space is said to be *irreducible* if it satisfies the following equivalent conditions.

(i) X cannot be expressed as a union $X_1 \cup X_2$, where $X_1, X_2 \subset X$ are proper closed subsets.

(ii) Any two nonempty open subsets $U_1, U_2 \subset X$ have nonempty intersection $U_1 \cap U_2 \neq \emptyset$. Otherwise X is said to be *reducible*.　　　　□

Note that two basic open sets $D(f), D(g) \subset \mathrm{Spm}\ R$ have intersection

$$D(f) \cap D(g) = D(fg).$$

If R is an integral domain, then $f \neq 0$ and $g \neq 0$ together imply $fg \neq 0$; so we obtain:

Proposition 3.15. *If R is an integral domain, then* $\mathrm{Spm}\ R$ *is irreducible.*　　□

In this case, just as for affine space $\mathbb{A}^n$, we can construct an elementary sheaf on $\mathrm{Spm}\ R$. Since R is an integral domain it has a field of fractions K. Given a nonzero element $f \in R$ we consider the subring $R[1/f] \subset K$ generated by R and $1/f$. This is the set of all elements of K expressible with denominator a power of f.

Proposition 3.16. *Let R be an integral domain. Then the map from nonempty open sets of $X = \mathrm{Spm}\ R$ to subrings of the field of fractions K of R:*

$$\mathcal{O}(X - V(\mathfrak{a})) = \bigcap_{0 \neq f \in \mathfrak{a}} R[1/f]$$

(or, equivalently, $\mathcal{O}(D(f)) = R[1/f]$ on the basic open sets) defines an elementary sheaf $\mathcal{O}$ of subrings of K on $X = \mathrm{Spm}\ R$.　　□

Lemma 3.17. *If the ideal $\mathfrak{a} \subset R$ is generated by nonzero elements $f_i \in \mathfrak{a}$, $i \in I$, then*

$$\bigcap_{0 \neq f \in \mathfrak{a}} R[1/f] = \bigcap_{i \in I} R[1/f_i].$$

Proof. The inclusion $\bigcap_{0 \neq f \in \mathfrak{a}} R[1/f] \subset \bigcap_{i \in I} R[1/f_i]$ is clear; we have to prove the converse. By hypothesis, an arbitrary element $f \in \mathfrak{a}$ can be written

$$f = \sum_{j \in J} a_j f_j,$$

where $a_j \in R$ and $J \subset I$ is a finite subset. Now suppose that $x \in K$ belongs to the right-hand intersection. Then for sufficiently large $n \in \mathbb{N}$ we have $f_j^n \in R$ for all $j \in J$. It follows that there exists some $N \in \mathbb{N}$ such that $f^N x \in R$, or, in other words, $x \in R[1/f]$. Since f is arbitrary, this shows that x is contained in the left-hand intersection. □

The following is clear.

Lemma 3.18. $\bigcap_{0 \neq f \in \mathfrak{a}} R[1/f] = \bigcap_{0 \neq f \in \sqrt{\mathfrak{a}}} R[1/f].$ □

Proof of Proposition 3.16. First note that if $V(\mathfrak{a}) = V(\mathfrak{b})$, then, by Hilbert's Nullstellensatz, $\sqrt{\mathfrak{a}} = \sqrt{\mathfrak{b}}$, so that by Lemma 3.18 the mapping $\mathcal{O}$ is well defined. Now suppose that an open set $U = X - V(\mathfrak{a})$ is a union $\bigcup_{i \in I} U_i$, where $U_i = X - V(\mathfrak{a}_i)$. Then

$$V(\mathfrak{a}) = \bigcap_{i \in I} = V\left(\sum_{i \in I} \mathfrak{a}_i\right).$$

The ideal $\sum_{i \in I} \mathfrak{a}_i$ is generated by the $\mathfrak{a}_i$, so by Lemma 3.17:

$$\mathcal{O}(U) = \bigcap_{i \in I} \mathcal{O}(U_i),$$

which shows that $\mathcal{O}$ is an elementary sheaf, as asserted. □

Definition 3.19. Let R be a finitely generated integral k-algebra. Then the collection of data consisting of:

(i) the set Spm R,
(ii) the Zariski topology on Spm R,
(iii) the elementary sheaf $\mathcal{O}$ over Spm R,

is called an *affine algebraic variety*. $\mathcal{O}$ is called the *structure sheaf* of the variety. Writing $X = $ Spm R, the k-algebra $k[X] := R$ is called the *coordinate ring* of X, and its field of fractions $k(X) := K$ is called the *algebraic function field* of X. □

Example 3.20. Let $R = k[x_1, \ldots, x_n]/\mathfrak{p}$, where $\mathfrak{p} \subset k[x_1, \ldots, x_n]$ is a prime ideal. Then Spm R is precisely the affine subvariety $V(\mathfrak{p}) \subset \mathbb{A}^n$ defined in the last section.

Conversely, Spm R is of this form for any finitely generated k-algebra R. To see this, suppose that $r_1, \ldots, r_n \in R$ are generators and consider the surjective homomorphism

$$k[x_1, \ldots, x_n] \to R \qquad x_i \mapsto r_i, \quad 1 \le i \le n.$$

The kernel $\mathfrak{p} \subset k[x_1, \ldots, x_n]$ of this map is a prime ideal, so that R is of the form above and Spm R is isomorphic to the affine subvariety $V(\mathfrak{p}) \subset \mathbb{A}^n$. This explains the terminology 'coordinate ring'. $\qquad\qquad\qquad\Box$

Choosing a different set of generators $r'_1, \ldots, r'_m \in R$ in this example represents Spm R as a different subvariety of affine space $V(\mathfrak{p}') \subset \mathbb{A}^m$ where $\mathfrak{p}' \subset k[x_1, \ldots, x_m]$. Thus Definition 3.19 improves on the ideas of part (a) by describing the points of an affine variety only in terms of an integral domain R (its coordinate ring), and independently of any particular ambient space $\mathbb{A}^n$ or $\mathbb{A}^m$.

We will explain next some important notions connected with this definition.

(c) Some important notions

Morphisms By Corollary 2.25, a homomorphism of finitely generated integral k-algebras

$$\phi : S \to R$$

induces a map of maximal spectra

$${}^t\phi : \text{Spm } R \to \text{Spm } S, \qquad \mathfrak{m} \mapsto \phi^{-1}(\mathfrak{m}).$$

With respect to the Zariski topology on each side, this map ${}^t\phi$ is continuous.

Write $X = \text{Spm } R$ and $Y = \text{Spm } S$, and consider in turn the structure sheaves $\mathcal{O}_X$ and $\mathcal{O}_Y$. If $U \subset Y$ is an open set, then there is a natural induced k-algebra homomorphism

$$\mathcal{O}_Y(U) \to \mathcal{O}_X(({}^t\phi)^{-1}U).$$

Such a continuous map ${}^t\phi : X \to Y$ together with its induced homomorphism of structure sheaves $\mathcal{O}_Y \to \mathcal{O}_X$ is called a *morphism* from X to Y.

If ϕ is surjective, then ${}^t\phi$ is a homeomorphism onto the closed subset $V(\mathfrak{p}) \subset$ Spm S determined by the ideal $\mathfrak{p} = \ker \phi \subset S$. In this case ${}^t\phi$ is called a *closed immersion*.

Proposition 3.21. *If $\phi : S \hookrightarrow R$ is an integral ring extension, then ${}^t\phi$ is surjective.*

Proof. Let $\mathfrak{m} \subset S$ be a maximal ideal. By Lemma 2.20, the ideal $\mathfrak{m}R$ is not the whole of R, and so it is contained in some maximal ideal $M \subset R$. Then $M \cap S = \mathfrak{m}$, showing that ${}^t\phi$ is surjective. $\qquad\square$

One should note that the image of a morphism between affine varieties is not itself necessarily an affine variety:

Example 3.22. The morphism

$$f : \mathbb{A}^2 \to \mathbb{A}^2, \qquad (x, y) \mapsto (x, xy)$$

has an image consisting of the union

$$(\mathbb{A}^2 - V(x)) \cup \{(0, 0)\}.$$

Of these sets, $\mathbb{A}^2 - V(x)$, the complement of the y-axis, is open and the second, the origin, is closed. Each is an affine subvariety of $\mathbb{A}^2$ but their union is not.

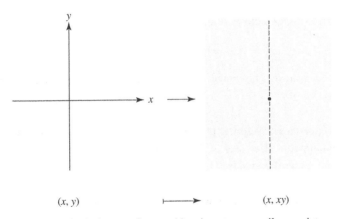

Figure 3.2: The image of a morphism is not necessarily a variety

$\qquad\square$

Products Let $X = V(\mathfrak{m}) \subset \mathbb{A}^n$ and $Y = V(\mathfrak{q}) \subset \mathbb{A}^m$ be affine subvarieties defined by prime ideals $\mathfrak{p} \subset k[x_1, \ldots, x_n]$ and $\mathfrak{q} \subset k[y_1, \ldots, y_m]$. Then the product of X and Y is by definition the affine variety

$$X \times Y = \{(a_1, \ldots, a_n, b_1, \ldots, b_m) \mid f(a) = g(b)$$

$$= 0 \text{ for all } f \in \mathfrak{p} \text{ and } g \in \mathfrak{q}\} \subset \mathbb{A}^{n+m}.$$

Equivalently, $X \times Y = \mathrm{Spm}\ R$, where R is the quotient ring

$$R = k[x_1, \ldots, x_n, y_1, \ldots, y_m]/(\mathfrak{p} + \mathfrak{q}).$$

This is precisely the tensor product of the coordinate rings of X and Y:

$$R = k[x_1, \ldots, x_n]/\mathfrak{p} \otimes k[y_1, \ldots, y_m]/\mathfrak{q}.$$

(We leave to the reader the verification that the tensor product of integral k-algebras is again integral.)

The product $X \times X$ of X with itself contains a distinguished closed subvariety:

Example 3.23. The diagonal. For any k-algebra R the tensor product $R \otimes_k R$ becomes a k-algebra with multiplication law

$$(a \otimes b)(c \otimes d) = ac \otimes bd.$$

With this ring structure there is a surjective homomorphism

$$m : R \otimes_k R \to R, \qquad a \otimes b \mapsto ab,$$

and this determines a closed immersion

$$^t m : X \to X \times X,$$

where $X = \mathrm{Spm}\ R$. The image of this map is called the *diagonal*, denoted by $\Delta \subset X \times X$. $\qquad\square$

General spectra and nilpotents The construction above of a topological space $\mathrm{Spm}\ R$ equipped with a sheaf of rings can be carried out even when the algebra R is not an integral domain.

Definition 3.24. A ring R in which $ab = 0$, $a, b \in R$, implies that either $a = 0$ or b is nilpotent is called *primary*. $\qquad\square$

If R is a primary ring over the field k, then the space $X = \mathrm{Spm}\ R$ is irreducible. To see this, we observe that the subset $\mathfrak{n} \subset R$ consisting of all nilpotent elements is a prime ideal, so the residue ring

$$R_{\mathrm{red}} := R/\mathfrak{n}$$

is an integral domain. As a topological space, X is exactly the same as $X_{\mathrm{red}} := \mathrm{Spm}\ R_{\mathrm{red}}$; that is, the quotient map $R \to R_{\mathrm{red}}$ induces a homeomorphism

$X_{\text{red}} := \text{Spm } R_{\text{red}} \xrightarrow{\sim} X$. However, the structure sheaves $\mathcal{O}_X$ and $\mathcal{O}_{X_{\text{red}}}$ are distinct. $\mathcal{O}_X$ is defined exactly as in Proposition 3.16, and for each open set $U \subset X$ there is a surjective ring homomorphism

$$\mathcal{O}_X \twoheadrightarrow \mathcal{O}_{X_{\text{red}}}.$$

In this sense X_{red} is a closed subvariety of X, and one thinks of X as a 'fattening' of X_{red}. $\mathcal{O}_X$ is the *total fraction ring* of R (see Section 8.2(a)).

Example 3.25. The simplest example of a primary ring that is not integral is $R = k[\epsilon]/(\epsilon^2)$: the residue class of ϵ is nonzero but its square is zero. Thus $R_{\text{red}} \cong k$ and $\text{Spm } R$ consists of the single point space $\text{Spm } R_{\text{red}}$ equipped with the ring R. Similarly, and more generally, one can consider the ring $k[\epsilon]/(\epsilon^n)$. □

Example 3.26. Let $\mathfrak{p} \subset R$ be a prime ideal. For each natural number $n \in \mathbb{N}$ the variety $\text{Spm } R/\mathfrak{p}^n$ has an underlying space equal to the closed subvariety $Y = \text{Spm } R/\mathfrak{p}$ of $X = \text{Spm } R$; but, as in the previous example, its coordinate ring $R/\mathfrak{p}^n$ is a primary ring containing nilpotents. This is called the $(n-1)$-*st infinitesimal neighbourhood* of $Y \subset X$. □

Spectra of this kind, and their (formal) limits

$$\lim_{n \to \infty} \text{Spm } k[\epsilon]/(\epsilon^n), \qquad \lim_{n \to \infty} \text{Spm } R/\mathfrak{p}^n,$$

are very important in deformation theory; the second limit is called the *formal neighbourhood* of $Y \subset X$.

More generally, one can construct the structure sheaf $\mathcal{O}_X$ on $X = \text{Spm } R$ for more general k-algebras R in a similar manner by using the primary decomposition of R.

Dominant morphisms Let ${}^t\phi : X \to Y$ be a morphism of affine varieties determined by a k-algebra homomorphism $\phi : S \to R$. Recall that by definition ${}^t\phi$ is a closed immersion when ϕ is surjective.

Definition 3.27. If ϕ is injective, then ${}^t\phi : X \to Y$ is called a *dominant morphism*. □

Let $\mathfrak{p} \subset S$ be the kernel of ϕ. This is a prime ideal, and by the isomorphism theorem ϕ decomposes as a composition:

$$S \twoheadrightarrow S/\mathfrak{p} \cong \phi(S) \hookrightarrow R. \tag{3.7}$$

Correspondingly, the morphism ${}^t\phi$ decomposes as

$$X \twoheadrightarrow W = \mathrm{Spm}\ S/\mathfrak{p} \hookrightarrow Y, \tag{3.8}$$

where the first map is dominant and the second is a closed immersion.

Theorem 3.28. *Let $f : X \to Y$ be a morphism of affine varieties, and let $Z \subset Y$ be the Zariski closure of the image $f(X)$. Then $f(X)$ contains a nonempty open subset of Z.*

Proof. First note that by the decomposition (3.8) it is enough to consider the case when f is dominant, that is, when $\phi : S \to R$ is injective. By Proposition 2.26 there exists some residue ring $\overline{R} = R/I$ for which the composition $S \hookrightarrow R \to \overline{R}$ is an algebraic ring extension. We can therefore assume that R is algebraic over S. Then, by Lemma 2.22, there is a nonzero element $a \in S$ such that $R[1/a]$ is integral over $S[1/a]$. So by Proposition 3.21 the image of f contains the open set $D(a) \subset Y$. $\qquad\square$

Open immersions Let R be an extension of S contained in the field of fractions K of S. When the morphism ${}^t\phi : \mathrm{Spm}\ R \to \mathrm{Spm}\ S$ induced by the inclusion $\phi : S \hookrightarrow R$ is a homeomorphism to an open subset, ${}^t\phi$ is called an *open immersion*. The following example is typical.

Example 3.29. Let

$$R = S\left[\frac{1}{s_1}, \ldots, \frac{1}{s_m}\right] \subset K$$

be the extension obtained by adjoining the inverses of nonzero elements $s_1, \ldots, s_m \in S$. Then ${}^t\phi$ is a homeomorphism,

$$\mathrm{Spm}\ R \to \mathrm{Spm}\ S - (V(s_1) \cup \cdots \cup V(s_m)),$$

and so is an open immersion. $\qquad\square$

Local properties Suppose that $a_1, \ldots, a_n \in R$ generate R as a module over itself. In other words, there exist $b_1, \ldots, b_n \in R$ such that

$$1 = b_1 a_1 + \cdots + b_n a_n.$$

Then the sets $D(a_i)$ form an open cover of $X = \mathrm{Spm}\, R$. (The set $\{a_1, \ldots, a_n\}$ is called a *partition of unity*. See Definition 8.26.)

A given property $\mathfrak{P}$ of X (or of R) is said to hold *locally* if, even if not satisfied by X (or R) itself, there exists an open cover $\{D(a_1), \ldots, D(a_n)\}$ of the above form such that $\mathfrak{P}$ holds for each open set $D(a_i)$ (or for each $R[1/a_i]$).

3.2 Algebraic varieties

Just as a differentiable manifold, or a complex manifold, is obtained by gluing together copies of $\mathbb{R}^n$ or $\mathbb{C}^n$, so an algebraic variety is an object obtained by gluing together affine varieties. We are going to explain this gluing process next.

(a) Gluing affine varieties

Definition 3.30. Let K be a field, finitely generated over k, and let X be a topological space equipped with an elementary sheaf $\mathcal{O}_X$ of k-subalgebras of K. Then the pair $(X, \mathcal{O}_X)$ is called an *algebraic variety* – also a *model of the algebraic function field* $k(X) := K$ – if there exists an open cover $\{U_i\}_{i \in I}$ of X with the following properties.

(i) Each U_i is an affine variety with function field K.
(ii) For each pair $i, j \in I$, the intersection $U_i \cap U_j$ is an open subset of each of U_i, U_j. $\qquad \square$

In general, algebraic varieties are constructed by a gluing construction in the sense of the following definition.

Definition 3.31. Let $(A, \mathcal{O}_A)$ and $(B, \mathcal{O}_B)$ be two affine varieties with the same algebraic function field K.

(i) If there exists an affine variety C and open immersions

$$\iota_A : C \hookrightarrow A, \qquad \iota_B : C \hookrightarrow B,$$

then we shall say that A and B have a common open set C and denote by $A \cup_C B$ the topological space obtained as the quotient of $A \cup B$ by the equivalence relation $\iota_A(x) \sim \iota_B(x)$ for $x \in C$.

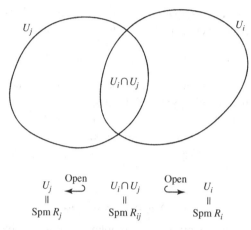

$$U_j \xrightarrow[]{\text{Open}} U_i \cap U_j \xrightarrow{\text{Open}} U_i$$

$$\parallel \qquad\qquad \parallel \qquad\qquad \parallel$$

$$\text{Spm } R_j \qquad \text{Spm } R_{ij} \qquad \text{Spm } R_i$$

Figure 3.3: Gluing affine varieties

(ii) Given an affine variety C as in (i), we define an elementary sheaf $\mathcal{O} = \mathcal{O}_{A \cup_C B}$ of subalgebras of K by $\mathcal{O}(U) = \mathcal{O}_A(U \cap A) \cap \mathcal{O}_B(U \cap B) \subset K$ for nonempty open sets $U \subset A \cup_C B$.

The ringed space $(A \cup_C B, \mathcal{O})$ is called the *gluing* of A and B along C. □

The space $(A \cup_C B, \mathcal{O})$ thus constructed is an algebraic variety. A case that often arises is the following.

Definition 3.32. Let R and S be integral k-algebras with a common field of fractions K, and for nonzero elements $a_1, \ldots, a_n \in R$ and $b_1, \ldots, b_m \in S$ suppose that

$$R\left[\frac{1}{a_1}, \ldots, \frac{1}{a_n}\right] = S\left[\frac{1}{b_1}, \ldots, \frac{1}{b_m}\right] =: T \subset K.$$

Then the induced maps $\text{Spm } T \hookrightarrow \text{Spm } R$ and $\text{Spm } T \hookrightarrow \text{Spm } S$ are open immersions, and the resulting gluing of $\text{Spm } R$ and $\text{Spm } S$ is called *simple* and written $\text{Spm } R \cup_T \text{Spm } S$. □

Clearly we do not obtain a new variety when $R = S = T$, and we shall disregard this trivial case.

Definition 3.33.

(i) If two affine varieties A, B are glued along a common open set C, then the image of the diagonal inclusion

$$C \to A \times B$$

is called the *graph* of the gluing. When this is a closed subset of the product the gluing is said to be *separated*.

(ii) An algebraic variety X is *separated* if it is covered by affine open sets $\{U_i\}_{i \in I}$ such that for each $i, j \in I$ the union $U_i \cup U_j$ is a separated gluing. □

Remarks 3.34.

(i) Separatedness of an algebraic variety is the analogue of the Hausdorff condition for a topological space; it is not related to the separatedness of a field extension!

(ii) An algebraic variety is separated if and only if its algebraic function field can be expressed as the field of fractions of a local ring. □

The following criterion allows one to check Definition 3.33(i) algebraically.

Proposition 3.35. *For a simple gluing* Spm $R \cup_T$ Spm S *the following conditions are equivalent:*

(i) the gluing is separated;
(ii) the subalgebra $T \subset K$ is generated by R and S. □

Note that condition (ii) here is not enough on its own to guarantee a gluing, as the following example shows.

Example 3.36. $R = k[x, xy]$ and $S = k[xy, y]$ together generate $T = k[x, y]$, but we do not get a gluing. Indeed, each of Spm R, Spm S and Spm T is isomorphic to $\mathbb{A}^2$, and the morphism Spm $T \to$ Spm R is that of Example 3.22. □

Example 3.37. Let K be the rational function field $k(t)$ in one variable. We can consider two different models of K, each obtained by gluing two copies of the affine line $\mathbb{A}^1$.

(1) Take $R = S = k[t]$ and $T = k[t, 1/t]$, the subalgebra of Laurent polynomials.

(2) Again each of R, S is the polynomial ring in one variable, but this time $R = k[t]$, $S = k[1/t]$ and $T = k[t, 1/t]$.

The graphs of these gluings can be pictured as in Figure 3.4.

What one sees is that in case (1) the origin is missing from the graph, which therefore fails to be a closed set. This is therefore a nonseparated gluing. In case (2), on the other hand, the graph is closed and the gluing is separated. Both cases illustrate Proposition 3.35.

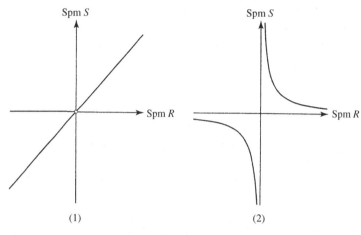

Figure 3.4

In fact the algebraic variety that one obtains in case (2) in the projective line $\mathbb{P}^1$ (see below). As already mentioned, separatedness corresponds to the fact that (over a subfield $k \subset \mathbb{C}$) this is Hausdorff in its Euclidean topology (though not in the Zariski topology!). (See Exercise 3.5.) Case (1), on the other hand, gives an algebraic variety consisting of two lines identified at all points except their respective origins. This space therefore fails to be Hausdorff even in the Euclidean topology.   □

Product varieties $X \times Y$ can be defined by gluing together the products of their affine charts. One should note that the Zariski topology on a product of algebraic varieties is not the same as the product topology (as the example $\mathbb{A}^1 \times \mathbb{A}^1$ shows).

The notion of separatedness of an algebraic variety in part (ii) of Definition 3.33 does not depend on the choice of open cover; in fact, the following holds.

Proposition 3.38. *For an algebraic variety X the following conditions are equivalent:*

(i) X is separated;
(ii) the diagonal map $X \to X \times X$ is a closed immersion.   □

Finally, we need to define morphisms of algebraic varieties.

Definition 3.39. For algebraic varieties X, Y, a *morphism* $f : X \to Y$ consists of a continuous map of the underlying topological spaces together with a homomorphism $\mathcal{O}_Y \to \mathcal{O}_X$ of the structure sheaves – that is, for every open set $U \subset Y$ a ring homomorphism

$$\mathcal{O}_Y(U) \to \mathcal{O}_X(f^{-1}U),$$

such that:

(i) there exist affine open covers $\{V_i\}$ of X and $\{U_j\}$ of Y with the property that each image $f(V_i)$ is contained in some U_j; and
(ii) the restrictions

$$f|_{V_i} : V_i \to U_j$$

are morphisms of affine varieties. □

In other words, just as algebraic varieties are constructed by gluing together affine varieties, so morphisms between algebraic varieties are defined by gluing together morphisms between their affine charts.

(b) Projective varieties

We begin with a fundamental example of the gluing construction just described.

Example 3.40. Projective space. Let K be the rational function field in n variables $k(x_1, \ldots, x_n)$, and introduce $n + 1$ indeterminates $X_0, X_1, \ldots, X_n$ such that $x_i = X_i / X_0$ for $i = 0, 1, \ldots, n$. Thus K is a subfield of $k(X_0, X_1, \ldots, X_n)$. Now, for each $i = 0, 1, \ldots, n$ let

$$R_{X_i} = k \left[\frac{X_0}{X_i}, \frac{X_1}{X_i}, \ldots, \frac{X_{i-1}}{X_i}, \frac{X_{i+1}}{X_i}, \ldots, \frac{X_n}{X_i} \right].$$

Each polynomial ring $R_{X_0}, R_{X_1}, \ldots, R_{X_n}$ is a subalgebra of K with K as its field of fractions. Note that each affine variety Spm R_{X_i} is isomorphic to $\mathbb{A}^n$.

For each i, j let $R_{X_i X_j} = R_{X_i} R_{X_j} \subset K$ be the subalgebra generated by R_{X_i} and R_{X_j}. Then Spm R_{X_i} and Spm R_{X_j} have Spm $R_{X_i X_j}$ as a common open set, and

$$\text{Spm } R_{X_i} \cup_{R_{X_i X_j}} \text{Spm } R_{X_j}$$

is a separated simple gluing. The algebraic variety obtained by gluing all of these affine spaces is the *projective space* $\mathbb{P}^n$. □

We now generalise this example to construct an algebraic variety Proj R from any graded ring R. Let $R = \bigoplus_{e \in \mathbb{Z}} R_{(e)}$ be a graded ring which is an integral domain and suppose also that $R_{(e)} = 0$ whenever $e < 0$. An element f/g of the field of fractions K of R is said to be *homogeneous* if each of f, g is homogeneous in R; in this case we define $\deg f/g = \deg f - \deg g$.

Definition 3.41.

(i) We denote by $K_0 \subset K$ the subfield consisting of elements of degree 0, including zero:

$$K_0 = \left\{ \frac{f}{g} \mid f, g \in R, \ g \neq 0, \ \deg f = \deg g \right\} \cup \{0\}.$$

(ii) Given a nonzero homogeneous element $h \in R$, we denote by $R_{h,0} \subset K_0$ the subalgebra

$$R_{h,0} = \left\{ \frac{f}{h^n} \mid f \in R, \ \deg f = n \deg h \right\} \cup \{0\}.$$

$\square$

Lemma 3.42.

(i) For every nonzero homogeneous element $h \in R$, the field of fractions of $R_{h,0}$ is K_0.

(ii) Given two homogeneous elements $h, l \in R$, the gluing $\mathrm{Spm}\ R_{h,0} \cup_T$ $\mathrm{Spm}\ R_{l,0}$, where $T = R_{h,0}R_{l,0} \subset K_0$, is simple and separated.

Proof. Part (i) is obvious, and part (ii) follows from Proposition 3.35 once we have shown that the gluing is simple. In other words, we have to show that $T = R_{h,0}[1/p]$ for some $p \in R_{h,0}$ (and similarly that $T = R_{l,0}[1/q]$ for some $q \in R_{l,0}$). Let $e \in \mathbb{N}$ be the lowest common multiple of $\deg h$ and $\deg l$, so that $e = a \deg h = b \deg l$, say. A general element of T is of the form $f/(h^r l^s)$, or $F/(h^a l^b)^m$ for sufficiently large $m \in \mathbb{N}$ and suitable homogeneous $F \in R$ with $\deg F = 2me$. This can be written

$$\frac{F}{(h^a l^b)^m} = \frac{F}{h^{2ma}} \left(\frac{h^a}{l^b} \right)^m \in R_{h,0}[h^a/l^b],$$

and hence $T = R_{h,0}[1/p]$, where $p = l^b/h^a \in R_{h,0}$, as required. $\square$

Definition 3.43.

(i) Given a finitely generated graded ring $R = \bigoplus_{e=0}^{\infty} R_{(e)}$, the algebraic variety with function field K_0 obtained by gluing the maximal spectra $\mathrm{Spm}\ R_{h,0}$ for all nonzer homogeneous elements $h \in R$ is denoted by $\mathrm{Proj}\ R$.

(ii) If $R_0 = k$, then the variety $\mathrm{Proj}\ R$ is said to be *projective*. $\square$

Proposition 3.44. *If $h_1, \ldots, h_m \in R$ are homogeneous generators, then* Proj R *is covered by the open sets* Spm $R_{h_1,0}, \ldots,$ Spm $R_{h_m,0}.$

Proof. We have to show, given a homogeneous element $h \in R$, that

$$\text{Spm } R_{h,0} \subset \bigcup_{i=1}^{m} \text{Spm } R_{h_i,0}.$$

Note that $h_1, \ldots, h_m$ are also generators of the ideal $R_+ = \bigoplus_{e>0} R_{(e)} \subset R$. Let $e \in \mathbb{N}$ be the lowest common multiple of the degrees of h and $h_1, \ldots, h_m$, so that

$$e = a_0 \deg h = a_1 \deg h_1 = \cdots = a_m \deg h_m$$

for some $a_0, \ldots, a_m \in \mathbb{N}$. Then h^{a_0} is contained in the radical of the ideal $(h_1^{a_1}, \ldots, h_m^{a_m}) \subset R$, so we can write $h^{a_0 N} = f_1 h_1^{a_1} + \cdots + f_m h_m^{a_m}$ for some $N \in \mathbb{N}$ and homogeneous elements $f_1, \ldots, f_m \in R$. In other words,

$$1 = f_1 \left(\frac{h_1^{a_1}}{h^{a_0 N}} \right) + \cdots + f_m \left(\frac{h_m^{a_m}}{h^{a_0 N}} \right),$$

and this says that the terms in brackets form a partition of unity in the ring $R_{h,0}$ (Section 3.1(c)). It implies that

$$\text{Spm } R_{h,0} = \bigcup_{i=1}^{m} \text{Spm } R_{h_i,0} \left[\frac{h_i^{a_i}}{h^{a_0 N}} \right],$$

where each Spm $R_{h_i,0} \left[h_i^{a_i} / h^{a_0 N} \right]$ is an open set in Spm $R_{h_i,0}.$ $\qquad\square$

Example 3.45. Let $R = k[X_0, X_1, \ldots, X_n]$, graded as usual by degree of polynomials. Then Proj R is nothing other than the n-dimensional projective space $\mathbb{P}^n$ constructed in Example 3.40. (We will return to this fundamental example in Section 6.1(a).) $\qquad\square$

Example 3.46. Weighted projective space. Given natural numbers $a_0, a_1, \ldots,$ $a_n \in \mathbb{N}$, we take $R = k[X_0, X_1, \ldots, X_n]$ to be a polynomial ring graded by $\deg X_i = a_i$ for each $i = 0, 1, \ldots, n$. So $R = \bigoplus_{e=0}^{\infty} R_{(e)}$, where $R_{(e)}$ is the vector space spanned by all monomials $X_0^{m_0} X_1^{m_1} \cdots X_n^{m_n}$ with $\sum a_i m_i = e$. We write

$$\text{Proj } R = \mathbb{P}(a_0 : a_1 : \ldots : a_n),$$

called a *weighted projective space*. Moreover, if the weights $a_0, a_1, \ldots, a_n$ have a common divisor, then an isomorphic projective variety is obtained by dividing

$a_0, a_1, \ldots, a_n$ through by their common divisor, that is:

$$\mathbb{P}(a_0 : a_1 : \ldots : a_n) \cong \mathbb{P}(a_0/b : a_1/b : \ldots : a_n/b),$$

$$\text{where } b = \gcd(a_0, a_1, \ldots, a_n).$$

(See also Example 3.72.)   □

The following fact, which we shall need later, is a consequence of Proposition 3.5.

Proposition 3.47. *A subvariety of* $\mathbb{P}^n$ *of codimension 1 is defined by the vanishing of a single homogeneous equation.*   □

3.3 Functors and algebraic groups

(a) A variety as a functor from algebras to sets

An affine algebraic variety X, at the most elementary level, is a subset of k^n defined as the set of common zeros of a system of polynomial equations

$$F_1(x_1, \ldots, x_n) = F_2(x_1, \ldots, x_n) = \cdots = F_m(x_1, \ldots, x_n) = 0. \qquad (3.9)$$

What is more essential to X, of course, is not the particular choice of equations (3.9) but rather the ideal that they generate. More than this, one can even view the particular choice of field k in which one looks for solutions as inessential to the system and define X as the solution set 'for any number system in which the system makes sense'.

To make this idea more precise it is convenient to use the language of functors. Let R be any k-algebra and consider the set

$$\underline{X}(R) = \{(x_1, \ldots, x_n) \in R^{\oplus n} \text{ satisfying the system of equations (3.9).}\}$$

We call $\underline{X}(R)$ the set of *R-valued points* of the variety X. This is no more than a set, of course, and so on its own does not have very much meaning. However, the point of view that follows turns out to be extraordinarily powerful. Namely, suppose that we have a homomorphism $f : R \to S$ of k-algebras. It is plain to see that, if $(x_1, \ldots, x_n)$ is an R-valued point of X, then $(f(x_1), \ldots, f(x_n))$ is an S-valued point. In other words, f determines a set mapping $\underline{X}(R) \to \underline{X}(S)$. Let us denote this mapping by $\underline{X}(f)$. If $f : R \to S$ and $g : S \to T$ are two k-algebra homomorphisms, then clearly

$$\underline{X}(g \circ f) = \underline{X}(g) \circ \underline{X}(f).$$

This observation can be expressed in formal language:

X is a (covariant) functor from the category of k-algebras to the category of sets.

We want to think of an algebraic variety not just in itself, but as a 'program' which takes as input a k-algebra and outputs a set.

It is also important to note that this functor determined by an algebraic variety does not depend on how we choose coordinates. Let $X = \mathrm{Spm}\, A$, where

$$A = k[x_1, \ldots, x_n]/(F_1, \ldots, F_m)$$

is the quotient ring by the ideal generated by the system of equations (3.9). Given an R-valued point $(a_1, \ldots, a_n) \in \underline{X}(R)$ we have a homomorphism of k-algebras

$$k[x_1, \ldots, x_n] \to R$$

obtained by mapping each $x_i \mapsto a_i$. The kernel is just the ideal generated by $F_1, \ldots, F_m$, and so we get a homomorphism $A \to R$. In this way we can identify

$$\underline{X}(R) = \mathrm{Hom}_k(A, R)$$

or, equivalently:

$$\underline{X}(R) = \mathrm{Mor}_k(\mathrm{Spm}\, R, X). \tag{3.10}$$

Moreover, if $f : R \to S$ is a homomorphism of k-algebras, then it induces a morphism $\overline{f} : \mathrm{Spm}\, S \to \mathrm{Spm}\, R$, and the set mapping

$$\underline{X}(f) : \mathrm{Mor}_k(\mathrm{Spm}\, R, X) \to \mathrm{Mor}_k(\mathrm{Spm}\, S, X) \tag{3.11}$$

is just given by composition $g \mapsto g \circ \overline{f}$:

$$\mathrm{Spm}\, S \xrightarrow{\ \overline{f}\ } \mathrm{Spm}\, R$$

$$\searrow \qquad \swarrow g$$

$$X$$

We can extend the functor $\underline{X}$ from affine varieties to arbitrary algebraic varieties. Since a variety X is obtained by patching together affine varieties U_i, for $i \in I$, some index set, we obtain a functor $\underline{X}$ by 'gluing' the functors $\underline{U_i}$; by definition of a variety this functor will not depend on the particular choice of our open cover $\{U_i\}_{i \in I}$. Moreover, given k-algebras R, S and a homomorphism $f : R \to S$, the interpretation of the set $\underline{X}(R)$ and the set mapping $\underline{X}(f)$ are exactly the same as (3.10) and (3.11).

This point of view will be needed in Chapter 11.

(b) Algebraic groups

Let us now look at a case where this functorial point of view is particularly useful. This is the definition of an algebraic group. Consider the most basic case of the special linear group $G = SL(n)$. Taking as coordinates the matrix entries $(x_{ij})_{1 \le i, j \le n}$, this is the degree n hypersurface in n^2-dimensional affine space:

$$G : \quad \det |x_{ij}| - 1 = 0.$$

Given a k-algebra R, the set of R-valued points is

$$\underline{G}(R) = \{(a_{ij}) \in R^{n^2} \mid \det |a_{ij}| = 1\} = SL(n, R).$$

In other words, $\underline{G}(R)$ is just the group of special linear matrices whose entries are in R. Moreover, given an algebra homomorphism $f : R \to S$, the induced map

$$\underline{G}(f) : SL(n, R) \to SL(n, S)$$

is a homomorphism of groups. So we see that $G = SL(n)$ is in fact a functor from the category of k-algebras to the category of *groups*.

Generalising, we say that:

an algebraic variety G is an algebraic group if $\underline{G}$ is a (covariant) functor from the category of k-algebras to the category of groups.

An example of an algebraic group which is not affine is the famous group law on a plane cubic curve. (See, for example, Cassels [1] §7.) In the case when G is affine, one can make the following definition.

Definition 3.48. Let A be a finitely generated k-algebra. Then $G = \mathrm{Spm}\, A$ is called an *affine algebraic group* if there exist k-algebra homomorphisms

$$\mu : A \to A \otimes_k A \quad \text{(coproduct)}$$
$$\epsilon : A \to k \qquad\quad \text{(coidentity)}$$
$$\iota : A \to A \qquad\quad \text{(coinverse)}$$

satisfying the following three conditions.

(i) The following diagram commutes:

$$
\begin{array}{ccc}
A & \xrightarrow{\;\mu\;} & A \otimes_k A \\[1em]
\mu \downarrow & & \downarrow 1_A \otimes \mu \\[1em]
A \otimes_k A & \xrightarrow{\mu \otimes 1_A} & A \otimes_k A \otimes_k A.
\end{array}
$$

(ii) Both of the compositions

$$
\begin{array}{ccc}
 & k \otimes_k A & \\
 & \epsilon \otimes 1_A \nearrow \qquad \searrow & \\
A \xrightarrow{\;\mu\;} A \otimes_k A & & A \\
 & 1_A \otimes \epsilon \searrow \qquad \nearrow & \\
 & A \otimes_k k &
\end{array}
$$

are equal to the identity.

(iii) The composition (where the last map m is multiplication in the algebra)

$$
A \xrightarrow{\;\mu\;} A \otimes_k A \xrightarrow{\;1_A \otimes \iota\;} A \otimes_k A \xrightarrow{\;m\;} A
$$

coincides with ϵ. □

These three requirements correspond, respectively, to associativity and the existence of an identity element and (right) inverses. The homomorphisms μ, ϵ, ι induce natural transformations $\underline{\mu}, \underline{\epsilon}, \underline{\iota}$, and these together with the axioms above guarantee that the functor

$$
\underline{G} : \{\text{algebras over } k\} \to \{\text{sets}\}
$$

actually takes values in the category of groups.

Example 3.49. If G is any group, the *group ring* $k[G]$ of G is the vector space (finite-dimensional if G is a finite group) with basis $\{[g]\}_{g \in G}$ and bilinear product defined (on basis vectors) by

$$
m : k[G] \times k[G] \to k[G]
$$
$$
([g], [h]) \mapsto [gh].
$$

This makes $k[G]$ into a k-algebra; we let A be its dual as a vector space, given the structure of a k-algebra by componentwise addition and multiplication. Spm A now becomes an affine algebraic group by taking

$$
\mu = m^\vee : A \to A \otimes A,
$$

the dual of the multiplication map on $k[G]$, ϵ to be evaluation at $1 \in G \subset k[G]$, and ι to be the pull-back of linear forms under the involution of $k[G]$ defined by inversion in G.

For example, when G is the cyclic group of order n we obtain

$$
A = k[t]/(t^n - 1),
$$

with $\mu(t) = t \otimes t$, $\epsilon(t) = 1$ and $\iota(t) = t^{n-1}$. As a functor, Spm A assigns to a k-algebra R the group

$$\underline{\text{Spm } A}\ (R) = \{a \in R \mid a^n = 1\}.$$

□

Example 3.50. The spectrum of a polynomial ring in one variable $A = k[s]$ can be given the structure of an affine algebraic group by defining $\mu(s) = s \otimes 1 + 1 \otimes s$, $\epsilon(s) = 0$ and $\iota(s) = -s$. This group is denoted by $\mathbb{G}_a$. As a functor, it assigns to a k-algebra R the additive group R itself. As a variety, of course, it is isomorphic to $\mathbb{A}^1$.   □

Example 3.51. The spectrum of the ring of Laurent polynomials $A = k[t, t^{-1}]$ is made into an affine algebraic group by defining $\mu(t) = t \otimes t$, $\epsilon(t) = 1$ and $\iota(t) = t^{-1}$. This group is denoted by $\mathbb{G}_m$. As a functor, it assigns to a k-algebra R the multiplicative group $R^{\times}$ of invertible elements in R. As a variety, it has an open immersion in $\mathbb{A}^1$ (as the complement of the origin) corresponding to the subring $k[t] \hookrightarrow k[t, t^{-1}]$.   □

Example 3.52. Let $A = k[x_{ij}, (\det x)^{-1}]$ be the polynomial ring in n^2 variables x_{ij}, $1 \le i, j \le n$, with the inverse of their determinant adjoined. Then Spm A is an open set in affine space $\mathbb{A}^{n^2}$ and becomes an affine algebraic group by

$$\mu(x_{ij}) = \sum_{l=1}^{n} x_{il} \otimes x_{lj}, \quad \epsilon(x_{ij}) = \delta_{ij}, \quad \iota(x_{ij}) = (\det x)^{-1}(\text{adj}\, x)_{ij}.$$

This is, of course, none other than the general linear group $GL(n)$. In the case $n = 1$ it is precisely the multiplicative group $\mathbb{G}_m$ of Example 3.51. As a functor it assigns to a k-algebra R the group $GL(n, R)$ of invertible $n \times n$ matrices with entries in R.   □

Example 3.53. The ring $A_0 = k[x_{ij}]/(\det x - 1)$ is the quotient of A (from Example 3.52) by the ideal $(\det x - 1) \subset A$. Observing that this ideal is contained in ker ϵ and is preserved by μ and ι, we see that these maps induce maps $\epsilon_0, \mu_0, \iota_0$ on A_0, making Spm A_0 into an affine algebraic group. This is a subgroup of Spm $A = GL(n)$ and is precisely the special linear group $SL(n)$.   □

The action of an algebraic group on an affine variety can also be viewed functorially in the same spirit. For future use we give the precise definition.

Definition 3.54. An *action* of an affine algebraic group $G = \text{Spm } A$ on an affine variety $X = \text{Spm } R$ is a morphism $G \times X \to X$ defined by a k-algebra homomorphism

$$\mu_R : R \to R \otimes_k A$$

satisfying the following two conditions.

(i) The composition

$$R \xrightarrow{\mu_R} R \otimes_k A \xrightarrow{1_R \otimes \epsilon} R \otimes_k k \xrightarrow{\sim} R$$

is equal to the identity.

(ii) The following diagram commutes:

$$
\begin{array}{ccc}
R & \xrightarrow{\mu_R} & R \otimes_k A \\
\mu_R \downarrow & & \downarrow \mu_R \otimes 1_A \\
R \otimes_k A & \xrightarrow{1_R \otimes \mu_A} & R \otimes_k A \otimes_k A.
\end{array}
$$
$\square$

3.4 Completeness and toric varieties

(a) Complete varieties

The *Hopf fibration* is a continuous surjection of the $2n + 1$-dimensional sphere

$$S^{2n+1} = \{(z_0, z_1, \ldots, z_n) \in \mathbb{C}^{n+1} \mid |z_0|^2 + |z_1|^2 + \cdots + |z_n|^2 = 1\}$$

onto a complex projective space $\mathbb{P}^n_{\mathbb{C}}$ (with the Euclidean topology):

$$S^1 \quad \text{circle fibre}$$

$$\downarrow$$

$$S^{2n+1} \subset \mathbb{R}^{2n+2} = \mathbb{C}^{n+1}$$

$$\downarrow \quad \text{Hopf map}$$

$$\mathbb{P}^n_{\mathbb{C}}$$

In particular, this implies that $\mathbb{P}^n_{\mathbb{C}}$ with its Euclidean topology is compact: that is, any infinite sequence of points in $\mathbb{P}^n_{\mathbb{C}}$ contains a convergent subsequence. The aim of this section is to discuss how this property should be formulated for an algebraic variety.

Definition 3.55.

(i) A map $f : X \to Y$ of topological spaces is called *closed* if for every closed
    set $Z \subset X$ the image $f(Z) \subset Y$ is closed.
(ii) An algebraic variety X is said to be *complete* if for any algebraic variety Y
     the projection morphism $X \times Y \to Y$ is closed.   □

Note that condition (i) is much stronger for the Zariski topology, with which
we are concerned, than for the Euclidean topology. For example, if Y is an
irreducible curve, then the image $f(Z)$ of a closed set $Z \subset X$ must be either
finite or the whole of Y. More generally, if $Y = \mathbb{A}^n$, then $f(Z)$ must be the
zero-set of some system of polynomials.

On the other hand, it is easy to see that products and closed subsets of complete
algebraic varieties are again complete.

Example 3.56. The affine line $\mathbb{A}^1$ is not complete, since the projection $\mathbb{A}^1 \times$
$\mathbb{A}^1 \to \mathbb{A}^1$, $(x, y) \mapsto x$, is not a closed map. For example, the image of the
closed set $V(xy - 1) \subset \mathbb{A}^1 \times \mathbb{A}^1 = \mathbb{A}^2$ is the punctured line $\mathbb{A}^1 - \{0\}$.   □

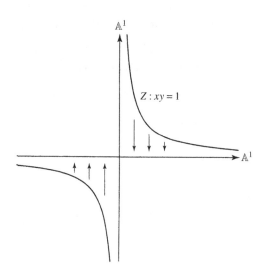

Figure 3.5: The affine line is not complete

We can use the same idea to make the following general observation.

Proposition 3.57. *The only complete affine variety is the single point*
Spm k.   □

Proof. Suppose that X is a complete affine variety realised as a closed subvariety $X \subset \mathbb{A}^n$, and let $x_1, \ldots, x_n$ be the coordinates on $\mathbb{A}^n$. We then consider the projection of the product $X \times \mathbb{A}^1 \to \mathbb{A}^1$, $(x, t) \mapsto t$. The image under this projection of the closed set

$$V(x_1 t - 1) \cap (X \times \mathbb{A}^1)$$

does not contain $0 \in \mathbb{A}^1$, and so by completeness it is either empty or a single point.

If the image is a single point, we let its coordinate be $t = a \neq 0$. This means that

$$X \subset V\left(x_1 - \frac{1}{a}\right) \subset \mathbb{A}^n.$$

On the other hand, the image can be empty only if

$$X \subset V(x_1).$$

In other words, we have fixed the value of the first coordinate x_1 for all points of $X \subset \mathbb{A}^n$. Similarly, we can determine the values of the other coordinates $x_2, \ldots, x_n$, so that X is a single point. $\qquad\square$

There is a general criterion for determining completeness of a variety X. One should think of this as being analogous to the criterion for compactness that any infinite sequence has a convergent subsequence. We make use of the notion of R-valued points, that is, of the functor $\underline{X}$ from k-algebras to sets, introduced in the previous section.

Valuative Criterion for Completeness 3.58. *Let X be a separated algebraic variety and suppose that for an arbitrary valuation ring V over k, with field of fractions K_V, the natural map*

$$\underline{X}(V) \to \underline{X}(K_V)$$

is surjective. Then X is complete.

Proof. Since the property of the projection $X \times Y \to Y$ being closed is local, it is enough to assume that $Y := \mathrm{Spm}\, S$ is an affine variety. Let $Z \subset X \times Y$ be a closed set. Without loss of generality we may assume that Z is irreducible. Replacing Y by the Zariski closure of the image of the second projection $\pi_2|_Z : Z \to Y$, to assume that this projection is a dominant morphism, we have to show that $\pi_2|_Z$ is surjective.

Let $y \in Y$ be an arbitrary point, and let $\mathfrak{m} \subset S$ be its corresponding maximal ideal. The projection $\pi_2|_Z$ induces an embedding of fields

$$k(Y) \hookrightarrow k(Z),$$

and hence by Theorem 2.37 there exists a valuation ring $(V, \mathfrak{m})$ in $k(Z)$ dominating $(S, \mathfrak{m})$. Then there is an injection $S \hookrightarrow V$ such that $\mathfrak{m}_V \cap S = \mathfrak{m}$.

The first projection $\pi_1|_Z : Z \to X$ determines a $k(Z)$-valued point of X and hence a K_V-valued point of X. By hypothesis this is induced by a V-valued point

$$\mathrm{Spm}\, V \to X$$

or, in other words, a homomorphism $R \to V$, where $\mathrm{Spm}\, R \subset X$ is some affine neighbourhood. Tensoring this with the inclusion $S \hookrightarrow V$ and composing with multiplication in V gives a map

$$R \otimes_k S \to V \otimes_k V \to V,$$

and the composition with $V \to V/\mathfrak{m}_V \cong k$ determines a point of $X \times Y$. By construction this point belongs to the closed set Z, and hence y is in the image of Z. $\qquad\square$

Our first application of the Valuative Criterion is the following.

Proposition 3.59. *The projective space $\mathbb{P}^n$ is complete.*

Proof. Let V be a valuation ring with valuation $v : K_V^* \to \Lambda$, and let $(a_0 : a_1 : \ldots : a_n)$ be the homogeneous coordinates of a K_V-valued point $a \in \mathbb{P}^n$ (see Example 3.40). Not all a_i are zero, and among the nonzero homogeneous coordinates we shall suppose that the valuation $v(a_j) \in \Lambda$ is minimal.

We have seen in Example 3.40 that $\mathbb{P}^n$ is covered by $n + 1$ affine spaces $U_0, U_1, \ldots, U_n \subset \mathbb{P}^n$, where each $U_i = \mathrm{Spm}\, R_i \cong \mathbb{A}^n$. Moreover, $a \in U_j$ with coordinates

$$\left(\frac{a_0}{a_j}, \frac{a_1}{a_j}, \ldots, \frac{a_{j-1}}{a_j}, \frac{a_{j+1}}{a_j}, \ldots, \frac{a_n}{a_j} \right).$$

Since $v(a_i/a_j) = v(a_i) - v(a_j) \geq 0$ for each i for which $a_i \neq 0$, it follows that all coordinates are in V, so that a is a V-valued point of $U_j \subset \mathbb{P}^n$. $\qquad\square$

Corollary 3.60. *Every closed subvariety of $\mathbb{P}^n$ is complete.* $\qquad\square$

(See also Corollary 3.73 at the end of the chapter.) Finally, we shall need the following fact later on (Section 9.1).

Proposition 3.61. *If $f : X \to Y$ is a morphism of algebraic varieties, where X is complete and Y is separated, then f is a closed map.*

Proof. The graph $\Gamma \hookrightarrow X \times Y$ of f is a closed immersion since by definition it is the pull-back of the diagonal $\Delta \subset Y \times Y$ under the map $f \times$ id. Then f can be expressed as the composition

$$X \xrightarrow{\sim} \Gamma \hookrightarrow X \times Y \to Y,$$

and since each of these maps is closed, so is f. □

(b) Toric varieties

We are now going to show how a complete algebraic variety, called a *toric variety*, can be obtained from data consisting of a partition of the real vector space $\mathbb{R}^n$ into convex rational polyhedral cones. (For a good introduction to this subject see Fulton [37].)

A point of $\mathbb{R}^n$ whose coordinates are all integers will be called a *lattice point*, and we denote the set of these by $N = \mathbb{Z}^n \subset \mathbb{R}^n$. A linear form $f : \mathbb{R}^n \to \mathbb{R}$ determines a closed half-space $\mathbb{H}_f = \{x \in \mathbb{R}^n | f(x) \geq 0\}$. By a *(convex) polyhedral cone* we mean a finite intersection

$$\sigma = \mathbb{H}_{f_1} \cap \cdots \cap \mathbb{H}_{f_m} \subset \mathbb{R}^n.$$

The forms $f_1, \ldots, f_m$ are called *supporting functions* of the cone σ, and the intersections of σ with the subspaces

$$V(f_{i_1}) \cap \cdots \cap V(f_{i_a}) \subset \mathbb{R}^n, \qquad 1 \leq i_1 < \cdots < i_a \leq m,$$

are called the *faces* of σ.

For simplicity, we will always assume that a polyhedral cone in $\mathbb{R}^n$ is n-dimensional – that is, not contained in any linear hyperplane of $\mathbb{R}^n$. A polyhedral cone will be called *nondegenerate* if it contains no lines (1-dimensional vector subspaces) of $\mathbb{R}^n$. It is *rational* if it is spanned by rays (1-dimensional faces) passing through lattice points.

Definition 3.62. A finite set $\Sigma = \{\sigma_1, \ldots, \sigma_t\}$ of rational convex polyhedral cones in $\mathbb{R}^n$ is called a *fan* if every intersection $\sigma_i \cap \sigma_j$ for $i \neq j$ is a face of σ_i and σ_j. □

Let $M \subset (\mathbb{R}^n)^\vee$ be the dual lattice of N, that is,

$$M = \{f \in (\mathbb{R}^n)^\vee \mid f(x) \in \mathbb{Z} \text{ for all } x \in N\}.$$

Corresponding to the fan Σ there is a *dual fan* in $(\mathbb{R}^n)^\vee$ whose shared rays we will denote by $l_1, \ldots, l_s \subset (\mathbb{R}^n)^\vee$. These define the faces of $\sigma_1, \ldots, \sigma_t$. In the examples following we will identify $\mathbb{R}^n$ with $(\mathbb{R}^n)^\vee$ via the standard inner product and draw $\sigma_1, \ldots, \sigma_t$ and $l_1, \ldots, l_s$ in the same picture.

The group algebra $k[M]$ of M (see Example 3.49) is a finitely generated algebra over k. Indeed, identifying $M \cong \mathbb{Z}^n$ by choosing a basis identifies $k[M]$ with the ring of Laurent polynomials in n variables:

$$k[M] \overset{\sim}{\to} k[x_1, \ldots, x_n, x_1^{-1}, \ldots, x_n^{-1}], \qquad (m_1, \ldots, m_n) \mapsto x_1^{m_1} \ldots x_n^{m_n}.$$

In other words, elements of M become Laurent monomials and we extend by linearity. Moreover, the coproduct

$$\mu : k[M] \to k[M] \otimes k[M], \qquad x^{\mathbf{m}} \mapsto x^{\mathbf{m}} \otimes x^{\mathbf{m}}$$

(where we write $x^{\mathbf{m}} = x_1^{m_1} \ldots x_n^{m_n}$ for $\mathbf{m} \in M$) makes the spectrum

$$T = \operatorname{Spm} k[M]$$

into an affine algebraic group isomorphic to the n-fold product $\mathbb{G}_m \times \cdots \times \mathbb{G}_m$.

Now consider a fan Σ. Each $\sigma \in \Sigma$ determines a semigroup

$$M_\sigma = \{\mathbf{m} \in M \mid \langle \mathbf{m}, \alpha \rangle \geq 0 \text{ for all } \alpha \in \sigma\},$$

where $\langle \, , \, \rangle$ denotes the natural nondegenerate pairing $M \times N \to \mathbb{Z}$. This in turn determines a subalgebra $k[M_\sigma] \subset k[M]$ of the Laurent polynomial ring. The assumption that σ is nondegenerate implies that the field of fractions of $k[M_\sigma]$ is $k(x_1, \ldots, x_n)$ – or, more precisely, that the inclusion $k[M_\sigma] \hookrightarrow k[M]$ induces an open immersion

$$T \hookrightarrow \operatorname{Spm} k[M_\sigma].$$

Moreover, the natural map

$$\mu_\sigma : k[M_\sigma] \to k[M_\sigma] \otimes k[M], \qquad x^{\mathbf{m}} \mapsto x^{\mathbf{m}} \otimes x^{\mathbf{m}}$$

for $\mathbf{m} \in M_\sigma$ defines an action of the group T on the affine variety $\operatorname{Spm} k[M_\sigma]$.

Lemma 3.63. *Given a fan* $\Sigma = \{\sigma_1, \ldots, \sigma_t\}$, *for each* $1 \leq i < j \leq t$ *the gluing*

$$\operatorname{Spm} k[M_{\sigma_i}] \cup_T \operatorname{Spm} k[M_{\sigma_j}]$$

is simple and separated.   □

This is a consequence of the defining condition on the intersection of the cones in a fan of Definition 3.62.

Definition 3.64. Given a fan $\Sigma = \{\sigma_1, \ldots, \sigma_t\}$ in $\mathbb{R}^n$, the gluing

$$X(\Sigma) := \operatorname{Spm} k[M_{\sigma_1}] \cup_T \cdots \cup_T \operatorname{Spm} k[M_{\sigma_t}]$$

is a separated algebraic variety with algebraic function field $k(x_1, \ldots, x_n)$, called a *toric variety*.

Note that the toric variety $X(\Sigma)$ contains the torus T as an open set common to all the affine charts $\operatorname{Spm} k[M_{\sigma_i}]$, and that $X(\Sigma)$ thus constructed carries a natural action of the group T.

Example 3.65. $n = 1$. The only possibility here is $s = t = 2$, with

$$\sigma_1 = l_1 = \mathbb{R}_{\geq 0},$$
$$\sigma_2 = l_2 = \mathbb{R}_{\leq 0}.$$

In this case, $k[M_1] = k[x]$, $k[M_2] = k[x^{-1}]$, and the spectrum of each is the affine line $\mathbb{A}^1$. The gluing is that of Example 3.37(2), and so

$$X(\Sigma) = \mathbb{P}^1.$$

Note also that $T = \mathbb{G}_m$, and this acts on $X(\Sigma) = \mathbb{P}^1$ by $(x_0 : x_1) \mapsto (tx_0 : t^{-1}x_1)$ for $t \in \mathbb{G}_m$. □

Example 3.66. $n = 2$. Let $\Sigma = \{\sigma_1, \sigma_2, \sigma_3, \sigma_4\}$ be the fan obtained by partitioning $\mathbb{R}^2$ into its four quadrants: $s = t = 4$ and $l_1, \ldots, l_4$ the rays spanned by $(\pm 1, 0), (0, \pm 1)$. So

$$
\begin{aligned}
k[M_1] &= k[x, y] \\
k[M_2] &= k[x^{-1}, y] \\
k[M_3] &= k[x^{-1}, y^{-1}] \\
k[M_4] &= k[x, y^{-1}].
\end{aligned}
$$

This is a simple gluing which gives the product

$$X(\Sigma) = \mathbb{P}^1 \times \mathbb{P}^1.$$

□

Example 3.67. $n = 2$. Let $s = t = 3$ and l_1, l_2, l_3 be the rays spanned by $(1, 0), (0, 1), (-1, -1)$, respectively. The cones $\sigma_1, \sigma_2, \sigma_3$ are the three regions shown in Figure 3.6.

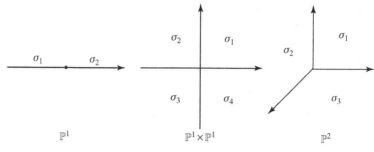

Figure 3.6

We see that

$$\begin{aligned}
k[M_1] &= k[x, y] \\
k[M_2] &= k[x^{-1}, x^{-1}y] \\
k[M_3] &= k[y^{-1}, xy^{-1}].
\end{aligned}$$

The toric variety obtained in this case is the projective plane (by defining homogeneous coordinates $(X_0 : X_1 : X_2) = (1 : x : y)$)

$$X(\Sigma) = \mathbb{P}^2.$$

□

A homomorphism $\mathbb{G}_m \to T$ is called a *1-parameter subgroup* of T. The set of these forms an abelian group and, in fact, is none other than the lattice $N \subset \mathbb{R}^n$. Note that as an algebraic variety $\mathbb{G}_m$ is isomorphic to the projective line $\mathbb{P}^1$ minus two points; more precisely, it is the intersection of the two affine charts $\mathbb{A}^1$ of $\mathbb{P}^1$, and these two charts are exchanged by the automorphism $t \leftrightarrow t^{-1}$ of $\mathbb{G}_m$.

The torus T acts on the toric variety $X(\Sigma)$, and via this action any 1-parameter subgroup $\lambda : \mathbb{G}_m \to T$ induces an action of $\mathbb{G}_m$ on $X(\Sigma)$. Let us use the same symbol λ to denote the composition

$$\lambda : \mathbb{G}_m \to T \to X(\Sigma).$$

If this morphism extends to a morphism $\mathbb{G}_m \subset \mathbb{A}^1 \to X(\Sigma)$, then we shall say that the 1-parameter subgroup λ *has a limit in* $X(\Sigma)$. We are now ready to state the main theorem of this section.

Theorem 3.68. *Let Σ be a fan in $\mathbb{R}^n$ and let $X(\Sigma) = \bigcup_{\sigma \in \Sigma} \mathrm{Spm}\, k[M_\sigma]$ be its associated toric variety. Then the following three conditions are equivalent.*

(i) $X(\Sigma)$ is complete.

(ii) *Every 1-parameter subgroup* $\lambda : \mathbb{G}_m \to T$ *of the torus* $T \subset X(\Sigma)$ *has a limit in* $X(\Sigma)$.

(iii) Σ *covers the space* $\mathbb{R}^n$,

$$\mathbb{R}^n = \bigcup_{\sigma \in \Sigma} \sigma.$$

(c) Approximation of valuations

Our aim in the rest of this section is to prove Theorem 3.68. To begin, the following lemma is well known.

Lemma 3.69. *If* $P \subset \mathbb{R}^n$ *is a convex cone not equal to* $\mathbb{R}^n$, *then there exists a half-space*

$$\mathbb{H}_f = \{x \in \mathbb{R}^n \mid f(x) \geq 0\}$$

for some $f \in (\mathbb{R}^n)^\vee$, *such that* $P \subset \mathbb{H}_f$. $\square$

Let $v : K^* \to \Lambda$ be a valuation of the rational function field $K = k(x_1, \ldots, x_n)$.

Lemma 3.70. *There exists a linear form* $f : M \to \mathbb{R}$ *(unique up to a scalar multiple) with the property that*

$$f(\mathbf{m}) \begin{cases} > 0 \\ < 0 \end{cases} \implies v(x^{\mathbf{m}}) \begin{cases} > 0 \\ < 0. \end{cases}$$

Proof. Let

$$M_v = \{\mathbf{m} \in M \mid v(x^{\mathbf{m}}) \geq 0\}.$$

This is a saturated subsemigroup of the lattice $M \subset (\mathbb{R}^n)^\vee$, which means that $M_v = P \cap M$ for some convex cone $P \subset (\mathbb{R}^n)^\vee$. If $P = (\mathbb{R}^n)^\vee$, then $v(x^{\mathbf{m}}) \geq 0$ for all Laurent monomials $x^{\mathbf{m}}$, which by inversion implies that $v(x^{\mathbf{m}}) = 0$ for all Laurent monomials. In this trivial case, therefore, it is enough to take $f \equiv 0$.

If P is a proper subset, take f to be a form as given by Lemma 3.69. This has the property that

$$v(x^{\mathbf{m}}) \geq 0 \implies f(\mathbf{m}) \geq 0$$

and hence, replacing $\mathbf{m}$ by $-\mathbf{m}$, that

$$v(x^{\mathbf{m}}) \leq 0 \implies f(\mathbf{m}) \leq 0.$$

The lemma follows from these two statements. $\square$

From this we deduce the following:

Approximation Theorem 3.71. *For any finite collection* $\mathbf{m}_1, \ldots, \mathbf{m}_s \in M$ *there exists a linear form* $g : M \to \mathbb{Z}$ *with the property that for each* $i = 1, \ldots, s$:

$$v(x^{\mathbf{m}_i}) \begin{cases} > 0 \\ = 0 \\ < 0 \end{cases} \iff g(\mathbf{m}_i) \begin{cases} > 0 \\ = 0 \\ < 0. \end{cases}$$

Proof. We argue by induction on the rank of M, noting that the result is trivial on a sublattice of rank 1.

We consider the linear form $f : M \to \mathbb{R}$ given by Lemma 3.70, whose kernel

$$M_0 = \ker f \subset M$$

is a sublattice of strictly lower rank than M. We re-order $\mathbf{m}_1, \ldots, \mathbf{m}_s$, if necessary, so that $\mathbf{m}_1, \ldots, \mathbf{m}_r \in M_0$ and $\mathbf{m}_{r+1}, \ldots, \mathbf{m}_s \notin M_0$. Then f is $\mathbb{R}$-valued but otherwise satisfies the requirements of the proposition for the elements $\mathbf{m}_{r+1}, \ldots, \mathbf{m}_s$. Moreover, since the rational numbers $\mathbb{Q}$ are dense in $\mathbb{R}$, we can perturb f, preserving the conditions of the proposition, to take values in $\mathbb{Q}$ (we just have to perturb the values taken on a basis). Multiplying by a suitable positive integer to clear denominators we obtain a $\mathbb{Z}$-valued form possessing the property of the proposition on the elements $\mathbf{m}_{r+1}, \ldots, \mathbf{m}_s$. Let us denote this by $g_{\text{main}} : M \to \mathbb{Z}$.

If $r = 0$, that is, if none of the elements $\mathbf{m}_1, \ldots, \mathbf{m}_s$ are in M_0, then we are done. So if $r \neq 0$, let us turn our attention to the elements $\mathbf{m}_1, \ldots, \mathbf{m}_r \in M_0$.

By the inductive hypothesis there exists a form $h : M_0 \to \mathbb{Z}$ such that

$$v(x^{\mathbf{m}_i}) \begin{cases} > 0 \\ = 0 \\ < 0 \end{cases} \iff h(\mathbf{m}_i) \begin{cases} > 0 \\ = 0 \\ < 0 \end{cases}$$

for each $i = 1, \ldots, r$. We let $g_{\text{sub}} : M \to \mathbb{Z}$ be an arbitrary extension of h to the whole of M. Then it is clear that for a sufficiently large integer $N \in \mathbb{Z}$ the form

$$g = N g_{\text{main}} + g_{\text{sub}}$$

has the required property.   □

Proof of Theorem 3.68. (i) $\Longrightarrow$ (ii) Consider the graph of $\lambda : \mathbb{G}_m \to T$ in the product $\mathbb{A}^1 \times X(\Sigma)$:

$$\Gamma \subset \mathbb{G}_m \times T \subset \mathbb{A}^1 \times X(\Sigma).$$

Denote its Zariski closure by $\overline{\Gamma} \subset \mathbb{A}^1 \times X(\Sigma)$ and consider the projection $\overline{\Gamma} \to \mathbb{A}^1$. By completeness the image is a closed set, and by construction it contains $\mathbb{G}_m$. The projection is therefore surjective and an isomorphism away from the origin of $\mathbb{A}^1$. But the local ring $\mathcal{O}_{\mathbb{A}^1,0}$ at the origin is a valuation ring, so by Example 2.31 and Exercise 2.9 the projection $\overline{\Gamma} \to \mathbb{A}^1$ is an isomorphism. Its inverse morphism is then the graph of a morphism $\overline{\lambda} : \mathbb{A}^1 \to X(\Sigma)$ extending λ.

(ii) $\Longrightarrow$ (iii) It is enough to show, since the lattice directions are dense, that the cones of the fan cover the lattice $N \subset \mathbb{R}^n$. For $x \in N$ let $\lambda_x : \mathbb{G}_m \to T$ be the corresponding 1-parameter subgroup. This has a limit $\overline{\lambda}_x : \mathbb{A}^1 \to X(\Sigma)$, and the limit point $\overline{\lambda}_x(0)$ is contained in Spm $k[M_\sigma]$ for some cone $\sigma \in \Sigma$. But this means precisely that $x \in \sigma$.

(iii) $\Longrightarrow$ (i) Let $l_1, \ldots, l_s$ be the rays of the dual fan of Σ spanned by lattice points $\mathbf{m}_1, \ldots, \mathbf{m}_s \in M$. We will apply the Valuative Criterion 3.58. Choose an arbitrary valuation ring V with $v : K_V^* \to \Lambda$, and a morphism $f : \text{Spm } K_V \to X(\Sigma)$, which we can suppose without loss of generality maps into the torus part $T \subset X(\Sigma)$ (otherwise we apply the argument instead to the lower dimensional toric strata).

Let $g \in N = \text{Hom}(M, \mathbb{Z})$ be the linear form given by the Approximation Theorem 3.71 applied to $\mathbf{m}_1, \ldots, \mathbf{m}_s$ and the valuation v. Since, by hypothesis, the fan Σ covers $\mathbb{R}^n$, it follows that g is contained in some cone $\sigma_i \in \Sigma$. This means that $g(M_{\sigma_i}) \geq 0$. The corresponding coordinate ring $k[M_{\sigma_i}]$ is generated by some subset of the Laurent monomials $x^{\mathbf{m}_1}, \ldots, x^{\mathbf{m}_s}$, and for members of this subset we have $g(\mathbf{m}_j) \geq 0$ and hence $v(x^{\mathbf{m}_j}) \geq 0$. Hence $k[M_{\sigma_i}]$ is contained in the valuation ring $V \subset K_V$, and so we obtain a V-valued point of $X(\Sigma)$. This extends the morphism f to

$$\text{Spm } K_V \subset \text{Spm } V \xrightarrow{f} X(\Sigma),$$

as required. $\qquad\qquad\square$

Example 3.72. Weighted projective space. Let $a = (a_0, a_1, \ldots, a_n)$ be a primitive element of the lattice $\widetilde{N} \cong \mathbb{Z}^{n+1}$, that is,

$$\gcd(a_0, a_1, \ldots, a_n) = 1.$$

Then the quotient $N = \widetilde{N}/\mathbb{Z}a$ is a free abelian group of rank n. If in addition all $a_i > 0$, then a partitions the first quadrant of $\widetilde{N} \otimes \mathbb{R} = \mathbb{R}^{n+1}$ into $n+1$ polyhedral cones of a fan with 1-dimensional faces spanned by a and the standard basis vectors. This fan projects to a fan covering $\mathbb{R}^n = N \otimes \mathbb{R}$.

The complete toric variety associated to this fan in $\mathbb{R}^n$ is the weighted projective space $\mathbb{P}(a_0 : a_1 : \ldots : a_n)$ of Example 3.46. When $a_0 = a_1 = \cdots = a_n = 1$, for example, it is nothing other than the n-dimensional projective space $\mathbb{P}^n$.

Just as for $\mathbb{P}^n$, the weighted projective space $\mathbb{P}(a_0 : a_1 : \ldots : a_n)$ can be described using $n + 1$ coordinates $X_0, X_1, \ldots, X_n$. Then $k[M]$ has a basis of Laurent monomials

$$X_0^{m_0} X_1^{m_1} \ldots X_n^{m_n} \qquad \text{such that } a_0 m_0 + a_1 m_1 + \cdots + a_n m_n = 0.$$

For each $i = 0, 1, \ldots, n$, the corresponding subalgebra $k[M_i] \subset k[M]$ is spanned by those monomials for which

$$m_j \geq 0 \qquad \text{for all } j \neq i.$$

□

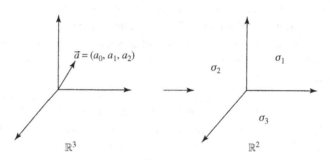

Figure 3.7

Corollary 3.73. *Every projective variety* Proj R *is complete.*

Proof. Let $x_0, x_1, \ldots, x_n$ be homogeneous generators of R with degrees $a_0, a_1, \ldots, a_n$, and consider the surjective graded homomorphism

$$k[X_0, X_1, \ldots, X_n] \to R$$

mapping $X_i \mapsto x_i$, where $\deg X_i = a_i$. Denote the kernel by $\mathfrak{a} \subset k[X_0, X_1, \ldots, X_n]$; then Proj R is isomorphic to the closed subvariety of the weighted projective space $\mathbb{P}(a_0 : a_1 : \ldots : a_n)$ defined by the ideal $\mathfrak{a}$. The weighted projective space is complete by Example 3.72, and hence so is Proj R. □

Remark 3.74. Note that this proof shows that a choice of homogeneous generators in the graded ring R, of degrees $a_0, \ldots, a_n$, is equivalent to specifying a closed immersion of Proj R in the weighted projective space $\mathbb{P}(a_0 : \ldots : a_n)$. In particular, a closed subvariety of $\mathbb{P}^n$ is precisely Proj R for some graded ring R generated in degree 1. We will return to Proj in Section 6.1(a).

Exercises

1. Show that an n-dimensional Euclidean space $\mathbb{R}^n$, for $n \geq 1$, is reducible in the sense of Definition 3.14. Show that this space is not Noetherian.

2. Show that a Noetherian topological space can be (uniquely) expressed as a finite union of irreducible closed sets.

3. Show that in a primary ring (Definition 3.24) the set of nilpotent elements is a prime ideal.

4. Let $S = \mathbb{Z}[\sqrt{-5}]$, and let R be the ring obtained by adjoining $(1 + \sqrt{-5})/2$. Show that Spm $R \to$ Spm S is an open immersion.

5. Let X be a topological space, and let $X \times X$ be the Cartesian product equipped with the product topology. Show that the diagonal subset $\Delta \subset X \times X$ is closed if and only if X is Hausdorff.

6. Consider the plane $\mathbb{R}^2$ with topology coming from the usual Euclidean metric. Then the map

$$\phi : \mathbb{R}^2 \to \mathbb{R}^2$$
$$(x, y) \mapsto (2x, y/2)$$

is a homeomorphism. When the cyclic group generated by ϕ acts on the punctured plane $\mathbb{R}^2 - \{0\}$, show that the quotient topological space is non-Hausdorff.

7. Give a direct proof that the weighted projective space $\mathbb{P}(a_0 : a_1 : \ldots : a_n)$ is complete, by the method of Proposition 3.59.

4

Algebraic groups and rings of invariants

In general it is hard to construct rings of invariants – that is, to determine explicitly a set of generators and relations. However, this is not actually necessary in order to say that a moduli space exists as an algebraic variety. For this one would like to understand – in the precise manner of a Galois theory, so to speak – the relationship between the invariant ideals in a ring and the ideals in its subring of invariants. What we need here is that the group that is acting is *linearly reductive*: this is the central notion in this chapter.

We begin by giving a careful definition of a representation of an algebraic group. Various important properties can be deduced only by following closely to this definition; for example, it allows us to deduce that all representations are locally finite-dimensional. The set of local distributions supported at the identity in an algebraic group G admits a convolution product, making it into a (noncommutative) algebra $\mathcal{H}(G)$, called the *distribution algebra*. The tangent space of G at the identity element, called the *Lie space* $\mathfrak{g} = \mathrm{Lie}\, G$, is a vector subspace of $\mathcal{H}(G)$. As is well known, it inherits a Lie algebra structure, although we will not use this in this book. As well as the Lie space, $\mathcal{H}(G)$ also contains a distinguished element Ω, called the Casimir element, constructed using an invariant inner product on the Lie space (Section 4.2). In Section 4.3 we use the Casimir element to prove the linear reductivity of $SL(n)$. We then prove Hilbert's Theorem 4.53 that if a linearly reductive algebraic group acts on a finitely generated algebra, then the invariant subalgebra is finitely generated. The key ingredient in the proof of this is Hilbert's Basis Theorem.

In Section 4.4 we determine the Hilbert series of the rings of classical binary invariants. Using the relation $e \star f - f \star e = h$ in the distribution algebra, we prove the dimension formula for (invariants of) $SL(2)$. As an application we derive the Cayley-Sylvester formula for the Hilbert series of the classical invariants for binary forms.

116

In the final section of this chapter we give an alternative proof of linear reductivity for $SL(2)$. This yields, in addition, a proof of geometric reductivity of $SL(2)$ over a ground field of positive characteristic.

4.1 Representations of algebraic groups

Let $G = \mathrm{Spm}\, A$ be an affine algebraic group over the field k.

Definition 4.1. An *(algebraic) representation* of the group G (or of the algebra A) is a pair consisting of a vector space V over k and a linear map $\mu_V : V \to V \otimes_k A$ satisfying the following conditions.

(i) The composition

$$V \xrightarrow{\mu_V} V \otimes_k A \xrightarrow{1 \otimes \epsilon} V$$

is the identity, where $\epsilon : A \to k$ is the coidentity.

(ii) The following diagram commutes, where $\mu_A : A \to A \otimes_k A$ is the coproduct.

$$
\begin{array}{ccc}
V & \xrightarrow{\mu_V} & V \otimes_k A \\
\mu_V \downarrow & & \downarrow \mu_V \otimes 1_A \\
V \otimes_k A & \xrightarrow{1_V \otimes \mu_A} & V \otimes_k A \otimes_k A
\end{array}
$$

$\square$

Example 4.2. The coordinate ring A of the group G, together with the coproduct μ_A, is itself an algebraic representation. $\square$

Remark 4.3. Let us check that this is equivalent to the usual definition of a representation as a linear action of G on V. If $\rho : G \to GL(n)$ is such a representation, then write $k[X_{ij}, (\det X)^{-1}]$ for the coordinate ring of $GL(n)$ and let $f_{ij} = \rho^* X_{ij} \in A$ be the pull-back of X_{ij} under ρ. In other words, the f_{ij} are the entries of the $n \times n$ matrix representation, viewed as functions on the group G. We then obtain an algebraic representation of G, in the sense of Definition 4.1, by taking an n-dimensional vector space V, a basis $\{e_i\}$, and the linear map given by

$$\mu_V : e_i \mapsto \sum_j e_j \otimes f_{ji}.$$

Conversely, suppose that $\mu : V \to A \otimes V$ is a finite-dimensional algebraic representation, and let $x_1, \ldots, x_n \in V$ be a basis. Then μ extends naturally to a homomorphism of polynomial rings

$$k[x_1, \ldots, x_n] \to A[x_1, \ldots, x_n].$$

This is just the same thing as a linear action of G on the dual space of V, viewed as an affine space, and this construction is inverse to the first. $\qquad\square$

All of the usual notions concerning group representations can be defined in the spirit of Definition 4.1. In what follows we shall often drop the subscript and write $\mu = \mu_V$ when there is no risk of confusion.

Definition 4.4. Given a representation $\mu : V \to V \otimes A$ of a group $G = \operatorname{Spm} A$:

(i) a vector $x \in V$ is said to be G-invariant if $\mu(x) = x \otimes 1$;

(ii) a subspace $U \subset V$ is called a subrepresentation if $\mu(U) \subset U \otimes A$. $\qquad\square$

Remark 4.5. In characteristic zero the coordinate ring A of a connected algebraic group is an integral domain. It follows from this that the above definitions are in this case equivalent to the usual notions for a rational representation $\rho : G \to GL(V)$. (See Exercise 4.8.)

The above definitions have some immediate consequences. The first says, in the language of Remark 4.3, that in any infinite dimensional representation only finitely many of the matrix entries f_{ij} are nonzero:

Proposition 4.6. *Every representation V of G is locally finite-dimensional. In other words, every $x \in V$ is contained in a finite-dimensional subrepresentation of the group.*

Proof. We can write $\mu(x)$ as a finite sum $\sum_i x_i \otimes f_i$ for some $x_i \in V$ and linearly independent elements $f_i \in A$. The linear span $U \subset V$ of the vectors x_i is then exactly what we require. First, it follows from Definition 4.1(i) that

$$x = \sum_i \epsilon(f_i) x_i,$$

so that $x \in U$. Second, the commutative diagram in Definition 4.1(ii) says that

$$\sum_i \mu(x_i) f_i = \sum_i x_i \mu_A(f_i) \in U \otimes A \otimes A.$$

Since the f_i are linearly independent, this implies that $\mu(x_i) \in U \otimes A$ for each i, so $U \subset V$ is indeed a (finite-dimensional) subrepresentation. $\qquad\square$

For the multiplicative group $\mathbb{G}_m$ of Example 3.51, the representations are particularly simple to describe. Given a vector space V and an integer $m \in \mathbb{Z}$, consider the map

$$V \to V \otimes k[t, t^{-1}], \qquad v \mapsto v \otimes t^m.$$

This defines an algebraic representation of $\mathbb{G}_m$, called its representation of *weight m*. By taking direct sums of these representations we get all representations of $\mathbb{G}_m$.

Proposition 4.7. *Every representation V of $\mathbb{G}_m$ is a direct sum $V = \sum_{m \in \mathbb{Z}} V_{(m)}$, where each $V_{(m)} \subset V$ is a subrepresentation of weight m.*

Given such a representation $V = \bigoplus V_{(m)}$ of $\mathbb{G}_m$, a vector $v \in V_{(m)}$ is said to be *homogeneous of weight m*.

Proof. For each integer $m \in \mathbb{Z}$ define

$$V_{(m)} = \{v \in V \mid \mu(v) = v \otimes t^m\}.$$

It is easy to verify that this is a subrepresentation of V (see Exercise 4.4), and by construction it has weight m. The proof that $V = \bigoplus_{m \in \mathbb{Z}} V_{(m)}$ is very similar to the proof of Proposition 4.6: begin by writing, for an arbitrary $v \in V$,

$$\mu(v) = \sum_{m \in \mathbb{Z}} v_m \otimes t^m \in V \otimes k[t, t^{-1}].$$

It follows from Definition 4.1(i) that $v = \sum v_m$, so it just remains to check that each $v_m \in V_{(m)}$. This will prove the direct sum decomposition since obviously $V_{(m)} \cap V_{(n)} = 0$ whenever $m \neq n$. However, Definition 4.1(ii) tells us that

$$\sum_{m \in \mathbb{Z}} \mu(v_m) t^m = \sum_{m \in \mathbb{Z}} v_m \otimes t^m \otimes t^m \in V \otimes A \otimes A,$$

and so by linear independence of the $t^m \in A$ it follows that $\mu(v_m) = v_m \otimes t^m$ for each $m \in \mathbb{Z}$; hence $v_m \in V_{(m)}$. $\qquad\square$

It is also easy to classify the representations of the additive group $\mathbb{G}_a$ (Example 3.50). Note, incidentally, that our assumption that the field k has characteristic zero is essential in the next proposition, as well as in the two examples which follow.

Proposition 4.8. *Every representation V of $\mathbb{G}_a = \mathrm{Spm}\, k[s]$ is given by*

$$\mu(v) = \sum_{n=0}^{\infty} f^n(v) \otimes \frac{s^n}{n!}$$

for some endomorphism $f \in \mathrm{End}\, V$ which is locally nilpotent (that is, every vector is eventually killed by iterates of f).

Proof. We have a sequence of linear maps $\delta_n : V \to V$ defined by

$$\mu(v) = \sum_{n=0}^{\infty} \delta_n(v) \otimes s^n \in V \otimes k[s].$$

By Definition 4.1 we see that $\delta_0(v) = v$ and

$$\sum_{n=0}^{\infty} \mu(\delta_n(v)) \otimes s^n = \sum_{n=0}^{\infty} \delta_n(v) \otimes (s \otimes 1 + 1 \otimes s)^n,$$

from which it follows that

$$\delta_m \circ \delta_n = \binom{m+n}{m} \delta_{m+n}.$$

The map $f = \delta_1$ therefore has the properties stated in the proposition. □

In the previous chapter (see Definition 3.54) we defined an action of a group $G = \mathrm{Spm}\, A$ on an affine variety $X = \mathrm{Spm}\, R$. We can now interpret this as simply a representation

$$\mu_R : R \to R \otimes_k A$$

which is also a ring homomorphism. The subset of G-invariants

$$R^G = \{ f \in R \mid \mu_R(f) = f \otimes 1 \}$$

is a subring of R.

Example 4.9. An action of the multiplicative group $\mathbb{G}_m = \mathrm{Spm}\, k[t, t^{-1}]$ on $X = \mathrm{Spm}\, R$ is equivalent to specifying a grading

$$R = \bigoplus_{m \in \mathbb{Z}} R_{(m)}, \qquad R_{(m)} R_{(n)} \subset R_{(m+n)}.$$

The invariants of $\mathbb{G}_m$ are then the homogeneous elements of weight 0 under this grading. Moreover, the linear endomorphism of R which rescales each summand $R_{(m)}$ by m,

$$E : R \to R, \qquad \sum f_m \mapsto \sum m f_m,$$

is a derivation of R. E is called the *Euler operator*. Trivially, the $\mathbb{G}_m$-invariants in R are the elements killed by E, that is, $R^{\mathbb{G}_m} = \ker E$. $\square$

Example 4.10. An action of the additive group $\mathbb{G}_a = \mathrm{Spm}\, k[s]$ on $X = \mathrm{Spm}\, R$ is equivalent to specifying a locally nilpotent derivation $D \in \mathrm{End}\, R$ of the function ring R (see Definition 4.15 below). The action $\mu_R : R \to R \otimes k[s]$ is then given by

$$\mu_R(f) = \sum_{n=0}^{\infty} D^n(f) \otimes \frac{s^n}{n!}.$$

The $\mathbb{G}_a$-invariants in R are the elements killed by D, that is, $R^{\mathbb{G}_a} = \ker D$. $\square$

We will later need to consider semiinvariants of group representations as well as invariants (see Chapter 6), and for these we make the next two definitions.

Definition 4.11. Let $G = \mathrm{Spm}\, A$ be an affine algebraic group. A *(1-dimensional) character* of G is a function $\chi \in A$ satisfying

$$\mu_A(\chi) = \chi \otimes \chi, \qquad \iota(\chi)\chi = 1.$$

$\square$

Note that the characters of G are invertible elements of the function ring A, and in fact they form a multiplicative subgroup of these.

Lemma 4.12. *The characters of the general linear group* $GL(n) = \mathrm{Spm}\, k[X_{ij}, (\det X)^{-1}]$ *are precisely the integer powers of the determinant* $(\det X)^n$, $n \in \mathbb{Z}$.

Proof. This is trivial: since $\det(X_{ij})$ is an irreducible polynomial, the only invertible elements of $k[X_{ij}, (\det X)^{-1}]$ are, up to multiplication by a scalar, the powers $(\det X)^n$. For every $n \in \mathbb{Z}$ and scalar $\lambda \in k$, moreover, $\lambda(\det X)^n$ is a character precisely when $\lambda = 1$. $\square$

Definition 4.13. Let χ be a character of an affine algebraic group G, and let V be a representation of G. A vector $x \in V$ satisfying

$$\mu_V(x) = x \otimes \chi$$

is called a semiinvariant of G with weight χ. The semiinvariants of V belonging to a given character χ of G form a subrepresentation (see Exercise 4.4), which we shall denote by $V_\chi \subset V$. □

An algebraic group $T = \text{Spm } A$ which is isomorphic to a direct product of copies of $\mathbb{G}_m$ is called an *algebraic torus*. In this case the set $X(T)$ of characters of T is a basis over k of the algebra A. The following fact follows from Proposition 4.7.

Proposition 4.14. *Let T be an algebraic torus and let $X(T)$ be its set of characters. Then every representation V of T is the direct sum of all its semiinvariant subrepresentations:*

$$V = \bigoplus_{\chi \in X(T)} V_{(\chi)}.$$

4.2 Algebraic groups and their Lie spaces

In this section we will define the Casimir operator associated to a representation of an algebraic group.

(a) Local distributions

Definition 4.15. Let R be a commutative ring over k and M an R-module (see Chapter 8). An *M-valued derivation* is a k-linear map

$$D : R \to M$$

satisfying the Leibniz rule $D(xy) = xD(y) + yD(x)$ for $x, y \in R$. □

An R-valued derivation $D : R \to R$ will simply be referred to as a derivation of R.

Example 4.16. Let $R = k[t_1, \ldots, t_n]$. The first examples of derivations are the partial derivatives

$$\frac{\partial}{\partial t_i} : R \to R.$$

Fixing $a_1, \ldots, a_n \in k$, a second example is the evaluation

$$\alpha : R \to k, \qquad f \mapsto \frac{\partial}{\partial t_i}(a_1, \ldots, a_n).$$

This is a derivation with values in the R-module $k = R/(t_1 - a_1, \ldots, t_n - a_n)$. □

One can generalise this example to any affine variety. Let p be a point of $X = \mathrm{Spm}\, R$, with maximal ideal $\mathfrak{m}_p \subset R$. Then a $k = R/\mathfrak{m}_p$-valued derivation is a linear map

$$\alpha : R \to k$$

with the property that

$$\alpha(fg) = f(p)\alpha(g) + g(p)\alpha(f)$$

for all $f, g \in R$. We shall sometimes refer to α as a *derivation of X at the point* $p \in X$. Note, in particular, that such a derivation vanishes on $\mathfrak{m}_p^2 \subset R$. It is this idea that we want to generalise next.

Definition 4.17. Let p be a point of $X = \mathrm{Spm}\, R$ with maximal ideal $\mathfrak{m}_p \subset R$. A *local distribution with support at $p \in X$* is a k-linear map $\alpha : R \to k$ with the property that $\alpha(\mathfrak{m}_p^N) = 0$ for sufficiently large $N \in \mathbb{N}$. $\qquad\square$

The degree $\deg \alpha$ of a local distribution supported at p is the minimum $d \in \mathbb{N}$ such that $\alpha(\mathfrak{m}_p^{d+1}) = 0$. Every local distribution of degree 0 is a scalar multiple of the evaluation map

$$ev_p : R \to k, \qquad f \mapsto f(p).$$

Lemma 4.18. *For a k-linear map $\alpha : R \to k$, the following are equivalent:*

(i) α is a derivation of $X = \mathrm{Spm}\, R$ at the point $p \in X$;
(ii) α is a local distribution, supported at $p \in X$, of degree 1 and satisfying $\alpha(1) = 0.$

Proof. (i) $\Longrightarrow$ (ii) It has already been observed that α is a local distribution of degree 1, while $\alpha(1) = \alpha(1 \cdot 1) = \alpha(1) + \alpha(1)$. Hence $\alpha(1) = 0$.

(ii) $\Longrightarrow$ (i) If $f, g \in R$, then $f - f(p), g - g(p) \in \mathfrak{m}_p$, so that $\deg \alpha = 1$ implies

$$\alpha((f - f(p))(g - g(p))) = 0.$$

Expanding and using the fact that $\alpha(f(p)g(p)) = f(p)g(p)\alpha(1) = 0$ gives

$$\alpha(fg) = \alpha(f)g(p) + \alpha(g)f(p),$$

as required. $\qquad\square$

It follows from the lemma that the vector space $\mathrm{Der}_k(R, R/\mathfrak{m}_p)$ of derivations of $X = \mathrm{Spm}\, R$ at $p \in X$ is isomorphic to the dual of $\mathfrak{m}_p/\mathfrak{m}_p^2$.

Definition 4.19. The k-vector space $(\mathfrak{m}_p/\mathfrak{m}_p^2)^\vee \cong \mathrm{Der}_k(R, R/\mathfrak{m}_p)$ is called the *Zariski tangent space* of X at the point p. □

Remark 4.20. The dimension of the Zariski tangent space at $p \in X$ is greater than or equal to the dimension of the variety X at p; X can be defined to be nonsingular at p if and only if the two dimensions are equal. (This is equivalent to Definition 9.44 in Section 9.3(b).) Over a field k of characteristic zero an algebraic group is always nonsingular. □

We now have a vector space isomorphism

$$\{\text{local distributions with degree} \leq d\} \cong (R/\mathfrak{m}_p^{d+1})^\vee,$$

and when $d = 1$ this decomposes into

$$k \oplus (\mathfrak{m}_p/\mathfrak{m}_p^2)^\vee,$$

where, by Lemma 4.18, the two summands are spanned by the evaluation map ev_p and by derivations at p, respectively.

More generally, for each $d \leq e$ the natural projection $R/\mathfrak{m}_p^{e+1} \to R/\mathfrak{m}_p^{d+1}$ induces an injection

$$(R/\mathfrak{m}_p^{d+1})^\vee \hookrightarrow (R/\mathfrak{m}_p^{e+1})^\vee.$$

There is therefore an ascending sequence:

$$k \subset (R/\mathfrak{m}_p^2)^\vee \subset (R/\mathfrak{m}_p^3)^\vee \subset \cdots \subset (R/\mathfrak{m}_p^{d+1})^\vee \subset \cdots.$$

The space of local distributions supported at p can thus be identified with the limit (that is, the union) of this sequence.

(b) The distribution algebra

If $G = \mathrm{Spm}\, A$ is an affine algebraic group with coordinate ring A, we will denote by $\mathcal{H}(G)$ the vector space of distributions $\alpha : A \to k$ supported at the identity element $e \in G$. The Zariski tangent space of G at this point is called the *Lie space* of G and is denoted by $\mathfrak{g} = \mathrm{Lie}(G) \subset \mathcal{H}(G)$.

Remark 4.21. The vector space $\mathfrak{g}$ acquires from $\mathcal{H}(G)$ (together with its convolution product, which we are about to define) the structure of a *Lie algebra*.

This is outlined in Exercise 4.3. However, we are not going to use the Lie algebra structure in this book. □

Let $\mu = \mu_A : A \to A \otimes A$ be the coproduct on G.

Definition 4.22. If $\alpha, \beta \in \mathcal{H}(G)$ are distributions supported at the identity, the *convolution product* $\alpha \star \beta$ of α and β is the composition

$$A \xrightarrow{\mu} A \otimes A \xrightarrow{\alpha \otimes \beta} k \otimes k \xrightarrow{\sim} k.$$

□

Lemma 4.23. *The convolution product of* $\alpha, \beta \in \mathcal{H}(G)$ *is again a distribution supported at the identity* $\alpha \star \beta \in \mathcal{H}(G)$*, and*

$$\deg \alpha \star \beta \leq \deg \alpha + \deg \beta.$$

Proof. Since $(e, e) \mapsto e$ under the group operation $G \times G \to G$, we have

$$\mu(\mathfrak{m}) \subset \mathfrak{m} \otimes A + A \otimes \mathfrak{m},$$

where $\mathfrak{m} = \mathfrak{m}_e$. Since μ is a ring homomorphism, this implies, for $a, b \in \mathbb{N}$,

$$\mu(\mathfrak{m}^{a+b+1}) \subset \sum_{i+j=a+b+1} \mathfrak{m}^i \otimes \mathfrak{m}^j.$$

Taking $a = \deg \alpha$ and $b = \deg \beta$, it follows from $\alpha(\mathfrak{m}^{a+1}) = 0$ and $\beta(\mathfrak{m}^{b+1}) = 0$ that $\alpha \star \beta(\mathfrak{m}^{a+b+1}) = 0$, which proves the lemma. □

Evaluation at the identity $\epsilon = ev_e \in \mathcal{H}(G)$ is an identity element for the convolution product. Moreover, it follows from the associative law for the coproduct μ (Definition 3.48(i)) that $\star$ is associative and thus makes $\mathcal{H}(G)$ into an associative algebra, called the *distribution algebra* of the algebraic group G.

Remark 4.24. In general, the distribution algebra is infinite-dimensional and noncommutative. If k has characteristic zero, it is a theorem of Cartier that $\mathcal{H}(G)$ is the universal enveloping algebra of a Lie algebra. □

Example 4.25. Consider the multiplicative group $G = \mathbb{G}_\mathrm{m} = \mathrm{Spm}\, k[t, t^{-1}]$. The vector space of distributions supported at the identity is, by definition,

$$\mathcal{H}(\mathbb{G}_\mathrm{m}) = \lim_{n \to \infty} \left(k[t]/(t-1)^n \right)^\vee.$$

As an algebra this is isomorphic to the polynomial ring $k[E]$ where

$$E = \frac{d}{dt}\bigg|_{t=1} : k[t, t^{-1}] \to k,$$

and $\mathrm{Lie}(\mathbb{G}_m)$ is the 1-dimensional linear span of E. In fact, by definition $E \star E$ takes $f(t) \in k[t, t^{-1}]$ to

$$\frac{\partial^2 \cdot}{\partial t \partial t'} f(tt')\bigg|_{t=t'=1} = \left(\frac{d}{dt}\right)^2 f(t)\bigg|_{t=1}.$$

Similarly, the n-th power $E^{\star n} = E \star \cdots \star E$ is equal to $(\partial^2/\partial t^n)|_{t=1}$.   □

Example 4.26. Similarly to the previous example, the additive group $\mathbb{G}_a = \mathrm{Spm}\, k[s]$ has distribution algebra

$$\mathcal{H}(\mathbb{G}_a) = \lim_{n \to \infty} \left(k[s]/(s^n)\right)^{\vee} = k[D].$$

This is the polynomial ring generated by

$$D = \frac{d}{ds}\bigg|_{s=0} : k[s] \to k.$$

□

A homomorphism of algebraic groups $G \to G'$ induces a ring homomorphism $\mathcal{H}(G) \to \mathcal{H}(G')$ and a linear map of Lie spaces $\mathrm{Lie}(G) \to \mathrm{Lie}(G')$. (This is also a homomorphism of Lie algebras in the sense of Exercise 4.3.) Note that the induced homomorphism of function rings is in the reverse direction, $A' \to A$, but our constructions dualise this once more so that both functors $\mathcal{H}$ and Lie are covariant.

We return now to the representations of $G = \mathrm{Spm}\, A$. Let $\mu_V : V \to V \otimes A$ be an algebraic representation, with an associated linear representation $\rho : G \to GL(V)$. For each distribution $\alpha \in \mathcal{H}(G)$ consider the k-linear map

$$V \xrightarrow{\mu} V \otimes A \xrightarrow{1 \otimes \alpha} V \otimes k \xrightarrow{\sim} V.$$

We will denote this composition by $\tilde{\rho}(\alpha) \in \mathrm{End}\, V$. From the associativity of μ (Definition 4.1(ii)) we obtain:

Lemma 4.27. *The map $\tilde{\rho} : \mathcal{H}(G) \to \mathrm{End}\, V$ is a ring homomorphism.*   □

In other words, the representation V is a (noncommutative) $\mathcal{H}(G)$-module. A vector $v \in V$ is G-invariant if and only if

$$\widetilde{\rho}(\alpha)v = \alpha(1)v$$

for all $\alpha \in \mathcal{H}(G)$. In particular (by Lemma 4.18(i)), $\widetilde{\rho}(\alpha)v = 0$ if $\alpha \in \mathfrak{g} = \mathrm{Lie}(G)$. That is, the Lie space $\mathfrak{g}$ kills all of the G-invariants in V.

Example 4.28. Let V be a representation of $\mathbb{G}_m$, and decompose a vector $v \in V$ into its homogeneous components $v = \sum_{m \in \mathbb{Z}} v_m$ under the action of $\mathbb{G}_m$. Then

$$\mu(v) = \sum v_m \otimes t^m,$$

and so the generating distribution

$$E = \left. \frac{d}{dt} \right|_{t=1} \in \mathcal{H}(\mathbb{G}_m) = k[E]$$

acts by $\widetilde{\rho}(E) : v \mapsto \sum m v_m$. In other words, it coincides with the Euler operator of Example 4.9. $\square$

Consider the action of G on itself by conjugation:

$$G \times G \to G, \qquad (x, g) \mapsto gxg^{-1}.$$

This induces an action of G on its coordinate ring A. Since the identity element $e \in G$ is fixed under conjugation, the action preserves the maximal ideal $\mathfrak{m} = \mathfrak{m}_e$, and in particular it induces a k-linear action on each quotient $A/\mathfrak{m}^n$ and on its dual space. It follows that $\mathcal{H}(G)$ becomes a linear representation of G. On the other hand, given a representation $\rho : G \to GL(V)$, the space $\mathrm{End}\, V$ also becomes a linear representation, by conjugation $T \mapsto \rho(g)T\rho(g)^{-1}$ for $T \in \mathrm{End}\, V$ and $g \in G$. With respect to this action we have:

Lemma 4.29. *The map* $\widetilde{\rho} : \mathcal{H}(G) \to \mathrm{End}\, V$ *is a homomorphism of G-representations.* $\square$

In particular, the Lie space $\mathfrak{g}$ is a subrepresentation of $\mathcal{H}(G)$; this is called the *adjoint representation* and is denoted by

$$\mathrm{Ad} : G \to GL(\mathfrak{g}).$$

Example 4.30. Consider the general linear group $GL(n)$ with coordinate ring $A = k[X_{ij}, (\det X)^{-1}]$. The Lie space $\mathfrak{gl}(n)$ of $GL(n)$ is the k-vector space with basis

$$E_{ij} = \left. \frac{\partial}{\partial X_{ij}} \right|_{X=I_n} : k[X_{ij}, (\det X)^{-1}] \to k$$

for $1 \le i, j \le n$. (Note that this generalises Example 4.25.) This space is isomorphic to the vector space of $n \times n$ matrices over k, by mapping E_{ij} to the matrix with a 1 in the (i, j)-th entry and 0s elsewhere. With this identification the adjoint representation is

$$\mathrm{Ad}(g) : \mathfrak{gl}(n) \to \mathfrak{gl}(n) \qquad M \mapsto gMg^{-1}.$$

Example 4.31. The special linear group $SL(n)$ has coordinate ring $A = k[X_{ij}]/(\det X - 1)$. We can identify its Lie space $\mathfrak{sl}(n)$ as follows. It is the Zariski tangent space $(\mathfrak{m}/\mathfrak{m}^2)^\vee$, where $\mathfrak{m} \subset A$ is the ideal at the identity matrix, and a tangent vector is therefore a ring homomorphism $f : A \to k[t]/(t^2)$ for which the composition $A \xrightarrow{f} k[t]/(t^2) \to k[t]/(t) = k$ coincides with the map $A \to A/\mathfrak{m}$. In other words, if we write ϵ for the residue class $t \bmod t^2$ and $k[\epsilon] = k[t]/(t^2)$, then a tangent vector is a matrix $I + \epsilon M$ which satisfies

$$1 = \det(I + \epsilon M) = 1 + \epsilon \operatorname{tr} M,$$

since $\epsilon^2 = 0$. Hence $\mathfrak{sl}(n) \subset \mathfrak{gl}(n)$ is the vector space of $n \times n$ matrices over k with trace zero.   □

(c) The Casimir operator

Consider now an inner product on the Lie space, that is, a symmetric and nondegenerate bilinear form

$$\kappa : \mathfrak{g} \times \mathfrak{g} \to k.$$

We will assume that κ is invariant under the adjoint representation $\mathrm{Ad} : G \to GL(\mathfrak{g})$.

Definition 4.32. Let κ be a G-invariant inner product on $\mathfrak{g}$, as above. Let $X_1, \ldots, X_N \in \mathfrak{g}$ be a basis of $\mathfrak{g}$ and let $X'_1, \ldots, X'_N \in \mathfrak{g}$ be its dual basis with respect to κ. The distribution

$$\Omega = X_1 \star X'_1 + \cdots + X_N \star X'_N \in \mathcal{H}(G)$$

is called the *Casimir element* over G with respect to κ.   □

Proposition 4.33. *The Casimir element Ω is independent of the choice of basis* $\{X_1, \ldots, X_N\}$.

Proof. A second basis $\{Y_1, \ldots, Y_N\}$, with dual basis $\{Y'_1, \ldots, Y'_N\}$ is related to the first by

$$Y_i = \sum_{j=1}^{N} a_{ij} X_j, \quad Y'_i = \sum_{j=1}^{N} a'_{ij} X'_j, \quad i = 1, \ldots, N,$$

where $A = (a_{ij})$, $A' = (a'_{ij})$ are some matrices satisfying $A^t A' = I_N$. So we compute

$$\sum_{i=1}^{N} Y_i \star Y'_i = \sum_{i=1}^{N} \left(\sum_{j=1}^{N} a_{ij} X_j \right) \star \left(\sum_{k=1}^{N} a'_{ik} X'_k \right)$$

$$= \sum_{j,k} \left(\sum_{i=1}^{N} a_{ij} a'_{ik} \right) X_j \star X'_k$$

$$= \sum_{j,k} \delta_{jk} X_j \star X'_k = \Omega,$$

as required. $\qquad\square$

For the Casimir element for $SL(n)$, see Example 4.48 below.

Since the inner product κ is assumed to be G-invariant, for each $g \in G$ the sets

$$\{ \mathrm{Ad}(g)(X_1), \ldots, \mathrm{Ad}(g)(X_N) \}, \qquad \{ \mathrm{Ad}(g)(X'_1), \ldots, \mathrm{Ad}(g)(X'_N) \},$$

are again dual bases. We therefore deduce:

Corollary 4.34. *The Casimir element $\Omega \in \mathcal{H}(G)$ is invariant under the action of G on the distribution algebra.* $\qquad\square$

Let $\rho : G \to GL(V)$ be a representation of G. This is an $\mathcal{H}(G)$-module via the homomorphism $\tilde{\rho} : \mathcal{H}(G) \to \mathrm{End}\, V$ of Lemma 4.27. In particular, the Casimir element Ω determines a linear endomorphism of V,

$$\tilde{\rho}(\Omega) : V \to V,$$

called the *Casimir operator* (with respect to the inner product κ). By Lemma 4.29 and Corollary 4.34 this is invariant under the conjugation action of G on End V: that is, it commutes with the action of G. Moreover, since $\mathfrak{g}$ kills the G-invariants $V^G \subset V$, so does the Casimir operator. In other words:

Corollary 4.35. *The Casimir operator is an endomorphism of each representation V of G, and*

$$V^G \subset \ker \left(\widetilde{\rho}(\Omega) \right).$$

4.3 Hilbert's Theorem

(a) Linear reductivity

Definition 4.36. An algebraic group G is said to be *linearly reductive* if, for every epimorphism $\phi : V \to W$ of G representations, the induced map on G-invariants $\phi^G : V^G \to W^G$ is surjective. $\qquad\square$

There are various equivalent definitions (see also Lemma 4.74 below):

Proposition 4.37. *The following conditions are all equivalent.*

(i) *G is linearly reductive.*

(ii) *For every epimorphism $\phi : V \to W$ of finite-dimensional representations the induced map on G-invariants $\phi^G : V^G \to W^G$ is surjective.*

(iii) *If V is any finite-dimensional representation and $v \in V$ is G-invariant modulo a proper subrepresentation $U \subset V$, then the coset $v + U$ contains a nontrivial G-invariant vector.*

Proof. (i) implies (ii) trivially. Applying (ii) to the quotient map $V \to V/U$ gives (iii), so we just have to show that (iii) implies (i). We suppose that $\phi : V \to W$ is an epimorphism of G representations and that $\phi(v) = w \in W^G$ for some $v \in V$. By local finite dimensionality (Proposition 4.6) there exists a finite-dimensional subrepresentation $V_0 \subset V$ containing v. Now $v \in V_0$ is G-invariant modulo the subrepresentation $U_0 = V_0 \cap \ker \phi$, so by property (iii) there exists a G-invariant vector $v' \in V_0$ such that $v' - v \in U_0$. Since $\phi(v') = w$, we have shown that $\phi^G : V^G \to W^G$ is surjective. $\qquad\square$

Proposition 4.38. *Every finite group is linearly reductive.*

Proof. Suppose that V is a finite-dimensional representation and $v \in V$ is a vector invariant modulo a subrepresentation $U \subset V$, and set

$$v' = \frac{1}{|G|} \sum_{g \in G} g \cdot v.$$

Clearly v' is G-invariant, while

$$v' - v = \frac{1}{|G|} \sum_{g \in G} (g \cdot v - v)$$

is contained in U. So we have verified condition (iii) of Proposition 4.37. $\quad\square$

Remark 4.39. The homomorphism $R : V \to V^G$ used in this proof, $R = \frac{1}{|G|} \sum_{g \in G} g$, is called a *Reynolds operator* and corresponds to *Cayley's Ω-process* in the work of Hilbert. (See Sturmfels [28].) One could, alternatively, prove the proposition by using R to verify Definition 4.36 directly, but we have used the criterion of Proposition 4.37 because this is the approach that we will take to prove the linear reductivity of $SL(n)$ (Theorem 4.43). $\quad\square$

Direct products of linearly reductive algebraic groups are linearly reductive; moreover, if $H \subset G$ is a normal subgroup and G is linearly reductive, then so is the quotient G/H. Conversely, if both H and G/H are linearly reductive, then G is linearly reductive.

Example 4.40. If G is an algebraic group whose connected component at the identity is linearly reductive, then G is linearly reductive. $\quad\square$

Proposition 4.41. *Every algebraic torus $(\mathbb{G}_m)^N$ is linearly reductive.*

Proof. It is enough to prove this for $T = \mathbb{G}_m$; again, we shall check condition (iii) of Proposition 4.37. By Proposition 4.7, a representation V and a subrepresentation U have weight decompositions

$$V = \bigoplus_{m \in \mathbb{Z}} V_{(m)}, \qquad U = \bigoplus_{m \in \mathbb{Z}} U_{(m)},$$

with $U_{(m)} \subset V_{(m)}$. T-invariance of an element $v = \sum v_{(m)}$ modulo U means that $v_{(m)} \in U$ for all $m \neq 0$. It follows that $v_{(0)}$ is a T-invariant element of the coset $v + U$, as required. $\quad\square$

Example 4.42. An example of a group which is not linearly reductive is the additive group $\mathbb{G}_a \cong k$. To see this, consider the 2-dimensional representation given by

$$\mathbb{G}_a \to GL(2), \qquad t \mapsto \begin{pmatrix} 1 & t \\ 0 & 1 \end{pmatrix}.$$

Then restriction to the x-axis,

$$V := k[x, y] \to W := k[x, y]/(y) = k[x],$$

is a surjective homomorphism of $\mathbb{G}_a$-representations. But $V^{\mathbb{G}_a} = k[y]$, $W^{\mathbb{G}_a} = k[x]$, so the induced homomorphism on invariants is not surjective. $\qquad\square$

Our aim in the remainder of this section is to prove:

Theorem 4.43. *The special linear group $SL(n)$ is linearly reductive.*

The general linear group $GL(n)$ is generated by its centre, which is isomorphic to $\mathbb{G}_m$, and the subgroup $SL(n)$. It can therefore be expressed as a quotient $GL(n) = (\mathbb{G}_m \times SL(n))/\mathbb{Z}_n$, and so we obtain:

Corollary 4.44. *$GL(n)$ is linearly reductive.* $\qquad\square$

Remark 4.45. The proof of Theorem 4.43 will be modelled on that of Proposition 4.38, using a Reynold's operator $R : V \to V^{SL(n)}$ (Remark 4.39). In this case R will be constructed using the Casimir operator for the representation V. For the case $n = 2$, on the other hand, we will give an alternative and more direct proof in Section 4.5. $\qquad\square$

Let U be any finite-dimensional vector space. The Lie space of $GL(U)$ is canonically isomorphic to End U, and the adjoint representation is the conjugation action of $GL(U)$ on this space (Example 4.30). Associating to a pair of endomorphisms of U the trace of their composite

$$\kappa : \text{End } U \times \text{End } U \to k, \qquad (f, g) \mapsto \text{tr } fg, \qquad (4.1)$$

defines a nondegenerate inner product (symmetric bilinear form) on End U. We will write $\kappa(f) = \kappa(f, f)$. The following is clear.

Lemma 4.46. *κ is invariant under the adjoint action of $GL(U)$. In other words,*

$$\kappa(\alpha f \alpha^{-1}) = \kappa(f)$$

is satisfied for all $f \in \text{End } U$ and $\alpha \in GL(U)$. $\qquad\square$

The Lie space $\mathfrak{sl}(U)$ of the special linear group $SL(U)$, as a subgroup of $GL(U)$, is the subalgebra of End U consisting of trace zero endomorphisms

(Example 4.31). We shall denote the restriction of the inner product κ to $\mathfrak{sl}(U)$ by the same symbol.

Lemma 4.47. *κ is a nondegenerate inner product on $\mathfrak{sl}(U)$ invariant under the adjoint action of $SL(U)$.*

Proof. κ is nondegenerate on End U, and with respect to κ the subspace $\mathfrak{sl}(U)$ is the orthogonal complement of the identity element I_U. Since $\kappa(I_U) = \dim U \neq 0$, it follows that κ is nondegenerate on $\mathfrak{sl}(U)$. $\qquad\square$

Example 4.48. Let us calculate the Casimir element for $SL(n)$ using this inner product. Let e_{ij} denote the sparse $n \times n$ matrix with a single 1 in the i-th row and the j-th column. Let $f_{ij} = e_{ji}$ and, for $i = 1, \ldots, n-1$, let

$$h_i = e_{ii} - e_{i+1,i+1}, \qquad m_i = e_{11} + \cdots + e_{ii} - \frac{i}{n}(e_{11} + \cdots + e_{nn}).$$

Then the Lie space $\mathfrak{sl}(n)$ has dual bases

$$\{e_{ij}\}_{i<j} \cup \{f_{ij}\}_{i<j} \cup \{h_i\}_i, \qquad \{f_{ij}\}_{i<j} \cup \{e_{ij}\}_{i<j} \cup \{m_i\}_i.$$

By Definition 4.32, we now compute

$$\Omega = \sum_{i<j}(e_{ij} \star f_{ij} + f_{ij} \star e_{ij}) + e_{11}^2 + \cdots + e_{nn}^2 - \frac{1}{n}(e_{11} + \cdots + e_{nn})^2.$$

For example, in the case $n = 2$, the Casimir element is

$$\Omega = e \star f + f \star e + \tfrac{1}{2}h \star h,$$

where

$$e = \begin{pmatrix} 0 & 1 \\ 0 & 0 \end{pmatrix}, \quad f = \begin{pmatrix} 0 & 0 \\ 1 & 0 \end{pmatrix}, \quad h = \begin{pmatrix} 1 & 0 \\ 0 & -1 \end{pmatrix}.$$

(See also Section 4.4(a).) $\qquad\square$

Proposition 4.49. *For a representation $\rho : SL(n) \to GL(V)$ the following are equivalent.*

(i) *The representation is trivial.*
(ii) *$\widetilde{\rho}(\Omega) = 0$.*
(iii) *tr $\widetilde{\rho}(\Omega) = 0$.*

Proof. (i) implies (ii) by Corollary 4.35, and this implies (iii) trivially; so we just have to show that (iii) implies (i).

Let $T \subset SL(n)$ be the torus subgroup of diagonal matrices and $\mathfrak{h} \subset \mathfrak{sl}(n)$ its Lie space. Under the action of T the representation V has, by Proposition 4.14, a character space decomposition

$$V = \bigoplus_{\chi \in X(T)} V_\chi.$$

Each $\chi : T \to \mathbb{G}_m$ corresponds to a linear form with integer coefficients $\overline{\chi} : \mathfrak{h} \to k$, and for each $h \in \mathfrak{h}$ we have

$$\operatorname{tr} \widetilde{\rho}(h)^2 = \sum_{\chi \neq 1} (\dim V_\chi) \overline{\chi}(h)^2.$$

It follows that if $\operatorname{tr} \widetilde{\rho}(h) = 0$ for all $h \in \mathfrak{h}$, then $\dim V_\chi = 0$ for all nontrivial characters $\chi \in X(T)$ – or, in other words, V is a trivial representation of the torus T. Since diagonalisable elements are dense in $SL(n)$, this implies that V is also trivial as a representation of $SL(n)$.

The proposition is proved, therefore, if we can show that $\operatorname{tr} \widetilde{\rho}(\Omega) = 0$ implies $\operatorname{tr} \widetilde{\rho}(h) = 0$ for all $h \in \mathfrak{h}$. We will do this just for $SL(2)$; the general case is similar. We have seen that the Casimir element is

$$\Omega = e \star f + f \star e + \tfrac{1}{2} h \star h \in \mathcal{H}(SL(2))$$

$$= \tfrac{1}{2}(e+f)^2 + \tfrac{1}{2}(\sqrt{-1}e - \sqrt{-1}f)^2 + \tfrac{1}{2}h^2,$$

where the matrices

$$e + f = \begin{pmatrix} 0 & 1 \\ 1 & 0 \end{pmatrix}, \quad \sqrt{-1}(e-f) = \begin{pmatrix} 0 & \sqrt{-1} \\ -\sqrt{-1} & 0 \end{pmatrix}, \quad h = \begin{pmatrix} 1 & 0 \\ 0 & -1 \end{pmatrix}$$

are conjugate in $\mathfrak{sl}(2)$. Hence

$$\operatorname{tr} \widetilde{\rho}(\Omega) = \frac{3}{2} \operatorname{tr} \widetilde{\rho}(h)^2,$$

and we are done. □

In particular, if the Casimir operator $\rho(\Omega)$ of a representation V is nilpotent, then the representation is trivial. Applying this to the subrepresentation $\ker \rho(\Omega)^n$ shows that $\ker (\widetilde{\rho}(\Omega))^m \subset V^{SL(n)}$ for any integer $m \geq 0$. Combining this with Corollary 4.35, we conclude:

Corollary 4.50. $V^{SL(n)} = \bigcup_{m \geq 0} \ker (\widetilde{\rho}(\Omega))^m.$ □

Proof of Theorem 4.43. Given a finite-dimensional representation V of $SL(n)$ we shall construct a Reynolds operator $R : V \to V^{SL(n)}$ analogous to the averaging operator $R = 1/|G| \sum_{g \in G} g$ used in the proof of Proposition 4.38 for the case of a finite group. This is constructed from the Casimir operator $C_V := \tilde{\rho}(\Omega) \in \operatorname{End} V$ as follows. Let $N = \dim V$, and let

$$\chi_V(t) = t^N + c_1 t^{N-1} + \cdots + c_{N-m} t^m$$

be the characteristic polynomial of C_V, where $c_{N-m} \neq 0$. The Reynolds operator is constructed by substituting the Casimir into the polynomial

$$P(t) := \frac{1}{c_{N-m} t^m} \chi_V(t) = \frac{1}{c_{N-m}} (t^{N-m} + c_1 t^{N-m-1} + \cdots + c_{N-m}).$$

That is, $R := P(C_V)$. This is a homomorphism of $SL(n)$-representations, and by the Cayley-Hamilton theorem it satisfies

$$C_V^m P(C_V) = 0.$$

It follows from Corollary 4.50 that the image of $P(C_V)$ is contained in $V^{SL(n)}$.

We can now follow the proof of Proposition 4.38 and apply the criterion in Proposition 4.37(iii) for linear reductivity. Suppose that $U \subset V$ is a subrepresentation and that $v \in V$ is $SL(n)$-invariant modulo U. This means that $\tilde{\rho}(\mathfrak{sl}(n))v \subset U$, and hence $C_V v \in U$. It follows from the way we have defined $P(t)$ that $P(C_V)v - v \in U$, and so the vector $v' = P(C_V)v$ satisfies the requirements of Proposition 4.37(iii). □

(b) Finite generation

Let G be an algebraic group acting on a polynomial ring S, preserving the grading.

Theorem 4.51 (Hilbert [19]). *If G is linearly reductive, then the ring of invariant polynomials S^G is finitely generated.*

Proof. We will essentially follow the original reasoning of Hilbert. The invariant ring is graded by

$$S^G = \bigoplus_{e \geq 0} S^G \cap S_e.$$

Let $S_+^G \subset S^G$ be the span of the invariants of positive degree and denote by $J \subset S$ the ideal generated in S by S_+^G. By Theorem 2.2, in fact, J is generated by finitely many polynomials $f_1, \ldots, f_N \in S_+^G$. In other words, the S-module homomorphism

$$\phi : S \oplus \cdots \oplus S \to J \qquad (h_1, \ldots, h_N) \mapsto \sum_{i=1}^N h_i f_i$$

is surjective.

Claim: S^G is generated by $f_1, \ldots, f_N$.

We pick an arbitrary homogeneous invariant $h \in S^G$. To show that h belongs to $k[f_1, \ldots, f_N]$ we shall use induction on $\deg h$. If $\deg h = 0$, then h is a constant, so this is clear. If $\deg h > 0$, then h belongs to the homogeneous ideal J, and therefore to the invariant subspace J^G, where we can view J as a representation of G. The map ϕ above is a surjective homomorphism of G-representations, so by linear reductivity the induced map of invariants $S^G \oplus \cdots \oplus S^G \to J^G$ is surjective. There therefore exist invariant polynomials $h_1', \ldots, h_N' \in S^G$ such that

$$h = \sum_{i=1}^N h_i' f_i.$$

The f_i all have positive degree, so $\deg h_i' < \deg h$. By the inductive hypothesis we may therefore assume that each $h_i' \in k[f_1, \ldots, f_N]$ and hence $h \in k[f_1, \ldots, f_N]$ also. □

Note that the last part of this proof is really just Proposition 2.41.

We turn now to the general case of G acting on a finitely generated k-algebra R. We shall show that, again, the invariant subring R^G is finitely generated, by reducing to the case of a polynomial ring as above.

Lemma 4.52. *Suppose that an algebraic group acts on a finitely generated k-algebra R. Then there exists a set of generators $r_1, \ldots, r_N$ of R whose k-linear span $\langle r_1, \ldots, r_N \rangle \subset R$ is a G-invariant vector subspace.*

Proof. Let $s_1, \ldots, s_M \in R$ be any set of generators. By local finiteness (Proposition 4.6) each s_i is contained in a finite-dimensional subrepresentation $V_i \subset R$ of G. It therefore suffices to extend $s_1, \ldots, s_M$ to a basis $r_1, \ldots, r_N$ of the finite-dimensional subspace $\sum_i V_i \subset R$. □

Geometrically, this lemma says that an affine algebraic variety acted on by an algebraic group can always be equivariantly embedded in an affine space $\mathbb{A}^N$ on which G acts linearly.

Theorem 4.53. *If a linearly reductive algebraic group G acts on a finitely generated k-algebra R, then the invariant ring R^G is finitely generated.*

Proof. Pick generators $r_1, \ldots, r_N$ of R as in Lemma 4.52. Then there is a surjective k-algebra homomorphism

$$S = k[x_1, \ldots, x_N] \to R$$

given by mapping $x_i \mapsto r_i$ for each $i = 1, \ldots, N$. Via this map G acts on the ring S, and by Theorem 4.51 the invariant ring S^G is finitely generated, while by linear reductivity the induced map $S^G \to R^G$ is surjective. It follows that R^G is finitely generated. $\qquad\square$

4.4 The Cayley-Sylvester Counting Theorem

In order to gain a concrete understanding of any invariant ring it is essential to be able to compute its Hilbert series. In this section we shall describe some methods for determining the Hilbert series for the case of classical binary invariants.

(a) SL(2)

Let us write a general unimodular 2×2 matrix as

$$X = \begin{pmatrix} a & b \\ c & d \end{pmatrix} \in SL(2).$$

The Lie space $\mathfrak{sl}(2)$ of $SL(2)$ has a basis consisting of three derivations:

$$e = \left.\frac{\partial}{\partial b}\right|_{X=I}, \quad f = \left.\frac{\partial}{\partial c}\right|_{X=I}, \quad h = \left.\left(\frac{\partial}{\partial a} - \frac{\partial}{\partial d}\right)\right|_{X=I}.$$

These correspond to the three subgroups of $SL(2)$

$$N^+ = \left\{ \begin{pmatrix} 1 & s \\ 0 & 1 \end{pmatrix} \mid s \in k \right\} \cong \mathbb{G}_a,$$

$$N^- = \left\{ \begin{pmatrix} 1 & 0 \\ s & 1 \end{pmatrix} \mid s \in k \right\} \cong \mathbb{G}_a,$$

$$T = \left\{ \begin{pmatrix} t & 0 \\ 0 & t^{-1} \end{pmatrix} \mid t \in k^\times \right\} \cong \mathbb{G}_m.$$

We can represent these basis elements, in the manner of Example 4.30, as matrices

$$e = \begin{pmatrix} 0 & 1 \\ 0 & 0 \end{pmatrix}, \quad f = \begin{pmatrix} 0 & 0 \\ 1 & 0 \end{pmatrix}, \quad h = \begin{pmatrix} 1 & 0 \\ 0 & -1 \end{pmatrix}.$$

The adjoint action of $SL(2)$ on $\mathfrak{sl}(2)$ is by conjugation, and its restriction to $T \subset SL(2)$ is therefore given by

$$\mathrm{Ad}\begin{pmatrix} t & 0 \\ 0 & t^{-1} \end{pmatrix} : \begin{cases} e \mapsto t^2 e \\ f \mapsto t^{-2} f \\ h \mapsto 0. \end{cases} \tag{4.2}$$

The invariant inner product (4.1) from Section 4.3(a) is given by the symmetric matrix

$$\begin{pmatrix} 0 & 1 & 0 \\ 1 & 0 & 0 \\ 0 & 0 & 2 \end{pmatrix}$$

with respect to the basis $e, f, h \in \mathfrak{sl}(2)$. We may therefore construct an orthonormal basis

$$\frac{e+f}{\sqrt{2}}, \quad \frac{e-f}{\sqrt{-2}}, \quad \frac{h}{\sqrt{2}}$$

and, as we have seen in Example 4.48, Casimir element

$$\Omega = e \star f + f \star e + \tfrac{1}{2} h \star h \in \mathcal{H}(SL(2)). \tag{4.3}$$

We now consider the basic representation $S = k[x, y]$ of $SL(2)$, and the action on S of the Lie space $\mathfrak{sl}(2) \subset \mathcal{H}(SL(2))$ and of the Casimir element on S, via the homomorphism $\tilde{\rho} : \mathcal{H}(SL(2)) \to \mathrm{End}\, S$. For simplicity we will usually drop the tilde and write just $\rho : \mathcal{H}(SL(2)) \to \mathrm{End}\, S$. This should not cause any confusion.

Example 4.54. $SL(2)$ acts on the right on the polynomial algebra $S = k[x, y]$ by

$$f(x, y) \cdot \begin{pmatrix} \alpha & \beta \\ \gamma & \delta \end{pmatrix} = f(\alpha x + \beta y, \gamma x + \delta y).$$

Let us compute the derivative at the identity of the restriction of this action to the subgroup $N^+ \subset SL(2)$:

$$f(x, y) \cdot \begin{pmatrix} 1 & s \\ 0 & 1 \end{pmatrix} = f(x + sy, y),$$

so we obtain

$$\frac{d}{ds}\Big|_{s=0}\left\{f(x,y)\cdot\begin{pmatrix}1 & s\\ 0 & 1\end{pmatrix}\right\} = y\frac{\partial}{\partial x}f(x,y).$$

Similarly, for $N^- \subset SL(2)$ we find

$$\frac{d}{ds}\Big|_{s=0}\left\{f(x,y)\cdot\begin{pmatrix}1 & 0\\ s & 1\end{pmatrix}\right\} = x\frac{\partial}{\partial y}f(x,y).$$

For the subgroup $T \subset SL(2)$ the restriction of the adjoint action is given by

$$f(x,y)\cdot\begin{pmatrix}t & 0\\ 0 & t^{-1}\end{pmatrix} = f(tx, t^{-1}y),$$

and we must differentiate at $t = 1$:

$$\frac{d}{dt}\Big|_{t=1}\left\{f(x,y)\cdot\begin{pmatrix}t & 0\\ 0 & t^{-1}\end{pmatrix}\right\} = x\frac{\partial f}{\partial x} - y\frac{\partial f}{\partial y}.$$

We conclude that the representation of the Lie space $\rho : \mathfrak{sl}(2) \to \mathrm{End}\, S$ is given by

$$\rho(e) = y\frac{\partial}{\partial x}, \quad \rho(f) = x\frac{\partial}{\partial y}, \quad \rho(h) = x\frac{\partial}{\partial x} - y\frac{\partial}{\partial y}.$$

From (4.3) the Casimir operator is

$$\rho(\Omega) = \rho(e)\rho(f) + \rho(f)\rho(e) + \tfrac{1}{2}\rho(h)^2$$

$$= E + \tfrac{1}{2}E^2,$$

where $E = x\frac{\partial}{\partial x} + y\frac{\partial}{\partial y}$. $\qquad\qquad\square$

In this example, for each $d \in \mathbb{N}$ the binary forms of degree d give a subrepresentation $V_d \subset S$. Note that by Euler's Theorem the Casimir operator on V_d is the scalar $d + d^2/2$.

Remark 4.55. The following inhomogeneous description of the representation V_d will reappear at the end of this chapter. Namely, V_d can be viewed as the $d + 1$-dimensional vector subspace of $k[x]$ consisting of polynomials of degree at most d. Then $SL(2)$ acts on the right on V_d by:

$$f(x)\cdot\begin{pmatrix}\alpha & \beta\\ \gamma & \delta\end{pmatrix} = (\gamma x + \delta)^d f(\frac{\alpha x + \beta}{\gamma x + \delta}).$$

(b) The dimension formula for SL(2)

If V is a finite-dimensional representation of $SL(2)$, then under the action of the torus $T \subset SL(2)$ it has a weight-space decomposition (see Proposition 4.7)

$$V = \bigoplus_{m \in \mathbb{Z}} V_{(m)}. \tag{4.4}$$

The following follows from (4.2):

Proposition 4.56 (Weight shift). *Under* $\rho : \mathfrak{sl}(2) \to \text{End } V$ *we have*

$$\rho(e) : V_{(m)} \to V_{(m+2)}, \qquad \rho(f) : V_{(m)} \to V_{(m-2)}.$$

Lemma 4.57. *In the distribution algebra* $\mathcal{H}(SL(2))$ *the following relation holds:*

$$e \star f - f \star e = h.$$

Proof. We fix independent variables a, b, c, d and a', b', c', d' as the entries of matrices

$$X = \begin{pmatrix} a & b \\ c & d \end{pmatrix}, \quad X' = \begin{pmatrix} a' & b' \\ c' & d' \end{pmatrix}.$$

By definition of the convolution product, the element $e \star f$ evaluated on a polynomial $F(X)$ in a, b, c, d gives

$$\left. \frac{\partial^2 F(XX')}{\partial b \partial c'} \right|_{X=X'=I}.$$

One easily checks that this is equal to

$$\left. \frac{\partial^2 F(X)}{\partial b \partial c} \right|_{X=I} + \left. \frac{\partial F(X)}{\partial a} \right|_{X=I}.$$

Similarly, the value of $f \star e$ at the polynomial $F(X)$ is

$$\left. \frac{\partial^2 F(X)}{\partial b \partial c} \right|_{X=I} + \left. \frac{\partial F(X)}{\partial d} \right|_{X=I}.$$

Subtracting these two expressions yields the identity in the lemma. $\quad\square$

Corollary 4.58. *The Casimir element of* $SL(2)$ *is*

$$\Omega = e \star f + f \star e + \tfrac{1}{2} h \star h = 2e \star f - h + \tfrac{1}{2} h \star h = 2f \star e + h + \tfrac{1}{2} h \star h.$$

$\square$

This allows us to locate explicitly the invariants of $SL(2)$ in the representation V:

Corollary 4.59. $V^{SL(2)} = \ker \{e : V_{(0)} \to V_{(2)}\}$.

Proof. Clearly $V^{SL(2)} \subset V_{(0)}$, and indeed it is contained in $\ker e$ since it is killed by the Lie space $\mathfrak{sl}(2)$. But conversely, every $v \in \ker e$ is killed by the Casimir operator, by Corollary 4.58. By Proposition 4.50, this implies that $v \in V^{SL(2)}$. $\qquad\square$

The dimension formula for $V^{SL(2)}$ will follow once we show that the map in Corollary 4.59 is surjective. In fact:

Lemma 4.60. *The composition* $e^2 : V_{(-2)} \xrightarrow{e} V_{(0)} \xrightarrow{e} V_{(2)}$ *is an isomorphism.*

Proof. First let us show that e^2 is injective. If $v \in \ker e^2$, then the vector $e(v) \in V_{(0)}$ is $SL(2)$-invariant by Corollary 4.59. In particular, $fe(v) = 0$. It follows that

$$\rho(\Omega)v = \rho(2f \star e + h + \tfrac{1}{2}h \star h)v = 0,$$

so that, by Proposition 4.50, $v \in V^{SL(2)}$. But v lies in the -2-weight space, which contains no nonzero invariants; hence $v = 0$.

By a similar argument the homomorphism $f^2 : V_{(2)} \to V_{(-2)}$ is injective. It follows that $\dim V_{(-2)} = \dim V_{(2)}$, and hence that e^2 is an isomorphism. $\quad\square$

It follows from this that $e : V_{(0)} \to V_{(2)}$ is surjective, and we deduce:

Dimension Formula 4.61. *If V is any finite-dimensional representation of $SL(2)$, with weight-space decomposition (4.4) with respect to the torus $T \subset SL(2)$, then*

$$\dim V^{SL(2)} = \dim V_{(0)} - \dim V_{(2)}.$$

$\qquad\square$

The generating function

$$\mathrm{ch}_V(q) = \sum_{m \in \mathbb{Z}} \dim V_{(m)} q^m \in \mathbb{Z}[q, q^{-1}]$$

of the weight-space decomposition (4.4) is called the *(formal) character* of the representation V. For example, the space V_d of binary forms of degree d

has character

$$\mathrm{ch}_{V_d}(q) = q^{-d} + q^{-d+2} + \cdots + q^{d-2} + q^d = \frac{q^{d+1} - q^{-d-1}}{q - q^{-1}}.$$

Corollary 4.62. $\dim V^{SL(2)} = -\operatorname*{Res}_{q=0}(q - q^{-1})\mathrm{ch}_V(q)$. $\square$

(c) A digression: Weyl measure

By Cauchy's integral formula we can re-express Corollary 4.62:

$$\dim V^{SL(2)} = -\frac{1}{2\pi i} \oint (q - q^{-1})\mathrm{ch}_V(q) dq,$$

where the integral is taken with winding number 1 around the origin. Taking the unit circle with parametrisation $q = e^{i\theta}$ transforms the integral to

$$\dim V^{SL(2)} = \frac{1}{\pi} \int_0^\pi (1 - \cos 2\theta)\mathrm{ch}_V(e^{i\theta}) d\theta. \tag{4.5}$$

(We have used here the Weyl symmetry $\mathrm{ch}_V(q) = \mathrm{ch}_V(q^{-1})$. This can be seen from the definition of the character, using conjugation by the element $\begin{pmatrix} 0 & 1 \\ -1 & 0 \end{pmatrix} \in SL(2)$.)

Note that every conjugacy class of the maximal compact subgroup $SU(2) \subset SL(2, \mathbb{C})$ has a unique representative of the form

$$A(\theta) = \begin{pmatrix} e^{i\theta} & 0 \\ 0 & e^{-i\theta} \end{pmatrix} \qquad \text{for } 0 \le \theta \le \pi.$$

Thus, if we define a *Weyl measure*

$$\mu(\theta) = 1 - \cos 2\theta = 2\sin^2\theta,$$

then, noting that $\int_0^\pi \mu(\theta) d\theta = \pi$, we see that the dimension formula (4.5) has the form of an average of the class function $\mathrm{ch}_V(\theta) = \mathrm{ch}_V(e_{i\theta})$ with respect to the measure μ.

Let us compare this situation with the Dimension Formula 1.11 for representations of a finite group G. If we denote by $\mathfrak{c}_1, \ldots, \mathfrak{c}_k \subset G$ its conjugacy classes, then, noting that the character $\chi : G \to \mathbb{C}$ of the representation is a class function, the Dimension Formula 1.11 can be written:

$$\dim V^G = \frac{1}{|G|} \sum_{i=1}^k |\mathfrak{c}_i| \chi(\mathfrak{c}_i).$$

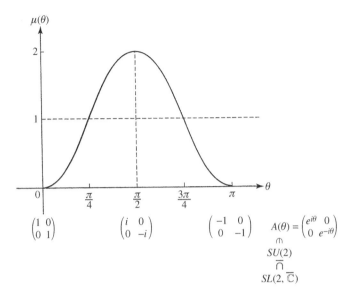

Figure 4.1: *Weyl measure* $\mu(\theta) = 2 \sin^2 \theta$

In other words (again noting that $\sum_{i=1}^{k} \mu(\mathfrak{c}_i) = |G|$), the dimension is given as the average of the character with respect to the measure $\mu(\mathfrak{c}) = |\mathfrak{c}|$.

In conclusion, then, in both cases of G a finite group or the compact Lie group $SU(2)$ we see that the dimension formula can be interpreted as the average of the character over the conjugacy classes of the group, with respect to cardinality or the Weyl measure, respectively.

(d) The Cayley-Sylvester Formula

As an application of the dimension formula we are now going to compute the Hilbert series of the classical binary invariant ring. If V is any n-dimensional representation of $SL(2)$, we consider the induced action of $SL(2)$ on the polynomial ring $S(V) = k[x_1, \ldots, x_n]$ of functions on V. Let $a_1, \ldots, a_n \in \mathbb{Z}$ be the weights (not necessarily distinct) of the torus $T \subset SL(2)$ (see Section 4.4(a)) occuring in the weight-space decomposition of the representation V (Proposition 4.14). The function

$$P(q;t) = \frac{1}{(1 - q^{a_1}t)(1 - q^{a_2}t) \cdots (1 - q^{a_n}t)} = \det\left(I_V - t\begin{pmatrix} q & 0 \\ 0 & q^{-1} \end{pmatrix}_V\right)^{-1}$$

is called the q-*Hilbert series* of the representation. Then Molien's Theorem 1.10 for finite groups has the following analogue for $SL(2)$.

Proposition 4.63. *The invariant ring $S(V)^{SL(2)}$ has Hilbert series*

$$P(t) = -\operatorname*{Res}_{q=0} (q - q^{-1})P(q;t).$$

Equivalently, if $P(q;t) = \sum_{m \in \mathbb{Z}} c_m(t)q^m$, then $P(t) = c_0(t) - c_2(t)$.

Proof. This is similar to the proof of Molien's Theorem. First, by making a linear change of coordinates we can assume that $x_1, \ldots, x_n$ diagonalise the action of $T \cong \mathbb{G}_m$. Then we note that the power series expansion of

$$R(x_1, \ldots, x_n) = \frac{1}{(1 - x_1)(1 - x_2) \cdots (1 - x_n)}$$

lists once each all the monomials of the ring S, and that the action on this expression of

$$\begin{pmatrix} q & 0 \\ 0 & q^{-1} \end{pmatrix} \in T$$

yields

$$R(q^{a_1}x_1, \ldots, q^{a_n}x_n) = \frac{1}{(1 - q^{a_1}x_1)(1 - q^{a_2}x_2) \cdots (1 - q^{a_n}x_n)}.$$

When this is expanded, the sum of the coefficients in degree e is precisely the formal character $\operatorname{ch}_{V_e}(q)$ of the representation V_e. It follows from this that

$$\sum_{e=0}^{\infty} \operatorname{ch}_{V_e}(q)t^e = R(q^{a_1}t, \ldots, q^{a_n}t) = P(q;t).$$

From Corollary 4.62

$$\dim S^{SL(2)} \cap V_e = -\operatorname*{Res}_{q=0} (q - q^{-1})\operatorname{ch}_{V_e}(q),$$

and hence the Hilbert series is

$$P(t) = -\operatorname*{Res}_{q=0} \sum_{e=0}^{\infty} (q - q^{-1})\operatorname{ch}_{V_e}(q)t^e = -\operatorname*{Res}_{q=0} (q - q^{-1})P(q;t).$$

$\square$

In particular, consider the representation $V = V_d$ of binary forms of degree d (see Example 4.54). The $d+1$ monomials $x^d, x^{d-1}y, \ldots, y^d$ are already a basis diagonalising the action of T and are transformed to

$$q^d x^d, q^{d-2} x^{d-1} y, \ldots, q^{-d} y^d$$

by the element $\begin{pmatrix} q & 0 \\ 0 & q^{-1} \end{pmatrix}$. The q-Hilbert series is therefore

$$P(q;t) = \prod_{i=0}^{d} \frac{1}{1 - q^{d-2i}t}.$$

Definition 4.64. The q-analogue of an integer d, of its factorial and of the binomial coefficients are, respectively:

(i) $\quad [d]_q = q^{d-1} + q^{d-3} + \cdots + q^{-d+3} + q^{-d+1} = \frac{q^d - q^{-d}}{q - q^{-1}},$

(ii) $$[d]_q! = \prod_{i=1}^{d} [i]_q,$$

(iii) $$\begin{bmatrix} d+e \\ e \end{bmatrix}_q = \frac{[d+e]_q!}{[d]_q! [e]_q!}.$$

$\square$

The corresponding classical notions are obtained by letting $q \to 1$. In this sense the following proposition is the q-analogue of the binomial theorem

$$\frac{1}{(1-t)^{d+1}} = \sum_{e \geq 0} \binom{d+e}{e} t^e.$$

Proposition 4.65.

$$\prod_{i=0}^{d} \frac{1}{1 - q^{d-2i}t} = \sum_{e \geq 0} \begin{bmatrix} d+e \\ e \end{bmatrix}_q t^e.$$

Proof. Denote the left-hand side by

$$\phi(q, t) = \prod_{i=0}^{d} \frac{1}{1 - q^{d-2i}t},$$

and its power series expansion in t by

$$\phi(q, t) = \sum_{e \geq 0} a_e(q) t^e, \qquad a_0(q) = 1.$$

Note that $\phi(q, t)$ satisfies the functional equation

$$\phi(q, q^2 t) = \frac{1 - q^{-d} t}{1 - q^{d+2} t} \phi(q, t).$$

Comparing terms on both sides, this gives

$$a_e(q) q^{2e} - a_{e-1} q^{2e+d} = a_e(q) - a_{e-1} q^{-d}.$$

Rearranging, we obtain the recurrence relation

$$a_e(q) = \frac{q^{e+d} - q^{-e-d}}{q^e - q^{-e}} a_{e-1}(q)$$

for the coefficients of $\phi(q, t)$, from which the proposition follows. $\square$

Returning to the representation V_d of $SL(2)$, with basis $x^d, x^{d-1} y, \ldots, y^d$, let $\xi_0, \ldots, \xi_d$ be the dual basis of $V_d^\vee$. From Propositions 4.63 and 4.65 we deduce:

Theorem 4.66. *The Hilbert series of the classical invariant ring $k[\xi_0, \ldots, \xi_d]^{SL(2)}$ for binary forms of degree d is given by:*

$$P^{(d)}(t) = -\sum_{e \geq 0} \left\{ \operatorname*{Res}_{q=0} (q - q^{-1}) \begin{bmatrix} d + e \\ e \end{bmatrix}_q \right\} t^e.$$

For the purpose of computing it is convenient to make a change of variable $u = q^2$. Then

$$\begin{bmatrix} d + e \\ e \end{bmatrix}_q = \frac{[e + 1]_q [e + 2]_q \cdots [e + d]_q}{[1]_q [2]_q \cdots [d]_q}$$

$$= q^{-de} \frac{(1 - u^{e+1})(1 - u^{e+2}) \cdots (1 - u^{e+d})}{(1 - u)(1 - u^2) \cdots (1 - u^d)}$$

and (note that the denominator begins with the quadratic factor!)

$$-(q - q^{-1}) \begin{bmatrix} d + e \\ e \end{bmatrix}_q = q^{-de-1} \frac{(1 - u^{e+1})(1 - u^{e+2}) \cdots (1 - u^{e+d})}{(1 - u^2) \cdots (1 - u^d)}. \qquad (4.6)$$

For a formal power series $f(u) \in \mathbb{Z}[[u]]$ we shall denote the coefficient of u^n by $[f(u)]_n \in \mathbb{Z}$.

Cayley-Sylvester Formula 4.67. *The vector space* $k[\xi_0, \ldots, \xi_d]_e^{SL(2)}$ *of classical invariants of degree e for binary d-ics has dimension*

$$m(d, e) = \begin{cases} \left[\dfrac{(1 - u^{e+1})(1 - u^{e+2}) \cdots (1 - u^{e+d})}{(1 - u^2) \cdots (1 - u^d)} \right]_{de/2} & \text{if } de \text{ is even,} \\[2em] 0 & \text{if } de \text{ is odd.} \end{cases}$$

Proof. The dimension $m(d, e)$ is equal to the residue appearing in Theorem 4.66, and we may note that, if de is odd, then this residue vanishes since the expansion of (4.6) contains only even powers of q. We shall therefore assume that de is even. Writing

$$R(u) = \frac{(1 - u^{e+1})(1 - u^{e+2}) \cdots (1 - u^{e+d})}{(1 - u^2) \cdots (1 - u^d)},$$

we have

$$m(d, e) = \frac{1}{2\pi i} \oint q^{-de-1} R(q^2) dq,$$

where the path of integration is a small circle around $q = 0$. Under the change of variable $u = q^2$, $du = 2q dq$, this is equal to

$$\frac{1}{2\pi i} \oint u^{-de/2} \frac{u^{-1}}{2} R(u) du,$$

where the contour now has winding number 2 about $u = 0$ and is therefore

$$m(d, e) = \frac{1}{2\pi i} \oint u^{-de/2-1} R(u) du$$

$$= \operatorname*{Res}_{u=0} u^{-de/2-1} R(u)$$

$$= [R(u)]_{de/2}.$$

$\square$

Corollary 4.68 (Hermite reciprocity). $m(d, e) = m(e, d)$. $\square$

(e) Some computational examples

Proposition 4.69. *For* $2 \le d \le 6$, *the Hilbert series* $P^{(d)}(t)$ *of the classical invariant ring for binary d-ics is given by the following table:*

d	$P^{(d)}(t)$
2	$\dfrac{1}{1-t^2}$
3	$\dfrac{1}{1-t^4}$
4	$\dfrac{1}{(1-t^2)(1-t^3)}$
5	$\dfrac{1+t^{18}}{(1-t^4)(1-t^8)(1-t^{12})}$
6	$\dfrac{1+t^{15}}{(1-t^2)(1-t^4)(1-t^6)(1-t^{10})}$

Proof. We shall just do the cases $d = 4, 5$, leaving the others to the reader. From the Cayley-Sylvester formula,

$$m(4, e) = \left[\frac{(1 - u^{e+1})(1 - u^{e+2})(1 - u^{e+3})(1 - u^{e+4})}{(1 - u^2)(1 - u^3)(1 - u^4)} \right]_{2e}.$$

In this expression we can expand the numerator, ignoring terms of degree greater than $2e$:

$$m(4, e) = \left[\frac{1 - u^{e+1} - u^{e+2} - u^{e+3} - u^{e+4}}{(1 - u^2)(1 - u^3)(1 - u^4)} \right]_{2e}$$

$$= \left[\frac{1}{(1 - u)(1 - u^{3/2})(1 - u^2)} \right]_e - \left[\frac{u + u^2 + u^3 + u^4}{(1 - u^2)(1 - u^3)(1 - u^4)} \right]_e$$

$$= \left[\frac{1 + u^{3/2}}{(1 - u)(1 - u^2)(1 - u^3)} \right]_e - \left[\frac{u}{(1 - u)(1 - u^2)(1 - u^3)} \right]_e$$

(where for the last term we have simply factorised $1 - u^4$). Exchanging $u^{3/2}$ and u between the two numerators, and noting that the second then has no integer powers in its expansion, we see that

$$m(4, e) = \left[\frac{1}{(1 - u^2)(1 - u^3)} \right]_e ,$$

and hence that $P^{(4)}(t) = 1/(1 - t^2)(1 - t^3)$.

We now turn to the case $d = 5$. Since $m(5, e) = 0$ whenever e is odd, it is enough to consider even values $e = 2a$. Then

$$m(5, 2a) = \left[\frac{(1 - u^{2a+1})(1 - u^{2a+2})(1 - u^{2a+3})(1 - u^{2a+4})(1 - u^{2a+5})}{(1 - u^2)(1 - u^3)(1 - u^4)(1 - u^5)} \right]_{5a} .$$

Expanding the numerator and ignoring terms of degree greater than $5a$, we obtain:

$$m(5, 2a) = \left[\frac{1}{(1 - u^2)(1 - u^3)(1 - u^4)(1 - u^5)} \right]_{5a}$$

$$- \left[\frac{u + u^2 + u^3 + u^4 + u^5}{(1 - u^2)(1 - u^3)(1 - u^4)(1 - u^5)} \right]_{3a} \qquad (4.7)$$

$$+ \left[\frac{u^3 + u^4 + 2u^5 + 2u^6 + 2u^7 + u^8 + u^9}{(1 - u^2)(1 - u^3)(1 - u^4)(1 - u^5)} \right]_a .$$

We deal with each of these three brackets. The first can be rewritten:

$$\left[\frac{1}{(1 - u^2)(1 - u^3)(1 - u^4)(1 - u^5)} \right]_{5a}$$

$$= \left[\frac{(1 + u^2 + u^4 + u^6 + u^8)(1 + u^3 + u^6 + u^9 + u^{12})(1 + u^4 + u^8 + u^{12} + u^{16})}{(1 - u^{10})(1 - u^{15})(1 - u^{20})(1 - u^5)} \right]_{5a}$$

$$= \left[\frac{1 + u^5 + 4u^{10} + 5u^{15} + 7u^{20} + 4u^{25} + 3u^{30}}{(1 - u^5)(1 - u^{10})(1 - u^{15})(1 - u^{20})} \right]_{5a} ,$$

where, in the last step, the numerator has been expanded ignoring terms that are not divisible by 5.

The second bracket in (4.7) can be rearranged similarly:

$$\left[\frac{u+u^2+u^3+u^4+u^5}{(1-u^2)(1-u^3)(1-u^4)(1-u^5)}\right]_{3a}$$

$$=\left[\frac{u(1+u^2+u^4)(1+u^4+u^8)(1+u+u^2)}{(1-u^6)(1-u^3)(1-u^{12})(1-u^3)}\right]_{3a}$$

$$=\left[\frac{2u^3+2u^6+3u^9+u^{12}+u^{15}}{(1-u^3)^2(1-u^6)(1-u^{12})}\right]_{3a}.$$

The third bracket is:

$$\left[\frac{u^3+u^4+2u^5+2u^6+2u^7+u^8+u^9}{(1-u^2)(1-u^3)(1-u^4)(1-u^5)}\right]_{a}$$

$$=\left[\frac{u^3(1+u+u^2+u^3+u^4)(1+u^2)}{(1-u^2)(1-u^3)(1-u^4)(1-u^5)}\right]_{a}$$

$$=\left[\frac{u^3}{(1-u)(1-u^2)^2(1-u^3)}\right]_{a}.$$

It follows from these computations that the Hilbert series is given by

$$P^{(5)}(\sqrt{t})=\frac{1+t+4t^2+5t^3+7t^4+4t^5+3t^5}{(1-t)(1-t^2)(1-t^3)(1-t^4)}-\frac{2t+2t^2+3t^3+t^4+t^5}{(1-t)^2(1-t^2)(1-t^4)}$$

$$+\frac{t^3}{(1-t)(1-t^2)^2(1-t^3)}$$

$$=\frac{1+t^9}{(1-t^2)(1-t^4)(1-t^6)}.$$

$\square$

In the cases $d=2,3$, the discriminant $D(\xi)\in k[\xi_0,\ldots,\xi_d]^{SL(2)}$ has degree 2, 4, respectively (Examples 1.21 and 1.22).

For the case $d=4$, we have constructed in Chapter 1 (see Section 1.3(b)) invariants $g_2(\xi)$, $g_3(\xi)\in k[\xi_0,\ldots,\xi_4]^{SL(2)}$ of degrees 2, 3, respectively. In fact, g_2 and g_3 are algebraically independent: this can be seen by restricting to the subspace $\xi_0=\xi_4$, $\xi_1=\xi_3=0$, on which the invariants reduce to

$$g_2(\xi)=\xi_0^2+3\xi_2^2,\qquad g_3(\xi)=(\xi_0^2-\xi_2^2)\xi_2.$$

These determine a map $k^2 \to k^2$, $(\xi_0, \xi_2) \mapsto (g_2, g_3)$ which is clearly surjective (since one can separate variables), so g_2, g_3 cannot satisfy any polynomial identity. From Propositions 4.69 and 1.9 we can therefore conclude:

Corollary 4.70. *The ring of invariants* $k[\xi_0, \dots, \xi_d]^{SL(2)}$ *is generated by the discriminant* $D(\xi)$ *when* $d = 2, 3$, *and by* $g_2(\xi)$, $g_3(\xi)$ *when* $d = 4$. $\qquad\Box$

The higher degree cases are less simple, but by constructing the invariant rings explicitly the following results are known. (See also Schur [26].)

Example 4.71. For $d = 5$, the Hilbert series can be written as

$$P^{(5)}(t) = \frac{1 - t^{36}}{(1 - t^4)(1 - t^8)(1 - t^{12})(1 - t^{18})},$$

and, indeed, the invariant ring $k[\xi_0, \dots, \xi_5]^{SL(2)}$ has four generators of degrees $4, 8, 12, 18$ satisfying a single relation of degree 36. Similar for $d = 6$,

$$P^{(6)}(t) = \frac{1 - t^{30}}{(1 - t^2)(1 - t^4)(1 - t^6)(1 - t^{10})(1 - t^{15})},$$

and the invariant ring has five generators of degrees $2, 4, 6, 10, 15$ and a single relation of degree 30. $\qquad\Box$

Example 4.72 (Shioda [27]). For $d = 8$, the Hilbert series is

$$P^{(8)}(t) = \frac{1 + t^8 + t^9 + t^{10} + t^{18}}{(1 - t^2)(1 - t^3)(1 - t^4)(1 - t^5)(1 - t^6)(1 - t^7)}$$

$$= \frac{1 + \sum_{a=16}^{20} t^a + \sum_{b=25}^{29} t^b - t^{45}}{(1 - t^2)(1 - t^3)(1 - t^4)(1 - t^5)(1 - t^6)(1 - t^7)(1 - t^8)(1 - t^9)(1 - t^{10})},$$

where the first expression is obtained from the Cayley-Sylvester formula, and the second is a convenient rearrangement. In this case the ring $k[\xi_0, \dots, \xi_8]^{SL(2)}$ is generated by nine invariants $J_2(\xi), \dots, J_{10}(\xi)$. These satisfy five relations of degrees $16, \dots, 20$, which in turn satisfy five syzygies. More precisely, the relations can be expressed as the Pfaffians of the five principal 4×4 minors of a skew-symmetric matrix

$$\begin{pmatrix} 0 & f_6(J) & f_7(J) & f_8(J) & f_9(J) \\ & 0 & f_8(J) & f_9(J) & f_{10}(J) \\ & & 0 & f_{10}(J) & f_{11}(J) \\ & - & & 0 & f_{12}(J) \\ & & & & 0 \end{pmatrix},$$

where each $f_i(J)$ is a weighted homogeneous polynomial of degree i, with $\deg J_m = m$. □

Remark 4.73. In fact, a Gorenstein ring of codimension 3 is always defined by an odd number $2k + 1$ of relations in its generators, and, by a theorem of Buchsbaum and Eisenbud [22], these relations can always be expressed as the Pfaffians of the principal $2k \times 2k$ minors of a skew-symmetric matrix, as above. □

The case $d = 7$ we prefer quietly to omit, though the interested reader may like to compute $P^{(7)}(t)$ for him or herself, or consult Dixmier and Lazard [24].

4.5 Geometric reductivity of $SL(2)$

We shall give in this section an alternative proof of linear reductivity in the special case of $SL(2)$. We begin by extending Proposition 4.37.

Lemma 4.74. *For an algebraic group G the following conditions are equivalent.*

(i) G is linearly reductive.
(ii) Given a finite-dimensional representation V of G and a surjective G-invariant linear form $f : V \to k$ there exists an invariant vector $w \in V^G$ such that $f(w) \neq 0$.
(iii) Given a finite-dimensional representation V of G and an invariant vector $w \in V^G$ there exists an G-invariant linear form $f : V \to k$ such that $f(w) \neq 0$.

Proof. (i) implies (ii) immediately from Definition 4.36, with G acting trivially on $W = k$. Conversely, (ii) implies (i) using the formulation of Proposition 4.37(ii). If $v \in W^G$, then we can decompose W as a representation of G, as $W = k\{v\} \oplus W'$ (Exercise 4.6). Then by condition (ii) the composition $V^G \subset V \to W \to k\{v\}$ is surjective.

(ii) is equivalent to (iii) by replacing V by its dual $V^\vee$ and noting that the space of G-invariant forms is $\mathrm{Hom}_G(V^\vee, k) = \mathrm{Hom}_G(k, V) = V^G$, where G acts trivially on k. □

It is the formulation of linear reductivity given by part (iii) of the lemma that we shall verify for $SL(2)$. First, by linear reductivity of $\mathbb{G}_m \cong T \subset SL(2)$, we can find a T-invariant linear form $e : V \to k$ such that $e(w) = 1$, and using this form we can define a map

$$\phi : V \to k[SL(2)]$$

by $\phi(x)(g) = e(g \cdot x)$ for $x \in V$ and $g \in SL(2)$. Equivalently, ϕ is the composition of the group action (see Definition 4.1) with $e \otimes 1$:

$$V \xrightarrow{\mu_V} V \otimes k[SL(2)] \xrightarrow{e \otimes 1} k \otimes k[SL(2)] = k[SL(2)].$$

The following properties of ϕ are easy to check.

Lemma 4.75.

(i) $\phi(w)$ is the constant function 1.

(ii) For all $x \in V$ the function $\phi(x)$ is invariant under the right-action of T, that is, $\phi(V) \subset k[SL(2)]^T$.

(iii) ϕ is a homomorphism of SL(2) representations.

We will give an explicit description of the invariant ring $k[SL(2)]^T$. The coordinate ring of $SL(2)$ is

$$k[SL(2)] = k[x, y, z, t]/(xt - yz - 1),$$

and this carries left- and right-actions of the group by left- and right-translation:

$$\begin{pmatrix} x & y \\ z & t \end{pmatrix} \mapsto \begin{pmatrix} a & b \\ c & d \end{pmatrix} \begin{pmatrix} x & y \\ z & t \end{pmatrix} \begin{pmatrix} a' & b' \\ c' & d' \end{pmatrix}.$$

We consider the right-action restricted to the torus $T \subset SL(2)$ and the left-action of $SL(2)$ on the invariant subalgebra $k[SL(2)]^T$. Since

$$\begin{pmatrix} x & y \\ z & t \end{pmatrix} \begin{pmatrix} q & 0 \\ 0 & q^{-1} \end{pmatrix} = \begin{pmatrix} qx & q^{-1}y \\ qz & q^{-1}t \end{pmatrix},$$

we have

$$k[x, y, z, t]^T = k[xy, xt, zy, zt].$$

The polynomial $xt - yz - 1$ is itself T-invariant, and so

$$k[SL(2)]^T = k[xy, xt, zy, zt]/(xt - yz - 1).$$

We can give this ring another description. For each natural number $n \in \mathbb{N}$ let R_n be the following set of rational functions in variables u, v:

$$R_n = \left\{ \frac{f(u, v)}{(u - v)^n} \mid \deg_u f(u, v) \le n, \ \deg_v f(u, v) \le n \right\}.$$

This is a vector space of dimension $(n + 1)^2$, and $R_n \subset R_{n+1}$. The union

$$R = \bigcup_{n \geq 0} R_n = \lim_{n \to \infty} R_n$$

is a subalgebra of $k[u, v, 1/(u - v)] \subset k(u, v)$, while the function field $k(u, v)$ is a representation of $SL(2)$ via

$$u \mapsto \frac{au + b}{cu + d}, \quad v \mapsto \frac{av + b}{cv + d}, \quad \begin{pmatrix} a & b \\ c & d \end{pmatrix} \in SL(2).$$

Note that under this action

$$u - v \mapsto \frac{au + b}{cu + d} - \frac{av + b}{cv + d} = \frac{u - v}{(cu + d)(cv + d)},$$

from which it follows that $R_n \subset k(u, v)$ is a subrepresentation. More precisely, R_n is isomorphic to the tensor product $V_n \otimes V_n$, where V_n is the $(n + 1)$-dimensional irreducible representation of $SL(2)$ as described in Remark 4.55.

Lemma 4.76. *There exists an isomorphism*

$$R \xrightarrow{\sim} k[SL(2)]^T = k[xy, xt, zy, zt]/(xt - yz - 1)$$

induced by mapping $u \mapsto x/z$, $v \mapsto y/t$, $1/(u - v) \mapsto zt$.   □

The proof of this is easy and is left to the reader.

Second proof of Theorem 4.43 for $n = 2$. We have to construct an invariant linear form $f \in \mathrm{Hom}_{SL(2)}(V, k)$ such that $f(w) \neq 0$, and we begin with $\phi : V \to k[SL(2)]$ constructed above. By Lemmas 4.75 and 4.76, the image of $\phi : V \to k[SL(2)]^T \cong R$ is contained in the finite-dimensional vector space R_n for some n. By definition, a general element is of the form

$$\frac{f(u, v)}{(u - v)^n} \quad \text{where } f(u, v) = \sum_{0 \leq i, j \leq n} a_{ij} u^i v^j.$$

Taking the determinant of the coefficient matrix $(a_{ij})_{0 \leq i, j \leq n}$ defines a function $\det : R_n \to k$ which is homogeneous of degree $n + 1$ and is $SL(2)$-invariant. Note that at the constant function

$$1 = \frac{(u - v)^n}{(u - v)^n} = \frac{u^n - nu^{n-1}v + \binom{n}{2}u^{n-2}v^2 - \cdots + (-v)^n}{(u - v)^n} \in R_n \subset k[SL(2)]^T$$

det takes the value

$$
\det \begin{vmatrix} & & & 1 \\ & & & -n \\ & & \binom{n}{2} & \\ & \cdots & & \\ & (-1)^{n-1}n & & \\ (-1)^n & & & \end{vmatrix} = \prod_{i=0}^{n}\binom{n}{i}.
$$

Let $h : R_n \to k$ be the formal differential of det at $1 \in R_n$, that is, the linear coefficient in the expansion of

$$
\det\left(1 + \epsilon\,\frac{f(u,v)}{(u-v)^n}\right).
$$

Then h is an $SL(2)$-invariant linear map given by

$$
\frac{f(u,v)}{(u-v)^n} \mapsto \sum_{i=0}^{n}(-1)^{n-i}a_{i,n-i}\prod_{j\neq i}\binom{n}{j}.
$$

In particular,

$$
h(1) = (-1)^n(n+1)\prod_{j=0}^{n}\binom{n}{j} \neq 0. \tag{4.8}
$$

Hence (using Lemma 4.75(i)) the composition

$$
f = h \circ \phi : V \to R_n \to k
$$

is an invariant linear form $f \in \mathrm{Hom}_{SL(2)}(V, k)$ with the required property. $\square$

Notice that in this proof the only place where we have used the assumption that the field k has characteristic zero is in the final step (4.8). In positive characteristic, even if we cannot use the differential h, we can nevertheless move the goalposts and use the function det to modify the condition of Lemma 4.74(iii):

Definition 4.77. An algebraic group G is *geometrically reductive* if, given a finite-dimensional representation V of G and an invariant vector $w \in V^G$, there exists a G-invariant homogeneous polynomial function $f : V \to k$ satisfying $f(w) = 1$. $\square$

Theorem 4.78 (Seshadri [76]). *If char $k = p > 0$, then $SL(2)$ is geometrically reductive.*

Proof. In the proof above we may take $n = p^\nu - 1$ for sufficiently large ν. Then

$$(u - v)^n = \frac{(u - v)^{p^\nu}}{u - v} = \frac{u^{p^\nu} - v^{p^\nu}}{u - v} = \sum_{i=0}^{n} u^{n-i} v^i.$$

In particular, the form $\det : R_n \to k$ takes the value 1 at $1 \in R_n$. So in this case the composition

$$f = \det \circ \phi : V \to R_n \to k$$

has exactly the properties asserted in the theorem.   □

Remarks 4.79.

(i) Obviously linear reductivity implies geometric reductivity, and over a field k of characteristic zero the converse is also true and the two conditions are equivalent. In positive characteristic, however, geometric reductivity is a strictly weaker condition, and in fact the only connected linearly reductive groups are tori. In particular, $SL(n)$ for $n \geq 2$ is geometrically reductive but not linearly reductive.

(ii) It is actually the property of geometric reductivity that the construction of quotient varieties depends upon. It turns out that both the finite-generatedness of the invariant ring and the separation of orbits by the invariants follow from the geometric reductivity of the group.

Exercises

1. If D and D' are derivations of a ring R, show that their commutator $[D, D'] = DD' - D'D$ is also a derivation.

2. Let $\alpha \in \mathcal{H}(G)$ be a distribution supported at the identity of an algebraic group G, let $R = k[G]$, and denote by $D_\alpha \in \operatorname{End} R$ the k-linear endomorphism

$$R \xrightarrow{\mu} R \otimes_k R \xrightarrow{\alpha \otimes 1} k \otimes_k R \overset{\sim}{\to} R.$$

Show that the following two conditions are equivalent:
(i) $\alpha : R \to k$ is a $k = R/\mathfrak{m}$-valued derivation;
(ii) $D_\alpha : R \to R$ is a derivation of R.

3. Show that the set $\operatorname{Lie}(G)$ of derivations at the identity is closed in distribution algebra $\mathcal{H}(G)$ under the commutator $[\alpha, \beta] = \alpha \star \beta - \beta \star \alpha$. (The Lie space $\operatorname{Lie}(G)$ equipped with this commutator product is called the *Lie algebra* of the algebraic group G.)

4. Let $\chi \in k[G]$ be a character of an algebraic group G, and let V be a linear representation of G. Show that

$$V_\chi := \{v \in V \mid \mu_V(v) = v \otimes \chi\}$$

   is a subrepresentation.
5. Let $\mu_V : V \to V \otimes k[G]$ and $\mu_W : W \to W \otimes k[G]$ be two representations of an algebraic group G.
   (i) Show that the tensor product $U \otimes V$ is a representation of G via the composition

   $$U \otimes V \xrightarrow{\mu_U \otimes \mu_V} U \otimes k[G] \otimes V \otimes k[G] \xrightarrow{\sim} U \otimes V \otimes k[G] \otimes k[G]$$
   $$\xrightarrow{1 \otimes m} U \otimes V \otimes k[G],$$

   where $m : k[G] \otimes k[G] \to k[G]$ is multiplication in the ring.
   (ii) Show that the space $\mathrm{Hom}_k(U, V)$ of k-linear maps from U to V is also a representation of G.
   (iii) Show that $f \in \mathrm{Hom}_k(U, V)$ is G-invariant if and only if f is a G-module homomorphism.
6. Suppose that G is linearly reductive, and let W be a subrepresentation of a finite-dimensional representation V of G. Prove that V decomposes as a direct sum of representations $V = W \oplus W'$. *Hint:* Apply linear reductivity to the surjective map of representations $\mathrm{Hom}_k(V, W) \to \mathrm{Hom}_k(W, W)$.
7. Let

$$0 \to U \to V \to W \to 0$$

   be an exact sequence of representations of G. Show that the induced sequence of spaces of invariants

$$0 \to U^G \to V^G \to W^G$$

   is exact. (That is, the functor which takes invariants is left-exact.)
8. Let $\mu : V \to V \otimes k[G]$ be a representation of an algebraic group G. Let $g \in G(k)$ be a k-valued point and $\mathfrak{m}_g \subset k[G]$ the corresponding maximal ideal. Denote by $\rho(g) \in \mathrm{End}\, V$ the composition

$$V \xrightarrow{\mu} V \otimes k[G] \xrightarrow{\mathrm{mod}\ \mathfrak{m}_g} V \otimes k \xrightarrow{\sim} V.$$

   Show that, if the coordinate ring $k[G]$ is an integral domain, then a vector $v \in V$ such that $\rho(g)(v) = v$ for every $g \in G(k)$ is a G-invariant.

5

The construction of quotient varieties

Suppose that an algebraic group G acts on an affine variety $X = \text{Spm } R$ (where we assume throughtout that R is an integral domain over k). We will consider the map to affine space

$$\phi : X \to \mathbb{A}^n \qquad x \mapsto (f_1(x), \ldots, f_n(x)),$$

given by some G-invariant functions $f_1, \ldots, f_n \in R^G$. Since each f_i is G-invariant, it is constant on the G-orbits. It follows that the map ϕ sends each orbit to a single point. This suggests – although a slight oversimplification – the essential idea for constructing a quotient variety by means of the invariant functions:

Aim: By taking sufficiently many invariants $f_1, \ldots, f_n \in R^G$, can we obtain a 'quotient variety' X/G as the image $\phi(X) \subset \mathbb{A}^n$?

First of all, what do we mean by 'sufficiently many' here? If the group G is linearly reductive, then by Hilbert's Theorem 4.51 the invariant ring R^G is finitely generated. In this case it should be enough to take $f_1, \ldots, f_n$ to be a set of generators. Second, our use of the word 'quotient' begs the following question:

Question 1: Do two distinct G-orbits always map to distinct points under ϕ?

Given that the functions $f_1, \ldots, f_n$ used to define ϕ generate all the invariants, this can be rephrased: given two distinct orbits of the G-action, does there exist any invariant form taking different values on the two orbits?

Moreover, if we ask for a 'variety' as the quotient, then we cannot avoid the following question. We have seen in Example 3.22 that the image of a morphism of algebraic varieties need not itself be an algebraic variety.

Question 2: Is the image of ϕ in $\mathbb{A}^n$ an algebraic variety?

We shall examine these questions in this chapter, and to what extent it is valid and consistent to view the image of $\phi(X) \subset \mathbb{A}^n$ (or more fundamentally Spm R^G) as the quotient of X by G. It turns out that Question 1 can be completely answered when G is linearly reductive, although the question first has to be modified slightly. On the other hand, it is not always the case that the quotient space of an affine variety is an algebraic variety at all (in any natural way), and so the above strategy has its limits.

The contents of this chapter are as follows. When a linearly reductive algebraic group G acts on an affine variety $X = \text{Spm } R$ as above, the morphism $\Phi : X \to X /\!/ G := \text{Spm } R^G$ induced by the inclusion $R^G \subset R$ is called the *affine quotient map*. In Section 5.1 we show that this is surjective and determines a one-to-one correspondence between closed G-orbits in X and points of $X /\!/ G$. A point of X is called *stable* if it belongs to a closed G-orbit of the same dimension as G. The stable points form an open subset $X^s \subset X$, and the restriction of Φ to X^s is a 'geometric quotient' in the sense that its fibres are precisely the G-orbits in X^s.

In the second section we apply this general theory to the classical case of hypersurfaces in $\mathbb{P}^n$. The set $X = U_{n,d}$ of nonsingular hypersurfaces of degree d in $\mathbb{P}^n$ is an affine variety in which (for $d \geq 3$) every point is stable for the action of $G = GL(n+1)$. Thus X/G exists as an affine variety parametrising smooth hypersurfaces up to projective equivalence. This is an open subset in the projective variety Proj $R_{n,d}$, where $R_{n,d} = k[\mathbb{V}_{n,d}]^{SL(n+1)}$ is the ring of classical invariants. Proj $R_{n,d}$ is the moduli space of 'semistable' hypersurfaces of degree d in $\mathbb{P}^n$, though we postpone the question of what semistable means geometrically until Chapter 7.

This is an example of a group action of 'ray type', for which a projective variety appears in a natural way as the quotient. We will study these more systematically in Chapter 6.

5.1 Affine quotients

(a) Separation of orbits

We begin with an example.

Example 5.1. Consider the action of the multiplicative group $\mathbb{G}_m$ on the affine plane $\mathbb{A}^2$, given by

$$(x, y) \mapsto (tx, t^{-1}y) \qquad \text{for } t \in \mathbb{G}_m.$$

The orbits of this action are of three types:

(i) The origin $(0, 0)$ is an orbit consisting of one point.
(ii) For each nonzero $a \in k$ the hyperbola $xy = a$ is a single orbit.
(iii) The x-axis minus the origin and the y-axis minus the origin are orbits.

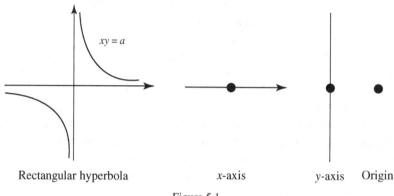

Rectangular hyperbola x-axis y-axis Origin

Figure 5.1

On the other hand, this action induces a $\mathbb{G}_m$-action on the coordinate ring $R = k[x, y]$, for which the invariant ring $R^{\mathbb{G}_m}$ is generated by the monomial xy. This generator determines, according to the recipe given in the introduction to this chapter, a map

$$\phi : \mathbb{A}^2 \to \mathbb{A}^1, \qquad (x, y) \mapsto xy.$$

This map separates the orbits (ii) but cannot distinguish the three orbits of (i) and (iii), all of which map to zero. (Compare this with Exercise 3.6.)   □

Even in this simple example, and even though the group $\mathbb{G}_m$ is linearly reductive, we have found a counterexample to Question 1 posed above. What is it that goes wrong here? The answer is quite simple: Question 1, as stated, overlooks the fact that the map ϕ is *continuous*. Every fibre of ϕ (that is, the preimage of each point) is a closed set. So if there is any orbit of the G-action which is not a closed set, then the answer to Question 1 must be negative. How can one get round this?

This example is quite typical and suggests the following general ideas.

(i) First, we could identify to a single point the three orbits which the invariants fail to separate.

(ii) Alternatively, note that the fixed point of the group action $(0, 0)$ is the ringleader of the troublemakers here: it is at just this point that the dimension of the stabiliser subgroup jumps. Removing this bad point levels up the dimensions of the stabilisers, and then we will get a good quotient.

(iii) On the other hand, even after carrying out (ii) the quotient that we obtain fails to be Hausdorff (or, in the language of algebraic geometry, separable). For this, we have to go a step further and remove, also, the x- and y-axes before taking the quotient. This done, we finally obtain a nice quotient of $\mathbb{A}^1 \times \mathbb{G}_m$ modulo $\mathbb{G}_m$, namely, $\mathbb{A}^1$.

Although the last approach (iii) has its merits, it is the first approach (i) that we shall follow in this chapter. (But see also the remarks at the beginning of Section 6.3.)

Definition 5.2. Two G-orbits $O, O' \subset X$ are said to be *closure-equivalent* if there exists between them a sequence of orbits

$$O = O_1, O_2, \ldots, O_{n-1}, O_n = O'$$

with the property that $\overline{O_i} \cap \overline{O_{i+1}} \neq \emptyset$ for each $i = 1, \ldots, n - 1$. $\qquad \square$

Invariant forms, because they are continuous, take the same value on closure-equivalent orbits. We should therefore modify Question 1 as follows.

Question 3: Do any two *closure-inequivalent* G-orbits map to distinct points under ϕ?

Equivalently, given any two closure-inequivalent orbits, does there exist an invariant form taking distinct values on the two orbits? And to this linear reductivity of the group gives us a complete answer:

Theorem 5.3 (Nagata, Mumford). *Suppose that a linearly reductive group G acts on an affine variety X. Given two orbits $O, O' \subset X$, the following three conditions are equivalent.*

(i) The closures of O, O' have a common point, $\overline{O} \cap \overline{O'} \neq \emptyset$.
(ii) O and O' are closure-equivalent.
(iii) O and O' fail to be separated by the G-invariants $k[X]^G$.

Proof. (i) $\Longrightarrow$ (ii) $\Longrightarrow$ (iii) is already clear. We shall prove the converse, (iii) $\Longrightarrow$ (i). In other words, we will show that if $\overline{O} \cap \overline{O'} = \emptyset$, then the orbits O, O' are separated by $k[X]^G$.

Step 1. Let $\mathfrak{a} \subset R := k[X]$ be the ideal of functions vanishing on the closure $\overline{O}$, and similarly let $\mathfrak{a}' \subset R$ be the ideal of $\overline{O'}$. We consider the ideal $\mathfrak{a} + \mathfrak{a}'$ generated by $\mathfrak{a}$ and $\mathfrak{a}'$. By hypothesis, the zero-set in X of the sum is empty, and so by Hilbert's Nullstellensatz 3.8 (take $f = 1$!) we have $\mathfrak{a} + \mathfrak{a}' = R$.

Step 2. The subsets $\overline{O}, \overline{O'} \subset X$ are preserved by the action of G, and this means that the ideals $\mathfrak{a}, \mathfrak{a}' \subset R$ are subrepresentations of G. Then the homomorphism of R-modules

$$\mathfrak{a} \oplus \mathfrak{a}' \to R, \qquad (a, a') \mapsto a + a',$$

is also a homomorphism of G-representations. By step 1 it is surjective, and so by linear reductivity (Definition 4.36) the map

$$(\mathfrak{a} \cap R^G) \oplus (\mathfrak{a}' \cap R^G) \to R^G$$

is also surjective. In particular, there exist invariants $f \in \mathfrak{a} \cap R^G$ and $f' \in \mathfrak{a}' \cap R^G$ satisfying $f + f' = 1$. The function f vanishes on the orbit O and takes the value 1 on the orbit O', so we are done.   □

Corollary 5.4. *If G is a linearly reductive group acting on an affine variety X, then distinct closed G-orbits are separated by the G-invariants $k[X]^G$.*   □

Corollary 5.5. *If G is a linearly reductive group acting on an affine variety, then each closure-equivalence class contains exactly one closed orbit. Moreover, this is contained in the closure of every orbit in the same equivalence class.*   □

Proof. By the previous corollary the closure-equivalence class can contain at most one closed orbit, so we just have to prove the existence of one. Let O be an orbit with minimal dimension in its equivalence class: we claim that this is a closed set. For if not, then the boundary $\overline{O} - O$ is a nonempty union of G-orbits which are both closure-equivalent to O and of smaller dimension, giving a contradiction. For the last part, if O' is an equivalent orbit, then $O \cap \overline{O'} = \overline{O} \cap \overline{O'}$ is nonempty. But by continuity $\overline{O'}$ is a union of orbits, so $O \subset \overline{O'}$.   □

It follows from Theorem 5.3 that the image $\phi(X) \subset \mathbb{A}^n$ parametrises the closure-equivalence classes of G-orbits in X, or, from Corollary 5.5, that it parametrises the closed G-orbits. On the other hand, for nonreductive groups the theorem is certainly not true:

Example 5.6. Consider the action of the additive group $\mathbb{G}_a$ on the affine plane by

$$(x, y) \mapsto (x, tx + y) \qquad \text{for } t \in \mathbb{G}_a.$$

The orbits are of the following two types:

(i) each vertical line away from the y-axis is an orbit;
(ii) each point on the y-axis is a single orbit.

In particular, we see that in this example every orbit is a closed set. (In fact, this is true whenever a unipotent group acts on an affine variety.) On the other hand, the invariant ring is $k[x] \subset k[x, y]$, which fails to separate any of the orbits of type (ii). Indeed, the group $\mathbb{G}_a$ is not linearly reductive (see Example 4.42). □

(b) Surjectivity of the affine quotient map

Next we consider Question 2 from the introduction to this chapter. We are interested in the map to affine space

$$\phi : X \to \mathbb{A}^n, \qquad x \mapsto (f_1(x), \dots, f_n(x)),$$

given by generators $f_1, \dots, f_n \in R^G$ of the ring of invariants. The Zariski closure of the image $Y \subset \mathbb{A}^n$ of this map is the set of all $(a_1, \dots, a_n) \in \mathbb{A}^n$ satisfying:

$$F(a_1, \dots, a_n) = 0 \text{ for all polynomial relations } F(f_1, \dots, f_n) \equiv 0$$
$$\text{among the generators } f_1, \dots, f_n \in R^G. \tag{5.1}$$

In other words, Y is the zero-set of the kernel $I \subset k[a_1, \dots, a_n]$ of the homomorphism

$$k[a_1, \dots, a_n] \to R^G \qquad a_i \mapsto f_i \quad \text{for } i = 1, \dots, n.$$

A priori ϕ maps X to Y, and we would like to know that it is surjective on this set.

Proposition 5.7. *If the group G is linearly reductive, then the image $\phi(X) \subset \mathbb{A}^n$ is equal to its Zariski closure $Y \subset \mathbb{A}^n$ above.*

Proof. The idea of the proof is similar to that of Hilbert's Theorem 4.51. Starting with a point $a = (a_1, \dots, a_n) \in Y$, that is, satisfying (5.1), we consider the homomorphism of R-modules

$$\pi : R \oplus \cdots \oplus R \to R, \qquad (b_1, \ldots, b_n) \mapsto \sum_{i=1}^{n} b_i(f_i - a_i).$$

Since each $f_i - a_i \in R$ is G-invariant, we see that π is in fact a homomorphism of G-representations. Also, we observe that the induced map π^G on G-invariants is not surjective: its image is the maximal ideal $\mathfrak{m}_a \subset R^G$ corresponding to the point $a \in Y$. Since G is linearly reductive, this implies that π itself cannot be surjective; its image is therefore contained in some maximal ideal $\mathfrak{m} \subset R$. By Corollary 2.25, the intersection $\mathfrak{m} \cap R^G$ is a maximal ideal in R^G, and therefore it coincides with the maximal ideal $\mathfrak{m}_a$. This shows that $a \in Y$ is the image of the point of X corresponding to $\mathfrak{m}$.   □

The closed subvariety $Y = \phi(X) \subset \mathbb{A}^n$ depends on the choice of generating invariants $f_1, \ldots, f_n$. In other words, $Y \cong \mathrm{Spm}\, [a_1, \ldots, a_n]/\sqrt{I}$. However, the ideal I is radical (that is, $\sqrt{I} = I$, since $R^G \subset R$ contains no nilpotents) and so Y is precisely the spectrum $\mathrm{Spm}\, R^G$.

Definition 5.8. We denote the affine variety $\mathrm{Spm}\, R^G$ by $X /\!/ G$. The inclusion $R^G \subset R$ determines a morphism of affine varieties

$$\Phi : X \to X /\!/ G,$$

which we shall call the *affine quotient map*.   □

What we have proved is the following.

Theorem 5.9. *If G is a linearly reductive group acting on an affine variety X, then the affine quotient map*

$$\Phi : X \to X /\!/ G = \mathrm{Spm}\, k[X]^G$$

is surjective and gives a one-to-one correspondence between points of $X /\!/ G$ and closure-equivalence classes of G-orbits in X.

In addition, Φ has the following property.

Proposition 5.10. *If $Z \subset X$ is a G-invariant closed subset, then its image $\Phi(Z)$ is also closed.*

Proof. Let $\mathfrak{a} \subset R$ be the ideal of Z. This is G-invariant, and so G acts on the residue ring $R/\mathfrak{a}$. By linear reductivity, the surjective ring homomorphism $R \to R/\mathfrak{a}$ restricts to a surjective map $R^G \to (R/\mathfrak{a})^G$. The kernel is $R^G \cap \mathfrak{a}$, and so we obtain an isomorphism

$$R^G/(R^G \cap \mathfrak{a}) \stackrel{\sim}{\to} (R/\mathfrak{a})^G.$$

Let us now interpret this geometrically. $R^G/(R^G \cap \mathfrak{a})$ is the coordinate ring of the closure in $X/\!/G$ of $\Phi(X)$, while $(R/\mathfrak{a})^G$ is that of $Z/\!/G$. Taking spectra, this therefore says that Φ induces an isomorphism of affine varieties

$$Z/\!/G \stackrel{\sim}{\to} \overline{\Phi(Z)}.$$

By Theorem 5.9, the affine quotient map $Z \to Z/\!/G$ is surjective, and this is just the restriction of Φ. $\qquad\qquad\square$

For any subset $A \subset X/\!/G$, the preimage $\Phi^{-1}A \subset X$ is G-invariant; if $\Phi^{-1}A$ is closed, then it follows from Proposition 5.10 that A is closed.

Corollary 5.11. *The affine quotient map Φ is a submersion. That is, if the preimage $\Phi^{-1}A \subset X$ of a subset $A \subset X/\!/G$ is open, then A is an open set.* $\qquad\square$

(c) Stability

Corollary 5.4 motivates the following fundamental notion.

Definition 5.12. Suppose that a linearly reductive group G acts on an affine variety X. A point $x \in X$ is said to be *stable* for the action of G if the following two conditions are satisfied.

(i) The orbit $Gx \subset X$ is a closed set.
(ii) The stabiliser subgroup $\mathrm{Stab}(x) = \{g \in G \mid gx = x\}$ is finite.

We denote the set of all stable points for the G-action by $X^s \subset X$. $\qquad\square$

Note that, given $x \in X$, the orbit Gx is the image of the map

$$\psi_x : G \to X, \qquad g \mapsto gx,$$

while the fibres of ψ_x are the left-cosets of $\mathrm{Stab}(x)$ in G. Thus the conditions (i) and (ii) of the definition are equivalent to requiring that ψ_x be a proper morphism (since the fibres are affine, so that ψ_x being proper implies that the fibres are complete and affine, and therefore finite).

Proposition 5.13. *Let* $Z \subset X$ *be the locus of points* $x \in X$ *for which* $\mathrm{Stab}(x)$ *is positive dimensional. Then* X^s *is the complement in* X *of* $\Phi^{-1}(\Phi(Z))$.

Proof. Suppose that $\Phi(x) \in \Phi(Z)$. If $x \in Z$, then condition 5.12(ii) fails; while if $x \notin Z$, then the fibre $\Phi^{-1}(\Phi(x))$ contains at least two G-orbits, and by Corollary 5.5 the orbit Gx (which does not have minimal dimension) cannot be closed – so condition 5.12(i) fails. This shows that $\Phi^{-1}(\Phi(Z)) \subset X - X^s$; the converse is similar. $\square$

Corollary 5.14. *All points are stable,* $X^s = X$, *if and only if all points of* X *have a finite stabiliser.* $\square$

By considering the map

$$G \times X \to X \times X, \qquad (g, x) \mapsto (gx, x),$$

determined by the group action, we see that $Z \subset X$ is a closed set. In fact, let $\widetilde{Z} \subset G \times X$ be the pull-back of the diagonal $\Delta \subset X \times X$; then Z is the locus along which the fibres of the projection $\widetilde{Z} \to X$ onto the second factor (that is, the stabiliser subgroups) have positive dimension. Since Z is also G-invariant, we deduce from Proposition 5.10:

Proposition 5.15. *The stable set* $X^s \subset X$ *and its image* $\Phi(X^s) \subset X /\!/ G$ *are open sets.*

Suppose that $x \in X$ is stable and that the orbits Gx and Gy, for another point $y \in X$, are closure-equivalent. Since Gx is a closed set, this implies that $Gx \subset \overline{Gy}$. Since $\dim Gx = \dim G \geq \dim \overline{Gy}$, it follows that $Gx = Gy$. From Theorem 5.3, therefore, we obtain:

Theorem 5.16. *Suppose that a linearly reductive group* G *acts on an affine variety* X, *and suppose that* $x \in X$ *is a stable point for the action. Then for any* $y \in X - Gx$ *there exists an invariant function* $f \in k[X]^G$ *such that* $f(x) \neq f(y)$. $\square$

Let us denote the image $\Phi(X^s)$ of the stable points by X^s/G. This is an open subset of $X//G$ and we have a commutative diagram:

$$
\begin{array}{ccc}
X^s & \subset & X \\
\downarrow & & \downarrow \Phi \\
X^s/G & \subset & X//G.
\end{array}
$$

Corollary 5.17. *The restriction* $X^s \to X^s/G$ *of the affine quotient map* Φ *gives a one-to-one correspondence between points of* X^s/G *and* G-*orbits in* X^s. $\square$

In this situation, one says that X^s/G is a *geometric quotient* of X^s by G. (For a more precise definition see Mumford et al. [30] chapter 4.)

5.2 Classical invariants and the moduli of smooth hypersurfaces in $\mathbb{P}^n$

We can apply the general theory of the last section to study the moduli of smooth hypersurfaces in $\mathbb{P}^n$. In Section 5.2(a) we show that the set of these is an affine variety, defined by the nonvanishing of the discriminant, and in Section 5.2(b) we show that every smooth hypersurface is stable for the action of the general linear group.

(a) Classical invariants and discriminants

A form of degree d in $n+1$ coordinates $x_0, x_1, \ldots, x_n$ can be written

$$
f(x_0, x_1, \ldots, x_n) = \sum_{|I|=d} a_I x^I.
$$

In this notation $I = (i_0, i_1, \ldots, i_n)$ is a multiindex, $0 \le i_\alpha \le n$ for each α, ranging through $\binom{n+d}{d}$ values for which $|I| := \sum i_\alpha = d$. For each multiindex, the monomial $x^I = x_0^{i_0} x_1^{i_1} \ldots x_n^{i_n}$ comes with coefficient $a_I \in k$.

We will denote by $V_{n,d}$ the vector space of homogeneous polynomials $f(x)$ of degree d and by $\mathbb{V}_{n,d}$ the associated affine space. (Thus $V_{1,d} = V_d$ in the notation of earlier chapters.) This space has dimension $\binom{n+d}{d}$, and $GL(n+1)$ acts on $\mathbb{V}_{n,d}$ on the right by $f(x) \mapsto f(gx)$. Equivalently, if we introduce $\binom{n+d}{d}$ independent variables ξ_I as coefficients for the generic form

$$
(\xi \lozenge x) = \sum_{|I|=d} \xi_I x^I \in k[x_0, \ldots, x_n; \ldots, \xi_I, \ldots],
$$

then a matrix $g = (a_{ij}) \in GL(n)$ transforms the form $(\xi \, \lozenge \, x)$ to

$$(\xi \, \lozenge \, gx) = \sum_{|I|=d} \xi_I \left(\sum a_{0j} x_j \right)^{i_0} \left(\sum a_{1j} x_j \right)^{i_1} \ldots \left(\sum a_{nj} x_j \right)^{i_n},$$

which, after expanding and gathering monomials x^I, can be written as

$$(\xi g \, \lozenge \, x) := \sum_{|I|=d} \xi_I(g) x^I.$$

The transformed coefficients $\xi_I(g)$ can be expressed as

$$\xi_I(g) = \sum_{|J|=d} g_I^J \xi_I$$

for some polynomials $g_I^J \in k[a_{ij}]$. (See Chapter 1 for the binary case $n = 1$). The following generalises Definition 1.24.

Definition 5.18. If a homogeneous polynomial $F(\xi)$ satisfies

$$F(\xi g) = F(\xi) \qquad \text{for all } g \in SL(n+1),$$

then F is called a *classical ($(n + 1)$-ary) invariant*. $\square$

Let $H_{n,d} = \mathbb{P}V_{n,d}$. In terms of $H_{n,d}$ we can interpret the classical invariants in more geometric language.

Proposition 5.19. *For a homogeneous polynomial $F(\xi) \in k[\mathbb{V}_{n,d}]$ the following two conditions are equivalent:*

(i) $F(\xi)$ is a classical invariant;
(ii) the subvariety $F(\xi) = 0$ in $H_{n,d}$ is $GL(n + 1)$-invariant.

Proof. It is enough to prove (ii) $\Longrightarrow$ (i), that is, that $F(\xi)$ is $SL(n+1)$-invariant when the projective subvariety $F(\xi) = 0$ is $GL(n + 1)$-invariant. We may also assume that $F(\xi)$ is irreducible. Then the ideal of the subvariety has a generator which is unique up to multiplication by a unit in the polynomial ring $k[\xi]$, that is, by a nonzero constant. This shows that the 1-dimensional subspace of $k[\xi]$ spanned by F is $GL(n + 1)$-invariant. By Lemma 4.12, every character of $GL(n + 1)$ is a power of the determinant, and it follows that $F(\xi)$ is invariant under the subgroup $SL(n + 1)$. $\square$

As in the case $n = 1$, the discriminant is a basic example of a classical invariant. Before we define it, let $X \subset \mathbb{P}^n$ be a hypersurface $f(x_0, x_1, \ldots, x_n) = 0$

in $\mathbb{P}^n$, where $f(x_0, x_1, \ldots, x_n)$ is a homogeneous polynomial of degree d. Recall that $p \in \mathbb{P}^n$ is a *singular point* of X if it is a solution of the simultaneous equations

$$\frac{\partial f}{\partial x_0} = \frac{\partial f}{\partial x_1} = \cdots = \frac{\partial f}{\partial x_n} = 0.$$

X is said to be *nonsingular* (or *smooth*) if these equations have no nonzero solutions. (See Section 1.4.)

We shall denote by $H_{n,d}^{\text{sing}} \subset H_{n,d}$ the set of all singular hypersurfaces $X \subset \mathbb{P}^n$. In order to analyse $H_{n,d}^{\text{sing}}$ we introduce the subset

$$Z = \{(p, X) \mid p \in \mathbb{P}^n \text{ is a singular point of } X \subset \mathbb{P}^n\}$$

of the product $\mathbb{P}^n \times H_{n,d}$. In coordinates Z is defined by $n + 1$ equations,

$$\frac{\partial}{\partial x_0} \sum_I \xi_I x^I = \cdots = \frac{\partial}{\partial x_n} \sum_I \xi_I x^I = 0. \tag{5.2}$$

In particular, $Z \subset \mathbb{P}^n \times H_{n,d}$ is a closed subvariety. Consider the projections to the two factors:

$$Z$$

$$\psi \swarrow \qquad \searrow \varphi$$

$$\mathbb{P}^n \qquad \qquad H_{n,d}$$

First, note that the image of φ is precisely $H_{n,d}^{\text{sing}}$. On the other side of the diagram, for any $p \in \mathbb{P}^n$ the fibre $\psi^{-1}(p)$ is the set of all hypersurfaces that are singular at the point p, and by (5.2) this set is defined by $n + 1$ linear equations in $H_{n,d}$. It follows that

$$\dim Z = \dim \mathbb{P}^n + \text{fibre dimension of } \psi$$

$$\geq n + \dim H_{n,d} - (n + 1)$$

$$= \dim H_{n,d} - 1.$$

In fact it is easy to see that equality holds: the image of ϕ is a proper subset since there exist nonsingular hypersurfaces (Exercise 5.1), and has the same dimension as Z because there exist hypersurfaces with exactly one singular point (Exercise 5.2). Hence $\dim Z \leq \dim H_{n,d} - 1$ and we conclude that

$$\dim H_{n,d}^{\text{sing}} = \dim Z = \dim H_{n,d} - 1.$$

Since the projective space $\mathbb{P}^n$ is complete, moreover, $H_{n,d}^{\text{sing}} \subset H_{n,d}$ is a closed subset, and by Proposition 3.47 this implies that it is defined by a single homogeneous equation.

Definition 5.20. The defining equation $D(\xi) \in k[\mathbb{V}_{n,d}]$ of the hypersurface $\dim H_{n,d}^{\text{sing}} \subset \dim H_{n,d}$ (determined up to multiplication by a scalar) is called the *discriminant* of forms of degree d on $\mathbb{P}^n$, or of degree d hypersurfaces in $\mathbb{P}^n$. (This generalises Definition 1.20 for the case $n = 1$.) $\square$

It is clear that $H_{n,d}^{\text{sing}}$ is invariant under $GL(n + 1)$, and so Proposition 5.19 implies:

Corollary 5.21. *The discriminant $D(\xi)$ is a classical invariant.* $\square$

Example 5.22. For degree $d = 2$, the space $\mathbb{V}_{n,2}$ of quadratic forms can be naturally identified with the vector space of $(n+1) \times (n+1)$ symmetric matrices (a_{ij}), $a_{ij} = a_{ji}$. Such a matrix determines the quadric hypersurface $Q \subset \mathbb{P}^n$ with equation

$$\sum_{i,j} a_{ij} x_i x_j = 0.$$

Then Q is singular if and only if $\det |a_{ij}| = 0$; thus the discriminant in this case is $D(a_{ij}) = \det |a_{ij}|$. $\square$

(b) Stability of smooth hypersurfaces

We need one more key fact for the construction of a moduli space for smooth hypersurfaces. This is due to Jordan [34] and Matsumura and Monsky [32].

Theorem 5.23. *Any homogeneous polynomial $f \in k[x_0, x_1, \ldots, x_n]$ with degree ≥ 3 is invariant under at most finitely many $g \in GL(n + 1)$.*

Proof. We shall prove the equivalent statement that f is invariant under no nonzero element of the Lie space $\mathfrak{gl}(n + 1)$. Recall from Example 4.30 that $\mathfrak{gl}(n + 1)$ is the vector space of all $(n + 1) \times (n + 1)$ matrices $A = (a_{ij})$, $0 \leq i, j \leq n$. Such a matrix determines a partial differential operator

$$\mathcal{D}_A = \sum_{i,j} a_{i,j} x_j \frac{\partial}{\partial x_i},$$

and the action of $\mathfrak{gl}(n+1)$ on polynomials induced by that of $GL(n+1)$ is

$$f \mapsto \mathcal{D}_A f = \sum_{i,j} a_{i,j} x_j \frac{\partial f}{\partial x_i}.$$

We have to show that the linear subspace $\{A \in \mathfrak{gl}(n+1) \mid \mathcal{D}_A(f) = 0\}$ is zero. Let $f_i = \frac{\partial f}{\partial x_i}$ for each $i = 0, 1, \ldots, n$. Then nonsingularity of the homogeneous polynomial $f(x_0, x_1, \ldots, x_n)$ means that the equations

$$f_0(x_0, x_1, \ldots, x_n) = f_1(x_0, x_1, \ldots, x_n) = \cdots = f_n(x_0, x_1, \ldots, x_n) = 0$$

have no nonzero solutions. This implies that the maximal ideal $\mathfrak{m} = (x_0, x_1, \ldots, x_n) \subset k[x_0, x_1, \ldots, x_n]$ is a minimal prime divisor of the ideal $(f_0, f_1, \ldots, f_n)$.

Claim: f_i is not a zero-divisor modulo the ideal $(f_0, \ldots, \widehat{f_i}, \ldots, f_n)$ for any $i = 0, 1, \ldots, n$. It is enough to consider the case $i = 0$. By Krull's Principal Ideal Theorem (see Atiyah and Macdonald [9] Chapter 11, or Eisenbud [61] Section 8.2.2) every minimal prime ideal containing $(f_1, \ldots, f_n)$ has height $\leq n$. Since $\mathfrak{m}$ has height $n + 1$, there is no $h \notin (f_1, \ldots, f_n)$ with $f_0 h \in (f_1, \ldots, f_n)$ – proving the claim.

Returning now to the equation $\mathcal{D}_A f = 0$, this is equivalent to an identity

$$\sum_{i=0}^{n} l_i(x) f_i(x) = 0 \qquad \text{for linear forms } l_i(x) = \sum_{j=0}^{n} a_{ij} x_j.$$

By the claim, this forces each $l_i(x)$ to be in the ideal $(f_0, \ldots, \widehat{f_i}, \ldots, f_n)$. But by hypothesis deg $f_i(x) \geq 2$, so we must have $l_i(x) \equiv 0$ for each i, and hence $A = 0$. □

The set of nonsingular homogeneous equations of hypersurfaces of degree d is an open subset of $\mathbb{V}_{n,d}$ defined by $D(\xi) \neq 0$. We shall denote this open set by $U_{n,d}$; it is an affine variety with coordinate ring

$$k[U_{n,d}] = k[\xi_I, D(\xi)^{-1}].$$

Then $GL(n+1)$ acts on this variety, and by Theorem 5.23 and Corollary 5.14 we have:

Corollary 5.24. *For* $d \geq 3$, *every point of* $U_{n,d} \subset \mathbb{V}_{n,d}$ *is stable for the action of* $GL(n+1)$. □

By Corollary 5.17, therefore, there exists a good quotient whose points parametrise precisely the $GL(n+1)$ orbits,

$$\Phi : U_{n,d} \to U_{n,d}/GL(n+1).$$

This quotient $U_{n,d}/GL(n+1)$ is called the *moduli space of smooth hypersurfaces of degree d in* $\mathbb{P}^n$; its points are in one-to-one correspondence with projective equivalence classes of such hypersurfaces.

The following example has already been seen in Chapter 1.

Example 5.25. Binary quartics. The variety $U_{1,4}$ is the set of all binary quartic forms (writing $\xi_0, \xi_1, \ldots, \xi_4$ instead of $\xi_{04}, \xi_{13}, \ldots, \xi_{40}$)

$$(\xi \mathbin{\vert} x, y) = \xi_0 x^4 + 4\xi_1 x^3 y + 6\xi_2 x^2 y^2 + 4\xi_3 x y^3 + \xi_4 y^4 \in \mathbb{V}_{1,4}$$

without repeated linear factors. By Corollary 4.70 and the fact that

$$D(\xi) = g_2(\xi)^3 - 27 g_3(\xi)^2,$$

the invariant ring $k[\xi_0, \ldots, \xi_4, D(\xi)^{-1}]^{GL(2)}$ is generated by $g_2(\xi)^3/D(\xi)$. Hence the affine quotient map is

$$\Phi : U_{1,4} = \left\{ \begin{matrix} \text{binary quartics without} \\ \text{repeated linear factors} \end{matrix} \right\} \to \mathbb{A}^1, \qquad \xi \mapsto g_2(\xi)^3/D(\xi).$$

The moduli space, parametrising $GL(2)$-equivalence classes of binary quartics, is in this case the affine line $\mathbb{A}^1$. The reader should compare this with the proof of Proposition 1.27. □

Example 5.26. Plane cubics. Ternary cubic forms $(\xi \mathbin{\vert} x, y, z)$ – or equivalently, plane cubic curves – live in the vector space

$$V_{2,3} = \langle x^3, y^3, z^3, x^2 y, y^2 z, z^2 x, x y^2, y z^2, z x^2, x y z \rangle.$$

As affine space $\mathbb{V}_{2,3} \cong \mathbb{A}^{10}$ this has coordinate ring

$$k[\mathbb{V}_{2,3}] = k[\xi_{300}, \xi_{030}, \xi_{003}, \xi_{210}, \xi_{021}, \xi_{102}, \xi_{120}, \xi_{012}, \xi_{201}, \xi_{111}],$$

and it is a classical result of Aronhold [21] that the ring $k[\mathbb{V}_{2,3}]^{SL(3)}$ is generated by two algebraically independent invariants $S(\xi)$ and $T(\xi)$, where $\deg S = 4$ and $\deg T = 6$. In particular, the discriminant $D(\xi)$ is in $k[S, T]$ and is computed to be

$$D = T^2 + 64 S^3.$$

(To give S, T explicitly one uses the canonical form of a cubic,

$$C: \quad x^3 + y^3 + z^3 + 6mxyz = 0. \tag{5.3}$$

Then $S = m - m^4$, $T = 1 - 20m^3 - 8m^6$ and $D = (1 + 8m^3)^3$.)

Restricting to the nonsingular cubics, the ring $k[\xi_I, D(\xi)^{-1}]^{GL(3)}$ is generated by a single invariant $S(\xi)^3/D(\xi)$. Hence the affine quotient map is

$$\Phi : U_{2,3} = \left\{ \begin{matrix} \text{nonsingular} \\ \text{plane cubics} \end{matrix} \right\} \rightarrow \mathbb{A}^1, \qquad \xi \mapsto \frac{S(\xi)^3}{D(\xi)}.$$

Thus the moduli space, as in the previous example, is the affine line $\mathbb{A}^1$.

For reference later on, let us say a bit more about this example. The $SL(3)$-invariance of S and T means that they have some projective geometric interpretation: the conditions $S = 0$ and $T = 0$ are projectively invariant properties of a cubic, and one can ask what they mean geometrically. The *Hessian* of a plane cubic $f(x, y, z) \in \mathbb{V}_{2,3}$ is the cubic

$$H(f)(x, y, z) = \begin{vmatrix} \partial^2 f/\partial^2 x & \partial^2 f/\partial x \partial y & \partial^2 f/\partial x \partial z \\ \partial^2 f/\partial y \partial x & \partial^2 f/\partial^2 y & \partial^2 f/\partial y \partial z \\ \partial^2 f/\partial z \partial x & \partial^2 f/\partial z \partial y & \partial^2 f/\partial^2 z \end{vmatrix} \in \mathbb{V}_{2,3}.$$

As a plane curve, $H(f)(x, y, z) = 0$ is the locus of points in $\mathbb{P}^2$ whose polar conic with respect to $f(x, y, z) = 0$ is a pair of lines. The interpretation of S, T is now the following (an exercise for the reader, using the canonical form (5.3)). First,

$$S = 0 \qquad \Longleftrightarrow \qquad H(f) \text{ factorises as three lines}$$

or, equivalently, if and only if f is a sum of three cubes $l_1^3 + l_2^3 + l_3^3$ of linear forms $l_i(x, y, z)$. On the other hand,

$$T = 0 \qquad \Longleftrightarrow \qquad H(H(f)) = f \text{ up to a scalar.}$$

For more details we refer the reader to Salmon [35] or Elliot [33].   □

(c) A moduli space for hypersurfaces in $\mathbb{P}^n$

We will now construct a complete variety which compactifies the affine moduli space $U_{n,d}/GL(n + 1)$ of the last section. This is intended to serve as a motivating example for the construction of the projective quotient in the next chapter, but nevertheless it is very classical and we will return to it in Chapter 7 after we have discussed the Hilbert-Mumford numerical criterion for stability and semistability.

We consider the ring of classical invariants

$$R_{n,d} = k[\mathbb{V}_{n,d}]^{SL(n+1)} = \bigoplus_{e=0}^{\infty} k[\xi_I]_e^{SL(n+1)},$$

graded by degree, and the projective variety Proj $R_{n,d}$ (Definition 3.43). By construction, this is covered by affine open sets Spm $R_{F,0}$ for homogeneous elements $F \in R = R_{n,d}$. By Hilbert's Theorem 4.51, the ring $R_{n,d}$ is finitely generated, and so Proj $R_{n,d}$ is covered by Spm $R_{F_1,0}, \ldots,$ Spm $R_{F_m,0}$ for some finite set of classical invariants $F_1(\xi), \ldots, F_m(\xi)$ generating the invariant ring $R_{n,d}$ (Proposition 3.44). On the other hand, it is easy to verify that

$$R_{F,0} = k\left[\xi_I, \frac{1}{F(\xi)}\right]^{GL(n+1)},$$

and hence we have:

Proposition 5.27. *For each classical invariant* $F(\xi) \in k[\mathbb{V}_{n,d}]^{SL(n+1)}$, *the affine variety* Spm $k[\xi_I, F(\xi)^{-1}]^{GL(n+1)}$ *is contained in* Proj $R_{n,d}$ *as an open set.*   □

Remarks 5.28.

(i) This proposition will be a special case of Remark 6.14(iv) in the next chapter. It says that the projective quotient is constructed by gluing together the affine quotients by $GL(n + 1)$ of localisations $\mathbb{V}_{n,d} - \{F = 0\}$, where F runs through all classical invariants.

(ii) The next proposition says that the function field of each affine variety Spm $R_{F,0}$ – that is, the field of fractions of $R_{F,0}$ – is equal to the field of invariant rational functions $k(\xi_I)^{GL(n+1)}$. This observation will be generalised in Proposition 6.16 in the next chapter.   □

Proposition 5.29. *The field of fractions of* $k[\xi_I, F(\xi)^{-1}]^{GL(n+1)}$ *is equal to the field of invariant rational functions* $k(\xi_I)^{GL(n+1)}$.

Proof. Let deg $F = h > 0$. An arbitrary rational function can be written as a ratio of polynomials $A(\xi)/B(\xi)$. Then $GL(n + 1)$-invariance forces A and B to be homogeneous of the same degree – let this be $e = \deg A = \deg B$. Then the two rational functions

$$a(\xi) = \frac{A(\xi)B(\xi)^{h-1}}{F(\xi)^e}, \qquad b(\xi) = \frac{B(\xi)^h}{F(\xi)^e}$$

are both $GL(n+1)$-invariant elements of $k[\xi_I, F(\xi)^{-1}]$, and their ratio is equal to $A(\xi)/B(\xi)$. $\qquad\square$

If in Proposition 5.27 we take F to be the discriminant, then from Corollary 5.24 we arrive at:

Corollary 5.30. *For $d \geq 3$, the moduli space $U_{n,d}/GL(n+1)$ of smooth hypersurfaces of degree d in $\mathbb{P}^n$ is contained as an open subset in* Proj $R_{n,d}$. $\qquad\square$

(d) Nullforms and the projective quotient map

We conclude this chapter by explaining the sense in which the projective variety Proj $R_{n,d}$ is itself a quotient of $\mathbb{V}_{n,d}$ by $GL(n+1)$ (Definition 5.36).

Definition 5.31. A form of degree d

$$(a \lozenge x_0, x_1, \ldots, x_n) = \sum_{|I|=d} a_I x^I \in \mathbb{V}_{n,d}$$

with the property that $F(a_I) = 0$ for every nonconstant classical invariant $F \in R_{n,d}$ was called by Hilbert a *nullform*. In Mumford's terminology, such a form is also called *unstable*. If a form is not unstable, it is called *semistable*. $\qquad\square$

Let $F(\xi) \in R_{n,d} = k[\xi_I]^{SL(n+1)}$ be an arbitrary invariant. This decomposes uniquely as a sum of homogeneous classical invariants

$$F(\xi) = F_{(0)} + F_{(1)}(\xi) + F_{(2)}(\xi) + \cdots + F_{(e)}(\xi),$$

where $\deg F_{(i)}(\xi) = i$. It follows that $a \in \mathbb{V}_{n,d}$ is a nullform if and only if $F(a) = F(0)$ for every $F \in R_{n,d}$. By Theorem 5.3 we deduce:

Proposition 5.32. *A form $a \in \mathbb{V}_{n,d}$ is a nullform if and only if the closure of its $SL(n+1)$-orbit contains the origin, that is, $0 \in \overline{SL(n+1) \cdot a}$.* $\qquad\square$

Together with Theorem 5.16, this implies:

Corollary 5.33. *If $a \in \mathbb{V}_{n,d}$ is a nullform, then it is not stable for the action of $SL(n+1)$.* $\qquad\square$

If $a \in \mathbb{V}_{n,d}$ is not a nullform, then it is semistable, so the corollary says:

$$\text{stable} \Longrightarrow \text{semistable}.$$

Note that if two forms $a, b \in \mathbb{V}_{n,d}$ are in the same $SL(n+1)$-orbit, then stability of one is equivalent to stability of the other (Definition 5.12). Similarly, since a classical invariant $F \in R_{n,d}$ is constant on $SL(n+1)$-orbits, its non-vanishing is independent of the $GL(n+1)$-action on forms. So we have:

Lemma 5.34. *If two forms $a, b \in \mathbb{V}_{n,d}$ are in the same $GL(n+1)$-orbit, then a is a nullform if and only if b is a nullform.* $\square$

Example 5.35. A binary quartic $a \in \mathbb{V}_{1,4}$ is a nullform if and only if the equation $(a \wr x, y) = 0$ has a root of multiplicity ≥ 3.

To see this, recall (see Corollary 4.70) that the ring $R_{1,4}$ of $SL(2)$-invariants is generated by

$$g_2(\xi) = \xi_0\xi_4 - 4\xi_1\xi_3 + 3\xi_2^2,$$
$$g_3(\xi) = \xi_0\xi_2\xi_4 - \xi_0\xi_3^2 - \xi_1^2\xi_4 + 2\xi_1\xi_2\xi_3 - \xi_2^3.$$

If $(a \wr x, y) = 0$ has a root of multiplicity ≥ 3, then it is $GL(2)$-equivalent to one of x^4, x^3y, each of which has $g_2(a) = g_3(a) = 0$ and is therefore a nullform. Conversely, if a is a nullform, then in particular its discriminant vanishes, $D(a) = 0$. This means that it has a multiple root, which up to $GL(2)$ we can take to be $x = 0$, so that

$$(a \wr x, y) = x^2(px^2 + qxy + ry^2).$$

The condition $g_2(a) = 0$ then implies $r = 0$, so that $x = 0$ has multiplicity ≥ 3. This example will be generalised in Proposition 7.9 and Example 7.10. $\square$

Consider now the affine quotient map for the action of $SL(n+1)$,

$$\Phi : \mathbb{V}_{n,d} \to \operatorname{Spm} R_{n,d}.$$

In this notation the set of nullforms is $\Phi^{-1}(\Phi(0))$, and we shall denote the complement of this set, the set of semistable forms, by $\mathbb{V}^{ss}_{n,d} \subset \mathbb{V}_{n,d}$. This is precisely the union of open sets

$$\mathbb{V}^{ss}_{n,d} = \{F_1(a) \neq 0\} \cup \cdots \cup \{F_r(a) \neq 0\}$$

for generators $F_1, \ldots, F_r \in R_{n,d}$. Then the affine quotient maps $\Phi_i : \{F_i(a) \neq 0\} \to U_i = \operatorname{Spm} k[\xi_I, F_i(\xi)^{-1}]^{GL(n+1)}$, for $i = 1, \ldots, r$, glue together to give a surjective morphism

$$\Psi : \mathbb{V}^{ss}_{n,d} \to \operatorname{Proj} R_{n,d}.$$

Definition 5.36. Ψ is called the *projective quotient map*, and

$$\mathbb{V}_{n,d}^{\mathrm{ss}} /\!\!/ GL(n+1) := \mathrm{Proj}\ R_{n,d}$$

is called the *moduli space of (semistable) hypersurfaces of degree d in* $\mathbb{P}^n$. $\square$

Example 5.37. Binary forms.

$\underline{d = 2, 3}$. Each of $R_{1,2}$ and $R_{1,3}$ is of the form $k[D]$, a polynomial ring in one variable generated by the discriminant. It follows that in each of these cases Proj R consists of a single point. (See Example 3.45.)

$\underline{d = 4}$. In this case $R_{1,4} = k[g_2, g_3]$, a polynomial ring in two variables with weights deg $g_2 = 2$, deg $g_3 = 3$. Hence Proj $R_{1,4} \cong \mathbb{P}(2:3)$, that is, the moduli space of binary quartics is a weighted projective line. In fact, this variety is isomorphic to $\mathbb{P}^1$. $\square$

Let R be the graded ring of Example 3.46, and let $F(X_0, X_1, \ldots, X_n) \in R_e$ be a homogeneous polynomial of degree e. Then the quotient ring $R/(F)$ is an integral domain which inherits a grading from R. The projective variety Proj $R/(F)$ can be identified with the zero set

$$\{F(X_0, X_1, \ldots, X_n) = 0\} \subset \mathbb{P}(a_0 : a_1 : \ldots : a_n)$$

and is called a *weighted hypersurface of degree e*.

Example 5.38. Binary quintics. As we have seen in Example 4.71, the invariant ring $R_{1,5}$ is of the form $k[X_0, X_1, X_2, X_3]/(F(X_0, X_1, X_2, X_3))$, where deg $F = 36$, and where

$$\deg X_0 = 4, \quad \deg X_1 = 8, \quad \deg X_2 = 12, \quad \deg X_3 = 18.$$

It follows that the moduli space of binary quintics is isomorphic to a weighted surface of degree 36 in the 3-dimensional weighted projective space $\mathbb{P}(4:8:12:18)$.

Binary sextics. Similarly, in the case $d = 6$ the moduli space Proj $R_{1,6}$ is isomorphic to a weighted hypersurface of degree 30 in $\mathbb{P}(2:4:6:10:15)$.

Binary octics. From Example 4.72, the moduli space of binary octics Proj $R_{1,8}$ is a 5-dimensional subvariety of $\mathbb{P}(2:3:4:5:6:7:8:9:10)$ defined by five 4×4 Pfaffians. $\square$

For forms in three variables – that is, curves in $\mathbb{P}^2$ – determining the ring of classical invariants $R_{2,d}$ is not easy, and we shall mention only the following result.

Example 5.39. Plane cubics. As we have seen in Example 5.26, the ring of classical invariants

$$R_{2,3} = k[\mathbb{V}_{2,3}]^{SL(3)}$$

is precisely the polynomial ring $k[S, T]$, where $\deg S = 4$ and $\deg T = 6$. Hence the moduli space of plane cubic curves is naturally isomorphic to the weighted projective line $\mathbb{P}(2 : 3)$. □

We next note that in this projective quotient variety there is a well-defined notion of stability.

Lemma 5.40. *For a nonzero form* $a \in \mathbb{V}_{n,d}$ *the following conditions are equivalent:*

(i) *a is stable for the action of* $SL(n + 1)$ *on* $\mathbb{V}_{n,d}$. *In other words, the orbit* $SL(n + 1)a \subset \mathbb{V}_{n,d}$ *is closed and the stabiliser* $\mathrm{Stab}(a) \subset SL(n + 1)$ *is finite.*

(ii) *a is stable for the action of* $GL(n+1)$ *on any open set* $\{F(\xi) \neq 0\} \subset \mathbb{V}_{n,d}$, *for* $F \in R_{n,d}$, *containing a.*

Proof. It is clear that finiteness of the stabiliser in (i) is equivalent to finiteness of the stabiliser in (ii); we have to show that closure of the orbits of a in (i) and (ii) are equivalent. First let us assume (i) – that is, that $SL(n+1) \cdot a \subset \mathbb{V}_{n,d}$ is closed. Then by the Nagata-Mumford Theorem 5.3 there exists some classical invariant $H(\xi)$ such that $H(a) \neq 0$. This determines a morphism $H : \mathbb{V}_{n,d} \to \mathbb{A}^1$, and we consider its restriction to the orbit of a:

$$\{H(\xi) \neq 0\}$$
$$\cup$$
$$H' : GL(n + 1) \cdot a \to \mathbb{A}^1 - \{0\}.$$

This map H' is surjective and its fibre is a disjoint union

$$\bigcup_{1 \leq i \leq N} \omega^i SL(n + 1) \cdot a, \qquad \omega^N = 1, \ N = \deg H. \tag{5.4}$$

By hypothesis this is closed in $\mathbb{V}_{n,d}$, and hence the orbit $GL(n + 1) \cdot a$ is a closed set.

Conversely, assume (ii). Then

$$GL(n + 1) \cdot a \cap \{F(\xi) = F(a)\}$$

is a disjoint union as in (5.4) and so $SL(n + 1) \cdot a$ is closed in $\mathbb{V}_{n,d}$.   □

Definition 5.41 (Mumford [30]). A nonzero form $a \in \mathbb{V}_{n,d}$, and the corresponding hypersurface in $\mathbb{P}^n$, are said to be *stable* if the conditions (i), (ii) of Lemma 5.40 are satisfied.   □

Example 5.42. Every nonsingular hypersurface of degree ≥ 3 (or, in the case $n = 1$, every binary form without repeated linear factors) is stable by Corollary 5.24.   □

Note that stability depends only on the $GL(n + 1)$-orbit of a form; note also that stability implies semistability. By Proposition 5.15, the set of stable forms $\mathbb{V}_{n,d}^s \subset \mathbb{V}_{n,d}$ and its image $\Psi(\mathbb{V}_{n,d}^s) \subset \text{Proj } R_{n,d}$ are open sets. By Corollary 5.17, $\Psi(\mathbb{V}_{n,d}^s)$ parametrises projective equivalence classes of stable hypersurfaces of degree d in $\mathbb{P}^n$ via the restricted map

$$\Psi : \mathbb{V}_{n,d}^s \rightarrow \Psi(\mathbb{V}_{n,d}^s) \subset \mathbb{V}_{n,d}^{ss} /\!/ GL(n + 1) = \text{Proj } R_{n,d}.$$

We write

$$\mathbb{V}_{n,d}^s / GL(n + 1) := \Psi(\mathbb{V}_{n,d}^s),$$

called the *moduli space of stable hypersurfaces.*

To summarise the constructions of this section:

$$\text{nonsingular} \Longrightarrow \text{stable} \Longrightarrow \text{semistable (not a nullform).}$$

Correspondingly, we have constructed moduli spaces:

$$\{D(\xi) \neq 0\}/GL(n+1) \underset{\text{open}}{\hookrightarrow} \mathbb{V}_{n,d}^s/GL(n+1) \underset{\text{open}}{\hookrightarrow} \mathbb{V}_{n,d}^{ss}/\!/GL(n+1)$$

$$\| \qquad\qquad\qquad \| \qquad\qquad\qquad \|$$

| $\left\{\begin{array}{l}\text{nonsingular hyper-}\\\text{surfaces up to pro-}\\\text{jective equivalence}\end{array}\right\}$ | $\left\{\begin{array}{l}\text{stable hypersurfaces}\\\text{up to projective}\\\text{equivalence}\end{array}\right\}$ | $\left\{\begin{array}{l}\text{semistable hypersur-}\\\text{faces up to closure}\\\text{equivalence}\qquad\text{of}\\GL(n+1)\text{-orbits}\end{array}\right\}$ |

It turns out that, even if the structure of the invariant ring $R_{n,d} = k[\xi_I]^{SL(n+1)}$ is not known, it is nevertheless possible to classify the stable forms and the nullforms. We will return to this question in Chapter 7 (Section 7.2(a)).

Exercises

1. For each n, d give an example of a nonsingular hypersurface $f(x_0, x_1, \ldots, x_n) = 0$.

2. For each n, d give an example of a hypersurface $f(x_0, x_1, \ldots, x_n) = 0$ with exactly one singular point.

6

The projective quotient

Let G be an algebraic group acting on an affine variety X. In the last chapter we discussed the following strategy for defining a quotient variety of X by G:

(A) For suitable G-invariant functions $f_1, \ldots, f_n \in k[X]^G$, consider the map to affine space

$$X \to \mathbb{A}^n, \qquad x \mapsto (f_1(x), \ldots, f_n(x)),$$

and take for the quotient the image of this map.

In this manner we constructed quite explicitly a moduli space for nonsingular hypersurfaces in $\mathbb{P}^n$; and by gluing together affine varieties obtained in this way (in other words, using Proj) we constructed a compactification of the moduli space.

Although at an elementary level this approach works well, the following improvement has a wider scope for applications:

(B) $f_0, f_1, \ldots, f_n$ need not be G-invariants, nor even regular functions on X, but their ratios f_i/f_j should all be G-invariant rational functions. (To be precise, the f_i should be G-invariant sections of a G-linearised invertible sheaf – see Section 6.2.) Thus chosen, we have a rational map (that is, a map defined on a nonempty open set in X) to projective space

$$X - - \to \mathbb{P}^n, \qquad x \mapsto (f_0(x) : f_1(x) : \ldots : f_n(x)),$$

and we take the image of this map.

Just as we arrive at an affine variety Spm $k[X]^G$ via (A), via (B) we arrive at a projective quotient, often called the GIT (geometric invariant theory) quotient. While (A) is the basic technique for constructing a quotient locally, (B) is generally more useful for constructing quotients globally. In fact, (A) is just the special case of (B) in which one takes $f_0 = 1$ and G-invariant regular functions $f_1, \ldots, f_n$. Moreover, the point of view of (B) is closely related to the so-called 'moment map', which allows one to construct symplectic reductions of symplectic manifolds and complex Kähler manifolds.

Note that the map to projective space in (B) fails to be defined at the common zeros of $f_0(x)$, $f_1(x)$, $\ldots$, $f_n(x)$. It is in this respect that the construction differs significantly from that of (A). By taking as many functions $f_0, f_1, \ldots, f_n$ as possible, one can reduce this common zero-set, but in general it will remain nonempty. Those points that remain in the zero-set for any choice of functions are called the *unstable points* for the group action (Hilbert's *nullforms* in the classical case of hypersurfaces studied in the last chapter (see Definition 5.31)).

Here is a summary of the chapter. For a linearly reductive group G acting on an affine variety $X = \mathrm{Spm}\, R$ and a choice of character $\chi \in \mathrm{Hom}(G, \mathbb{G}_{\mathrm{m}})$, we construct in Section 6.1 the *Proj quotient map* $\Phi_\chi : X -- \to X /\!/_\chi G :=$ Proj $\bigoplus_{m \in \mathbb{Z}} R^G_{\chi^m}$, which is a rational map defined on the open set X^{ss} of semistable points with respect to χ. This improves on the affine quotient $X /\!/ G$ (which is the case of the trivial character $\chi = 1$) and is obtained by gluing together affine quotient maps of covering open sets. A classical example is the moduli space Proj $R_{n,d}$ of semistable hypersurfaces in Chapter 5, where $R_{n,d}$ coincides with $\bigoplus_{m \geq 0} R^G_{\chi^m}$ with $G = GL(n+1)$, $\chi = \det$ and $R = k[\mathbb{V}_{n,d}]$.

In Section 6.2 we briefly discuss a generalisation $X /\!/_M G := \mathrm{Proj}\, S(M)^G$ in which M is an invertible GR-module and $S(M)$ is its symmetric tensor algebra. The quotient $X /\!/_\chi G$ is the case $M = R$ with G-action via $\chi : G \to \mathbb{G}_{\mathrm{m}}$. This generalisation allows one to answer fully the 'Italian problem' (for locally factorial X) of constructing a *birational quotient*: that is, there exists some M for which $X /\!/_M G$ has function field equal to $k(X)^G$.

When the character $\chi \in \mathrm{Hom}(G, \mathbb{G}_{\mathrm{m}})$ moves, the Proj quotient $X /\!/_\chi G$ undergoes a birational transformation. A flop is a special case of this. We examine in Section 6.3 some examples of such moving quotients of torus actions on affine space, which have natural descriptions as toric varieties. In complex geometry the different quotients are parametrised by the symplectic *moment map*.

6.1 Extending the idea of a quotient: from values to ratios

Throughout this section we work with an affine variety $X = \mathrm{Spm}\, R$. Recall that the quotient constructed in the last chapter followed the first of the three approaches introduced in the discussion following Example 5.1:

(i) *Identify to a single point orbits which the ring of invariants fails to separate.*

When an algebraic group G acts on the affine variety X, by definition it acts on its coordinate ring R and one considers the affine quotient map

$$\Phi : \operatorname{Spm} R \to \operatorname{Spm} R^G.$$

What we showed was that when G is linearly reductive this map expresses $\operatorname{Spm} R^G$ as a parameter space for the closure-equivalence classes of G-orbits in X (Theorem 5.9). Moreover, as we will see in Chapter 11 (see Example 11.8), $\operatorname{Spm} R^G$ is even a categorical quotient. Nevertheless, the quotient problem for affine varieties does not end here. We still need one key idea which concerns the invariant rational functions on X.

In ring-theoretic language, the action $G \curvearrowright X$ is written as a *coaction* (see Definition 3.54)

$$\mu_X : R \to R \otimes_k k[G]. \tag{6.1}$$

To say that a function $f \in R$ is G-invariant then means that $\mu_X(f) = f \otimes 1$.

Definition 6.1. We shall denote the field of fractions of R by $Q(R)$ or, alternatively, by $k(X)$. An element $a/b \in Q(R)$ is called *G-invariant* if it satisfies:

$$(a \otimes 1)\mu_X(b) = (b \otimes 1)\mu_X(a).$$

$\square$

Note that G-invariance of an element does not depend on how it is represented. Moreover, it is a property closed under addition, multiplication and division, and the set of invariant elements is therefore a subfield of $Q(R)$. This is called the *invariant field* and written $Q(R)^G$. Or, denoted $k(X)^G$, it is called the *invariant function field* of X under the group action.

The Italian problem: *If an algebraic variety X has a quotient under the action of G, does this quotient have function field equal to $k(X)^G$?*

We call this question 'Italian' in honour of the Italian school of algebraic geometry that left for posterity so much work on birational geometry. (See also the preface to the first edition of Mumford et al [30].) A variety satisfying this requirement is called a *birational quotient*. Stability (Definition 5.12) gives one solution to the problem of finding birational quotients.

Proposition 6.2. *Suppose that a linearly reductive algebraic group G acts on an affine variety X, and that there exists a stable point for the action. Then* *every invariant rational function can be expressed as a ratio of invariant regular functions. In other words, $k(X)^G$ coincides with the field of fractions of $k[X]^G$.*

Proof. The set of stable points is a (by hypothesis, nonempty) open set $X^s \subset X$. Let h/f, for $f, h \in R$, be an invariant rational function on $X = \mathrm{Spm}\, R$; we shall write $R_f = R[1/f]$ and $(R^G)' = R^G[h/f]$ (see Section 8.2(a) in the next chapter). Then, corresponding to the inclusions

$$R_f \supset (R^G)' \supset R^G,$$

we obtain a sequence of dominant morphisms (see Definition 3.27)

$$X \supset \{f(x) \neq 0\} \to Y' := \mathrm{Spm}\,(R^G)' \to Y := \mathrm{Spm}\, R^G.$$

By Theorem 3.28, the image of a dominant morphism contains a nonempty open set. Moreover, G-invariance of h/f implies that every orbit collapses to a single point under the map $\{f(x) \neq 0\} \to Y'$. We claim that this forces the element h/f to be algebraic over the field $Q(R^G)$. Suppose it were transcendental. In this case Y' would have dimension strictly greater than that of Y. But this contradicts the fact (see Corollary 5.17) that the fibres of the (nontrivial) quotient map $X^s \to Y$ are single orbits.

So h/f satisfies some irreducible polynomial equation over $Q(R^G)$, of degree n, say. This means that over some open set the morphism $Y' \to Y$ is n-to-one. By the same reasoning as above (that is, Corollary 5.17) we must have $n = 1$. Hence $h/f \in Q(R^G)$.   □

It follows from the proposition that, if $\mathrm{Spm}\, R$ contains any stable points for the action of G, then $\mathrm{Spm}\, R^G$ satifies the Italian condition. However, there are many examples in which this is not the case.

Example 6.3. Let the multiplicative group $\mathbb{G}_m$ act on affine space $\mathbb{A}^{n+1}$ by rescaling coordinates:

$$(x_0, x_1, \ldots, x_n) \mapsto (tx_0, tx_1, \ldots, tx_n), \qquad t \in \mathbb{G}_m.$$

Expressed ring-theoretically, $\mathbb{G}_m$ acts on the polynomial ring $k[x_0, x_1, \ldots, x_n]$ by simultaneously multiplying every variable x_i by $t \in \mathbb{G}_m$. The only invariants in $k[x_0, x_1, \ldots, x_n]$ are the constants, so the affine quotient map is the trivial morphism

$$\Phi : \mathbb{A}^{n+1} \to \mathrm{Spm}\, k,$$

collapsing the whole space to a single point.

Geometrically, one might find it unreasonable that the quotient of an $n + 1$-dimensional variety by a 1-dimensional group should be a single point. However, the orbits of the action $\mathbb{G}_m \curvearrowright \mathbb{A}^{n+1}$ are of two kinds:

(i) lines through the origin, and
(ii) the origin itself.

Of these, only the second is a closed orbit. Thus all orbits are closure-equivalent, so this example does not contradict Theorem 5.9.

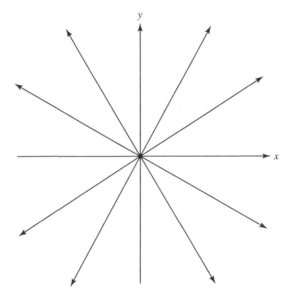

Figure 6.1: The orbits in Example 6.3 ($n = 1$)

On the other hand, we can observe that although there are no (nonconstant) invariants, there are plenty of invariant rational functions. These are all quotients $f(x)/g(x)$, where $f, g \in k[x_0, x_1, \ldots, x_n]$ are homogeneous polynomials of the same degree $\deg f = \deg g$. Equivalently, they are all the rational functions of the ratios $x_1/x_0, \ldots, x_n/x_0$, and so the invariant function field is

$$k(x_0, x_1, \ldots, x_n)^{\mathbb{G}_m} = k\left(\frac{x_1}{x_0}, \ldots, \frac{x_n}{x_0}\right).$$

This has dimension (that is, transcendence degree over k) equal to $n = \dim \mathbb{A}^{n+1} - \dim \mathbb{G}_m$. As the reader can guess, the 'correct' quotient satisfying Italian condition in this example is projective space $\mathbb{P}^n$, which parametrises all of the orbits away from the origin. We shall justify this in what follows.   □

(a) The projective spectrum

We are going to construct a new quotient of the form Proj R (see Section 3.2(b)). First, let us consider again the construction of projective space $\mathbb{P}^n$ by gluing together affine spaces in Example 3.40. It could be defined alternatively by the following four steps.

(i) As a set, $\mathbb{P}^n$ consists of all ratios $(a_0 : a_1 : \ldots : a_n)$, or, in other words,

$$\mathbb{P}^n = (k^{n+1} - \{0\})/k^\times.$$

(ii) As a topological space, $\mathbb{P}^n$ is given the Zariski topology. That is, we take as a basis of open sets the complements $U_f \subset \mathbb{P}^n$ of zero-sets of homogeneous polynomials $f(x_0, x_1, \ldots, x_n)$,

$$U_f = \{(a_0 : a_1 : \ldots : a_n) \mid f(a_0 : a_1 : \ldots : a_n) \neq 0\}.$$

(iii) As a variety, $\mathbb{P}^n$ has algebraic function field (see Definition 3.30)

$$K_0 = k\left(\frac{x_1}{x_0}, \ldots, \frac{x_n}{x_0}\right).$$

(iv) The structure sheaf $\mathcal{O}_{\mathbb{P}^n}$ is the elementary sheaf of subrings of K_0 (see Definition 3.2) given on the basic open sets by

$$\mathcal{O}_{\mathbb{P}^n} : U_f \mapsto \left\{ \frac{g}{f^m} \;\middle|\; f, g \text{ are homogeneous polynomials with} \right.$$

$$\left. \deg g = m \deg f, m \geq 0 \right\}.$$

In fact, what we have described here is nothing other than the projective spectrum Proj R (Definition 3.43) of a graded integral domain over k,

$$R = \bigoplus_{m \geq 0} R_{(m)}.$$

Let us consider those homogeneous ideals of R which are maximal among homogeneous ideals, called *maximal homogeneous ideals*. Of these there are two kinds:

(H0) Sums $J + R_+$, where

$$R_+ = \bigoplus_{m > 0} R_{(m)}$$

is the 'irrelevant ideal' and J is a maximal ideal in $R_{(0)}$.

(H1) Those not containing the irrelevant ideal R_+.

Maximal homogeneous ideals of type (H0) are actually maximal ideals of R, but those of type (H1) are not. In case (H1), the residue ring is isomorphic to a polynomial ring in one variable, in fact.

By definition, a homogeneous ideal is just an ideal which is invariant under the natural action of $\mathbb{G}_m$ on R. Consequently, maximal homogeneous ideals correspond geometrically to $\mathbb{G}_m$-invariant closed sets in Spm R which are minimal among such sets. Those of type (H0) are fixed points under the action $\mathbb{G}_m \curvearrowright$ Spm R, while in case (H1) they are the closures of 1-dimensional orbits.

Given a homogeneous element $a \in R$ we can define a ring

$$R_{a,0} = \left\{ \frac{b}{a^n} \mid a, b \text{ are homogeneous polynomials with} \right.$$

$$\left. \deg b = m \deg a, m \geq 0 \right\}.$$

(See Definition 3.41.) Given also a maximal homogeneous ideal $\mathfrak{m} \subset R$ not containing a we obtain a maximal ideal

$$\left\{ \frac{b}{a^n} \mid b \in \mathfrak{m} \right\} \subset R_{a,0}.$$

Conversely, any maximal ideal in $R_{a,0}$ determines, by the set of numerators of its elements, a maximal ideal in R. In this way we arrive at the following description of Proj R.

Proposition 6.4. *The ringed space given by the following four properties is an algebraic variety, and is isomorphic to the variety* Proj R *of Definition 3.43 constructed by gluing affine varieties.*

  (i) Set: The underlying set is that of maximal homogeneous ideals in R of type (H1).
 (ii) Topology: The set of these ideals is equipped with the Zariski topology. That is, a basis of open sets consists of

$$U_a = \{ \mathfrak{m} \mid a \notin \mathfrak{m} \}$$

  for homogeneous elements $a \in R$.
(iii) Function field: This is the field of ratios of homogeneous elements of R of equal degree (together with zero),

$$K_0 = \left\{ \frac{a}{b} \mid a, b \in R, \quad \deg a = \deg b \right\} \cup \{0\}.$$

(iv) *Structure sheaf:* $\mathcal{O}$ is the sheaf of subalgebras of K_0 defined on basic open sets by

$$\mathcal{O}(U_a) = R_{a,0}.$$

$\square$

One should note that Proj R contains less information than the graded ring R. To see this, let j be a natural number and consider the grade subring generated in degrees divisible by j,

$$R^{[j]} := \bigoplus_{n \geq 0} R_{(nj)}.$$

Replacing R by $R^{[j]}$ does not change the collection of coordinate rings $R_{a,0} \cong (R^{[j]})_{a^j,0}$ and the gluing data they determine, and it follows that restriction of maximal homogeneous ideals defines an isomorphism of algebraic varieties

$$\text{Proj } R \xrightarrow{\sim} \text{Proj } R^{[j]}, \qquad \mathfrak{m} \mapsto \mathfrak{m} \cap R^{[j]}. \tag{6.2}$$

Next, note that for each maximal homogeneous ideal $\mathfrak{m} \subset R$ the restriction to elements of degree zero $\mathfrak{m} \cap R_{(0)}$ is a maximal ideal of $R_{(0)}$, and so there is a mapping

$$\phi : \text{Proj } R \to \text{Spm } R_{(0)}, \qquad \mathfrak{m} \mapsto \mathfrak{m} \cap R_{(0)}. \tag{6.3}$$

To see that this is a morphism of algebraic varieties, consider its restriction to the affine varieties Spm $R_{a,0}$ from which Proj R is obtained by gluing. Each $R_{a,0}$ contains $R_{(0)}$ as a subring and so has an induced morphism of spectra Spm $R_{a,0} \to$ Spm $R_{(0)}$. This is nothing other than the restriction of (6.3).

The map ϕ is called the *structure morphism* of Proj R. The following example is trivial but will be needed later.

Example 6.5. Let $R_{(0)}$ be a finitely generated algebra over k. The polynomial ring in one variable $R_{(0)}[u]$ is a graded ring by assigning $\deg R_{(0)} = 0, \deg u = 1$. In this case the structure morphism

$$\text{Proj } R_{(0)}[u] \xrightarrow{\sim} \text{Spm } R_{(0)}$$

is an isomorphism of varieties. $\square$

The degree zero component $R_{(0)}$, as well as being a subring of R, is also the residue ring modulo the irrelevant ideal R_+. Corresponding to the projection $R \to R_{(0)}$, then, there is a closed subvariety $F \subset$ Spm R which is the image of

Spm $R_{(0)}$. As a set this consists of the maximal (homogeneous) ideals of type (H0) above; and in fact the map Spm $R_{(0)} \to F \subset \text{Proj } R$ is an isomorphism whose inverse is the restriction of the structure morphism. As already noted, if we let $\mathbb{G}_m$ act on Spm R via its graded action on R, F is just the set of fixed points. Maximal homogeneous ideals of type (H1) are in one-to-one correspondence with the $\mathbb{G}_m$ orbits in Spm $R - F$. In other words, as a set:

$$\text{Proj } R = (\text{Spm } R - F)/k^{\times}.$$

Remark 6.6. We need to make a technical remark which will be needed later. A graded ring R need not be an integral domain for Proj R to be defined as an algebraic variety. More generally, let R_S be the localisation of R by the multiplicative set $S \subset R$ of nonzero homogeneous elements of positive degree. (For the definition of localisation, see Section 8.2.) Assume, first, that R_S is an integral domain or, equivalently, that R_h is an integral domain for every $h \in S$. Then the field of fractions of R_h does not depend on $h \in S$. Gluing their spectra Spm R_h, we obtain an algebraic variety which we also denote by Proj R. In fact, this is isomorphic to Proj $\overline{R}$, where $\overline{R}$ is the image of the natural homomorphism $R \to R_S$ and is an integral domain.

Our assumption that R_S is an integral domain is not essential here. Rather, it is enough that R_S is *locally integral*. This is satisfied, in particular, if Spm R is smooth away from the fixed-point set $F \subset \text{Spm } R$. Then Proj R can be defined, similarly, as a disjoint union of algebraic varieties.   □

(b) The Proj quotient

Suppose that our affine variety $X = \text{Spm } R$ is acted upon by the group $\mathbb{G}_m$. By Proposition 4.7, the ring R has a direct sum decomposition

$$R = \bigoplus_{m \in \mathbb{Z}} R_{(m)} \tag{6.4}$$

in which $R_{(m)}$ is the summand of weight m for the action. Note that the invariant ring $R^{\mathbb{G}_m}$ is the same as $R_{(0)}$. Note also that the group action preserves the algebra structure of R, and so R becomes a graded ring via this decomposition.

Definition 6.7. The action $\mathbb{G}_m \curvearrowright X$ is said to be of *ray type* if, in the decomposition (6.4), either $R_{(m)} = 0$ for all $m < 0$ or $R_{(m)} = 0$ for all $m > 0$. By exchanging t with t^{-1} in $\mathbb{G}_m = \text{Spm } k[t, t^{-1}]$, if necessary, it is enough to assume that $R_{(m)} = 0$ for all $m < 0$, so that

$$R = \bigoplus_{m \geq 0} R_{(m)}.$$

$\square$

As we have seen in the previous section, the closed set $F \subset X$ determined by the irrelevant ideal in R is the fixed point set for the action of $\mathbb{G}_m$, and Proj R can be viewed as the quotient by $\mathbb{G}_m$ of the complement $X - F$. Moreover, if we choose a nonzero homogeneous element $a \in R_+$, then $R_{a,0} = (R_a)_{(0)}$ and there is a commutative diagram, where Spm R_a is the basic open set $D(a)$ (see Chapter 3, (3.6)):

$$
\begin{array}{ccccc}
\text{Spm } R_a = & D(a) & \hookrightarrow & X - F & \subset X \\
& \downarrow & & \downarrow & \\
\text{Spm } R_{a,0} = & D(a)/\!/\mathbb{G}_m & \hookrightarrow & \text{Proj } R &
\end{array}
\tag{6.5}
$$

The left-hand vertical map is the affine quotient map for the action $\mathbb{G}_m \curvearrowright D(a) \subset X$. Since $X - F$ is covered by open sets of the form $D(a)$, this shows that the right-hand vertical map $X - F \to \text{Proj } R$ is a morphism of algebraic varieties and is locally an affine quotient map. This suggests the following terminology.

Definition 6.8. If $\mathbb{G}_m \curvearrowright X = \text{Spm } R$ is an action of ray type, then the projective spectrum Proj $R =: X /\!/\mathbb{G}_m$ is called the *Proj quotient* of X by $\mathbb{G}_m$. $\square$

Denoting by Φ the affine quotient map and by ϕ the structure morphism (6.3), the following diagram commutes:

$$
\begin{array}{ccc}
X - F & \hookrightarrow & X \\
\downarrow & & \downarrow \Phi \\
\text{Proj } R & \xrightarrow{\phi} & \text{Spm } R_{(0)} = \text{Spm } R^{\mathbb{G}_m}
\end{array}
\tag{6.6}
$$

Thus the Proj quotient is precisely (the base change of) an affine quotient away from the fixed point set F.

Example 6.9. When $\mathbb{G}_m$ acts on $\mathbb{A}^{n+1}$ by $x \mapsto tx$, $t \in \mathbb{G}_m$, the action on the polynomial ring $R = k[x_0, x_1, \ldots, x_n]$ is ray type, while $F = \{0\}$ and diagram (6.6) is:

$$\begin{array}{ccc} \mathbb{A}^{n+1} - \{0\} & \hookrightarrow & \mathbb{A}^{n+1} \\ \downarrow & & \downarrow \Phi \\ \mathbb{P}^n & \xrightarrow{\phi} & \mathrm{Spm}\, k = \text{point} \end{array}$$

□

Let us now consider an action on $X = \mathrm{Spm}\, R$ of a general algebraic group G. In many cases where the invariant regular functions and the invariant rational functions disagree the source of the disagreement, as in Example 6.3, is the multiplicative group $\mathbb{G}_m$. (Though see also Example 6.21 in the next section.) What we are going to do next, using the preceding discussion, is to build a projective quotient $X /\!/_\chi G$ associated to each 1-dimensional representation

$$\chi : G \to \mathbb{G}_m = \mathrm{Spm}\, k[t, t^{-1}].$$

Consider the function t pulled back to G via χ: we shall use the same symbol χ to denote this function. For the coproduct μ_G (Definition 4.11) this function satisfies $\mu_G(\chi) = \chi \otimes \chi$.

Definition 6.10. A function $f \in R$ on $X = \mathrm{Spm}\, R$ satisfying

$$\mu_X(f) = f \otimes \chi,$$

where μ_X is the coaction (6.1), is called a *semiinvariant of weight* χ for the G action. (See Definition 4.13.) This condition can be written set-theoretically as $f(g \cdot x) = \chi(g) f(x)$ for all $g \in G$, $x \in X$. □

Example 6.11. Let $G = GL(n)$ and $w \in \mathbb{Z}$ be an integer. Then the determinantal power $\chi = \det^w$ is a 1-dimensional character of G, and by Lemma 4.12 every character is of this form. A semiinvariant $f \in R$ with respect to $\chi = \det^w$ is called a *classical semiinvariant of weight w*. Set-theoretically this means that $f(g \cdot x) = (\det g)^w f(x)$ for all $g \in GL(n)$, $x \in X$. □

Obviously the set of semiinvariants of a given weight χ is a vector subspace of R, and we denote this space by R_χ^G. A product of semiinvariants of weights χ, χ' is again a semiinvariant of weight $\chi \chi'$. In particular, this means that the direct sum

$$\bigoplus_{m \in \mathbb{Z}} R^G_{\chi^m} \tag{6.7}$$

has the structure of a graded ring.

Definition 6.12. The action of G on $X = \text{Spm } R$ is of *ray type* with respect to $\chi \in \text{Hom}(G, \mathbb{G}_m)$ if either $R^G_{\chi^m} = 0$ for all $m < 0$ or for all $m > 0$. If $\text{Hom}(G, \mathbb{G}_m) \cong \mathbb{Z}$, then this definition is independent of the choice of χ and we will just say that the action $G \curvearrowright X$ is of ray type.   □

Just as for $\mathbb{G}_m$-actions, it is enough to assume that $R^G_{\chi^m} = 0$ for all $m < 0$, and then the ring of semiinvariants is

$$\bigoplus_{m \geq 0} R^G_{\chi^m}. \tag{6.8}$$

Notice that because of the isomorphism (6.2) the projective spectrum of this graded ring depends only on the ray $\mathbb{R}_+$ spanned by χ in the real vector space $\text{Hom}(G, \mathbb{G}_m) \otimes_{\mathbb{Z}} \mathbb{R}$. (The set of characters $\text{Hom}(G, \mathbb{G}_m)$ is a finitely generated free abelian group.)

Definition 6.13. Let $G \curvearrowright X = \text{Spm } R$ be any action and $\chi \in \text{Hom}(G, \mathbb{G}_m)$ a character.

(i) The projective spectrum

$$X /\!/_\chi G := \text{Proj} \bigoplus_{m \geq 0} R^G_{\chi^m}$$

of the graded ring (6.8) is called the *Proj quotient in direction χ* of the action $G \curvearrowright X$.

(ii) A point $x \in X$ satisfying $f(x) \neq 0$ for some semiinvariant $f \in R$ with weight equal to some positive power χ^n, $n > 0$, is said to be *semistable with respect to χ*; if no such f exists, then $x \in X$ is called *unstable*. The set of points semistable with respect to χ is an open set which we denote by $X^{ss}_\chi \subset X$.   □

How do we know that the ring of semiinvariants (6.8) is finitely generated? By Example 6.5 we can identify $X = \text{Spm } R$ with $\text{Proj } R[u]$. We let G act on the graded ring $R[u]$ with a twist by χ^{-1}, that is:

$$g \cdot (f \otimes u^m) = (g \cdot f) \otimes \chi(g)^{-m} u^m, \quad \text{for } g \in G, \ f \in R \text{ and } m \geq 0.$$

Then the ring (6.8) is precisely the invariant ring under this action, and hence it follows from Hilbert's Theorem 4.51 that, if G is linearly reductive, then the ring (6.8) is finitely generated. From the inclusion homomorphism

$$\bigoplus_{m \geq 0} R^G_{\chi^m} \hookrightarrow R[u]$$

we obtain a rational map

$$X - - \to X /\!/_\chi G.$$

In concrete terms this is given by $x \mapsto (f_0(x) : f_1(x) : \ldots : f_n(x)) \in \mathbb{P}(a_0 : a_1 : \ldots : a_n)$, where $f_0, \ldots, f_n \in \bigoplus_{m \geq 0} R^G_{\chi^m}$ are generating semiinvariants of degrees $a_0, \ldots, a_n$. The rational map is therefore defined on the open set X^{ss}_χ, and the morphism

$$\Phi_\chi : X^{ss}_\chi \to X /\!/_\chi G \tag{6.9}$$

is called the *Proj quotient map in direction χ*.

Remarks 6.14.

(i) First of all, suppose that $\chi = 1$ is the trivial character. The constant function $f = 1 \in R$ is a (semi)invariant satisfying $f(x) \neq 0$ for every $x \in X$, so all points of X are semistable. On the other hand, for every $m \geq 0$ the space $R^G_{\chi^m}$ is nothing but the ring of G-invariants, and so the graded ring (6.8) reduces to the polynomial ring $R^G[u]$. Thus by Example 6.5 its Proj is isomorphic to Spm R^G, and so the Proj quotient coincides with the affine quotient. *In this sense the Proj quotient extends the idea of the affine quotient.*

(ii) When the character χ is nontrivial, the semiinvariant ring (6.8) is the same as the ring of invariants in R under the action of the kernel

$$G_\chi := \ker \{G \xrightarrow{\chi} \mathbb{G}_m\}.$$

More precisely, $R^{G_\chi} = \bigoplus (R^{G_\chi})_{(m)}$, where the grading is the weight-space decomposition under the action of $G/G_\chi \cong \mathbb{G}_m$, and then

$$(R^{G_\chi})_{(m)} = R^G_{\chi^m}.$$

Thus the Proj quotient $X /\!/_\chi G$ is obtained by taking the affine quotient of X by G_χ, but replacing Spm by Proj.

(iii) Generalising (6.6) there is a commutative diagram:

$$X^{ss}_\chi \quad \hookrightarrow \quad X$$

$$\downarrow \Phi_\chi \qquad\qquad \downarrow \Phi$$

$$\text{Proj } R^{G_\chi} = \text{Proj} \left(\bigoplus_{m \geq 0} R^G_{\chi^m} \right) = X /\!\!/_\chi G \xrightarrow{\ \phi\ } X /\!\!/ G = \text{Spm } R^G$$

$$(6.10)$$

Thus on the open set $X^{ss}_\chi \subset X$ the Proj quotient is obtained by base change from the affine quotient.

(iv) If $f \in R$ is a semiinvariant in direction χ, then the following diagram commutes:

$$G \curvearrowright \quad D(f) \quad \hookrightarrow \quad X^{ss}_\chi \quad \subset X$$

$$\downarrow \qquad\qquad \downarrow \qquad\qquad (6.11)$$

$$\text{Spm } (R[1/f])^G \quad = \quad D(f) /\!\!/ G \quad \hookrightarrow \quad X /\!\!/_\chi G$$

Here the left-hand vertical map is the affine quotient and the right-hand vertical map is the Proj quotient. In this way one sees that the Proj quotient map is an affine quotient map locally and indeed is obtained by gluing such maps. *Hence $X /\!\!/_\chi G$ is a moduli space for closure-equivalence classes of G-orbits in the semistable set X^{ss}_χ.* (But note that this is not the same as closure-equivalence in X.)

(v) The motivating example for all of this is the moduli space of hypersurfaces of degree d in $\mathbb{P}^n$ discussed in the last chapter. Here $X = \mathbb{V}_{n,d}$, the affine space of forms of degree d acted on by $G = GL(n+1)$, and we use the character $\chi = \det$, for which $G_\chi = SL(n+1)$. The semiinvariant ring is none other than $R_{n,d} = k[\mathbb{V}_{n,d}]^{SL(n+1)}$ and the moduli space is Proj $R_{n,d} = \mathbb{V}_{n,d} /\!\!/_{\det} GL(n+1)$.

(vi) Finally, note that, in view of Remark 6.6, under suitable conditions we can define the Proj quotient $X /\!\!/_\chi G$ as a disjoint union of algebraic varieties even if the semiinvariant ring is not an integral domain. In particular, it is enough that the semistable set X^{ss}_χ is smooth.

Remark 6.14(iv) allows us to generalise Proposition 6.2. First we need to define stability with respect to a character χ; note that for the trivial character $\chi = 1$ the following definition agrees with Definition 5.12 of the previous chapter.

Definition 6.15. If $x \in X$ is a semistable point with respect to character χ, then x is *stable with respect to* χ if the orbit $G_\chi \cdot x \subset X$, where $G_\chi = \ker \chi$, is a closed set and the stabiliser subgroup $\{g \in G \mid g \cdot x = x\}$ is finite.   □

Proposition 6.16. *Suppose that a linearly reductive algebraic group acts on an affine variety X and that X contains stable points with respect to a character χ of G. Then every invariant rational function can be expressed as a ratio of semiinvariants of weight χ. In particular, the algebraic function field of $X /\!/_\chi G$ is $k(X)^G$.*   □

The quotient $X /\!/_\chi G$ therefore satisfies the Italian condition. In addition, it follows from diagram (6.11) that all of the results of Section 5.1(c) can be extended unchanged to the present situation, and in particular we obtain:

Proposition 6.17. *If all χ-semistable point are stable, $X_\chi^{ss} = X_\chi^s$, then the fibres of the map $X_\chi^s \rightarrow X /\!/_\chi G$ are closed orbits.*   □

In this last situation we denote the quotient by $X /_\chi G$, often called the *geometric quotient*. (See Mumford et al [30] Chapter 4.)

(c) The Proj quotient by a $GL(n)$-action of ray type

The classical case of the above construction arises when G is the general linear group $GL(n)$ and the character χ is the determinant

$$\det : GL(n) \rightarrow \mathbb{G}_m.$$

Indeed, the motivating example of the previous chapter was exactly of this form (Remark 6.14(ii)). Although we do not have anything new to add for this case, it will be used so often in what follows that it is worth restating the important points in this section.

First of all we take the ring of invariants of the kernel $SL(n)$ of det. This ring $R^{SL(n)}$ is acted upon by $\mathbb{G}_m$, identified with the quotient group $GL(n)/SL(n)$. The ring of invariants is graded by this action:

$$R^{SL(n)} = \bigoplus_{w \in \mathbb{Z}} R_{(w)}^{SL(n)}.$$

We note that $R_{(w)}^{SL(n)} = R_{(\det)^w}^{GL(n)}$, the semiinvariants of weight w for the action of $GL(n)$ (Example 6.11).

The action of $GL(n)$ on an affine variety $X = \mathrm{Spm}\, R$ is of ray type if and only if the induced action of $\mathbb{G}_m$ on $R^{SL(n)}$ is of ray type. Just as in Definition 6.7, it is enough to assume (by exchanging $g \leftrightarrow^t g^{-1}$ in the $GL(n)$ action if necessary) that $R^{SL(n)}_{(w)} = 0$ for all $w < 0$. In this case Proj $R^{SL(n)}$ is the Proj quotient of the action $GL(n) \curvearrowright X$.

Thus the Proj quotient $X /\!/ GL(n)$ is obtained by taking the affine quotient $X /\!/ SL(n)$ but replacing the affine spectrum with the projective spectrum, and it parametrises closure-equivalence classes of $GL(n)$-orbits in X^{ss}. (Note that it does *not* parametrise closure-equivalence classes in X, as the baby example 6.9 shows!) Diagram (6.10) now looks like:

$$X^{ss} \hookrightarrow X$$

$$\downarrow \qquad\qquad \downarrow \Phi$$

$$\text{Proj } R^{SL(n)} \xrightarrow{\phi} \text{Spm } R^{GL(n)}$$

Example 6.18. Binary forms revisited. $GL(2)$ acts, by transformation of binary forms $(\xi \wr x, y)$ of degree d, on the $d + 1$-dimensional affine space

$$\mathbb{V}_d = \{(\xi \wr x, y)\} = \text{Spm } k[\xi_0, \xi_1, \dots, \xi_d].$$

The graded ring of semiinvariants of $GL(2)$ is the same as the ring of invariants of $SL(2)$. More precisely, given a polynomial $F(\xi)$,

$$\begin{pmatrix} F(\xi) \text{ is } SL(2) \text{ invariant and} \\ \text{homogeneous of degree } m \end{pmatrix} \iff \begin{pmatrix} F(\xi) \text{ is } GL(2) \text{ semiinvariant} \\ \text{of weight } w = \frac{dm}{2} \end{pmatrix}.$$

In the case $d = 4$, the semiinvariant ring is $k[g_2, g_3]$ generated by $g_2(\xi)$ of weight 4 and $g_3(\xi)$ of weight 6, and the Proj quotient $\mathbb{V}_4 /\!/ GL(2)$ is isomorphic to $\mathbb{P}^1$ (Proposition 1.25 and Corollary 4.70).   □

A point $x \in X$ is semistable with respect to an action $GL(n) \curvearrowright X$ of ray type if there exists a semiinvariant $f \in R = k[X]$ of positive weight for which $f(x) \neq 0$. A semistable point $x \in X$ is stable if the orbit $SL(n) \cdot x \subset X$ is closed and the stabiliser subgroup is finite, and on the stable set $X^s \subset X^{ss}$ the Proj quotient is a moduli space for $GL(n)$-orbits.

This is a generalisation of the definition of a semistable form (Definition 5.31). Let $F \subset X$ be the fixed point set under the action of the scalar matrices $\mathbb{G}_m \subset GL(n)$; this will be called the *irrelevant set*. The following is then essentially Theorem 5.32.

Proposition 6.19. *The following conditions on a point $x \in X$ are equivalent:*

(i) $x \in X$ is semistable;

(ii) the closure of the orbit $SL(n) \cdot x$ does not intersect the irrelevant set $F \subset X$.

Proof. (i) $\Longrightarrow$ (ii) By semistability there exists a semiinvariant f of positive weight satisfying $f(x) \neq 0$. This f is therefore nonzero on the orbit $SL(n) \cdot x$. On the other hand, f is identically zero on F; so (ii) follows.

(ii) $\Longrightarrow$ (i) By Theorem 5.3, x and F are separated by some invariant: that is, there exists $f \in R^{SL(n)}$, $R = k[X]$, for which $f(x) \neq 0$ and $f|_F \equiv 0$. But vanishing on F implies that $f \in R_+^{SL(n)}$, and hence by taking homogeneous components of f we obtain (i). $\qquad\square$

In almost all applications one has $R_{(0)} = k$, and in this case the Proj quotient is a complete variety. (In general, it is *proper* over the affine quotient Spm R^G.) Here the fixed point set F is a single point, which we shall denote by $O \in X$.

Corollary 6.20. *When $R_{(0)} = k$, $x \in X$ is semistable for the action of $GL(n)$ if and only if $O \notin \overline{SL(n) \cdot x}$.* $\qquad\square$

In the moduli construction for hypersurfaces in $\mathbb{P}^n$ this was Proposition 5.32.

6.2 Linearisation and Proj quotients

Apart from Definition 6.23, most of this section will not be used afterwards in this book, but nonetheless we include it for completeness and for the sake of clarity. Beginners are invited to skip it. Taking Example 6.3 as point of departure, and also motivated by the case of projective hypersurfaces in Chapter 5, we have developed and extended the notion of quotient variety. However, as the next example shows, even in the absence of $\mathbb{G}_m$ it can happen that there are not enough invariants to make the theory work.

Example 6.21. First consider the quadric surface in $Y \subset \mathbb{A}^3$ with equation

$$AC - B^2 + \tfrac{1}{4} = 0.$$

If we identify Y with the set of symmetric 2×2 matrices with determinant $-1/4$, then it is acted upon by $SL(2)$ in the usual way:

$$\begin{pmatrix} A & B \\ B & C \end{pmatrix} \mapsto P \begin{pmatrix} A & B \\ B & C \end{pmatrix}^t P, \qquad P \in SL(2).$$

Note that the stabiliser subgroups are the conjugates of $T \subset SL(2)$ consisting of diagonal matrices $\begin{pmatrix} q & 0 \\ 0 & q^{-1} \end{pmatrix}$. Thus Y is the quotient variety $SL(2)/T$. (See Section 4.5.)

Next, consider the set L of matrices of the form

$$\begin{pmatrix} x & A & B - \frac{1}{2} \\ z & B + \frac{1}{2} & C \end{pmatrix}$$

and of rank 1. Then L is a 3-dimensional closed subvariety in $\mathbb{A}^5$. The map forgetting x, z,

$$L \to Y, \qquad (A, B, C, x, z) \mapsto (A, B, C),$$

is surjective and its fibre is a 1-dimensional vector space with

$$\begin{pmatrix} 0 & A & B - \frac{1}{2} \\ 0 & B + \frac{1}{2} & C \end{pmatrix} \tag{6.12}$$

as its origin. In other words, L is a line bundle over Y (in the geometric sense, rather than the algebraic sense of the next chapter). Moreover, the group $SL(2)$ also acts on L:

$$\begin{pmatrix} x & A & B - \frac{1}{2} \\ z & B + \frac{1}{2} & C \end{pmatrix} \mapsto \begin{pmatrix} ax + bz & A' & B' - \frac{1}{2} \\ cx + dz & B' + \frac{1}{2} & C' \end{pmatrix},$$

where

$$\begin{pmatrix} A' & B' \\ B' & C' \end{pmatrix} = P \begin{pmatrix} A & B \\ B & C \end{pmatrix}^t P, \qquad P = \begin{pmatrix} a & b \\ c & d \end{pmatrix} \in SL(2).$$

There are two orbits of this action: an open orbit isomorphic to $SL(2)$ and a closed orbit consisting of points (6.12) isomorphic to Y.

The variety we are interested in is the 4-dimensional fibre product $X :=$ $L \times_Y L$. In more concrete terms, this is the set of rank 1 matrices of the form

$$\begin{pmatrix} x_1 & x_2 & A & B - \frac{1}{2} \\ z_1 & z_2 & B + \frac{1}{2} & C \end{pmatrix}.$$

This is a closed subvariety of $\mathbb{A}^7$, and the map forgetting x_1, x_2, z_1, z_2 expresses X as a rank 2 vector bundle

$$X \to Y.$$

That is, each fibre is a 2-dimensional vector space.

Claim: Under the above action $SL(2) \curvearrowright X$, the only invariants are the constant functions.

For example, the element $\begin{pmatrix} q & 0 \\ 0 & q^{-1} \end{pmatrix} \in SL(2)$ maps

$$p = \begin{pmatrix} x_1 & x_2 & 1 & 0 \\ 0 & 0 & 0 & 0 \end{pmatrix} \mapsto \begin{pmatrix} qx_1 & qx_2 & 1 & 0 \\ 0 & 0 & 0 & 0 \end{pmatrix} \in X.$$

This has a limit as $q \to 0$, which is contained in the zero section of the vector bundle $X \to Y$. Similarly, at other points of X, we see that all orbits are closure-equivalent to points of the zero section. Thus, if f is an invariant, then by $SL(2)$-invariance and continuity its value at each point is equal to its value at a limiting point in the zero section. But the zero section is isomorphic to Y, on which $SL(2)$ acts transitively. Hence f is constant, proving the claim.

On the other hand, x_2/x_1 is an invariant rational function and generates the invariant function field $k(X)^G$. □

Even in cases of this sort, it is nevertheless possible to construct a birational quotient by using G-linearised invertible R-modules, which we define next.

Definition 6.22. Suppose as usual that G acts on an affine variety $X = \mathrm{Spm}\, R$. Then a GR-*module* is a representation of G which is also an R-module M for which the defining coaction

$$\mu_M : M \to M \otimes_k k[G]$$

is a homomorphism of R-modules, where R acts on $M \otimes_k k[G]$ via $\mu_R : R \to R \otimes_k k[G]$. In other words, $\mu_M(ar) = \mu_R(a)\mu_M(m)$ for $a \in R$, $m \in M$. □

Reversing the roles of G and R in this definition:

Definition 6.23. Given an R-module M, a G-*linearisation* of M is a coaction $\mu_M : M \to M \otimes_k k[G]$ making M into a GR-module. □

Suppose M is an invertible R-module (see Definition 8.60) and admits a G-linearisation. Then the tensor algebra of M, that is, the direct sum of tensor products

$$S(M) := \bigoplus_{m \geq 0} M^{\otimes m},$$

has the structure of a commutative algebra over R on which the group G acts. We can therefore consider its ring of invariants, which by Hilbert's Theorem is finitely generated. On the other hand, since M is locally isomorphic to R (by definition of an invertible R-module), $S(M)$ is locally isomorphic to the polynomial algebra $R[u]$. So it follows from Example 6.5 that the structure morphism Proj $S(M) \to$ Spm R is an isomorphism, and in the same manner as (6.9) we obtain a morphism

$$X_M^{ss} \to X /\!/_M G := \text{Proj } S(M)^G.$$

This is called the *Proj quotient* coming from the GR-module M.

Example 6.24. If χ is a 1-dimensional character of G, then the homomorphism

$$R \to R \otimes_k k[G], \qquad a \mapsto a \otimes \chi^{-1}$$

is a G-linearisation of R itself viewed as an R-module. Thus the pair (R, χ) defines a GR-module, and the corresponding Proj quotient is precisely $X /\!/_\chi G$, defined as the projective spectrum of the graded ring of χ-semiinvariants (Definition 6.13). □

Here is an example of an invertible GR-module which is not isomorphic to R.

Example 6.25. Let X be the 4-dimensional variety of Example 6.21, and consider the following subspace of the function field $k(X)$:

$$M = \{ f \in k(X) \mid x_1 f \in k[X], \ z_1 f \in k[X] \}.$$

The ring of invariants in $S(M)$ is generated by $1, x_2/x_1$. Moreover, the complement of the zero section in X is exactly the set of semistable points with respect to M. Thus the quotient is the projective line $\mathbb{P}^1$.

Note that $x_1 = z_1 = 0$ defines the codimension 1 subvariety $L \subset X$, and it is usual to denote the module M by $\mathcal{O}_X(-L)$. Since the rational function x_1/x_2 on X is $SL(2)$-invariant, so is its zero-set L, and hence M carries an $SL(2)$-linearisation. □

Examination of this example indicates another solution to the 'Italian problem' of constructing birational quotients.

Theorem 6.26. *Suppose that a linearly reductive group G acts on an affine variety X, where $k[X]$ is locally a unique factorisation domain (X is said to be locally factorial). Then there exists a GR-module M whose associated projective quotient $X /\!/_M G$ has a rational function field equal to the invariant function field $k(X)^G$.* □

Since we are not going to make any use of this result, we merely sketch the proof. Let $f_1, \ldots, f_N$ be generators of the field $k(X)^G$, and let $D \subset X$ be the sum of the polar divisors $(f_i)_\infty$. Then we let M be the set of rational functions with poles at most along D:

$$M = \{ f \in k(X) \mid (f) + D \geq 0 \}.$$

Local factoriality now implies that as a $k[X]$-module M is invertible. It is also a representation of G since D is G-invariant. Finally, the invariants of the tensor algebra $S(M)$ include the functions $1, f_1, \ldots, f_N$, and so $X /\!/_M G$ is a birational quotient.

6.3 Moving quotients

Taking the problem of invariant rational functions as our starting point, we have now generalised the idea of a quotient variety in the following way.

- First, we abandoned the hope of getting a quotient that classifies all orbits of the group action.
- After removing the unstable orbits, we considered the quotient problem restricted to as large as possible an open set.

In terms of the ideas suggested in Section 5.1(a), we have moved from approach (i) to (i)+(ii)+(iii). In this generalisation, if the group action is not of ray type, then we obtain different quotients depending on the choice of 1-dimensional character of G (or, more generally, the choice of a G-linearised invertible module). For example, an action of $\mathbb{G}_m$ which is not of ray type, corresponding to the identity, trivial and inverse characters $\mathbb{G}_m \to \mathbb{G}_m$ we obtain three quotients:

character		quotient	
identity	$t \mapsto t$	$X /\!/_+ \mathbb{G}_m$	$= \mathrm{Proj}\ R_+$
trivial	$t \mapsto 1$	$X /\!/_0 \mathbb{G}_m$	$= \mathrm{Proj}\ R_0$
inverse	$t \mapsto t^{-1}$	$X /\!/_- \mathbb{G}_m$	$= \mathrm{Proj}\ R_-$

(R_-, R_0 and R_+ denote the subspaces of the graded ring $R = \bigoplus_{m \in \mathbb{Z}} R_{(m)}$ determined by the $\mathbb{G}_m$-action with $m < 0$, $m = 0$ and $m > 0$, respectively.) We will denote these three quotients by X_-, X_0, X_+. As an example, we begin by considering a simple flop.

(a) Flops

Consider the affine space $\mathbb{A}^N$ on a finite-dimensional representation of the multiplicative group $\mathbb{G}_m$. By Proposition 4.41, the action of $\mathbb{G}_m$ can be diagonalised:

that is, one can find coordinates $x_1, \ldots, x_N$ with respect to which the action of $t \in \mathbb{G}_m$ is

$$(x_1, \ldots, x_N) \mapsto (t^{a_1} x_1, \ldots, t^{a_N} x_N)$$

for some fixed integers $a_1, \ldots, a_N \in \mathbb{Z}$. We may assume without loss of generality that all of the $a_1, \ldots, a_N$ are nonzero. What is important here is the distribution in $\mathbb{Z}$ of the values taken in this sequence; we shall consider here the simple case in which $+1$ occurs p times and -1 occurs q times for some $p + q = N$. We rename the last q coordinates $y_1, \ldots, y_q$; then (reordering if necessary) the action of $t \in \mathbb{G}_m$ is

$$(x_1, \ldots, x_p, y_1, \ldots, y_q) \mapsto (tx_1, \ldots, tx_p, t^{-1}y_1, \ldots, t^{-1}y_q). \qquad (6.13)$$

We now give to the coordinate ring $R = k[x_1, \ldots, x_p, y_1, \ldots, y_q]$ of $\mathbb{A}^N$ a grading by defining

$$\begin{aligned} \deg x_i &= 1 & i &= 1, \ldots, p, \\ \deg y_j &= -1 & j &= 1, \ldots, q. \end{aligned} \qquad (6.14)$$

In this graded ring $R = \bigoplus_{n \in \mathbb{Z}} R_{(n)}$, the component $R_{(0)}$ is generated by pq invariants $x_i y_j$ while the subalgebra R_+ is equal to $R_{(0)}[x_1, \ldots, x_p]$. Thus Proj R_+ is the variety constructed by gluing p affine open sets

$$\mathrm{Spm}\, R_+[1/x_1]_{(0)}, \quad \ldots, \quad \mathrm{Spm}\, R_+[1/x_p]_{(0)}.$$

Proj R_- is constructed similarly.

Theorem 6.27.

(i) X_0 is the subvariety of affine space $\mathbb{A}^{pq}$ consisting of $p \times q$ matrices of rank at most 1,

$$\mathrm{rank} \begin{pmatrix} z_{11} & \cdots & z_{1q} \\ & \cdots & \\ z_{p1} & \cdots & z_{pq} \end{pmatrix} \leq 1.$$

This is defined by $\binom{p}{2} \times \binom{q}{2}$ quadratic equations

$$\begin{vmatrix} z_{ij} & z_{ij'} \\ z_{i'j} & z_{i'j'} \end{vmatrix} = 0,$$

for $1 \leq i < i' \leq p$, $1 \leq j < j' \leq q$. Equivalently, X_0 is the affine cone over the Segre variety $\mathbb{P}^{p-1} \times \mathbb{P}^{q-1} \subset \mathbb{P}^{pq-1}$ (see Section 8.1(b)).

(ii) X_+ is the subvariety of $X_0 \times \mathbb{P}^{p-1}$ defined by

$$X_+ = \left\{ (Z, \mathbf{x} = (x_1 : \ldots : x_p)) \mid \mathrm{rank} \left({}^t\mathbf{x}, Z \right) \leq 1 \right\}.$$

(iii) X_- *is the subvariety of* $X_0 \times \mathbb{P}^{q-1}$ *defined by*

$$X_- = \left\{ \left(Z, \mathbf{y} = (y_1 : \ldots : y_q) \right) \mid \mathrm{rank} \begin{pmatrix} \mathbf{y} \\ Z \end{pmatrix} \leq 1 \right\}.$$

$\square$

The cases p or $q = 0$ we have already seen in Example 6.3. The cases p or $q = 1$ are exceptional. Of these, the case $p = q = 1$ is particularly exceptional, and in fact there are isomorphisms $X_\pm \xrightarrow{\sim} X_0$.

Example 6.28. In the case $p = q = 1$, the action of $\mathbb{G}_m$ on $\mathbb{A}^2$ is by

$$(x, y) \mapsto (tx, t^{-1}y).$$

This is the example we gave in Section 5.1(a) to illustrate the problem of separating orbits. The three quotients X_-, X_0, X_+ are all isomorphic to $\mathbb{A}^1$. $\square$

Example 6.29. The case $p = q = 2$ is well known and is the basic 3-dimensional *flop*. We write $p_{ij} = x_i y_j$; recall that these are the generating invariants of R. Thus

$$\begin{aligned} X_0 &= \mathrm{Spm}\, k[x_1 y_1, x_1 y_2, x_2 y_1, x_2 y_2] \\ &= \mathrm{Spm}\, k[p_{11}, p_{12}, p_{21}, p_{22}]/(p_{11} p_{22} - p_{12} p_{21}). \end{aligned}$$

This is a quadric hypersurface in $\mathbb{A}^4$ with an isolated singular point at the origin. X_+ and X_- are both resolutions of this singular point, and the exceptional sets (that is, the fibres over the origin) C_+ and C_- are both copies of $\mathbb{P}^1$. The two morphisms

$$X_+ - C_+ \to X_0 - \{0\} \leftarrow X_- - C_-$$

are both isomorphisms, but the composition $X_+ - C_+ \leftrightarrow X_- - C_-$ does not extend in either direction to an isomorphism of X_+ with X_- (it is a proper birational map). $\square$

As in this example, when $p, q \geq 2$ the three quotients X_-, X_0, X_+ all have the same rational function field, but the rational maps

$$X_- \leftarrow - \to X_+$$

cannot be extended in either direction to a morphism.

Remarks 6.30.

(i) Denote by $\widetilde{X}_0$ the blow-up of $X_0 \subset \mathbb{A}^{pq}$ at the origin. Then the morphism $\widetilde{X}_0 \to X_0$ resolves the singular point of X_0, with the exceptional set

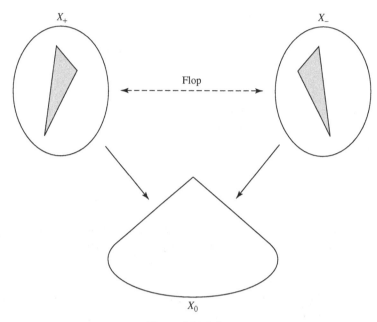

Figure 6.2: A flop

isomorphic to the product $\mathbb{P}^{p-1} \times \mathbb{P}^{q-1}$. In fact this mapping is via $X_\pm$, and there is a commutative diagram:

$$
\begin{array}{ccc}
\widetilde{X}_0 & \rightarrow & X_+ \\
\downarrow & & \downarrow \\
X_- & \rightarrow & X_0
\end{array}
$$

This is precisely the fibre product of the maps $X_\pm \rightarrow X_0$.

(ii) We have a pair of varieties $X_\pm$ together with subvarieties $C_+ \subset X_+, C_- \subset X_-$ whose complements are isomorphic. In this situation, suppose that (the restriction to $C_\pm$ of) the canonical line bundle $\mathcal{O}(K)$ (that is, the determinant line bundle $\det \Omega_X$ of the cotangent vector bundle, also denoted ω_X – see Definition 9.44 in Chapter 9) changes from negative to positive as we pass from X_- to X_+. Such a birational map is called a *flip* and is supposed to take one step towards a *minimal model* of the function field. In the present example we have

$$
\mathcal{O}(K)|_{C_-} = \mathcal{O}_{\mathbb{P}^{q-1}}(p-q), \qquad \mathcal{O}(K)|_{C_+} = \mathcal{O}_{\mathbb{P}^{p-1}}(q-p).
$$

Hence, if $p < q$, then the birational map $X_- - - \to X_+$ is a flip (and its inverse $X_+ - - \to X_-$ is an *inverse flip*). See Mori and Kollár [39].

(b) Toric varieties as quotient varieties

When the action of G on X is not of ray type with respect to $\chi \in \mathrm{Hom}(G, \mathbb{G}_\mathrm{m})$, the projective quotient $X /\!/_\chi G$ depends on the ray $\mathbb{R}_+ \cdot \chi$ in $\mathrm{Hom}(G, \mathbb{G}_\mathrm{m}) \otimes_\mathbb{Z} \mathbb{R}$, and not just on the direction $\mathbb{R} \cdot \chi$.

We begin with an example which in effect compactifies that of part (a) above. Consider two actions of $\mathbb{G}_\mathrm{m}$ on $X = \mathbb{A}^{p+q+1}$:

$$\lambda(s): \quad (x_1, \ldots, x_p, y_1, \ldots, y_q, z) \mapsto (sx_1, \ldots, sx_p, y_1, \ldots, y_q, sz)$$
$$\mu(t): \quad (x_1, \ldots, x_p, y_1, \ldots, y_q, z) \mapsto (x_1, \ldots, x_p, ty_1, \ldots, ty_q, tz),$$

where $s, t \in \mathbb{G}_\mathrm{m}$. The two actions commute and so define an action of the 2-dimensional torus $G = \mathbb{G}_\mathrm{m} \times \mathbb{G}_\mathrm{m}$. The characters of G correspond to pairs of integers $a, b \in \mathbb{Z}$,

$$\chi_{a,b}: G \to \mathbb{G}_\mathrm{m}, \qquad (s, t) \mapsto s^a t^b.$$

We shall sometimes denote this character simply by (a, b).

When $(a, b) = (0, 1)$ the action on X of $\ker \chi_{0,1}$ is λ and the associated ring of invariants is the polynomial ring $k[y_1, \ldots, y_q]$. Here we assign the grading coming from the action of $\chi_{0,1}(s, t) = t$, which is the standard grading with $\deg y_1 = \cdots = \deg y_q = 1$. By Remark 6.14(ii), the Proj quotient $X /\!/_{(0,1)} G$ is therefore $\mathbb{P}^{q-1}$. Similarly, the Proj quotient $X /\!/_{(1,0)} G$ is $\mathbb{P}^{p-1}$. Neither is a birational quotient.

Next, consider $(a, b) = (1, 1)$. The kernel of $\chi_{1,1}$ acts on X as $\lambda(s)\mu(s)^{-1}$, $s \in \mathbb{G}_\mathrm{m}$, and the ring of invariants is $k[x_i y_j, z]_{1 \le i \le p, \, 1 \le j \le q}$. Hence in this case the Proj quotient $X /\!/_{(1,1)} G$ is the affine cone over the Segre variety $\mathbb{P}^{p-1} \times \mathbb{P}^{q-1} \subset \mathbb{P}^{pq-1}$.

Now suppose that the character $\chi_{a,b}$ has a direction in between $(0, 1)$ and $(1, 1)$. Then the projective quotient $X /\!/_{(a,b)} G$ has the structure of a $\mathbb{P}^p$-bundle over $\mathbb{P}^{q-1}$. This bundle has a section $\cong \mathbb{P}^{q-1}$ which collapses to the vertex of the Segre cone as we move to $X /\!/_{(1,1)} G$. Similarly, the quotient in any direction between $(1, 1)$ and $(1, 0)$ is a $\mathbb{P}^q$-bundle over $\mathbb{P}^{p-1}$. The birational transformation of $X /\!/_{(a,b)} G$ as the ray in the direction (a, b) crosses over the ray $(1, 1)$ is a flop. This phenomenon is called *wall crossing*.

As another example, let us take a look at a sextic del Pezzo surface with a torus action. Think of $\mathbb{A}^6$ as the space of 2×3 matrices

$$\begin{pmatrix} x_1 & x_2 & x_3 \\ y_1 & y_2 & y_3 \end{pmatrix}.$$

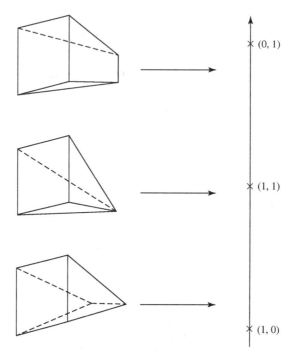

Figure 6.3: Wall crossing

The two tori $(\mathbb{G}_m)^2$ and $(\mathbb{G}_m)^3$ act on this space, on the left and right, respectively, as the diagonal matrices

$$\begin{pmatrix} s_1 & \\ & s_2 \end{pmatrix}, \qquad \begin{pmatrix} t_1 & & \\ & t_2 & \\ & & t_3 \end{pmatrix}.$$

These two actions commute, and the diagonal subgroup $\mathbb{G}_m$ of each has the same action. There is therefore a 1-dimensional subgroup of $\mathbb{G}_m{}^2 \times \mathbb{G}_m{}^3$ which acts trivially, and we let G be the quotient by this subgroup. This is a 4-dimensional torus $G \cong \mathbb{G}_m{}^4$ acting on $\mathbb{A}^6$. The group $\mathrm{Hom}(G, \mathbb{G}_m)$ of characters

$$(s_1, s_2 : t_1, t_2, t_3) \mapsto s_1^{a_1} s_2^{a_2} t_1^{b_1} t_2^{b_2} t_3^{b_3}$$

can be identified with the abelian group

$$\{(a_1, a_2 : b_1, b_2, b_3) \in \mathbb{Z}^5 \mid a_1 + a_2 = b_1 + b_2 + b_3\}.$$

The natural coordinates on $\mathbb{A}^6$ are semiinvariants for the G-action with weights given in the following table:

semiinvariant	x_1	x_2	x_3
weight $\in$ Hom$(G, \mathbb{G}_m)$	$(10 : 100)$	$(10 : 010)$	$(10 : 001)$
semiinvariant	y_1	y_2	y_3
weight $\in$ Hom$(G, \mathbb{G}_m)$	$(01 : 100)$	$(01 : 010)$	$(01 : 001)$

These weights are the six vertices of a triangular prism (the *effective poly-hedron*). The centroids of the five faces of this prism form the vertices of a 6-sided polyhedron (two tetrahedra glued along a triangular face), which we will denote by N (called the *nef polyhedron*).

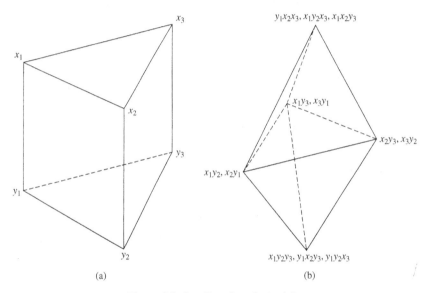

(a)   (b)

Figure 6.4: Semiinvariants by weight

We can describe how the Proj quotient $\mathbb{A}^6 /\!/_\chi G$ depends on the ray direction of a character

$$\chi = (a_1, a_2 : b_1, b_2, b_3) \in \text{Hom}(G, \mathbb{G}_m).$$

(i) If $\mathbb{R}_+ \cdot \chi$ lies outside N, then the quotient is either empty or a single point.
(ii) If $\mathbb{R}_+ \cdot \chi$ is the direction of the top vertex $(21 : 111)$ of N, then the semiinvariants are generated by

$$y_1 x_2 x_3, \quad x_1 y_2 x_3, \quad x_1 x_2 y_3.$$

Taking the projective spectrum we see that the quotient is $\mathbb{P}^2$. Similarly, the quotient at the bottom vertex $(12 : 111)$ is $\mathbb{P}^2$.

(iii) If $\mathbb{R}_+ \cdot \chi$ is the direction of one of the three remaining vertices $(11 : 011)$, $(11 : 101)$, $(11 : 110)$ of N, then the semiinvariant rings are $k[y_2x_3, x_2y_3]$, $k[y_3x_1, x_3y_1]$, $k[y_1x_2, x_1y_2]$, respectively. So these three quotients are each isomorphic to $\mathbb{P}^1$.

(iv) If $\mathbb{R}_+ \cdot \chi$ is the direction of an interior point of N, then the Proj quotient is a toric variety defined by the fan as shown in Figure 6.5. This is a surface obtained by gluing six copies of $\mathbb{A}^2$ (see Section 3.4(b)).

Particularly pretty is the point $\chi = (33 : 222)$. One can read off from the fan that the surface is $\mathbb{P}^2$ blown up at three points, or equivalently $\mathbb{P}^1 \times \mathbb{P}^1$ blown up in two points. (See, for example, Fulton [37].) In this case we get seven generating semiinvariants:

$$x_1x_2x_3y_1y_2y_3, \quad x_2^2x_3y_2^2y_3, \quad x_3^2x_1y_3^2y_1, \quad x_1^2x_2y_1^2y_2,$$
$$x_2x_3^2y_2y_3^2, \quad x_3x_1^2y_3y_1^2, \quad x_1x_2^2y_1y_2^2,$$

and these embed the quotient variety in $\mathbb{P}^6$. The image is called the *del Pezzo surface of degree 6*.

(v) The reader may like to examine the remaining cases for him- or herself. They are all toric varieties.

In fact, it is known that every toric variety can be represented as a quotient of affine space $\mathbb{A}^N$ by some torus action.

(c) Moment maps

What we have seen in this chapter is that, from the point of view of Proj quotients, it is most natural to think of quotient varieties as occuring in families. We can move from one to another in a suitable parameter space. In terms of differential geometry, and viewing complex varieties as symplectic manifolds, what is responsible for this phenomenon is a *moment map*. One can define precisely what is meant here, but we shall just indicate how it arises naturally in some of our examples. (But see Mumford et al [30] or McDuff and Salamon [40].) First consider the action $\mathbb{G}_m \curvearrowright \mathbb{A}^{n+1}$ of Example 6.3. We assume that the ground field is the field of complex numbers $\mathbb{C}$; thus $\mathbb{C}^*$ acts on the space $\mathbb{C}^{n+1}$ by scalar multiplication. In this case the moment map is

$$\mu : \mathbb{C}^{n+1} \to \mathbb{R}, \qquad (z_0, z_1, \ldots, z_n) \mapsto \sum_{i=0}^{n} |z_i|^2.$$

This map is invariant under the action of the subgroup $U(1) = \{z \in \mathbb{C}^* \mid |z| = 1\} \subset \mathbb{C}^*$, and for any real number $a \in \mathbb{R}$ one can consider the restricted action

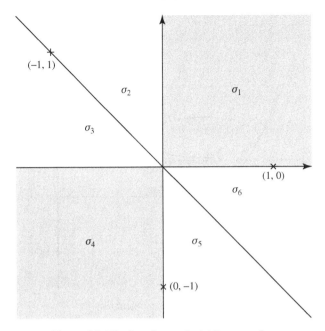

Figure 6.5: The fan of a sextic del Pezzo surface

of $U(1)$ on the fibre $\mu^{-1}(a)$ and its orbit space

$$X_a = \mu^{-1}(a)/U(1).$$

It is easy to see that we get:

	$\mu^{-1}(a)$	X_a
$a > 0$	S^{2n+1}	$\mathbb{CP}^n$
$a = 0$	point	point
$a < 0$	empty	empty

These quotients are parametrised by the real line, passing from $\mathbb{CP}^n$ to the empty set with a single point appearing as we cross the boundary. This corresponds to the three algebraically constructed quotients

$$\text{Proj } R_+ = \mathbb{P}^n, \quad \text{Proj } R_{(0)} = \text{point}, \quad \text{Proj } R_+ = \text{empty set}.$$

One can also view μ as a Morse function on $\mathbb{C}^{n+1}$, for which the origin is a critical point of index $(0, 2n)$, and the sphere S^{2n+1} is the level set for $a > 0$.

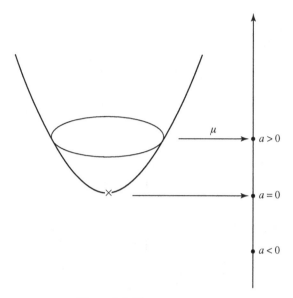

Figure 6.6: The moment map

The moment map of the action (6.13) of Section 6.3(a) is the map

$$\mu : \mathbb{C}^{p+q} \to \mathbb{R}, \quad (x_1, \ldots, x_p, y_1, \ldots, y_q) \mapsto \sum_{i=1}^{p} |x_i|^2 - \sum_{j=1}^{q} |y_j|^2.$$

Again, μ is a Morse function on $\mathbb{C}^{p+q} = \mathbb{R}^{2p+2q}$, with the origin as a critical point of index $(2p, 2q)$.

The moment map giving the sextic del Pezzo surface is

$$\mu : \mathbb{C}^6 \to \mathbb{R}^4 \subset \mathbb{R}^5,$$

$$\begin{pmatrix} x_1 & x_2 & x_3 \\ y_1 & y_2 & y_3 \end{pmatrix} \mapsto \left(\sum |x_i|^2, \sum |y_i|^2 : |x_1|^2 + |y_1|^2, |x_2|^2 + |y_2|^2, \right.$$

$$\left. |x_3|^2 + |y_3|^2 \right).$$

7

The numerical criterion and some applications

Our aim in this book is to study the Proj quotient, and some applications of this, when an algebraic group G acts on an affine variety X. In fact, as we have seen, this is not a quotient of X, but of its subset $X^{ss} \subset X$ of semistable points for the group action, and to get a good quotient we have to restrict further still to the set $X^s \subset X^{ss}$ of stable points. But in general this leaves unanswered the fundamental problem of determining whether or not a given point $x \in X$ is (semi)stable. Let us review very briefly how, in general, we will answer this question in some particular cases.

(1) In Chapter 5 we have already looked at the action of $GL(n + 1)$ on the affine space of homogeneous polynomials of degree d in $n + 1$ variables $f_d(x_0, x_1, \ldots, x_n)$, and we have seen that all nonsingular forms are stable (Corollary 5.24).

(2) In Chapter 8 we are going to consider the action of $GL(r)$ on the affine space of $r \times n$ matrices, and it will turn out that stability and semistability are both equivalent to the condition of having maximal rank (Proposition 8.1).

(3) In Chapter 10, under the action $GL(N) \curvearrowright \mathrm{Alt}_{N,2}(H^0(L))$, we will see that a point is semistable if and only if it is the Gieseker matrix of a semistable rank 2 vector bundle with determinant L (Propositions 10.69 and 10.70 and Lemma 10.81).

In each of these examples semistability is shown using some explicit semi-invariants – in case (1) the discriminant, in (2) the determinantal minors and in (3) the Pfaffian minors.

Nevertheless, it is possible to determine the (semi)stable points of a group action even without knowing the semiinvariants, and that is what we will discuss in this chapter. It should be regarded essentially as an interlude, though, as the numerical criterion will not be needed in later chapters for the moduli constructions for line bundles and vector bundles. For the classical examples of Section 7.2, on the other hand, it does give very explicit geometric information.

7.1 The numerical criterion

Although similar results can be shown for any linearly reductive group, we will restrict our attention in this chapter to the general linear group $GL(n)$; and we will restrict ourselves, moreover, to actions $GL(n) \curvearrowright X$ of ray type (Definitions 6.7 and 6.12). We will denote by $F \subset X$ the fixed point set under the multiplicative group $\mathbb{G}_m \subset GL(n)$ of scalar matrices, called the irrelevant set (see Section 6.1(c)).

(a) 1-parameter subgroups

Definition 7.1. Let G be any algebraic group. A nontrivial homomorphism $\lambda : \mathbb{G}_m \to G$ is called a *1-parameter subgroup* of G, or 1-PS for short.   □

If G acts on a variety X, then the group $\mathbb{G}_m$ acts on X via the 1-PS λ: that is, $x \mapsto \lambda(t) \cdot x, x \in X, t \in \mathbb{G}_m$. If we regard $\mathbb{G}_m \hookrightarrow \mathbb{A}^1$ by taking the spectrum of the inclusion of rings $k[t] \hookrightarrow k[t, t^{-1}]$, so that $\mathbb{G}_m = \mathbb{A}^1 - \{0\}$, then we can consider the limit $\lim_{t \to 0} \lambda(t) \cdot x$.

Definition 7.2. If a 1-PS $\lambda : \mathbb{G}_m \to X$ extends to a morphism $\mathbb{A}^1 \to X$, then the image of the origin $0 \in \mathbb{A}^1$ is called the *limit* of λ as $t \to 0$ and written $\lim_{t \to 0} \lambda(t) \cdot x$.   □

In what follows we shall always assume that the variety X is separated (Definition 3.33). This guarantees the uniqueness of limits. The following two theorems are together called the *Hilbert-Mumford Numerical Criterion*.

Theorem 7.3. The Hilbert-Mumford Numerical Criterion. *For an action* $GL(n) \curvearrowright X$ *of ray type and a point* $x \in X$ *the following conditions are equivalent.*

(i) $x \in X$ *is semistable.*
(ii) For every 1-PS in $SL(n) \subset GL(n)$ *the limit* $\lim_{t \to 0} \lambda(t) \cdot x$ *either does not exist, or it exists but it is not contained in the irrelevant set* $F \subset X$.

Theorem 7.4. *For an action* $GL(n) \curvearrowright X$ *of ray type and a point* $x \in X$ *the following conditions are equivalent.*

(i) $x \in X$ *is stable.*
(ii) $x \notin F$ *and the limit* $\lim_{t \to 0} \lambda(t) \cdot x$ *does not exist for any 1-PS in* $SL(n) \subset GL(n)$.   □

Given integers $r_1, \ldots, r_n \in \mathbb{Z}$, not all zero, with $r_1 + \cdots + r_n = 0$, we have a 1-PS, called a *diagonal* 1-PS:

$$\mathbb{G}_m \to SL(n), \quad t \mapsto \begin{pmatrix} t^{r_1} & & & \\ & t^{r_2} & & \\ & & \ddots & \\ & & & t^{r_n} \end{pmatrix}. \tag{7.1}$$

In fact, every 1-PS in $SL(n)$ is conjugate to a diagonal 1-PS. More precisely, the following is true.

Proposition 7.5. *For every 1-PS $\lambda : \mathbb{G}_m \to SL(n)$ there exist integers $r_1 \leq r_2 \leq \cdots \leq r_n$ for which λ is conjugate in $SL(n)$ to the diagonal 1-PS (7.1).*

The kind of group action $GL(n) \curvearrowright X$ that we often encounter is where X is a vector space V, viewed as an affine space, and the action $GL(n) \curvearrowright V$ is a linear representation. Typically, the centre $\mathbb{G}_m \subset GL(N)$ acts by $x \mapsto t^M x$, $x \in V, t \in \mathbb{G}_m$, for some positive integer $M \in \mathbb{N}$, and in this situation the irrelevant set F is just the origin $0 \in V$. It is for this case that we will prove the two theorems above. Namely:

$$x \in V \text{ is semistable} \iff \begin{array}{l} \lim_{t \to 0} \lambda(t) \cdot x \neq 0 \\ \text{for every 1-PS } \lambda : \mathbb{G}_m \to SL(n), \end{array} \tag{7.2}$$

$$x \in V \text{ is stable} \iff \begin{array}{l} \lim_{t \to 0} \lambda(t) \cdot x \text{ does not exist} \\ \text{for any 1-PS } \lambda : \mathbb{G}_m \to SL(n). \end{array} \tag{7.3}$$

(b) The proof

By definition of an action of an algebraic group, if R is any algebra over k, then $SL(n, R)$ acts on $V \otimes_k R$. In particular, if R is an integral domain with field of fractions K, then we get an action $SL(n, K) \curvearrowright V \otimes_k K$.

Proposition 7.6. *Suppose that $y \in V$ belongs to the closure of the orbit $SL(n) \cdot x \subset V$. Then there exist a (not necessarily discrete) valuation ring $(R, \mathfrak{m})$ and a matrix $\Xi \in SL(n, K)$ such that $R/\mathfrak{m} \cong k$ and satisfying the following two conditions.*

(a) $\Xi \cdot x \in V \otimes_k R$;
(b) $\Xi \cdot x \equiv y \mod \mathfrak{m}$.

Proof. We will write $G = SL(n)$. Let $W \subset V$ be the closure of the orbit $G \cdot x$. Then the dominant morphism $G \to W$, $g \mapsto g \cdot x$ corresponds to an injective homomorphisms of rings:

$$
\begin{array}{ccc}
k[W] & \hookrightarrow & k[G] \\
\cap & & \cap \\
k(W) & \hookrightarrow & k(G)
\end{array}
$$

Let $\mathfrak{n}$ be the maximal ideal of $y \in W$. By Theorem 2.37, there exists a valuation ring $(R, \mathfrak{m})$ dominating $(k[W], \mathfrak{n})$ with field of fractions $k(G)$ and residue field isomorphic to k.

Any morphism of affine varieties Spm $A \to SL(n)$ determines an element of $SL(n, A)$. In particular, the identity map $G \to SL(n)$ determines a matrix $\Xi \in SL(n, k[G])$, and we can view this as belonging to $SL(n, k(G))$. But then $\Xi \cdot x \in V \otimes_k k[W]$ and $\Xi \cdot x \equiv y \bmod \mathfrak{n}$, as required. $\qquad\square$

To understand what is going on in this proof we can give an analytical explanation as well. This captures the essential idea behind the more general case.

If $y \in W = \overline{G \cdot x}$, then we can find a holomorphic map $\phi : \Delta \to W$ of the disc $\Delta = \{z \in \mathbb{C} \mid |z| < 1\}$ such that $\phi(0) = y$ and $\phi(\Delta') \subset G \cdot x$, where $\Delta' = \Delta - 0$. Suppose that the restriction of ϕ to Δ' lifts to a holomorphic map $\psi : \Delta' \to G$. We then have a commutative diagram:

$$
\begin{array}{ccc}
\Delta' & \overset{\psi}{\longrightarrow} & G \\
\cap & & \downarrow \\
\Delta & \overset{\phi}{\longrightarrow} & W \subset V \cong \mathbb{C}^m
\end{array}
$$

Let $\mathbb{C}\{\{z\}\}$ be the ring of germs of holomorphic functions at the origin $0 \in \Delta$, and let K be its field of fractions. The map ψ then corresponds to an element $\Psi(z) \in SL(n, K)$. Since $\phi(0) = y$, it follows that

$$
\lim_{z \to 0} \Psi(z) \cdot x = y.
$$

So $R := \mathbb{C}\{\{z\}\}$ and $\Xi := \Psi(z)$ satisfy conditions (a) and (b) of the proposition.

In general, if $(R, \mathfrak{m})$ is a valuation ring and $\Xi \in SL(n, K)$ a matrix satisfying conditions (a) and (b) of Proposition 7.6, then we will write

$$
\lim_{R} \Xi \cdot x = y.
$$

There are now two essential lemmas. We let R be a valuation ring and K its field of fractions.

Lemma 7.7. *Any $n \times n$ matrix $\Xi \in \mathrm{Mat}_n(K)$ can be expressed as a product*

$$\Xi = ADB,$$

where $D = \mathrm{diag}(\xi_1, \ldots, \xi_n)$ is a diagonal matrix and $A, B \in SL(n, R)$.

Proof. If $\Xi = 0$, this is trivial; so we can assume that $\Xi = (\xi_{ij}) \neq 0$. We then consider the minimum $v_{\min}$ of the valuations $v(\xi_{ij})$ of the nonzero entries of Ξ. After multiplying on the left and right by permutation matrices we may assume that this minimum value is $v_{\min} = v(\xi_{11})$. We can now write

$$\begin{pmatrix} 1 & 0 & 0 & \cdots & 0 \\ y_2 & 1 & 0 & & \\ y_3 & 0 & 1 & & \\ \vdots & & & \ddots & \\ y_n & & & & 1 \end{pmatrix} \Xi \begin{pmatrix} 1 & z_2 & z_3 & \cdots & z_n \\ 0 & 1 & 0 & & \\ 0 & 0 & 1 & & \\ \vdots & & & \ddots & \\ 0 & & & & 1 \end{pmatrix} = \begin{pmatrix} \xi_{11} & 0 & 0 & \cdots & 0 \\ 0 & * & * & & * \\ 0 & * & * & & * \\ \vdots & & & \ddots & \\ 0 & * & * & & * \end{pmatrix},$$

where $y_i = -\xi_{i1}/\xi_{11}$ and $z_j = -\xi_{1,j}/\xi_{11}$. Note that both y_i, z_j belong to R by the way we have chosen ξ_{11}. Thus both of the matrices on the left-hand side belong to $SL(n, R)$. We now repeat the argument for the $(n-1) \times (n-1)$ submatrix on the right-hand side, until we obtain a diagonal matrix

$$D = \begin{pmatrix} \xi_1 & & & \\ & \xi_2 & & \\ & & \ddots & \\ & & & \xi_n \end{pmatrix}, \qquad v(\xi_1) \le v(\xi_2) \le \cdots \le v(\xi_n),$$

satisfying the requirements of the lemma. $\qquad\square$

Let Λ be the valuation group of R (see Section 2.4(b)). This is a totally ordered additive group. Given a finite set of elements $\chi_1, \ldots, \chi_n \in \Lambda$, the next lemma, which is essentially Proposition 3.71, says that linear inequalities on these elements can always be replaced by inequalities on a corresponding set of rational integers.

Proposition 7.8. *Let $\chi_1, \ldots, \chi_n \in \Lambda$ and let $\Phi \subset \mathbb{Z}^n$ be any finite subset. Then there exist integers $r_1, \ldots, r_n \in \mathbb{Z}$ with the property that, for all $(a_1, \ldots, a_n) \in \Phi$,*

$$a_1\chi_1 + \cdots + a_n\chi_n \begin{cases} > 0 \\ = 0 \\ < 0 \end{cases} \iff a_1 r_1 + \cdots + a_n r_n \begin{cases} > 0 \\ = 0 \\ < 0. \end{cases} \tag{7.4}$$

Outline of the proof. The elements $\chi_1, \ldots, \chi_n \in \Lambda$ partition $\mathbb{Z}^n$ into three subsemigroups:

$$
\begin{aligned}
C_+ &= \{(a_1, \ldots, a_n) \mid a_1\chi_1 + \cdots + a_n\chi_n > 0\}, \\
C_0 &= \{(a_1, \ldots, a_n) \mid a_1\chi_1 + \cdots + a_n\chi_n = 0\}, \\
C_- &= \{(a_1, \ldots, a_n) \mid a_1\chi_1 + \cdots + a_n\chi_n < 0\}.
\end{aligned}
$$

There now exists a hyperplane

$$
r_1 x_1 + \cdots + r_n x_n = 0
$$

in $\mathbb{R}^n$ which contains C_0 and which partitions $\mathbb{R}^n$ into half-spaces which intersect $\mathbb{Z}^n$ in $C_\pm$. (In the case $n = 2$, this is called a *Dirichlet section*.) Since Φ is a finite set, the real numbers $r_1, \ldots, r_n$ can be made rational by a small perturbation; they can then be assumed to be integers by multiplying through by their common denominator. (See Section 3.4(c).) □

Let us denote the second expression in (7.4) by $\langle \mathbf{a} \mid r_1, \ldots, r_n \rangle$.

Proof of (7.2). By Proposition 6.19 we just have to show that the following are equivalent:

(a) $0 \in \overline{SL(n) \cdot x}$.
(b) $\lim_{t \to 0} \lambda(t) \cdot x = 0$ for some 1-PS $\lambda : \mathbb{G}_m \to SL(n)$.

(b) $\Longrightarrow$ (a) is obvious, and we just need to show (a) $\Longrightarrow$ (b). By Proposition 7.6 (with $y = 0$) there exists a valuation ring R, with field of fractions K, and a matrix $\Xi \in SL(n, K)$ such that

$$
\lim_R \Xi \cdot x = 0. \tag{7.5}
$$

By Lemma 7.7 we can write

$$
\Xi = ADB, \qquad D = \begin{pmatrix} \xi_1 & & & \\ & \xi_2 & & \\ & & \ddots & \\ & & & \xi_n \end{pmatrix},
$$

where $A, B \in SL(n, R)$ and $\xi_1, \ldots, \xi_n \in K$. Let $T \subset SL(n)$ be the group of diagonal matrices; the action of T on V can be diagonalised using a suitable basis $e_1, \ldots, e_m \in V$. Each vector e_i then spans an eigenspace on which T acts with some weight

$$
\mathbf{a}_i = (a_{i1}, \ldots, a_{in}) \in \mathbb{Z}^n,
$$

in the sense that

$$
\begin{pmatrix}
t_1 & & & \\
& t_2 & & \\
& & \ddots & \\
& & & t_n
\end{pmatrix}
\cdot e_i = t_1^{a_{i1}} t_2^{a_{i2}} \dots t_n^{a_{in}} e_i, \qquad 1 \le i \le m. \tag{7.6}
$$

Let $\rho : SL(n) \to SL(V)$ be the linear representation giving the action that we are considering. (So $g \cdot x$ means $\rho(g)x$.) Then $\rho(\Xi) = \rho(A)\rho(D)\rho(B)$ and there exists a limit $\lim_R \rho(A)$ with determinant 1. This implies, by (7.5), that

$$
\lim_R \rho(D)(\rho(B)x) = 0. \tag{7.7}
$$

Using the basis above we will write

$$
B \cdot x = \rho(B)x = f_1 e_1 + \dots + f_m e_m. \tag{7.8}
$$

Since $B \in SL(n, R)$, the coefficients f_i belong to R, and (7.6) says that

$$
\rho(D)(\rho(B)x) = \xi^{\mathbf{a}_1} f_1 e_1 + \dots + \xi^{\mathbf{a}_m} f_m e_m,
$$

where $\xi^{\mathbf{a}_i} := \xi_1^{a_{i1}} \xi_2^{a_{21}} \dots \xi_n^{a_{in}}$. Since the limit of this expression is zero, we must have $v(\xi^{\mathbf{a}_i} f_i) > 0$ for each $i = 1, \dots, m$. Hence

$$
v(f_i) = 0 \quad \Longrightarrow \quad v(\xi^{\mathbf{a}_i}) > 0. \tag{7.9}
$$

We now consider the residue classes modulo the valuation ideal $\mathfrak{m} \subset R$ of B and f_i; we denote these by $\overline{B} \in SL(n, k)$ and $\overline{f}_i \in k$. Reducing (7.8) mod $\mathfrak{m}$ gives

$$
\overline{B} \cdot x = \overline{f}_1 e_1 + \dots + \overline{f}_m e_m \in V.
$$

By (7.9), $v(\xi^{\mathbf{a}_i}) > 0$ whenever $\overline{f}_i \neq 0$. It follows that (7.7) remains valid with B replaced by $\overline{B}$:

$$
\lim_R \rho(D)\left(\rho(\overline{B})x\right) = 0.
$$

We now apply Proposition 7.8 to the set of weights $\Phi = \{\mathbf{a}_i\}_{i=1}^m$ of the representation ρ and the values $v(\xi_1), \dots, v(\xi_n) \in \Lambda$. This tells us that there exist integers $r_1, \dots, r_n \in \mathbb{Z}$ such that for all $1 \le i \le m$:

$$
v(\xi^{\mathbf{a}_i}) > 0 \quad \Longleftrightarrow \quad \langle \mathbf{a}_i \mid r_1, \dots, r_n \rangle > 0.
$$

For this set of integers we have

$$\lim_{t \to 0} \begin{pmatrix} t^{r_1} & & & \\ & t^{r_2} & & \\ & & \ddots & \\ & & & t^{r_n} \end{pmatrix} \cdot \left(\rho(\overline{B})x \right) = 0.$$

But $\rho(\overline{B}) \in SL(n, k)$, and so we have constructed a 1-PS proving (b). $\square$

Proof of (7.3). The proof here is similar to that above; we will suppose that $x \in V$ is semistable but not stable, and show that in this case $\lim_{t \to 0} \lambda(t) \cdot x$ exists for some 1-PS λ. Let us write $G = SL(n)$. There are two possibilities that we have to consider:

(a) The case in which the orbit $G \cdot x$ is not closed.
(b) The case in which the stabiliser $G_x = \{g \in G \mid g \cdot x = x\}$ is not finite.

In case (a) we apply Proposition 7.6 to a point y in the boundary of the closure $W = \overline{G \cdot x}$. This gives a matrix $\Xi \in SL(n, K)$ such that

$$\lim_R \Xi \cdot x = y \notin G \cdot x,$$

and writing $\Xi = ADB$ we have

$$\lim_R D\, (B \cdot x) = y \notin G \cdot x.$$

Using the same basis as in the previous proof we have (7.8),

$$B \cdot x = f_1 e_1 + \cdots + f_m e_m, \qquad f_i \in R,$$

and modulo the maximal ideal $\mathfrak{n} \subset R$ at y,

$$\overline{B} \cdot x = \overline{f}_1 e_1 + \cdots + \overline{f}_m e_m.$$

Since the limit exists we have $v(\xi^{a_i} f_i) \geq 0$ for each $i = 1, \ldots, m$, and so by the same argument as before

$$\overline{f}_i \neq 0 \quad \Longrightarrow \quad v(\xi^{a_i}) \geq 0.$$

Since the limit is not contained in $G \cdot x$, the diagonal matrix D is not contained in $SL(n, R)$; so $v(\xi_i) < 0$ for some component ξ_i. We now apply Proposition 7.8 to the set of weights Φ consisting of the single vector $(0, \ldots, 0, 1, 0, \ldots, 0)$

(*i*-th entry) to obtain $(r_1, \ldots, r_n)$ with $r_i < 0$ and such that there exists a limit

$$\lim_{t \to 0} \begin{pmatrix} t^{r_1} & & & \\ & t^{r_2} & & \\ & & \ddots & \\ & & & t^{r_n} \end{pmatrix} \cdot \left(\rho(\overline{B})x \right).$$

This deals with case (a). For (b), we remark that $G = SL(n)$ is an affine variety and that the stabiliser $G_x \subset G$ is a closed subvariety and therefore is itself an affine variety. So it follows from Proposition 3.57 and the Valuative Criterion 3.58 that there exist a valuation ring R (with field of fractions K) and a matrix $\Xi \in SL(n, K)$ satisfying:

(i) $\Xi \cdot x = x$,   (ii) $\lim_{R} \Xi$ does not exist.

By Lemma 7.7 we can write $\Xi = ADB$. Now, by (ii) we must have $v(\xi_i) < 0$ for some component ξ_i of D; and since $\rho(A)(\rho(D)\rho(B)x) = x$ it follows that the limit $\lim_{R} \rho(D)\left(\rho(\overline{B})x \right) = x$ exists. The proof is now the same as for case (a). $\qquad\qquad\square$

7.2 Examples and applications

(a) Stability of projective hypersurfaces

We can apply the Hilbert-Mumford Criteria (7.2) and (7.3) to answer the question left open at the end of Chapter 5: what is the geometric interpretation of stability of homogeneous polynomials in $n + 1$ variables under the action of $GL(n + 1)$, or, equivalently, of hypersurfaces in $\mathbb{P}^n$ under the action of the projective group?

The first case $n = 1$ of binary forms is easy to deal with. In the case of any ray-type action of $GL(2)$, (semi)stability is determined by the limits under a single 1-PS (up to conjugacy in $SL(2)$)

$$\lambda : \mathbb{G}_m \to SL(2), \qquad t \mapsto \begin{pmatrix} t & 0 \\ 0 & t^{-1} \end{pmatrix}. \qquad (7.10)$$

Proposition 7.9.

(i) *A binary form* $(a \lozenge x, y)$ *of degree d is stable if and only if every linear factor has multiplicity* $< d/2$.

(ii) *A binary form* $(a \lozenge x, y)$ *of degree d is semistable if and only if every linear factor has multiplicity* $\leq d/2$.

Proof. We will deal with parts (i) and (ii) simultaneously and prove the 'only if' direction first.

Writing

$$(a \lozenge x, y) = a_0 x^d + d a_1 x^{d-1} y + \cdots + a_d y^d,$$

we may assume coordinates chosen (that is, move the form within its $GL(2)$-orbit) so that the multiple zero is $y = 0$. This means that

$$a_0 = a_1 = \cdots = a_{m-1} = 0.$$

We shall prove (ii) first. By hypothesis, $m > d/2$ in this case. Thus $(a \lozenge x, y)$ is a sum of terms $a_i \binom{d}{i} x^i y^{d-i}$ for $i > d/2$, so that under the action of the torus

$$\mathbb{G}_m \to T \subset SL(2) \qquad t \mapsto \begin{pmatrix} t & 0 \\ 0 & t^{-1} \end{pmatrix}$$

we have $\lim_{t \to 0}(a \lozenge tx, t^{-1}y) = 0$. This shows that $(a \lozenge x, y)$ is a nullform by Corollary 6.20.

For part (i) assume that $m \geq d/2$. If d is odd this implies $m > d/2$ so the result already follows from (ii); we may therefore assume that $d = 2s$ is even. Then, by hypothesis, $m \geq s$ and we can assume that equality holds, again by part (ii). Thus

$$a_0 = a_1 = \cdots = a_{s-1} = 0, \qquad a_s \neq 0.$$

So under the torus action we have

$$\lim_{t \to 0} (a \lozenge tx, t^{-1}y) = a_s x^s y^s.$$

Observe that the form $a_s x^s y^s \in \mathbb{V}_{1,d}$ is nonstable, since its stabiliser in $SL(2)$ contains T and is therefore positive-dimensional. If it is contained in the orbit $SL(2)a \subset \mathbb{V}_{1,d}$, then a also fails to be stable; otherwise, if it is not contained in the orbit $SL(2)a$, then this orbit is not closed, and so again a is nonstable.

To prove the converse, we assume that $(a \lozenge x, y)$ fails to be stable. By (7.3) this is equivalent to the existence of $\lim_{t \to 0} \lambda(t) \cdot a$ under some 1-PS, which by changing coordinates (that is, conjugating in $SL(2)$) we can take to be (7.10). This is equivalent to x dividing $(a \lozenge x, y)$ with multiplicity $\geq d/2$. This proves part (i). Similarly, $\lim_{t \to 0} \lambda(t) \cdot a = 0$ is equivalent to x dividing $(a \lozenge x, y)$ with multiplicity $> d/2$, and so part (ii) follows from (7.2). $\square$

Example 7.10. $d = 4$. A binary quartic is stable if and only if it has no repeated linear factors. $\square$

When we pass from two variables to three or more the number of 1-PSs that have to be considered increases rapidly. The following generalisation of the binary case is proved similarly, but we omit the details as they are illustrated by the examples below.

Proposition 7.11. *Let* $(a \wr x) \in \mathbb{V}_{n,d}$ *be a homogeneous form of degree d in n variables.*

(i) *$(a \wr x)$ is a nullform, that is, it fails to be semistable under the action $GL(n) \curvearrowright \mathbb{V}_{n,d}$, if and only if there exists a vector $(r_1, \ldots, r_n) \in \mathbb{Z}^n$ such that $a_I = 0$ for all I such that $\langle I \mid r_1, \ldots, r_n \rangle \geq 0$.*

(ii) *$(a \wr x)$ fails to be stable under the action $GL(n) \curvearrowright \mathbb{V}_{n,d}$ if and only if there exists a nonzero vector $(r_1, \ldots, r_n) \in \mathbb{Z}^n - \{0\}$ such that $a_I = 0$ for all I such that $\langle I \mid r_1, \ldots, r_n \rangle > 0$.* $\square$

This proposition has a simple geometric interpretation. Consider the case $n = 3$. (This is entirely representative of the general case.) Suppose we arrange the monomials of degree d in x, y, z in a triangle in the plane

$$\{(i_1, i_2, i_3) \in \mathbb{Z} \times \mathbb{Z} \times \mathbb{Z} \mid i_1 + i_2 + i_3 = d, \quad i_1, i_2, i_3 \geq 0\}.$$

Then a 1-PS (7.1) determined by a vector $\mathbf{r} = (r_1, r_2, r_3)$ corresponds to the line $\mathbf{r}^{\perp}$ through the centroid, and the condition for a form $(a \wr x, y, z)$ to be a nullform is that all the monomials occuring nontrivially (that is, with nonzero coefficient) in the form lie in the open half-plane strictly on one side of $\mathbf{r}^{\perp}$. Failure to be stable is equivalent to all monomials appearing nontrivially lying in the closed half-plane on one side of $\mathbf{r}^{\perp}$. (See Figure 7.2.)

Example 7.12. Plane cubics. The singular behaviour of a cubic in $\mathbb{P}^2$ is classified by the nine types shown in Figure 7.1.
    We claim:

(i) A plane cubic curve $C \subset \mathbb{P}^2$ is semistable if and only if its only singular points are ordinary double points.

(ii) $C \subset \mathbb{P}^2$ is stable if and only if it is nonsingular.

In Figure 7.1, in other words, case (1) is stable (this already follows from Corollary 5.24) and cases (2), (4) and (6) are semistable, while the others are all nullforms. Note, incidentally, that all the nullforms belong to closure-equivalent orbits.

Proof. The proofs of (i) and (ii) are entirely similar, and we will just prove (ii). We arrange the cubic monomials in x, y, z in a triangle with xyz at the centroid,

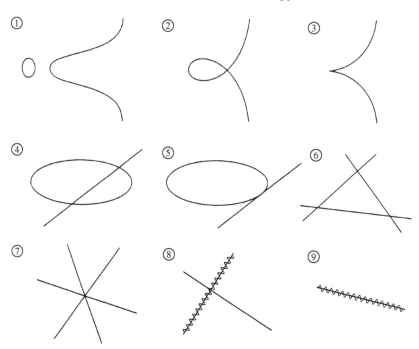

Figure 7.1: Classification of plane cubics

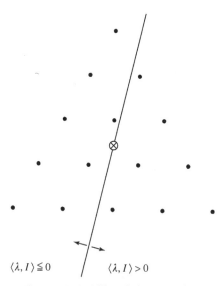

Figure 7.2: Stability of plane quartics

and let the line $\mathbf{r}^{\perp}$ (corresponding to a diagonal 1-PS) rotate through this point. As the line varies we examine the geometry of the cubic curve whose equation is supported in the closed half-plane on one side or other of the line.

For example, for $\mathbf{r} = (2, -1, -1)$ the equation of the curve lives in the half-plane:

$$
\begin{array}{cccc}
 & \times & & \\
 & \times & \times & \\
xy^2 & xyz & xz^2 & \\
y^3 & y^2z & yz^2 & z^3
\end{array}
\qquad\qquad
\begin{array}{cccc}
 & & 6 & \\
 & 3 & & 3 \\
0 & 0 & 0 & \\
-3 & -3 & -3 & -3
\end{array}
$$

(where the right-hand triangle shows the weight of the corresponding mono-mial). This is precisely the condition for $C \subset \mathbb{P}^2$ to have a singular point at $(1 : 0 : 0)$. Conversely, if C is singular, then by choosing homogeneous coordinates so that the singular point is at $(1 : 0 : 0)$ we see in this way that C is not stable.

A second possibility, corresponding to $\mathbf{r} = (1, -2, 1)$, is

$$
\begin{array}{cccc}
 & \times & & \\
x^2y & & \times & \\
xy^2 & xyz & & \times \\
y^3 & y^2z & yz^2 & \times
\end{array}
\qquad\qquad
\begin{array}{cccc}
 & & 3 & \\
 & 0 & & 3 \\
-3 & 0 & & 3 \\
-6 & -3 & 0 & 3
\end{array}
$$

This is equivalent to C containing the line $y = 0$ as a component; and conversely, again, if C contains a line, then we can assume, by changing coordinates, that it is this one.

Finally, we note that up to symmetry of x, y, z these two cases contain all possibilities as the line $\mathbf{r}^{\perp}$ rotates about the centroid. $\qquad\qquad\square$

For higher degree plane curves the enumeration of singular types quickly becomes horrendous, but nevertheless a similar analysis is possible. We will content ourselves with the statement for degree 4.

Example 7.13. Plane quartics.

(i) A plane quartic curve $C \subset \mathbb{P}^2$ is semistable if and only if it has no triple point, and is not the sum of a plane cubic and an inflectional tangent line.

(ii) A semistable plane quartic curve $C \subset \mathbb{P}^2$ is stable if and only if it has no tacnode.

(A tacnode is a double point with a single tangent line with contact of order 4. Its local canonical form is $y^2 = x^4$.)   □

Figure 7.2 shows the case of a nullform (with equation supported on the right-hand side of the line) which is the union of a plane cubic and an inflectional tangent. (See also Mumford [47].)

(b) Cubic surfaces

We turn now to cubic surfaces in $\mathbb{P}^3$. Our first aim is to show:

Theorem 7.14. *A cubic surface $S \subset \mathbb{P}^3$ is stable under the action $GL(4) \curvearrowright \mathbb{V}_{3,3}$ if and only if it has finitely many ordinary double points and no worse singularities.*

Remark 7.15. We should first say a few words about double points. Suppose P is a double point of a surface $S \subset \mathbb{A}^3$, and choose corrdinates so that $P = (0, 0, 0)$ is the origin. Then the equation of S decomposes into homogeneous polynomials

$$f_2(x, y, z) + f_3(x, y, z) + \cdots = 0,$$

where, by hypothesis, $f_2 \neq 0$. The *rank* of the double point is then defined to be the rank of the quadratic form $f_2(x, y, z)$. An *ordinary double point* is, by definition, a double point of rank 3. The tangent cone of S at P is the quadric cone $\{f_2(x, y, z) = 0\} \subset \mathbb{A}^3$ with vertex P. If $P \in S$ is a double point of rank 2, then the tangent cone is a pair of planes (that is, f_2 factorises as a product of linear forms) and the intersection of these planes is called the *axis* of the double point.   □

A 1-PS $\lambda : \mathbb{G}_m \to SL(4)$ will be called *normalised* if its image is in the torus

$$T = \{\mathrm{diag}(t_0, t_1, t_2, t_3) \mid t_0 t_1 t_2 t_3 = 1\} \subset SL(4)$$

and it is of the form

$$\lambda : t \mapsto \mathrm{diag}(t^{r_0}, t^{r_1}, t^{r_2}, t^{r_3}), \qquad \text{where } r_0 \geq r_1 \geq r_2 \geq r_3.$$

(Note that $r_0 + r_1 + r_2 + r_3 = 0$. In particular, $r_3 < 0$.) Every 1-PS is conjugate to a normalised 1-PS, and so when we apply the Hilbert-Mumford criterion it is sufficient to consider only normalised 1-PSs.

We want to look for cubic surfaces which have a limit as $t \to 0$ under such a 1-PS: that is, which fail to be stable. The equation of such a surface must be a linear combination of monomials from the set

$$M^{\oplus}(\lambda) := \{w^a x^b y^c z^d \mid ar_0 + br_1 + cr_2 + dr_3 \geq 0\}.$$

We will denote the set of all 20 cubic monomials in w, x, y, z by M.

Notation 7.16. We denote the monomial $m = w^a x^b y^c z^d$ by $\langle a, b, c, d|$, the 1-PS $\mu : t \mapsto \mathrm{diag}(t^{r_0}, t^{r_1}, t^{r_2}, t^{r_3})$ by $|r_0, r_1, r_2, r_3\rangle$, and the inner product $ar_0 + br_1 + cr_2 + dr_3$ by $\langle m \mid \mu \rangle = \langle a, b, c, d \mid r_0, r_1, r_2, r_3 \rangle$. ☐

Proposition 7.17. *If λ is any normalised 1-PS, then $M^{\oplus}(\lambda)$ is a subset of one of:*

(1) $M^{\oplus}(1, 1, 0, -2)$, (2) $M^{\oplus}(2, 0, -1, -1)$, (3) $M^{\oplus}(1, 0, 0, -1)$.

☐

We will prove this in a moment. We need to put a partial ordering on the monomial set M by:

$$m \geq m' \text{ for } m, m' \in M \qquad \Longleftrightarrow \qquad \begin{array}{l} \langle m \mid \lambda \rangle \geq \langle m' \mid \lambda \rangle \text{ for all} \\ \text{normalised 1-PSs } \lambda. \end{array} \qquad (7.11)$$

With this definition the following fact is easily checked.

Lemma 7.18.

$$\langle a, b, c, d| \geq \langle a', b', c', d'| \qquad \Longleftrightarrow \qquad \begin{cases} a \geq a' \\ a + b \geq a' + b' \\ a + b + c \geq a' + b' + c'. \end{cases}$$

☐

From this lemma we obtain Figure 7.3, where the right-hand column is $3a + 2b + c$. Now the subset $M^{\oplus}(\lambda) \subset M$ is an ideal with respect to the partial ordering, in the sense that, if $m \in M^{\oplus}(\lambda)$ and $m' \geq m$, then $m' \in M^{\oplus}(\lambda)$. Dually, its complement $M^-(\lambda) \subset M$ has the property that, if $m \in M^-(\lambda)$ and $m \geq m'$, then $m' \in M^-(\lambda)$. Because of these properties, in order to prove an inclusion relation $M^{\oplus}(\lambda) \subset M^{\oplus}(\mu)$ or, equivalently, that $M^{\oplus}(\lambda) \cap M^-(\mu) = \emptyset$, it is enough to check just the maximal elements of $M^-(\mu)$:

Proposition 7.19. *The following conditions are equivalent:*

(i) $M^{\oplus}(\lambda) \subset M^{\oplus}(\mu)$.
(ii) $\langle m \mid \lambda \rangle < 0$ for every maximal monomial $m \in M^-(\mu)$.

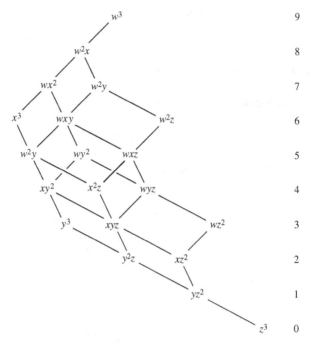

Figure 7.3: Partial ordering of the monomial set M

Proof. (i) is equivalent to $M^-(\mu) \cap M^\oplus(\lambda) = \emptyset$, while, if $M^-(\mu)$ has maximal monomials $m_1, \ldots, m_n$, then

$$M^-(\mu) = \bigcup_{i=1}^n \{m \in M \mid m_i \ge m\}.$$

This immediately implies that (i) is equivalent to (ii). □

Proof of Proposition 7.17. For the three sets (1), (2), (3) we find the following:

1-PS μ	monomials of $M^-(\mu)$	maximal elements
$\lvert 1, 1, 0, -2\rangle$	$(y, z)^2 z$, $\quad (w, x)(y, z)z$	wyz
$\lvert 2, 0, -1, -1\rangle$	$(x, y, z)^2(y, z)$	$x^2 y$
$\lvert 1, 0, 0, -1\rangle$	$(x, y, z)^2 z$, $\quad wz^2$	$x^2 z$, wz^2

We have to show that $M^-(\lambda)$ contains one of these sets, for any normalised 1-PS λ.

If either $wyz \in M^-(\lambda)$ or $x^2y \in M^-(\lambda)$, then we are done, so suppose not: that is,

$$\langle 1, 0, 1, 1 \mid \lambda \rangle \geq 0, \qquad \langle 0, 2, 1, 0 \mid \lambda \rangle \geq 0.$$

We now observe (using $r_0 + r_1 + r_2 + r_3 = 0$) that

$$2\langle 1, 0, 1, 1 \mid \lambda \rangle + \langle 0, 2, 0, 1 \mid \lambda \rangle = \langle 2, 2, 2, 3 \mid \lambda \rangle = r_3 < 0$$

and

$$\langle 1, 0, 1, 1 \mid \lambda \rangle + \langle 0, 2, 1, 0 \mid \lambda \rangle + \langle 1, 0, 0, 2 \mid \lambda \rangle = \langle 2, 2, 2, 3 \mid \lambda \rangle = r_3 < 0,$$

and these inequalities force $\langle 0, 2, 0, 1 \mid \lambda \rangle < 0$ and $\langle 1, 0, 0, 2 \mid \lambda \rangle < 0$, or in other words x^2z and $wz^2 \in M^-(\lambda)$. $\qquad \square$

Proof of Theorem 7.14. We first assume that S is not stable and examine its singularities. By the Hilbert-Mumford Criterion (7.3), instability means that the equation of S has a limit as $t \to 0$ under some 1-PS λ, which we can assume to be normalised. Existence of the limit, in turn, implies that the equation of S belongs to the linear span of the monomials $M^{\oplus}(\lambda)$, and by Proposition 7.17 it therefore belongs to the linear span of one of the sets (1),(2),(3). We consider each of these in turn.

A general cubic $S \subset \mathbb{P}^3$ spanned by $M^{\oplus}(1, 1, 0, -2)$ has the form

$$c(w, x, y) + zq(w, x) = 0,$$

where c is a cubic and q is a quadratic form. S therefore has a singular point at $P = (0 : 0 : 0 : 1)$. But the tangent cone at this point is $q(w, x) = 0$, which has rank 2, so that P is not an ordinary double point.

Next, $M^{\oplus}(2, 0, -1, -1)$ spans cubics of the form

$$wq(w, x, y, z) + ax^3 = 0,$$

where q is quadratic. Such a cubic contains the line $w = x = 0$, and on this line we can find a singular point whose tangent cone contains the plane $w = 0$, and is therefore not an ordinary double point.

Finally, a cubic spanned by $M^{\oplus}(1, 0, 0, -1)$ has the form

$$c(w, x, y, z) + l(w, x, y)wz = 0,$$

where c is cubic and l is linear. This has a rank 2 double point at $P = (0 : 0 : 0 : 1)$ whose tangent cone is the pair of planes $l(w, x, y)w = 0$.

We now have to show the converse, that, if $S \subset \mathbb{P}^3$ is stable, then it has only ordinary double points. But this is easy: if the point $P = (0 : 0 : 0 : 1)$ were a singular point worse than an ordinary double point, then the equation of S would (after a suitable choice of homogeneous coordinates) necessarily be of the form $M^{\oplus}(1, 1, 0, -2)$ above – and such a cubic is not stable.   □

Theorem 7.20 (Hilbert [20]). *A cubic surface $S \subset \mathbb{P}^3$ is semistable under the action $GL(4) \curvearrowright \mathbb{V}_{3,3}$ if and only if it has at most finitely many singularities of the following types:*

(1) ordinary double points,
(2) rank 2 double points whose axes are not contained in S.

Remark 7.21. One says that the surface has (at most) rational double points, of type A_1 in case (1) or type A_2 in case (2).   □

Theorem 7.20 is proved in the same way as Theorem 7.14, using the numerical criterion (7.2). Given a normalised 1-PS $\lambda = |r_0, r_1, r_2, r_3\rangle$ we define

$$M^+(\lambda) := \{ w^a x^b y^c z^d \mid ar_0 + br_1 + cr_2 + dr_3 > 0 \}.$$

A cubic surface unstable with respect to λ then has equation in the linear span of $M^+(\lambda)$.

Proposition 7.22. *If λ is any normalised 1-PS, then $M^+(\lambda)$ is a subset of one of:*

(1) $M^+(3, 1, 1, -5)$,   (2) $M^+(3, 3, -1, -5)$,   (3) $M^+(3, -1, -1, -1)$.

Proof. Exactly the same as the proof of Proposition 7.17. We begin by noting (where $M^{\ominus}$ denotes the complement of M^+ in M):

1-PS μ	monomials of $M^{\ominus}(\mu)$	maximal elements
$\lvert 3, 1, 1, -5\rangle$	$(w, x, y)(x, y)z$, $(w, x, y)z^2$, z^3	wxz
$\lvert 3, 3, -1, -5\rangle$	$(w, x)(y, z)z$, $(y, z)^3$	wyz, y^3
$\lvert 3, -1, -1, -1\rangle$	$(x, y, z)^3$	x^3

If $M^{\ominus}(\lambda)$ contains wxz or x^3, then it contains $M^{\ominus}(3, 1, 1, -5)$ or $M^{\ominus}(3, -1, -1, -1)$, respectively, and we are done. If not, then both of $\langle 1, 1, 0, 1 \mid \lambda \rangle$ and $\langle 0, 3, 0, 0 \mid \lambda \rangle$ are strictly positive. From the equalities

$$\langle 0, 3, 0, 0 \mid \lambda \rangle + 3\langle 1, 1, 0, 1 \mid \lambda \rangle = 0,$$
$$3\langle 1, 0, 1, 1 \mid \lambda \rangle + \langle 0, 3, 0, 0 \mid \lambda \rangle = 0,$$

therefore, it follows that $\langle 0, 3, 0, 0 \mid \lambda \rangle \leq 0$ and $\langle 1, 0, 1, 1 \mid \lambda \rangle \leq 0$, and hence that $M^{\ominus}(\lambda)$ contains $M^{\ominus}(3, 3, -1, -5)$. $\qquad\square$

Proof of Theorem 7.20. Suppose that $S \subset \mathbb{P}^3$ is unstable. Then its equation lives in the linear span of one of the sets (1), (2), (3) of the previous proposition, and we consider each of these in turn.

In $M^+(3, 1, 1, -5)$, a general cubic $S \subset \mathbb{P}^3$ has the form

$$c(w, x, y) + aw^2 z = 0.$$

This has a rank 1 double point at $(0 : 0 : 0 : 1)$ whose tangent cone is the double plane $w^2 = 0$. In $M^+(3, 3, -1, -5)$ a general cubic has the form

$$y^2 l(w, x) + y q_1(w, x) + z q_2(w, x) + c(w, x) = 0.$$

Again, this surface has a double point at $(0 : 0 : 0 : 1)$ with tangent cone $q_2(w, x) = 0$ and axis $w = x = 0$ which lies on the surface. Finally, a cubic in $M^+(3, -1, -1, -1)$ looks like

$$wq(w, x, y, z) = 0,$$

which is reducible (and in particular is singular along a plane conic).

From this analysis we conclude that, if the cubic surface $S \subset \mathbb{P}^3$ is unstable, then one of the following holds:

(a) S has a triple point.
(b) S has a double point of rank 1.
(c) S contains the axis of a rank 2 double point.
(d) S is reducible.

The converse is easy and we leave it to the reader. $\qquad\square$

Finally, we will classify the closed orbits of semistable points under the action $GL(4) \curvearrowright \mathbb{V}_{3,3}$. Of course, all stable orbits are closed; for the nonstable orbits the problem is, to which of the types (1), (2), (3) of Proposition 7.17 do they belong? Note that, if an orbit is closed, then it can be represented by the limit of a point under any 1-PS. As an example, let us see what happens when we take a cubic of type (1) and pass to the limit under the 1-PS $\lambda = |1, 1, 0, -2\rangle$. A monomial m remains in the limit if and only if $\langle m \mid \lambda \rangle = 0$, and the only monomials with this property are y^3 and wxz. The limiting cubic is therefore of the form $ay^3 + bwxz$, and semistability guarantees that both $a, b \neq 0$. We therefore have a candidate for a closed semistable and nonstable orbit:

$$SL(4) \cdot (y^3 - wxz)$$

In fact, if we apply the same reasoning to types (2) and (3), then we find (up to scalar) the same orbit again, and no others. We just have to check:

Proposition 7.23. $SL(4) \cdot (y^3 - wxz)$ *is a closed orbit.*

Proof. The surface $S: y^3 - wxz = 0$ is semistable since it has three singular points $(1 : 0 : 0 : 0), (0 : 1 : 0 : 0), (0 : 0 : 0 : 1)$ which are all rank 2 double points whose axes do not lie on S. (In fact, these axes are the edges of the tetrahedron of reference passing through the last vertex $(0 : 0 : 1 : 0)$.) Consider the closure of the orbit $SL(4) \cdot (y^3 - wxz)$. This must contain a closed orbit, and what we have seen above is that the latter must be projectively equivalent to the orbit of $y^3 - wxz$. But this means that $SL(4) \cdot (y^3 - wxz)$ is itself closed.   □

Theorem 7.24. *The moduli space of semistable cubic surfaces* $\mathbb{V}_{3,3}//GL(4)$ *is the one-point compactification of the geometric quotient* $\mathbb{V}_{3,3}^s/GL(4)$ *which parametrises projective equivalence classes of stable cubic surfaces.*   □

The single point that we add corresponds to the closure equivalence class of semistable surfaces which are not stable, and this class is represented by the unique closed orbit of the surface $y^3 = wxz$.

Remark 7.25. It is easy to see that the surface $S: y^3 = wxz$ is the image of the rational map
$$\mathbb{P}^2 - - \to \mathbb{P}^3, \qquad (a : b : c) \mapsto (a^2b : b^2c : abc : c^2a).$$
It is therefore isomorphic to the quotient of $\mathbb{P}^2$ by the cyclic group $\mathbb{Z}/3$ generated by the automorphism
$$\sigma : \mathbb{P}^2 \to \mathbb{P}^2, \qquad (x : y : z) \mapsto (y : z : x).$$

□

(c) Finite point sets in projective space

Let us consider finite (unordered) point sets in projective space $\Lambda = \{p_1, \ldots, p_d\} \subset \mathbb{P}^n$ and their stability under the action of the general linear group $GL(n + 1) \curvearrowright S^d\mathbb{P}^n$. Of course, the symmetric product $S^d\mathbb{P}^n$ is not an affine variety, so to define stability we need to embed it in a suitable projective space in such a way that $GL(n + 1)$ acts on its affine cone.

In the case $n = 1$ we can view $\{p_1, \ldots, p_d\} \subset \mathbb{P}^1$ as the zero-set of a binary form of degree d,

$$F_d(x, y) = \prod_{i=1}^{d} (b_1 x - a_i y),$$

where $p_1 = (a_1 : b_1), \ldots, p_d = (a_d : b_d)$. In other words, $S^d \mathbb{P}^1 \cong \mathbb{P}(V_d)$, and the quotient we are considering is $V_d /\!/ GL(2)$. Then, as we have seen in Proposition 7.9,

$$\begin{pmatrix} \text{the set } \{p_1, \ldots, p_d\} \subset \mathbb{P}^1 \\ \text{is semistable} \end{pmatrix} \quad \Longleftrightarrow \quad \begin{pmatrix} \text{no more than } [d/2] \text{ of the} \\ \text{points coincide} \end{pmatrix}.$$

This has the following generalisation. We can view the set $\Lambda = \{p_1, \ldots, p_d\} \subset \mathbb{P}^n$ as a hypersurface of degree d, reducible as a union of d hyperplanes, in the dual projective space $(\mathbb{P}^n)^\vee$, as follows. In terms of homogeneous coordinates we write

$$p_i = (a_0^{(i)} : a_1^{(i)} : \ldots : a_n^{(i)})$$

and the linear form with these coordinates as coefficents as

$$\pi_{p_i}(x_0, x_1, \ldots, x_n) := a_0^{(i)} x_0 + a_1^{(i)} x_1 + \cdots + a_n^{(i)} x_n.$$

We then associate to the unordered point set Λ the homogeneous polynomial

$$\pi_\Lambda(x_0, x_1, \ldots, x_n) = \prod_{i=1}^{d} \pi_{p_i}(x_0, x_1, \ldots, x_n),$$

and by unique factorisation (Theorem 2.10) the set $\Lambda \subset \mathbb{P}^n$ is completely determined by this polynomial.

Definition 7.26. The point set $\Lambda \subset \mathbb{P}^n$ will be called *(semi)stable* if the corresponding degree d form π_Λ is (semi)stable. $\square$

Proposition 7.27. *A point set $\Lambda \subset \mathbb{P}^n$ is stable under the action of $GL(n + 1)$ if and only if*

$$\frac{\sharp \, \Lambda \cap P}{\sharp \, \Lambda} < \frac{\dim P + 1}{n + 1}$$

for every projective subspace $P \subset \mathbb{P}^n$. If the inequality $\leq$ is satisfied, then Λ is semistable.

Proof. We will assume that the inequality holds and deduce that the form $\pi_\Lambda(x)$, and hence Λ, is stable, leaving the converse to the reader.

Let $\mu : \mathbb{G}_m \to SL(n+1)$ be an arbitrary 1-PS; we have to show that there is no limit $\lim_{t \to 0} \mu(t) \cdot \pi_\Lambda(x)$. We first choose homogeneous coordinates $(x_0 : x_1 : \ldots : x_n)$ for which μ is normalised so that

$$\mu : t \mapsto \operatorname{diag}(t^{r_0}, t^{r_1}, \ldots, t^{r_n}), \qquad \text{where } r_0 \le r_1 \le \cdots \le r_n.$$

Let $P_k \subset \mathbb{P}^n$ be the subspace defined by $x_0 = \cdots = x_{n-k-1} = 0$, and let a_k be the number of points in $\Lambda \cap P_k$. The form $\pi_\Lambda(x)$ contains (with nonzero coefficient) the monomial

$$x_0^{a_n - a_{n-1}} x_1^{a_{n-1} - a_{n-2}} \ldots x_{n-1}^{a_1 - a_0} x_n^{a_0}$$

with weight (with respect to μ)

$$w = (a_n - a_{n-1})r_0 + (a_{n-1} - a_{n-2})r_1 + \cdots + (a_1 - a_0)r_{n-1} + a_0 r_n.$$

The inequality in the proposition, applied to P_k, implies that

$$a_k < \frac{k+1}{n+1}d,$$

and hence

$$w = a_n r_0 - a_{n-1}(r_0 - r_1) - a_{n-2}(r_1 - r_2) - \cdots$$
$$-a_1(r_{n-2} - r_{n-1}) - a_0(r_{n-1} - r_n)$$

$$< dr_0 - \frac{nd}{n+1}(r_0 - r_1) - \frac{(n-1)d}{n+1}(r_1 - r_2) - \cdots$$
$$-\frac{2d}{n+1}(r_{n-2} - r_{n-1}) - \frac{d}{n+1}(r_{n-1} - r_n)$$

$$= \frac{d}{n+1}\left((n+1)r_0 - n(r_0 - r_1) - (n-1)(r_1 - r_2) - \cdots\right.$$
$$\left. -2(r_{n-2} - r_{n-1}) - (r_{n-1} - r_n)\right)$$

$$= 0.$$

This shows that $\mu(t) \cdot \pi_\Lambda(x)$ does not have a limit as $t \to 0$, as claimed. $\square$

Example 7.28. Five points in $\mathbb{P}^2$. In this case, stability and semistability are equivalent and a set is stable if and only if:

(1) the five points are all distinct, and
(2) no four of the points lie on a line. $\square$

Example 7.29. Six points in $\mathbb{P}^3$. Here a set is stable if and only if:

(1) the six points are all distinct,
(2) no three of the points lie on a line, and
(3) no five of the points lie on a plane.

Semistability is equivalent to conditions (1), (3), and

(2′) no four points lie on a line. □

8

Grassmannians and vector bundles

It is well known that the set of vector subspaces of a fixed dimension in a fixed vector space is a projective algebraic variety, called the Grassmannian. We are going to examine the Grassmannian as an example of a Proj quotient by a group action of ray type. In Section 8.1, using a construction of this variety by means of invariants, we shall study, in the case $\mathbb{G}(2, n)$, its coordinate ring. We compute its Hilbert series, its generators and their relations.

From Section 8.2 we shall review, as preparation for the chapters which follow, the theory of modules over a ring. We discuss localisation and gluings by partitions of unity, free modules, tensor products and flat modules. In Section 8.3 we define locally free modules and invertible modules, and the properties of these which follow from flatness.

The set of equivalence classes of invertible modules forms an abelian group under the tensor product, called the *Picard group* of the ring. For the ring of integers of an algebraic number field, for example, this coincides with the divisor class group. In Section 8.4 we calculate the Picard group explicitly in the cases of an imaginary quadratic field and of an affine hyperelliptic curve (that is, a quadratic extension of the polynomial ring $k[x]$). This paves the way for the later discussion of the Jacobian variety.

Just as one obtains an algebraic variety by gluing together affine spectra of algebras, so too one can glue invertible modules, or locally free modules, over a ring to form line bundles, or vector bundles, over algebraic varieties. The line bundles (vector bundles of rank 1) form a group Pic X under $\otimes$, which coincides with Pic R when $X = \text{Spm } R$ is affine.

In the final section we construct the tautological line bundle on a projective space and the universal vector bundle on a Grassmannian $\mathbb{G}(r, n)$, and use this to show that the Grassmannian represents the functor $\mathcal{G}r(r, n)$ which assigns to a ring R the set of locally free rank r submodules of $R^{\oplus n}$ up to isomorphism.

We compute the tangent space of $\mathbb{G}(r, n)$ at a point and of the Grassmannian functor which it represents.

8.1 Grassmannians as quotient varieties

The set of $r \times n$ matrices is an rn-dimensional vector space, and via multiplication on the left by $r \times r$ matrices this vector space becomes a representation of the general linear group $GL(r)$. We denote this space, viewed as an affine space, by $\mathrm{Mat}(r, n)$. We shall consider its quotient by the action of $GL(r)$.

First of all, let

$$X = \begin{pmatrix} x_{11} & x_{12} & \cdots & x_{1n} \\ & & \cdots & \\ x_{r1} & x_{r2} & \cdots & x_{rn} \end{pmatrix} \tag{8.1}$$

be a matrix of independent variables. Using these variables we identify $k[\mathrm{Mat}(r, n)]$ with $k[x_{ij}]$, on which $GL(r)$ acts. We consider the projective spectrum of the semiinvariant ring

$$k[\mathrm{Mat}(r, n)]^{SL(r)} = \bigoplus_{w \geq 0} k[\mathrm{Mat}(r, n)]^{SL(r)}_{(w)}. \tag{8.2}$$

The weight w of a homogeneous polynomial is $1/r$ times its degree; so in particular all components of negative weight are zero, and the only polynomials of weight zero are the constants. In other words, the action $GL(r) \curvearrowright \mathrm{Mat}(r, n)$ is of ray type (Definition 6.12). Moreover, the fixed point set F of the multiplicative group of scalar matrices $\mathbb{G}_{\mathrm{m}} \subset GL(r)$ is just the origin $O \in \mathrm{Mat}(r, n)$ (see Section 6.1(a)).

For each subset $I = \{i_1 < \cdots < i_r\} \subset \{1, \ldots, n\}$ we will denote by X_I the $r \times r$ submatrix of (8.1) constructed from columns $i_1, \ldots, i_r$. The minor $\det X_I$ is a homogeneous polynomial of degree r in $k[\mathrm{Mat}(r, n)]$ and is a semiinvariant of weight 1.

Proposition 8.1. *Given a matrix* $A \in \mathrm{Mat}(r, n)$, *the following are equivalent:*

(i) A is stable for the action of $GL(r)$.
(ii) A is semistable for the action of $GL(r)$.
(iii) A has rank r.

Proof. (i) $\Longrightarrow$ (ii) is trivial. For (ii) $\Longrightarrow$ (iii), we suppose rank $A < r$. Then, by moving A within its $GL(r)$ orbit we may assume that its first row is zero:

$$A = \begin{pmatrix} 0 & 0 & \cdots & \cdots & 0 \\ * & * & \cdots & \cdots & * \\ & & \cdots & \cdots & \\ * & * & \cdots & \cdots & * \end{pmatrix}.$$

The 1-parameter subgroup of the special linear group $\lambda : \mathbb{G}_m \to SL(r)$ defined by

$$\lambda(t) = \begin{pmatrix} t^{-r+1} & & & \\ & t & & \\ & & \ddots & \\ & & & t \end{pmatrix}$$

acts on the left on $\mathrm{Mat}(r, n)$ and sends A to

$$\begin{pmatrix} 0 & 0 & \cdots & \cdots & 0 \\ t* & t* & \cdots & \cdots & t* \\ & & \cdots & \cdots & \\ t* & t* & \cdots & \cdots & t* \end{pmatrix}.$$

As $t \to 0$ this tends to the origin O, which therefore lies in the closure of the $SL(r)$ orbit of A.

(iii) $\Longrightarrow$ (i) If A has rank r, then there is a submatrix A_I whose determinant is nonzero. Since $\det A_I$ is a semiinvariant of positive weight, it follows that A is semistable. Moreover, the set of matrices of rank r is an open set on which the stabiliser subgroup in $GL(r)$ is trivial. It then follows from Corollary 5.14 that all the orbits in this subset are closed. $\qquad\square$

We assume from now on that $r < n$. The projective spectrum of $k[\mathrm{Mat}(r, n)]^{SL(r)}$ then parametrises the set of stable $GL(r)$-orbits in $\mathrm{Mat}(r, n)$. (This follows from Theorem 5.3, though in the present case it is also easy to prove directly.) By associating to a matrix $A \in \mathrm{Mat}(r, n)$ the space spanned by its rows, this orbit space coincides with the set of r-dimensional subspaces of a fixed n-dimensional vector space.

Definition 8.2. The projective spectrum

$$\mathbb{G}(r, n) := \mathrm{Proj}\, k[\mathrm{Mat}(r, n)]^{SL(r)}$$

is called the *Grassmannian variety* of linear r-planes in k^n. $\qquad\square$

The unstable points in $\mathrm{Mat}(r, n)$ are the matrices for which all $r \times r$ minors $\det X_I$ are zero. Thus $\mathbb{G}(r, n)$ is covered by $\binom{n}{r}$ affine varieties $\{\det X_I \neq 0\}/GL(r)$. Moreover, each of these affine varieties is isomorphic to $\mathbb{A}^{r(n-r)}$. For example, when $I = \{1, 2, \ldots, r\}$ each orbit in the affine open set is uniquely represented by a matrix

$$(X_I)^{-1}X = \begin{pmatrix} 1 & 0 & \cdots & 0 & * & \cdots & * \\ 0 & 1 & \cdots & 0 & * & \cdots & * \\ & & \cdots & & & \cdots & \\ 0 & 0 & \cdots & 1 & * & \cdots & * \end{pmatrix} \tag{8.3}$$

in which the $r(n-r)$ entries $*$ serve as coordinates on $\mathbb{A}^{r(n-r)}$. (Note that each $*$ is a $GL(r)$-invariant rational form on $\mathrm{Mat}(r, n)$.)

Example 8.3. The case $r = 1$ is the construction of projective space $\mathbb{P}^n$ in Examples 3.40 and 3.45. □

In the remainder of this section we shall examine the semiinvariant ring (8.2) in the case $r = 2$.

(a) Hilbert series

As a representation of $SL(2)$, the vector space $\mathrm{Mat}(2, n)$ is the direct sum $V_1 \oplus \cdots \oplus V_1$ of n copies of the 2-dimensional irreducible representation V_1 (see Section 4.4(a)). The q-Hilbert series (see Section 4.4(d)) of V_1 is $P(q;t) = 1/(1 - qt)(1 - q^{-1}t)$, so the q-Hilbert series of $\mathrm{Mat}(2, n)$ is

$$P(q;t) = \frac{1}{(1 - qt)^n(1 - q^{-1}t)^n}.$$

It follows from Proposition 4.63 that the invariant ring $k[\mathrm{Mat}(2, n)]^{SL(2)}$ has Hilbert series

$$\sum_{k=0}^{\infty} \dim \left\{ \begin{matrix} SL(2) \text{ invariant} \\ \text{forms of degree } k \end{matrix} \right\} t^k = \operatorname*{Res}_{q=0} \left[\frac{q^{-1} - q}{(1 - qt)^n(1 - q^{-1}t)^n} \right].$$

The expression in square brackets is

$$\left(\frac{1}{q} - q \right) \sum_{j=0}^{\infty} \binom{n + j - 1}{j} q^j t^j \sum_{k=0}^{\infty} \binom{n + k - 1}{k} \frac{t^k}{q^k},$$

and by reading off the coefficient of q^{-1} and replacing t by $\sqrt{t}$ we obtain the following.

Proposition 8.4. *The Hilbert series of the semiinvariant ring (8.2), computing the dimensions of the spaces of weight w semiinvariants, is equal to:*

$$P_n(t) = \sum_{w=0}^{\infty} \dim \left(k[\text{Mat}(2, n)]_{(w)}^{SL(2)} \right) t^w$$

$$= \sum_{w=0}^{\infty} \left\{ \binom{n+w-1}{w}^2 - \binom{n+w}{w+1}\binom{n+w-2}{w-1} \right\} t^w.$$

□

The expression in braces, the dimension of the space of semiinvariants of weight w, is

$$H_n(w) :=$$
$$\frac{(w+n-1)(w+n-2)^2(w+n-3)^2\dots(w+3)^2(w+2)^2(w+1)}{(n-1)!(n-2)!},$$
(8.4)

which is a polynomial in w of degree $2n - 4$.

Example 8.5. For $n \leq 6$ these polynomials are

$$H_2(w) = 1$$

$$H_3(w) = \frac{(w+2)(w+1)}{2!}$$

$$= \binom{w+2}{2}$$

$$H_4(w) = \frac{(w+3)(w+2)^2(w+1)}{3!2!}$$

$$= \binom{w+4}{4} + \binom{w+3}{4}$$

$$H_5(w) = \frac{(w+4)(w+3)^2(w+2)^2(w+1)}{4!3!}$$

$$= \binom{w+6}{6} + 3\binom{w+5}{6} + \binom{w+4}{6}$$

$$H_6(w) = \frac{(w+5)(w+4)^2(w+3)^2(w+2)^2(w+1)}{5!4!}$$

$$= \binom{w+8}{8} + 6\binom{w+7}{8} + 6\binom{w+6}{8} + \binom{w+5}{8}.$$

With these coefficients we obtain Hilbert series:

$$P_2(t) = \frac{1}{1-t}$$

$$P_3(t) = \frac{1}{(1-t)^3}$$

$$P_4(t) = \frac{1+t}{(1-t)^5} \qquad = \frac{1-t^2}{(1-t)^6}$$

$$P_5(t) = \frac{1+3t+t^2}{(1-t)^7} \qquad = \frac{1-5t^2+5t^3-t^5}{(1-t)^{10}}$$

$$P_6(t) = \frac{1+6t+6t^2+t^3}{(1-t)^9} = \frac{1-15t^2+35t^3-21t^4-21t^5+35t^6-15t^7+t^9}{(1-t)^{15}}.$$

The cases $n = 2, 3, 4$ correspond, respectively, to the simplest Grassmannians $\mathbb{G}(2, 2) = \{\text{point}\}$, $\mathbb{G}(2, 3) = (\mathbb{P}^2)^\vee$ and the quadric $\mathbb{G}(2, 4) \subset \mathbb{P}^5$. (See Example 8.18). $\qquad\qquad\Box$

(b) Standard monomials and the ring of invariants

The following is the *first fundamental theorem of invariant theory*.

Theorem 8.6. *The ring of (semi)invariants* $k[\text{Mat}(r, n)]^{SL(r)}$ *is generated by the* $\binom{n}{r}$ *minors* $\det X_I$, $|I| = r$, *of the matrix (8.1).* $\qquad\qquad\Box$

We will give a proof of this for the case $r = 2$. By Proposition 8.1, the common zero-set in $\text{Mat}(2, n)$ of all the positive weight semiinvariants consists of the matrices of rank ≤ 1. The projectivisation of this set in $\mathbb{P}\text{Mat}(2, n) = \mathbb{P}^{2n-1}$ is the image of the *Segre map*

$$\mathbb{P}^1 \times \mathbb{P}^{n-1} \hookrightarrow \mathbb{P}^{2n-1}, \qquad ((a_1 : a_2), (b_1 : \ldots : b_n)) \mapsto \begin{pmatrix} a_1b_1 & \cdots & a_1b_n \\ a_2b_1 & \cdots & a_2b_n \end{pmatrix}.$$

Writing

$$X = \begin{pmatrix} x_1 & x_2 & \cdots & x_n \\ y_1 & y_2 & \cdots & y_n \end{pmatrix} \tag{8.5}$$

for the matrix of indeterminates (8.1), denote by $I_n \subset k[x_1, \ldots, x_n, y_1, \ldots, y_n]$ the ideal of polynomials vanishing identically on the Segre variety. This is the kernel of the homomorphism

$$
\begin{aligned}
k[x_1, \ldots, x_n, y_1, \ldots, y_n] \quad &\to \quad k[s_1, s_2, t_1, \ldots, t_n] \\
x_i \quad &\mapsto \quad s_1 t_i, \\
y_j \quad &\mapsto \quad s_2 t_j.
\end{aligned}
\tag{8.6}
$$

This kernel contains, in particular, the 2×2 minors

$$p_{ij}(x, y) := \begin{vmatrix} x_i & x_j \\ y_i & y_j \end{vmatrix}, \qquad 1 \le i < j \le n. \tag{8.7}$$

The case $r = 2$ of Theorem 8.6 reduces, using the claim made in the proof of Theorem 4.51, to the following statement.

Theorem 8.7. *The homogeneous ideal I_n of the Segre variety $\mathbb{P}^1 \times \mathbb{P}^n \hookrightarrow \mathbb{P}^{2n-1}$ is generated by the minors (8.7).*

It will be convenient to adopt the notation $\mathbf{x} = (x_1, \ldots, x_n), \mathbf{y} = (y_1, \ldots, y_n)$, $k[\mathbf{x}, \mathbf{y}] = k[x_1, \ldots, x_n, y_1, \ldots, y_n]$.

Definition 8.8. A *standard monomial* in $k[\mathbf{x}, \mathbf{y}]$ is a monomial

$$x_{i_1} x_{i_2} \ldots x_{i_a} y_{j_1} y_{j_2} \ldots y_{j_b}, \qquad i_1 \le \cdots \le i_a, \quad j_1 \le \cdots \le j_b,$$

for which $i_a \le j_i$. $\qquad\qquad\qquad\qquad\qquad\qquad\qquad\qquad \square$

Let $I'_n \subset k[\mathbf{x}, \mathbf{y}]$ be the ideal generated by the minors $p_{ij}(x, y)$. The idea, to show that $I_n \subset I'_n$, is to 'straighten' arbitrary monomials, modulo I'_n, into standard monomials.

Lemma 8.9 (Straightening quadratic monomials). *Every quadratic monomial is congruent to a standard monomial modulo I'_n.*

Proof. $x_i x_j$ and $y_i y_j$ are already standard monomials, as is $x_i y_j$ if $i \le j$. If $i > j$, we just note that $x_i y_j$ is congruent to $x_j y_i$, which is standard, modulo $p_{ij}(x, y)$. $\qquad\qquad\qquad\qquad\qquad\qquad\qquad\qquad \square$

Lemma 8.10 (Straightening higher monomials). *An arbitrary monomial in* $k[\mathbf{x}, \mathbf{y}]$ *is congruent to a standard monomial modulo* I'_n.

Proof. Consider a monomial $m = x_{i_1} \ldots x_{i_a} y_{j_1} \ldots y_{j_b}$ as in Definition 8.8 and suppose that $i_a > j_1$. We write $\alpha(m) = i_a - j_1$. Using the minor $p_{i_a j_1}(x, y)$ we can replace $x_{i_a} y_{j_1}$ by $x j_1 y_{i_a}$, and repeating this operation if necessary (since x_{i_a} or y_{j_1} may occur with multiplicity in m) we can replace m by another monomial m' in the same residue class modulo I'_n in which one or other of x_{i_a} or y_{j_1} does not appear. We then have $\alpha(m') < \alpha(m)$. Repeating this procedure we eventually obtain a monomial m'' for which $\alpha(m'') \leq 0$ and which is therefore standard. $\qquad\square$

The monomial in Definition 8.8 is of degree $d = a + b$. If we substitute $x_i \mapsto s_1 t_i$, $y_j \mapsto s_2 t_j$, then we obtain a monomial

$$s_1^a s_2^b t_{i_1} t_{i_2} \ldots t_{i_a} t_{j_1} t_{j_2} \ldots t_{j_b},$$

which has degree d in each of $\mathbf{s} := (s_1, s_2)$ and $\mathbf{t} := (t_1, \ldots, t_n)$. We will call this a monomial of bidegree (d, d) in $(\mathbf{s}, \mathbf{t})$.

Lemma 8.11. *The map induced by the substitution (8.6)*

$$\left\{ \begin{matrix} \text{standard monomials} \\ \text{of degree } d \end{matrix} \right\} \to \left\{ \begin{matrix} \text{degree } d \\ \text{monomials in } \mathbf{s} \end{matrix} \right\} \times \left\{ \begin{matrix} \text{degree } d \\ \text{monomials in } \mathbf{t} \end{matrix} \right\}$$

is bijective.

Proof. Any monomial of bidegree (d, d) in $(\mathbf{s}, \mathbf{t})$ can be written uniquely as $s_1^a s_2^b t_{i_1} \ldots t_{i_a} t_{j_1} \ldots t_{j_b}$ with $a + b = d$ and $1 \leq i_1 \leq \cdots \leq i_a \leq j_1 \leq \cdots \leq j_b$. But this is the image of the standard monomial $x_{i_1} \ldots x_{i_a} y_{j_1} \ldots y_{j_b}$. $\qquad\square$

Proof of Theorem 8.7. We have seen that $I'_n \subset I_n$, and it remains to show the reverse inclusion. As d varies, the monomials of bidegree (d, d) in $\mathbf{s}, \mathbf{t}$ form a basis of (the image of) $k[\mathbf{x}, \mathbf{y}]/I_n$. On the other hand, by Lemma 8.10, the standard monomials generate $k[\mathbf{x}, \mathbf{y}]/I'_n$. Hence the two ideals I_n, I'_n coincide. $\qquad\square$

(c) Young tableaux and the Plücker relations

A *Young diagram of size* $(k_1, \ldots, k_r)$ is an array of r rows of empty boxes, with k_i in the i-th row. Rectangular Young diagrams, for which $k_1 = \cdots = k_r = w$, say, provide a useful tool for describing the weight w summand of the semiinvariant ring $k[\mathrm{Mat}(r, n)]^{SL(r)}$. Here we restrict ourselves to the case $r = 2$ and will only be concerned with Young diagrams of size (w, w):

We are going to use these to determine the relations among the generating minors $p_{ij}(x, y)$ of the homogenous coordinate ring $k[\mathrm{Mat}(2, n)]^{SL(2)}$ of the Grassmannian $\mathbb{G}(2, n)$.

Definition 8.12.

(i) A Young diagram of size (w, w) in which each box contains an integer from $\{1, \ldots, n\}$ is called a *Young tableau* (for Mat$(2, n)$):

$$
\begin{array}{|c|c|c|c|}
\hline
i_1 & i_2 & \cdots & i_w \\
\hline
j_1 & j_2 & \cdots & j_w \\
\hline
\end{array}
\qquad 1 \le i_\alpha, j_\beta \le n.
$$

(ii) A *standard tableau* (for Mat$(2, n)$) is a Young tableau whose entries satisfy the two conditions:

(S1)
$$
\begin{aligned}
i_1 \le i_2 \le \cdots \le i_w \\
j_1 \le j_2 \le \cdots \le j_w
\end{aligned}
$$

(S2)
$$
\begin{array}{ccccc}
i_1 & i_2 & & i_w \\
\wedge & \wedge & \cdots & \wedge \\
j_1 & j_2 & & j_w
\end{array}
$$

□

Examples 8.13. $w = 1$. Here a standard tableau looks like $\begin{array}{|c|} \hline i \\ \hline j \\ \hline \end{array}$ with $i < j$.

$w = 2$. If $n = 3$, for example, there are exactly six standard tableaux:

$$
\begin{array}{|c|c|}\hline 1&1\\\hline 2&2\\\hline\end{array}
\quad
\begin{array}{|c|c|}\hline 1&1\\\hline 2&3\\\hline\end{array}
\quad
\begin{array}{|c|c|}\hline 1&2\\\hline 2&3\\\hline\end{array}
\quad
\begin{array}{|c|c|}\hline 1&1\\\hline 3&3\\\hline\end{array}
\quad
\begin{array}{|c|c|}\hline 1&2\\\hline 3&3\\\hline\end{array}
\quad
\begin{array}{|c|c|}\hline 2&2\\\hline 3&3\\\hline\end{array}
$$

□

Lemma 8.14. *The number of standard tableaux of size (w, w) for* Mat$(2, n)$ *is equal to*

$$
\binom{n+w-1}{w}^2 - \binom{n+w}{w+1}\binom{n+w-2}{w-1}.
$$

Proof. First, the number of Young tableaux of size (w, w) satisfying just condition (S1) is

$$\binom{n + w - 1}{w}^2.$$

(Choosing either row is equivalent to choosing w objects from $1, \ldots, n$, $R_1, \ldots, R_{w-1}$, where R_a denotes the rule 'let $i_a = i_{a+1}$'.) Within this set we will classify those Young tableaux that are not standard. Suppose T is a Young tableau satisfying condition (S1), and that the first column from the left in which condition (S2) is violated is the a-th. In other words, $i_1 < j_1, \ldots, i_{a-1} < j_{a-1}$, but $i_a \geq j_a$. In this situation we obtain a pair of nondecreasing sequences:

$$j_1 \leq \cdots \leq j_a \leq i_a \leq i_{a+1} \leq \cdots \leq i_w \qquad \text{of length } w + 1,$$
$$i_1 \leq \cdots \leq i_{a-1} \leq j_{a+1} \leq \cdots \leq j_w \qquad \text{of length } w - 1.$$

The set of such pairs has cardinality $\binom{n+w}{w+1}\binom{n+w-2}{w-1}$, by the same reasoning as above. It is therefore enough to show that the map

$$\begin{Bmatrix} \text{nonstandard} \\ \text{Young tableaux} \end{Bmatrix} \longrightarrow \begin{Bmatrix} \text{nondecreasing} \\ \text{sequences of} \\ \text{length } w + 1 \end{Bmatrix} \times \begin{Bmatrix} \text{nondecreasing} \\ \text{sequences of} \\ \text{length } w - 1 \end{Bmatrix} \qquad (8.8)$$

is a bijection. We will construct the inverse map. Given sequences

$$i_0 \leq i_1 \leq \cdots i_{w-1} \leq i_w,$$
$$j_1 \leq \cdots \leq j_{w-1},$$

we want to construct a nonstandard Young tableau N. First we compare i_0 and j_1. If $i_0 \leq j_1$, then

$$N = \begin{array}{|c|c|c|c|c|} \hline i_1 & i_2 & i_3 & \cdots & i_w \\ \hline i_0 & j_1 & j_2 & \cdots & j_{w-1} \\ \hline \end{array}$$

will do. If $i_0 > j_1$, then the first column of N will be

$$\begin{array}{|c|c|c|c|} \hline j_1 & & \cdots & \\ \hline i_0 & & \cdots & \\ \hline \end{array}$$

and we then compare i_1 and j_2. If $i_1 \leq j_2$, then we take

$$N = \begin{array}{|c|c|c|c|c|} \hline j_1 & i_2 & i_3 & \cdots & i_w \\ \hline i_0 & i_1 & j_2 & \cdots & j_{w-1} \\ \hline \end{array}$$

and we are done. If $i_1 > j_2$, then the first two columns are determined:

j_1	j_2		$\cdots$	
i_0	i_1		$\cdots$	

and we compare i_2 and j_3. Repeating this process we eventually obtain a nonstandard tableau and an inverse of the map (8.8). □

We now consider the ideal of relations among the 2×2 minors (8.7), that is, the kernel of the ring homomorpism

$$S = k[p_{ij}]_{1 \leq i,j \leq n} \to k[\mathrm{Mat}(2, n)]^{SL(2)}, \quad p_{ij} \mapsto p_{ij}(x, y).$$

We denote this kernel by $J_n \subset S$.

For distinct numbers $i, j, k, l \in \{1, \ldots, n\}$ consider the 4×4 determinant

$$\begin{vmatrix} x_i & x_j & x_k & x_l \\ y_i & y_j & y_k & y_l \\ x_i & x_j & x_k & x_l \\ y_i & y_j & y_k & y_l \end{vmatrix}.$$

Evaluating this by the Laplace expansion along the first two rows yields an identity

$$p_{ij}(x, y)p_{kl}(x, y) - p_{ik}(x, y)p_{jl}(x, y) + p_{jk}(x, y)p_{il}(x, y) = 0. \tag{8.9}$$

These $\binom{n}{4}$ relations are called the *Plücker relations* for the ring $k[\mathrm{Mat}(2, n)]^{SL(2)}$.

The following is a special case (for $r = 2$) of the *second fundamental theorem of invariant theory*.

Theorem 8.15. *The ideal $J_n \subset S = k[p_{ij}]$ is generated by the Plücker relations*

$$p_{ij}p_{kl} - p_{ik}p_{jl} + p_{jk}p_{il}, \qquad 1 \leq i < j < k < l \leq n.$$

□

Proof. To each monomial of degree w in S corresponds a Young tableau of size (w, w) by

$$p_{i_1 j_1} \cdots p_{i_w j_w} \mapsto \begin{array}{|c|c|c|c|} \hline i_1 & i_2 & \cdots & i_w \\ \hline j_1 & j_2 & \cdots & j_w \\ \hline \end{array}.$$

A monomial which corresponds in this way to a standard tableau is called a *standard monomial*. Let $J_n' \subset S$ be the ideal generated by the Plücker relations – so clearly $J_n' \subset J_n$. For $i < j < k < l$, the Plücker relation (8.9) can be interpreted as saying that the monomial

$$p_{jk} p_{il} \equiv p_{ik} p_{jl} - p_{ij} p_{kl} \quad \mathrm{mod}\ J_n'$$

is nonstandard but is expressed as a sum of standard monomials modulo J_n'. (This is 'quadratic straightening' – compare with Lemma 8.9.) Thus, by applying this process inductively one sees that an arbitrary monomial in S is congruent modulo J_n' to a standard monomial. In other words, the residue ring S/J_n' is spanned, as a vector space, by standard monomials. On the other hand, by Proposition 8.4 and Lemma 8.14, the number of standard monomials of degree w is equal to the dimension of $(S/J_n)_w$, and hence $\dim(S/J_n')_w \le \dim(S/J_n)_w$. The reverse inequality holds because $J_n' \subset J_n$, and so the two ideals are equal. $\square$

Remarks 8.16.

(i) It follows from this proof that for each $w \in \mathbb{N}$ the standard monomials of degree w form a basis of $(S/J_n)_w = k[\mathrm{Mat}(2, n)]_{(w)}^{SL(2)}$.

(ii) As well as an action of $GL(r)$ on the left, the space $\mathrm{Mat}(r, n)$ has an action on the right by $GL(n)$. Consequently, the semiinvariant ring $k[\mathrm{Mat}(r, n)]^{SL(r)}$ is a representation of the group $GL(n)$. Moreover, the weight w summand $k[\mathrm{Mat}(r, n)]_{(w)}^{SL(r)}$ is a finite-dimensional subrepresentation, and using the theory of characters one can show that it is irreducible. $\square$

(d) Grassmannians as projective varieties

By Theorem 8.6, the Grassmannian $\mathbb{G}(2, n) = \mathrm{Proj}\ k[\mathrm{Mat}(2, n)]^{SL(2)}$ has an embedding as a closed subvariety of $\binom{n}{2} - 1 = (n - 2)(n + 1)/2$-dimensional projective space:

$$\mathbb{G}(2, n) \hookrightarrow \mathbb{P}^{(n-2)(n+1)/2}.$$

This map is called the *Plücker embedding*. By Proposition 8.4, the polynomial $H_n(w)$ of degree $2n - 4$ (see (8.4)) is the Hilbert polynomial of $\mathbb{G}(2, n)$, which means the following.

Suppose that $S = \bigoplus_{w=0}^{\infty} S_w$ is a graded integral domain with $S_0 = k$ and generated over k by S_1. Then (see Remark 3.74) $X = \mathrm{Proj}\ S$ has a closed immersion in a projective space $\mathbb{P}^N$, where $N + 1 = \dim_k S_1$.

(i) There exists a polynomial $H_S(x) \in \mathbb{Q}[x]$ such that for some $w_0 \in \mathbb{N}$

$$\dim S_w = H_S(w) \quad \text{for all } w \ge w_0.$$

$H_S(x)$ is called the *Hilbert polynomial* of S.

(ii) The degree of $H_S(x)$ is equal to the dimension of X.

(iii) The leading coefficient of $H_S(x)$ is equal to $(\deg X)/m!$, where $m = \dim X$. That is,

$$H_S(x) = \frac{\deg X}{m!} x^m + \text{lower degree terms}, \quad m = \dim X.$$

By definition, the degree of X is the number of intersection points

$$X \cap H_1 \cap \cdots \cap H_m$$

with m general hyperplanes in $\mathbb{P}^N$.

From (8.4) we obtain the following. (And we will return to re-examine this degree in the next subsection.)

Proposition 8.17. $\mathbb{G}(2, n) \subset \mathbb{P}^{(n-2)(n+1)/2}$ *has degree* $\dfrac{1}{n-1}\dbinom{2n-4}{n-2}$. $\quad\square$

This number is called the $((n - 2)$-nd) *Catalan number* and has various interpretations in combinatorics. For low values of n it takes the following values.

dim $\mathbb{G}(2, n)$	2	4	6	8	10	12	14	16	18	20
deg $\mathbb{G}(2, n)$	1	2	5	14	42	132	429	1430	4862	16796

By Theorem 8.15, the Grassmannian $\mathbb{G}(2, n) \subset \mathbb{P}^{n(n-3)/2}$ is cut out (scheme-theoretically) by the $\binom{n}{4}$ quadrics determined by the Plücker relations of Theorem 8.15. Note, incidentally, that these quadrics are exactly the 4×4 Pfaffian minors of the $n \times n$ skew-symmetric matrix $P := (p_{ij})_{1 \le i, j \le n}$. (See Section 10.3(a).)

Examples 8.18.

$\mathbb{G}(2, 3)$ is isomorphic to the projective plane $\mathbb{P}^2$.

$\mathbb{G}(2, 4)$ is isomorphic to a nonsingular quadric hypersurface in $\mathbb{P}^5$.

$\mathbb{G}(2, 5) \subset \mathbb{P}^6$ has codimension 3 and degree five and is the zero-set of 5 quadrics.

$\mathbb{G}(2, 6) \subset \mathbb{P}^{14}$ has codimension 6 and degree 14 and is the zero-set of 15 quadrics. In this case the 15 quadrics are the partial derivatives, with respect to the 15 homogeneous coordinates, of the cubic Pfaffian of the 6×6 skew-symmetric matrix of Plücker coordinates. $\quad\square$

The first and second fundamental theorems of invariant theory say that the homogeneous coordinate ring of the Grassmannian $R = k[\text{Mat}(2, n)]^{SL(2)}$, as a module over the polynomial ring $S = k[p_{ij}]$, can be expressed in an exact sequence as follows:

$$0 \leftarrow R \leftarrow S \leftarrow \binom{n}{4} S(-2).$$

Here $S(e)$ denotes the graded S-module equal to S but with grading shifted by e. In this sense the arrows are all homomorphisms preserving the gradings. For the first few values of n one can use Example 8.5 to deduce that the exact sequence extends as follows:

$\mathbb{G}(2, 4):\quad 0 \leftarrow R \leftarrow S \leftarrow S(-2) \leftarrow 0$
$\mathbb{G}(2, 5):\quad 0 \leftarrow R \leftarrow S \leftarrow 5S(-2) \leftarrow 5S(-3) \leftarrow S(-5) \leftarrow 0.$
$\mathbb{G}(2, 6):\quad 0 \leftarrow R \leftarrow S \leftarrow 15S(-2) \leftarrow 35S(-3) \leftarrow 21S(-4) \oplus 21S(-5)$
$\qquad\qquad \leftarrow 35S(-6) \leftarrow 15S(-7) \leftarrow 8S(-8) \leftarrow 0.$

(e) A digression: the degree of the Grassmannian

We will briefly explain the degree appearing in Proposition 8.17 from another more topological point of view – and the reader can happily skip this at a first reading. Our main reason for including this is that the ideas, in particular the use of the Pascal triangle, will reappear in Chapter 12 in connection with the intersection numbers in the moduli spaces of vector bundles and parabolic bundles, and the case of the Grassmannian may serve as a useful preliminary example.

But first an even easier example. Throughout this section our field will be $k = \mathbb{C}$.

Example 8.19. What is the degree of the Segre variety

$$\mathbb{P}^n \times \mathbb{P}^n \hookrightarrow \mathbb{P}^N, \qquad (\mathbf{x}, \mathbf{y}) \mapsto \mathbf{x} \cdot \mathbf{y}^t,$$

where $N = n(n + 2)$? We have to compute in the cohomology ring $H^*(\mathbb{P}^n \times \mathbb{P}^n, \mathbb{Z})$, which is isomorphic to $\mathbb{Z}[x, y]/I$, where the generators x, y are the hyperplane classes $c_1(\mathcal{O}(1))$ pulled back from each of the two factors and $I = (x^{n+1}, y^{n+1})$ is the ideal of relations.

The class $x^n y^n$, of top degree, is Poincaré dual to a point, while the hyperplane class in the Segre space $\mathbb{P}^N$ is $x + y$. So the required degree is the integer $d \in \mathbb{Z}$ such that

$$(x + y)^{2n} \equiv d\, x^n y^n \quad \bmod I.$$

By simple binomial expansion this number is $d = \binom{2n}{n}$.

Notice that this binomial coefficient is computed from a Pascal triangle, truncated by the relations $x^{n+1} = y^{n+1} = 0$. For example, in the case $n = 4$, the degree $d = 70$ comes from:

$$
\begin{array}{ccccccccc}
 & & & & 1 & & & & \\
 & & & 1 & & 1 & & & \\
 & & 1 & & 2 & & 1 & & \\
 & 1 & & 3 & & 3 & & 1 & \\
1 & & 4 & & 6 & & 4 & & 1 \\
 & 5 & & 10 & & 10 & & 5 & \\
 & & 15 & & 20 & & 15 & & \\
 & & & 35 & & 35 & & & \\
 & & & & 70 & & & & \\
\end{array}
$$

Notice also that if we fiddle this Pascal triangle by replacing the left-hand central column with 0s, then the Catalan numbers appear in the central column, and the degree 14 at the bottom is (according to Proposition 8.17) that of the 8-dimensional Grassmannian $\mathbb{G}(2, 6)$!

$$
\begin{array}{c|cccccc}
 & & 1 & & & & \\
0 & & & 1 & & & \\
 & & 1 & & 1 & & \\
0 & & & 2 & & 1 & \\
 & & 2 & & 3 & & 1 \\
0 & & & 5 & & 4 & \\
 & & 5 & & 9 & & \\
0 & & & 14 & & & \\
 & 14 & & & & & \\
\end{array}
$$

It is this phenomenon that we are going to explain.   □

The Grassmannian $\mathbb{G}(r, n)$ carries a tautological vector subbundle and quotient bundle:

$$0 \to \mathcal{F} \to \mathcal{O}^{\oplus n} \to \mathcal{Q} \to 0.$$

(See Section 8.5(b).) If $\mathcal{F}$ has Chern classes $x_1, \ldots, x_r$ and $\mathcal{Q}$ has Chern classes $s_1, \ldots, s_{n-r}$, then the total Chern classes satisfy

$$c(\mathcal{F})c(\mathcal{Q}) = 1, \quad c(\mathcal{F}) = 1 + x_1 + \cdots + x_r, \quad c(\mathcal{Q}) = 1 + s_1 + \cdots + s_{n-r}.$$

$$(8.10)$$

In other words, the s_i are polynomials in the x_j defined by the formal power series expansion

$$\sum_{i \geq 0} s_i(x_1, \ldots, x_r) t^i \equiv \frac{1}{1 + x_1 t + \cdots + x_r t^r}. \tag{8.11}$$

It is known that the classes $x_1, \ldots, x_r, s_1, \ldots, s_{n-r}$ generate the cohomology ring $H^*(\mathbb{G}(r, n), \mathbb{Z})$ and that (8.10) generates all the relations among them. (See, for example, Bott and Tu [51].) In particular, it follows that the cohomology ring is generated by just $x_1, \ldots, x_r$ with r relations $s_{n-r+1} = \cdots = s_n = 0$:

$$H^*(\mathbb{G}(r, n), \mathbb{Z}) = \mathbb{Z}[x_1, \ldots, x_r]/(s_{n-r+1}, \ldots, s_n). \tag{8.12}$$

These relations are an obvious consequence of the fact that rank $\mathcal{Q} = n - r$.

Proof of Proposition 8.17. We now restrict our attention to the case $r = 2$ and the Grassmannian $\mathbb{G} = \mathbb{G}(2, n)$. We will write the cohomology ring as

$$H^*(\mathbb{G}, \mathbb{Z}) = \mathbb{Z}[A, B]/(s_{n-1}, s_n),$$

where $A = -x_1$, $B = x_2$ and the polynomials $s_i(A, B)$ are determined, via (8.11), by the recurrence relation

$$s_{i+1} - As_i + Bs_{i-1} = 0, \qquad s_0 = 1, \quad s_1 = A. \tag{8.13}$$

For example,

$$\begin{aligned} s_2 &= A^2 - B, \\ s_3 &= A^3 - 2AB, \\ s_4 &= A^4 - 3A^2B + B^2, \end{aligned}$$

and so on.

Note that $A = c_1(\mathcal{F}^\vee) = c_1(\det \mathcal{F}^\vee)$. In other words, it is the hyperplane class in the Plücker embedding. Our problem, therefore, is to determine the class A^N where $N = \dim \mathbb{G} = 2(n - 2)$.

On the other hand, B^{n-2} is Poincaré dual to a point. This is because B is the second Chern class of $\mathcal{F}^\vee$, and so is Poincaré dual to the zero-set of a global section of $\mathcal{F}^\vee$, that is, of the set of lines contained in a hyperplane of $\mathbb{P}^{n-1}$. So B^{n-2} is Poincaré dual to the set of lines contained in $n - 2$ general hyperplanes – that is, a $\mathbb{P}^1$.

Hence the degree of $\mathbb{G}(2, n) \subset \mathbb{P}^{(n-2)(n+1)/2}$ is the number $d \in \mathbb{Z}$ such that

$$A^{2(n-2)} \equiv d \, B^{n-2} \quad \mod (s_{n-1}, s_n).$$

Now, just as in Example 8.19, this number is determined by a Pascal triangle. Namely, spread out the monomials in A, B, s_i of top degree $\leq N = \dim \mathbb{G}$ in an array (illustrated here for $n = 6$):

$$B^4$$
$$AB^3s_1$$
$$A^2B^3 \qquad\qquad A^2B^2s_2$$
$$A^3B^2s_1 \qquad\qquad A^3Bs_3$$
$$A^4B^2 \qquad\qquad A^4Bs_2 \qquad\qquad A^4s_4$$
$$A^5Bs_1 \qquad\qquad A^5s_3$$
$$A^6B \qquad\qquad A^6s_2$$
$$A^7s_1$$
$$A^8$$

The recurrence relation (8.13) says precisely that this array is a Pascal triangle, with each entry obtained by adding those diagonally above it. From this it follows at once that the degree d is the bottommost entry in the Pascal triangle:

$$
\begin{array}{ccccccc}
1 & & & & & & \\
 & 1 & & & & & \\
1 & & 1 & & & & \\
 & 2 & & 1 & & & \\
2 & & 3 & & 1 & & \\
\vdots & & & \ddots & & & 1 \\
 & & & & n-2 & & \\
 & & & & & & \\
\vdots & & & & & & \\
 & d & & & & &
\end{array}
$$

Each entry in the array is the number of descending paths from the top (corresponding to B^4). In particular, the degree d is the Catalan number

$$\frac{(2n-4)!}{(n-1)!(n-2)!} = \frac{1}{n-1}\binom{2n-4}{n-2}.$$

(See, for example, Conway and Guy [53] p.105. Counting descending paths in the right-hand Pascal triangle is equivalent to counting 'mountain ranges'. Alternatively, see Stanley [58], where many combinatorial interpretations of Catalan numbers can be found in Ex. 6.19 (pp. 219–229).)　　　　□

8.2 Modules over a ring

Let R be a (commutative) ring. An action of R on an abelian group M is a map

$$R \times M \to M, \qquad (a, m) \mapsto am,$$

which is distributive, $a(m + m') = am + am'$, associative, $a(a'm) = (aa')m$, and satisfies $0m = 0$ and $1m = m$. Equivalently, we have a ring homomorphism $R \to \mathrm{Hom}(M, M)$. An abelian group M equipped with such an action of R is called an *R-module*. Homomorphisms of R-modules $M \to N$, submodules $N \subset M$, quotient modules M/N and direct sums $M \oplus N$ of R-modules are all defined in the usual way. We refer the reader to Atiyah and Macdonald [9] for a more systematic treatment than we can give here.

Examples 8.20.

(i) Every abelian group is a module over the ring $\mathbb{Z}$, while a module over a field k is the same as a vector space over k.
(ii) Any ring R is itself an R-module. In this case, the submodules are nothing other than the ideals of R.
(iii) Any ring homomorphism $\phi : R \to S$ makes S into an R module by the action $as := \phi(a)s$ for $a \in R, s \in S$. $\qquad\square$

If M is an R-module and $\mathfrak{a} \subset R$ is an ideal, then

$$\mathfrak{a}M := \{am \mid a \in \mathfrak{a}, \ m \in M\} \subset M$$

is a submodule. The ring R acts on the quotient $M/\mathfrak{a}M$, and the restriction of this action to $\mathfrak{a}$ is zero. Hence $M/\mathfrak{a}M$ is actually an $R/\mathfrak{a}$-module. This is called the *reduction of M modulo* $\mathfrak{a}$.

(a) Localisation

Localisation is a notion complementary to that of reduction $M/\mathfrak{a}M$. Let $S_0 \subset R$ be the subset consisting of elements which are not divisors of zero, and on the Cartesian product $R \times S_0$ define an equivalence relation

$$(x, y) \sim (x', y') \iff xy' = yx'. \tag{8.14}$$

We denote the equivalence class of (x, y) by x/y. The set of equivalence classes $(R \times S_0)/\sim$ can then be given a ring structure by the rules:

$$\begin{aligned}
\frac{x}{y} + \frac{x'}{y'} &= \frac{xy' + x'y}{yy'}, \\
\frac{x}{y} \times \frac{x'}{y'} &= \frac{xx'}{yy'}.
\end{aligned} \tag{8.15}$$

This ring is called the *total fraction ring* of R and is denoted by $Q(R)$. When R is an integral domain, $Q(R)$ is its field of fractions. The injection $R \hookrightarrow Q(R)$, $x \mapsto \frac{x}{1}$, identifies R with a subring of its total fraction ring.

We now generalise this construction. A subset $S \subset R - \{0\}$ containing $1 \in R$ is called *multiplicatively closed* if $x, y \in S$ implies $xy \in S$. We can then define an equivalence relation on the product $R \times S$ by

$$(x, y) \sim (x', y') \iff s(xy' - yx') = 0 \quad \text{for some } s \in S. \tag{8.16}$$

In the same way as (8.15) we now put a ring structure on the set of equivalence classes $(R \times S)/ \sim$. This ring is denoted $S^{-1}R$. Note that if $S = S_0$, and therefore contains no divisors of zero, then the two equivalence relations (8.14) and (8.16) coincide. Thus $S_0^{-1}R$ is equal to the total fraction ring $Q(R)$.

The map

$$R \to S^{-1}R, \quad x \mapsto \frac{x}{1} \tag{8.17}$$

is a ring homomorphism whose kernel is the set of $x \in R$ such that $sx = 0$ for some $s \in S$. In general, therefore, it is not injective.

Example 8.21. Localisation at one element. If $a \in R$ is not nilpotent, then the set $S = \{1, a, a^2, a^3, \ldots\}$ does not contain zero and is multiplicatively closed. In this case $S^{-1}R$ is denoted by R_a. $\qquad\square$

When R is an integral domain, R_a is the subring $R[1/a]$ of the field of fractions generated by R and $1/a$ which has already been used in discussing algebraic varieties, and in particular the construction of the structure sheaf (see Section 3.1).

Definition 8.22. The complement $S = R - \mathfrak{p}$ of a prime ideal $\mathfrak{p} \subset R$ is a multiplicatively closed set which excludes zero. In this case the ring $R_\mathfrak{p} := S^{-1}R$ is called the *localisation* of R at $\mathfrak{p}$. $\qquad\square$

The reason for the terminology is that $R_\mathfrak{p}$ is a *local ring*, that is, a ring containing a unique maximal ideal. This ideal consists of elements expressible as x/s, where $x \in \mathfrak{p}$, and is denoted by $\mathfrak{p}R_\mathfrak{p}$. By construction of $R_\mathfrak{p}$, all elements not contained in $\mathfrak{p}R_\mathfrak{p}$ are invertible, and it follows that $\mathfrak{p}R_\mathfrak{p} \subset R_\mathfrak{p}$ is maximal and, moreover, is the unique maximal ideal.

The following is a very important fact about local rings. The reader should compare the proof with that of Lemma 2.20.

Nakayama's Lemma 8.23. *Let M be a finitely generated module over a local ring* $(R, \mathfrak{m})$. *Then* $M = \mathfrak{m}M$ *only if* $M = 0$.

Proof. Let $m_1, \ldots, m_r$ be generators of M. Then the condition $M = \mathfrak{m}M$ says that

$$
\begin{pmatrix} m_1 \\ \vdots \\ m_r \end{pmatrix} = A \begin{pmatrix} m_1 \\ \vdots \\ m_r \end{pmatrix}
$$

for some $r \times r$ matrix with entries in $\mathfrak{m}$. Rewriting this as

$$
(I_r - A) \begin{pmatrix} m_1 \\ \vdots \\ m_r \end{pmatrix} = 0
$$

and multiplying on the left by the adjugate of the R valued matrix $I_r - A$, we obtain relations

$$
\det(I_r - A)m_i = 0 \quad \text{for each } i = 1, \ldots, r.
$$

But $\det(I_r - A)$ is of the form $1 + a$, for $a \in \mathfrak{m}$, and hence is not contained in $\mathfrak{m}$. Since $(R, \mathfrak{m})$ is a local ring, this implies that $\det(I_r - A)$ is an invertible element of R, and hence $m_1 = \cdots = m_r = 0$. □

Nakayama's lemma is often used in the following form, whose proof we leave as an exercise.

Corollary 8.24. *Let M be a finitely generated module over a local ring* $(R, \mathfrak{m})$. *Then elements* $m_1, \ldots, m_r$ *generate M if and only if their residue classes* $\overline{m}_1, \ldots, \overline{m}_r$ *span the quotient* $M/\mathfrak{m}M$ *as a vector space over the field* $R/\mathfrak{m}$. □

In this situation, if $\overline{m}_1, \ldots, \overline{m}_r \in M/\mathfrak{m}M$ are a basis over $R/\mathfrak{m}$, then we say that $m_1, \ldots, m_r$ form a *minimal system of generators of M*.

We can now generalise the above construction of fractions, with respect to a multiplicatively closed subset $S \subset R$, to any R-module M. First we put the same equivalence relation (8.16) on the product $M \times S$. Then we define an action of $S^{-1}R$ on the set of equivalence classes $(M \times S)/\sim$ by

$$
\frac{m}{s} + \frac{m'}{s'} = \frac{sm' + s'm}{ss'}, \qquad \frac{a}{s} \times \frac{m}{s'} = \frac{am}{ss'}. \tag{8.18}
$$

This defines an $S^{-1}R$-module which we denote by $S^{-1}M$. Analogously to (8.17), there is a natural map $M \to S^{-1}M$ with kernel

$$\ker\{M \to S^{-1}M\} = \{m \in M \mid sm = 0 \text{ for some } s \in S\}. \tag{8.19}$$

Corresponding to Example 8.21 and Definition 8.22, we can define localisations M_a at a nonnilpotent element $a \in R$, and $M_{\mathfrak{p}}$ at a prime ideal $\mathfrak{p} \subset R$.

In the case when S is the set S_0 of ring elements which do not divide zero, $Q(M) := S_0^{-1}M$ is called the *total fraction module*.

Definition 8.25. An R-module M is called a *torsion module* if $Q(M) = 0$. If the natural map $M \to Q(M)$ is injective, then M is said to be *torsion free*. □

(b) Local versus global

Let $\mathfrak{b} \subset R$ be any ideal and $a \in R$ any ring element. In the ring R_a the set of elements of the form x/a^n with $n \in \mathbb{N}$ and $x \in \mathfrak{b}$ is an ideal which we denote by $\mathfrak{b}R_a$. Note that, if $a \in \mathfrak{b}$, then $1 \in \mathfrak{b}R_a$; so we get a proper ideal only if $a \notin \mathfrak{b}$. If $\mathfrak{b} \subset R$ is a prime ideal, then $\mathfrak{b}R_a \subset R_a$ is also a prime ideal, and $\mathfrak{b}$ is the inverse image of $\mathfrak{b}R_a$ via the map $R \to R_a$.

For the set of maximal ideals, in particular, we therefore have a bijection:

$$\mathrm{Spm}\, R_a \cong D(a) := \{\mathfrak{m} \mid a \notin \mathfrak{m}\} \subset \mathrm{Spm}\, R.$$

(When R is an integral domain, we have already seen this in Section 3.2.) Note that, if $a \in R$ is nilpotent, then $R_a = 0$ (and more generally $M_a = 0$ for any R-module M), and so $D(a)$ is the empty set. (This is the converse of Theorem 2.27.)

Definition 8.26. Suppose that the ideal generated by $a_1, \dots, a_n \in R$ contains 1. That is, there exist $b_1, \dots, b_n \in R$ such that

$$a_1 b_1 + \cdots + a_n b_n = 1.$$

Then the set $\{a_1, \dots, a_n\}$ is called a *partition of unity*. In this case we have $\bigcup_{i=1}^n D(a_i) = \mathrm{Spm}\, R$. Given an R-module M, the collection of localisations $R_{a_1}, \dots, R_{a_n}$ and $M_{a_1}, \dots, M_{a_n}$ is called a *covering* of M. □

(See the paragraph on local properties in Section 3.1(c).) By expanding the relation $(a_1 b_1 + \cdots + a_n b_n)^N = 1$ for $N \in \mathbb{N}$ sufficiently large, we find:

Lemma 8.27. *If $\{a_1, \dots, a_n\}$ is a partition of unity, then for any natural number $k \in \mathbb{N}$ the set of powers $\{a_1^k, \dots, a_n^k\}$ is also a partition of unity.* □

Note that the coverings of an R-module M given by $\{a_1, \ldots, a_n\}$ and $\{a_1^k, \ldots, a_n^k\}$ are the same.

Proposition 8.28. *If $\{a_1, \ldots, a_n\}$ is a partition of unity and M is an R-module, then the natural homomorphism*

$$M \to M_{a_1} \oplus \cdots \oplus M_{a_n}, \quad m \mapsto (\frac{m}{1}, \ldots, \frac{m}{1})$$

is injective.

Proof. Suppose that $m \in M$ belongs to the kernel. That is, $m/1 \in M_{a_i}$ is zero for each $i = 1, \ldots, n$. By (8.19) this means that for some $k \in \mathbb{N}$ we have

$$a_1^k m = \cdots = a_n^k m = 0.$$

But by Lemma 8.27 this implies that $m = 0$. $\square$

Corollary 8.29. *If an R-module M admits a covering $\{M_{a_i}\}$ for which every $M_{a_i} = 0$, then $M = 0$.* $\square$

In other words, the property $M = 0$ holds locally, in the sense of Section 3.1(c), if and only if it holds globally. Such properties are common; in particular we shall often use the following.

Proposition 8.30. *Let $f : M \to N$ be a homomorphism of R-modules, and let $\{a_1, \ldots, a_n\}$ be a partition of unity of R. If all the localisations*

$$f_{a_i} : M_{a_i} \to N_{a_i}, \quad i = 1, \ldots, n,$$

are isomorphisms (or injective, surjective, zero), *then f is an isomorphism* (or injective, surjective, zero, respectively). $\square$

The 'localness' of the vanishing of a module can also be expressed 'pointwise':

Lemma 8.31. *The following properties of an R-module M are equivalent.*

(i) $M = 0$.
(ii) $M_{\mathfrak{m}} = 0$ at every maximal ideal $\mathfrak{m} \subset R$.

Proof. (i) implies (ii) is trivial; we shall prove that (ii) implies (i). Given $m \in M$, let

$$\text{Ann}(m) = \{a \in R \mid am = 0\}.$$

This subset is an ideal in R, called the annihilator of $m \in M$. We shall show that $\text{Ann}(m) = R$ for all $m \in M$; this will show that $M = 0$. Pick a maximal ideal $\mathfrak{m} \subset R$. Then (8.19) and the hypothesis that $S^{-1}M = M_{\mathfrak{m}} = 0$ imply that $\text{Ann}(m)$ contains the complement S of $\mathfrak{m}$. This shows that $\text{Ann}(m)$ is not contained in any maximal ideal $\mathfrak{m} \subset R$ and is therefore equal to R. $\square$

It is important to examine next the gluing principles by which a module M is reconstructed from a covering $\{M_{a_i}\}$. First note that for each i, j the compositions

$$M \to M_{a_i} \xrightarrow{r} M_{a_i a_j} \quad \text{and} \quad M \to M_{a_j} \xrightarrow{l} M_{a_i a_j}$$

agree.

Proposition 8.32. *The sequence*

$$M \to \bigoplus_i M_{a_i} \underset{l}{\overset{r}{\rightrightarrows}} \bigoplus_{i,j} M_{a_i a_j}$$

is exact in the sense that, given any collection of elements $\{x_i \in M_{a_i}\}$ obeying the compatibility condition $x_i = x_j \in M_{a_i a_j}$ for all i, j, there exists a (unique) element $m \in M$ such that $x_i = m/1$ for each i.

Proof. Replacing the partition of unity $\{a_1, \dots, a_n\}$ by $\{a_1^k, \dots, a_n^k\}$ if necessary, it is enough to assume that each $x_i = m_i/a_i$ for some $m_i \in M$. By hypothesis, $(a_i a_j)^p(a_j m_i - a_i m_j) = 0$ for some $p \in \mathbb{N}$, and we can take p to be the same for all i, j. By Lemma 8.27 there exist elements $b_1, \dots, b_n \in R$ such that $\sum_j a_j^{p+1} b_j = 1$. If we take $m = \sum_j a_j^p b_j m_j$, then for each i we find

$$a_i^{p+1} m = \sum_j (a_i a_j)^p a_i b_j m_j = \sum_j (a_i a_j)^p a_j b_j m_i = a_i^p m_i.$$

Hence in the module M_{a_i} we have $m/1 = a_i^p m_i/a_i^{p+1} = x_i$. The uniqueness of m follows from Proposition 8.28. $\square$

Let $\text{Hom}(S, R)$ be the set of ring homomorphisms $S \to R$.

Proposition 8.33. *Given a covering $\{R_{a_i}\}$ of the ring R, the following sequence is exact:*

$$\text{Hom}(S, R) \to \prod_i \text{Hom}(S, R_{a_i}) \underset{l}{\overset{r}{\rightrightarrows}} \prod_{i,j} \text{Hom}(S, R_{a_i a_j}).$$

$\square$

Proposition 8.32 allows one to reconstruct the module M itself by gluing. (This is a special case of *descent under a faithful flat morphism*.)

Proposition 8.34. *Suppose that a ring R admits a covering $\{R_{a_i}\}$ and data consisting of the following.*

(i) For each i an R_{a_i}-module M_i,
(ii) for each pair i, j an isomorphism $f_{ij} : (M_i)_{a_j} \xrightarrow{\sim} (M_j)_{a_i}$ satisfying:
(iii) for each triple i, j, k, the cocycle condition $f_{ij} f_{jk} = f_{ik} : (M_i)_{a_j a_k} \xrightarrow{\sim} (M_k)_{a_j a_i}$.

Then there exists, uniquely up to isomorphism, an R-module M such that $M_{a_i} \cong M_i$ for each i. □

To prove this it is enough to take for M the kernel of the homomorphism

$$\bigoplus_i M_{a_i} \to \bigoplus_{i,j} M_{a_i a_j}, \qquad (m_i) \mapsto \left(\frac{f_{ij}(m_i)}{1} - \frac{m_j}{1} \right)_{ij}.$$

We leave the details to the reader.

(c) Free modules

The following generalises the notion of a basis in a vector space.

Definition 8.35. A *free basis* in an R-module is a subset $B = \{m_i\}_{i \in I}$ satisfying the following conditions.

(i) B generates M over R.
(ii) $\sum_i a_i m_i = 0$ for $a_i \in R$ only if all $a_i = 0$. □

An R-module M which admits a free basis is said to be *free*. Equivalently, M is isomorphic to a direct sum of copies of R.

Lemma 8.36. *Let M be a free R-module with a free basis B.*

(i) For any maximal ideal $\mathfrak{m} \subset R$, the quotient module $M/\mathfrak{m}M$ has dimension equal to $|B|$ as a vector space over the field $R/\mathfrak{m}$.
(ii) The cardinality $|B|$ is independent of the choice of basis and depends only on the module M.

Proof. (i) follows simply from the fact that the set $\{\overline{m} \mid m \in B\} \subset M/\mathfrak{m}M$ is a vector space basis. Part (ii) follows from (i). □

This cardinality is called the *rank* of the free module M. When R is an integral domain, it is also equal to the dimension of $Q(M)$ as a vector space over the field of fractions $Q(R)$.

Every free module is torsion free (Definition 8.25). In special cases the converse is also true (see Exercise 7.2):

Proposition 8.37. *If R is a principal ideal domain, then every finitely generated torsion free R-module M is free.*   □

The case $R = \mathbb{Z}$ is well known. By Theorem 2.4 the proposition also applies to the case $R = k[x]$. In the following chapters we shall often use the following (see Exercise 8.3):

Corollary 8.38. *If R is a discrete valuation ring (Section 2.4(b)), then every finitely generated torsion free R-module is free.*   □

We now suppose that R contains the field k. If $B = \{m_i\}_{i \in I}$ is a free basis, then the expression of an element $m \in M$ as $m = \sum_i a_i m_i$ determines a linear map

$$M \to R \otimes_k M, \quad m \mapsto \sum_i a_i \otimes m_i.$$

If we view the vector space $R \otimes_k M$ as an R-module by multiplication on the first factor, then this map is a homomorphism of R-modules. This homomorphism can be used to characterise free modules.

Lemma 8.39. *Suppose R contains the field k and there exists a homomorphism of R-modules $f : M \to R \otimes_k M$ (that is, $f(am) = (a \otimes 1)f(m)$ for $a \in R$ and $m \in M$) satisfying the following two conditions:*

1. the composition of f with the map

$$R \otimes_k M \to M, \quad a \otimes m \mapsto am$$

is the identity map;

2. the following diagram commutes:

$$
\begin{array}{ccc}
M & \xrightarrow{\ f\ } & R \otimes_k M \\
f \downarrow & & \downarrow 1_R \otimes f \\
R \otimes_k M & \to & R \otimes_k R \otimes_k M \\
a \otimes m & \mapsto & a \otimes 1 \otimes m
\end{array}
$$

Then M is a free R-module. Moreover, any basis of the k-vector space

$$M_0 = \{m \in M \mid f(m) = 1 \otimes m\}$$

is a free basis for M.

Proof. $M_0 \subset M$ is a vector subspace over k, and so it is enough to show that the natural homomorphism of R-modules

$$\phi : R \otimes_k M_0 \to M, \qquad a \otimes m \mapsto am$$

is an isomorphism. Suppose that $\sum_j b_j m_j = 0$ for some $b_j \in R$ and $m_j \in M_0$. Applying f to this relation:

$$0 = \sum_j f(b_j m_j) = \sum_j (b_j \otimes 1) f(m_j) = \sum_j (b_j \otimes 1)(1 \otimes m_j) = \sum_j b_j \otimes m_j.$$

This shows that ϕ is injective.

Next, let $\{b_j\}_{j \in J}$ be a basis for R as a vector space over k. Given any $m \in M$, we can write $f(m) = \sum_j b_j \otimes m_j$ for some $m_j \in M$. Applying the commutative diagram (2) we have

$$\sum_j b_j \otimes 1 \otimes m_j = \sum_j b_j \otimes f(m_j).$$

But $\{b_j\}$ was chosen to be a basis, so we conclude that $1 \otimes m_j = f(m_j)$ for each j, which means that $m_j \in M_0$. By condition (1), $m = \sum_j b_j \otimes m_j$, and hence ϕ is surjective. $\qquad\square$

(d) Tensor products and flat modules

The tensor product of two R-modules is a notion that unifies those of reduction modulo an ideal $M/\mathfrak{a}M$ on the one hand, and localisation $S^{-1}M$ on the other.

Let $\{m_i\}_{i \in I}$ be a set of generators of M. Then the R-module homomorphism

$$R^{\oplus I} \to M, \qquad (a_i)_{i \in I} \mapsto \sum a_i m_i$$

is called a *free cover* of M, and we denote its kernel by $K_M \subset R^{\oplus I}$. Now let N be another R-module. Then the subset

$$\{(a_i n)_{i \in I} \mid (a_i)_{i \in I} \in K_M, \, n \in N\} \subset N^{\oplus I}$$

generates a submodule which we denote by $K_M N \subset N^{\oplus I}$.

Definition 8.40.

(i) The *tensor product* of M and N over R is defined to be the quotient R-module

$$M \otimes_R N := N^{\oplus I}/K_M N.$$

(ii) Given $m = \sum a_i m_i \in M$ and $n \in N$, the residue class of $(a_i n)_{i \in I}$ in $N^{\oplus I}/K_M N$ is independent of the choice of generators $\{m_i\}_{i \in I}$ and is denoted by

$$m \otimes n \in M \otimes_R N.$$

$\square$

The R-module $M \otimes_R N$ is independent of the choice of generators, up to isomorphism. Many of the important operations on modules can be expressed as tensor products.

Examples 8.41.

(i) Let $\mathfrak{a} \subset R$ be an ideal. Then the module $M = R/\mathfrak{a}$ is generated by a single element, so that $K_M = \mathfrak{a}$ and $R/\mathfrak{a} \otimes_R N = N/\mathfrak{a}N$. In particular, note that $R \otimes_R N = N$.

(ii) The tensor product is distributive over direct sums:

$$(M_1 \oplus M_2) \otimes_R N \cong (M_1 \otimes_R N) \oplus (M_2 \otimes_R N).$$

In particular, if M is a free module of rank r, then $M \otimes_R N \cong N^{\oplus r}$.

(iii) Let $M = S^{-1}R$, where $S \subset R$ is a multiplicatively closed subset. Taking $\{1/s\}_{s \in S}$ as a system of generators, we see that K_M is generated by differences $[1/s] - t[1/st] \in R^{\oplus S}$. Thus the tensor product $S^{-1}R \otimes_R N$ is isomorphic to $S^{-1}N$.

$\square$

One can now easily verify commutativity and associativity of the tensor product:

$$\begin{aligned} M \otimes_R N &\cong N \otimes_R M, \\ L \otimes_R (M \otimes_R N) &\cong (L \otimes_R M) \otimes_R N. \end{aligned} \tag{8.20}$$

Moreover, by definition of the tensor product, any R-module homomorphism $f : N_1 \to N_2$ induces an R-module homomorphism

$$f_M : M \otimes_R N_1 \to M \otimes_R N_2, \quad m \otimes n \mapsto m \otimes f(n).$$

In other words, $M \otimes_R$ is a functor from the category of R-modules to itself.

We now come to a very important notion.

Definition 8.42. An R-module is *flat* if for every homomorphism $f : N_1 \to N_2$ of R modules,

$$f : N_1 \to N_2 \text{ injective} \implies f : M \otimes_R N_1 \to M \otimes_R N_2 \text{ is injective.}$$

$\square$

Examples 8.43.

(i) Every free module is flat.

(ii) Let $S \subset R$ be a multiplicatively closed subset. It follows easily from (8.19) that, if $f : N_1 \to N_2$ is an injective R-module homomorphism, then $f : S^{-1}N_1 \to S^{-1}N_2$ is also injective. By Example 8.41(iii), this implies that $S^{-1}R$ is a flat R-module. $\square$

From part (ii) together with Example 8.41(iii) and the fact that the tensor product of two flat modules is again flat we deduce the following.

Lemma 8.44. *A module M is flat over R if and only if the localisation $M_{\mathfrak{m}}$ is flat over $R_{\mathfrak{m}}$ for every maximal ideal $\mathfrak{m} \subset R$.* $\square$

The first indication of the importance of flatness is the following.

Proposition 8.45. *Over a local ring $(R, \mathfrak{m})$ every finitely generated flat module is free.*

Proof. Suppose that M is an R-module with a minimal system of generators $m_1, \ldots, m_r \in M$. We will show that this system is a free basis. Let

$$a_1 m_1 + \cdots + a_r m_r = 0$$

be a linear relation among the generators, and let $\mathfrak{a} \subset R$ be the ideal generated by the coefficients $a_1, \ldots, a_r \in R$. Then the element

$$\alpha := a_1 \otimes m_1 + \cdots + a_r \otimes m_r \in \mathfrak{a} \otimes_R M$$

is in the kernel of the R-module homomorphism

$$\mathfrak{a} \otimes_R M \to M = R \otimes_R M,$$

and therefore $\alpha = 0$ if M is flat. We want to deduce from this that $a_1 = \cdots = a_r = 0$. We consider the vector space over $R/\mathfrak{m}$,

$$(\mathfrak{a}/\mathfrak{m}\mathfrak{a}) \otimes_{R/\mathfrak{m}} (M/\mathfrak{m}M).$$

Note that this is a quotient module of $\mathfrak{a} \otimes_R M$ and that $\sum_i \overline{a}_i \otimes \overline{m}_i = \overline{\alpha} = 0$. But by definition of a minimal system of generators this implies that all the (generating) elements $\overline{a}_1, \ldots, \overline{a}_r \in \mathfrak{a}/\mathfrak{m}\mathfrak{a}$ are zero. By Nakayama's Lemma 8.23 this implies that $\mathfrak{a} = 0$, and we are done. $\square$

8.3 Locally free modules and flatness

(a) Locally free modules

Unlike the vanishing $M = 0$, the properties $M \cong R$ and of being free are not local properties.

Definition 8.46. An R-module M is *locally free* if it admits a covering $\{M_{a_i}\}$ by some partition of unity $a_1, \ldots, a_n \in R$, for which each M_{a_i} is a free R_{a_i}-module. $\square$

Given some mild conditions, local freeness can be characterised in terms of the localisations at maximal ideals:

Proposition 8.47. *Suppose that R is a Noetherian ring and that M is a finitely generated R-module. Then the following are equivalent.*

(i) M is locally free.
(ii) For every maximal ideal $\mathfrak{m} \subset R$ the localisation $M_\mathfrak{m}$ is a free $R_\mathfrak{m}$-module.

Exercise 8.5 shows that the hypothesis that M is finitely generated cannot be relaxed.

Proof. (i) $\Longrightarrow$ (ii) Let $\{a_1, \ldots, a_n\} \subset R$ be a partition of unity as in Definition 8.46. Then the multiplicatively closed set $R - \mathfrak{m}$ contains some a_i. By hypothesis, M_{a_i} is a free R_{a_i}-module, and hence $M_\mathfrak{m}$ is a free $R_\mathfrak{m}$-module.

(ii) $\Longrightarrow$ (i) Let $\mathfrak{m} \subset R$ be a maximal ideal and let $m_1/s, \ldots, m_r/s \in M_\mathfrak{m}$ be a free basis, where $m_i \in M$ and $s \in R$, $s \notin \mathfrak{m}$. We then consider the homomorphism of R-modules

$$R_s \oplus \cdots R_s \to M_s, \qquad (a_1, \ldots, a_r) \mapsto \sum_{i=1}^{r} \frac{a_i m_i}{s}.$$

Denote the kernel and cokernel by K, C. These are both R-modules whose localisation at $\mathfrak{m}$ is zero; they are finitely generated, and so there exists a ring element $t \notin \mathfrak{m}$ such that $tK = tC = 0$. Therefore, taking $a = st$, the localisation M_a is a free R_a-module.

What we have shown is that for every maximal ideal $\mathfrak{m} \subset R$ there exists $a_\mathfrak{m} \in R - \mathfrak{m}$ for which the localisation $M_{a_\mathfrak{m}}$ is free. The ideal generated by the $a_\mathfrak{m}$ as $\mathfrak{m}$ ranges through all maximal ideals is the whole of R. Since R is Noetherian, a finite subset of $a_\mathfrak{m}$ can be taken to give a partition of unity, and hence M is locally free. $\qquad\square$

Proposition 8.48. *A finitely generated module M over a Noetherian ring R is locally free if and only if it is flat.*

Proof. If M is locally free, then each localisation $M_\mathfrak{m}$ at a maximal ideal $\mathfrak{m} \subset R$ is a free $R_\mathfrak{m}$-module, and therefore a flat $R_\mathfrak{m}$-module by Example 8.43(1). By Lemma 8.44, this implies that M is flat over R. Conversely, if M is flat, then each localisation $M_\mathfrak{m}$ is flat over $R_\mathfrak{m}$ and therefore free by Proposition 8.45. Hence M is locally free by Proposition 8.47. $\qquad\square$

If M is a locally free module, then by Lemma 8.36 the set of maximal ideals for which $\dim_{R/\mathfrak{m}}(M/\mathfrak{m}M)$ equals some value r is an open subset of Spm R. Hence:

Proposition 8.49. *If M is any locally free R-module, then $\dim_{R/\mathfrak{m}}(M/\mathfrak{m}M)$ is constant on connected components of* Spm R. $\qquad\square$

The *rank* of a locally free module (at a maximal ideal $\mathfrak{m} \subset R$) is defined to be this dimension $\dim_{R/\mathfrak{m}}(M/\mathfrak{m}M)$. By Lemma 8.36 it is equal to the rank of the free localisations of M.

Remarks 8.50.

(i) Partitions of Spm R into two disjoint open subsets correspond to (nontrivial) *idempotents* $e \in R$, $e^2 = e$. See Exercise 7.6.

(ii) If R has no nontrivial nilpotents, then the converse of Proposition 8.49 is also true. See Exercise 8.7. $\qquad\square$

Many of the linear algebra constructions that are familiar for vector spaces carry over in a similar manner for locally free modules.

Proposition 8.51. *If M, N are locally free R-modules, then the following hold.*

(i) *The direct sum $M \oplus N$ is locally free and* rank $M \oplus N$ = rank M + rank N.

(ii) *The R-module* $\mathrm{Hom}_R(M, N)$ *is locally free and* rank $\mathrm{Hom}_R(M, N)$ = rank M × rank N.

(iii) *The tensor product* $M \otimes_R N$ *is locally free and* rank $M \otimes_R N$ = rank M × rank N.

Proof. We will prove (iii). By hypothesis there are partitions of unity $\{a_1, \ldots, a_n\}$ and $\{b_1, \ldots, b_m\}$ giving coverings $\{M_{a_i}\}$ and $\{N_{b_j}\}$ by free modules. Then the collection $\{a_i b_j\}$ is also a partition of unity and each $M_{a_i b_j}$, $N_{a_i b_j}$ is free. Therefore the tensor products $M_{a_i b_j} \otimes N_{a_i b_j}$ are free (tensor products of free modules are free by Example 8.41(ii)); and since $\otimes$ commutes with localisation, it follows that $M \otimes_R N$ is locally free.   □

The special case $N = R$ of part (ii) of this proposition is called the *dual* of the locally free module M, and denoted

$$M^{\vee} := \mathrm{Hom}_R(M, R).$$

Proposition 8.52. *If M is any locally free module, then* $(M^{\vee})^{\vee} \cong M$.

Proof. The evaluation map

$$M \times \mathrm{Hom}_R(M, R) \to R, \quad (m, f) \mapsto f(m)$$

determines an R-module homomorphism

$$M \to \mathrm{Hom}_R(\mathrm{Hom}_R(M, R), R) = (M^{\vee})^{\vee}.$$

If M is a free module, this is an isomorphim. But Hom_R commutes with localisation, and so by Proposition 8.30 it is also an isomorphism for any locally free module M.   □

The next fact will be needed in Section 8.5(b) later on.

Lemma 8.53. *If M is a locally free R-module, and $R \to S$ is any ring homomorphism, then $M \otimes_R S$ is a locally free S-module.*   □

(b) Exact sequences and flatness

A sequence of R-module homomorphisms

$$\cdots \longrightarrow N_{i-1} \xrightarrow{f_{i-1}} N_i \xrightarrow{f_i} N_{i+1} \longrightarrow \cdots$$

is *exact* if, at each term, Im $f_{i-1} = \ker f_i \subset N_i$. Of particular importance is the case

$$0 \to N_1 \to N_2 \to N_3 \to 0, \tag{8.21}$$

in which the map $N_1 \to N_2$ is injective and the map $N_2 \to N_3$ is surjective. This is called a *short exact sequence*.

Proposition 8.54. *For the short exact sequence (8.21):*

(i) if N_1, N_3 are flat, then N_2 is flat;
(ii) if N_2, N_3 are flat, then N_1 is flat. □

Before proving this we need some preliminary facts. The first of these is really the background to the definition (8.42) of flatness.

Lemma 8.55 (Right exactness of $\otimes_R$). *If the sequence*

$$N_1 \to N_2 \to N_3 \to 0$$

is exact and M is any R-module, then the sequence

$$M \otimes_R N_1 \to M \otimes_R N_2 \to M \otimes_R N_3 \to 0$$

is exact.

Proof. We consider first the case $M = R/\mathfrak{a}$, where $\mathfrak{a} \subset R$ is an ideal. We have a commutative diagram with exact rows:

$$
\begin{array}{ccccccc}
N_1 & \xrightarrow{f} & N_2 & \xrightarrow{g} & N_3 & \longrightarrow & 0 \\
 & & \cup & & \cup & & \\
 & & \mathfrak{a}N_2 & \longrightarrow & \mathfrak{a}N_3 & \longrightarrow & 0
\end{array}
$$

A residue class $\bar{n} \in N_2/\mathfrak{a}N_2$ maps to zero under $g/\mathfrak{a}g$ if and only if $n \in N_2$ lies in $f(N_1) + \mathfrak{a}N_2$, and this in turn is equivalent to saying that $\bar{n}$ is in the image of $N_1/\mathfrak{a}N_1$. This proves the lemma for $M = R/\mathfrak{a}$, using Example 8.41(1).

For the general case we return to Definition 8.40. If $M = R^{\oplus I}/K_M$, then the tensor product is $M \otimes_R N = N^{\oplus I}/K_M N$. We now consider the diagram

$$
\begin{array}{ccccccc}
N_1^{\oplus I} & \longrightarrow & N_2^{\oplus I} & \longrightarrow & N_3^{\oplus I} & \longrightarrow & 0 \\
 & & \cup & & \cup & & \\
 & & K_M N_2 & \longrightarrow & K_M N_3 & \longrightarrow & 0
\end{array}
$$

and apply the same reasoning as in the first case. □

It follows from this that if M is a flat R-module, then the functor $M \otimes_R$ takes short exact sequences to short exact sequences. Such a functor is said to be *exact*.

A flat module need not be free, and one can discuss whether there exists a free basis.

Lemma 8.56. *Suppose that M, N are R-modules and M is flat. Suppose also that elements $m_1, \ldots, m_r \in M$ and $n_1, \ldots, n_r \in N$ satisfy $\sum m_i \otimes n_i = 0 \in M \otimes_R N$. Then it is possible to express*

$$\begin{pmatrix} m_1 \\ \vdots \\ m_r \end{pmatrix} = \begin{pmatrix} a_{11} \\ \vdots \\ a_{r1} \end{pmatrix} m_1' + \cdots + \begin{pmatrix} a_{1s} \\ \vdots \\ a_{rs} \end{pmatrix} m_s'$$

for some $m_1', \ldots, m_s' \in M$ and ring elements a_{ij} satisfying

$$(n_1, \ldots, n_r) \begin{pmatrix} a_{11} & \cdots & a_{1s} \\ \vdots & & \vdots \\ a_{r1} & \cdots & a_{rs} \end{pmatrix} = 0.$$

Proof. We consider the homomorphism

$$f : R \oplus \cdots \oplus R \to N, \qquad \mathbf{a} = \begin{pmatrix} a_1 \\ \vdots \\ a_r \end{pmatrix} \mapsto \sum_{i=1}^{r} a_i n_i,$$

and tensor with M. Writing $K = \ker f$, the flatness of M implies that we obtain an exact sequence

$$M \otimes_R K \to M \oplus \cdots \oplus M \xrightarrow{1 \otimes f} M \otimes_R N.$$

Since $\mathbf{m} = (m_1, \ldots, m_r) \in \ker 1 \otimes f$, it follows that $\mathbf{m}$ is equal to the image of $\sum_{j=1}^{s} m_j' \otimes \mathbf{a}_j$ for some $m_1', \ldots, m_s' \in M$ and $\mathbf{a}_1, \ldots, \mathbf{a}_s \in K$. $\square$

We are now moving towards the proof of Proposition 8.54. The following is well known.

Snake Lemma 8.57. *Let*

$$\begin{array}{ccccccccc}
0 & \longrightarrow & U & \longrightarrow & V & \longrightarrow & W & \longrightarrow & 0 \\
  &   & f \downarrow &   & g \downarrow &   & h \downarrow &   &   \\
0 & \longrightarrow & U' & \longrightarrow & V' & \longrightarrow & W' & \longrightarrow & 0
\end{array}$$

be a commutative diagram of modules in which each row is exact. Then there is an exact sequence

$$0 \to \ker f \to \ker g \to \ker h \xrightarrow{\delta} \operatorname{coker} f \to \operatorname{coker} g \to \operatorname{coker} h \to 0$$

where the connecting map $\delta : \ker h \to \operatorname{coker} f$ *is defined by* $\delta : w \mapsto g(v) \in U' \bmod f(U)$, *where* $v \in V$ *is a lift of* w. $\qquad\square$

Lemma 8.58. *Suppose*

$$0 \longrightarrow N_1 \xrightarrow{f} N_2 \longrightarrow N_3 \longrightarrow 0$$

is a short exact sequence of R-modules and that N_3 is flat. Then for any R-module M the homomorphism

$$1_M \otimes f : M \otimes_R N_1 \to M \otimes_R N_2$$

is injective.

Proof. As in the proof of Lemma 8.55, we write $M = R^{\oplus I}/K_M$. Then, applying the Snake Lemma to the commutative diagram

$$
\begin{array}{ccccccc}
K_M \otimes_R N_2 & \to & N_2^{\oplus I} & \to & M \otimes_R N_2 & \to 0 \\
\uparrow 1_{K_M} \otimes f & & \uparrow f^{\oplus I} & & \uparrow 1_M \otimes f & \\
K_M \otimes_R N_1 & \to & N_1^{\oplus I} & \to & M \otimes_R N_1 & \to 0
\end{array}
$$

we obtain an exact sequence

$$
\begin{array}{ccc}
K_M \otimes_R N_3 & \to & N_3^{\oplus I} \\
\| & & \| \\
\ker f^{\oplus I} \;\; \to \;\; \ker 1_M \otimes f \;\; \to \;\; \operatorname{coker} 1_{K_M} \otimes f \;\; \to \;\; \operatorname{coker} f^{\oplus I}.
\end{array}
$$

Since f is injective, $\ker f^{\oplus I} = 0$. On the other hand, $K_M \to R^{\oplus I}$ is injective and N_3 is flat, so $\operatorname{coker} 1_{K_M} \otimes f \to \operatorname{coker} f^{\oplus I}$ is also injective. Hence $\ker 1_M \otimes f = 0$. $\qquad\square$

Proof of Proposition 8.54. Let $f : A \to B$ be an injective homomorphism of R-modules and consider the commutative diagram:

$$N_1 \otimes_R B \quad \to \quad N_2 \otimes_R B \quad \to \quad N_3 \otimes_R B$$

$$\uparrow f_1 \qquad\qquad \uparrow f_2 \qquad\qquad \uparrow f_3$$

$$N_1 \otimes_R A \quad \xrightarrow{\alpha} \quad N_2 \otimes_R A \quad \to \quad N_3 \otimes_R A$$

(i) If f_1 and f_3 are injective, then f_2 is injective.
(ii) By Lemma 8.58 the lower left map α is injective. If f_2 is injective, then this implies that f_1 is injective. □

One can summarise the results of this section as follows.

Theorem 8.59. *Let R be a Noetherian ring and*

$$N_1 \xrightarrow{f} N_2 \longrightarrow N_3 \longrightarrow \cdots \longrightarrow N_{a-1} \longrightarrow N_a \longrightarrow 0$$

an exact sequence of flat R-modules. If the R-module $\ker f$ *is finitely generated, then it is locally free.* □

8.4 The Picard group

Definition 8.60.

(i) A locally free R-module L of rank 1 is called an *invertible R-module*. Equivalently, L is an R-module locally isomorphic to R.
(ii) The set of all isomorphism classes of invertible R-modules is denoted by $\mathrm{Pic}\,R$. □

By Proposition 8.51, the tensor product of two invertible modules is again invertible. Moreover, if L is an invertible module, then tensoring with its dual gives

$$L \otimes_R L^{\vee} \cong \mathrm{Hom}_R(L, L) \cong R.$$

For this reason we write $L^{\vee} = L^{-1}$ in this case, and $\mathrm{Pic}\,R$ becomes a group under the operation $\otimes_R$, called the *Picard group* of R. By (8.20) this is an abelian group.

(a) Algebraic number fields

An ideal $\mathfrak{a} \subset R$ is called an *invertible ideal* if there exists an ideal $\mathfrak{b} \subset R$ such that $\mathfrak{a}\mathfrak{b} = cR$ for some $c \in R$ not dividing zero. An invertible ideal is an invertible R-module (Exercise 8.8), and historically these were the first invertible modules to be studied.

Example 8.61. In the integral domain $R = \mathbb{Z}[\sqrt{-7}]$ the ideal $\mathfrak{a} = (2, 1+\sqrt{-7})$ is invertible since

$$\mathfrak{a}^2 = \mathfrak{a}\bar{\mathfrak{a}} = (2, 1 + \sqrt{-7})(2, 1 - \sqrt{-7}) = (4).$$

In this example $\mathfrak{a}$ is locally free but not free. (Compare this with Exercise 2.1.) □

Recall that a root of a monic polynomial with coefficients in R is said to be integral over R (Definition 2.19).

Lemma 8.62. *Let R be a subring of a field K.*

(i) *An element $b \in K$ is integral over R if and only if the subring $R[b] \subset K$ is finitely generated as an R-module.*

(ii) *The set of $b \in K$ that are integral over R is a subring of K, called the integral closure of R in K.*

(iii) *Suppose that $b \in K$ is a root of an equation*

$$f(x) = a_0 x^n + a_1 x^{n-1} + \cdots + a_{n-1}x + a_n = 0,$$

whose coefficients $a_i \in K$ are all integral over R. Then the coefficients of $f(x)/(x - b)$ are also integral over R.

(iv) *Given polynomials $f(x) = \sum_{i=0}^n a_i x^{n-i}$ and $g(x) = \sum_{j=0}^m b_j x^{m-j} \in K[x]$, suppose that $f(x)g(x) \in R[x]$. Then the products $a_i b_j \in K$ are all integral over R.*

Proof.

(i) is well known and is an application of the determinant trick used in the proof of Lemma 2.20 and Nakayama's Lemma 8.23, to the action of b on $R[b]$.

(ii) follows from (i).

(iii) Multiplying the equation by a_0^{n-1} shows that $a_0 b$ is integral over R. So if we write $f(x) = a_0(x - b)x^{n-1} + g(x)$, then the polynomial $g(x)$ has the same properties as $f(x)$ but has degree one less. The result therefore follows by induction on the degree.

(iv) We can assume that $a_0 b_0 \neq 0$. Let $\alpha_1, \ldots, \alpha_{m+n}$ be the roots of $f(x)g(x) = 0$ in some algebraically closed field containing K. Then for each subset $I \subset \{1, \ldots, m+n\}$ it follows from (iii) that $a_0 b_0 \prod_{i \in I} \alpha_i$ is integral over R.

Using the relations between the roots of an equation and its coefficients, each $a_i b_j$ can be expressed as some sum of such products, and by (ii) it is therefore integral over R. ☐

Let K be an algebraic number field, that is, a finite extension of the rational numbers $\mathbb{Q}$, and let $\mathcal{O}_K$ be the ring of algebraic integers in K,

$$\mathcal{O}_K := (\text{integral closure of } \mathbb{Z} \text{ in } K) \subset K.$$

Proposition 8.63. *Every nonzero ideal* $\mathfrak{a} \subset \mathcal{O}_K$ *is invertible.*

Proof. Suppose that $\mathfrak{a}$ has generators $a_1, \ldots, a_n$ and consider the polynomial

$$f(x) = a_1 x^{n-1} + \cdots + a_{n-1} x + a_n \in \mathcal{O}_K[x].$$

If K has degree $d = [K : \mathbb{Q}]$, then we can construct d polynomials $f^{(1)} = f, f^{(2)}, \ldots, f^{(d)} \in \mathbb{C}[x]$, whose coefficients are the conjugates in $\mathbb{C}$ of $a_1, \ldots, a_n \in K$ over $\mathbb{Q}$. Let $g(x) = f^{(2)} \ldots f^{(d)}$; then $f(x)g(x) \in \mathbb{Z}[x]$ and $g(x) \in \mathcal{O}_K[x]$. Let m be the greatest common divisor of the coefficients of $f(x)g(x)$ and $\mathfrak{b} \subset \mathcal{O}_K$ the ideal generated by the coefficients of $g(x)$. Clearly $m \in \mathfrak{a}\mathfrak{b}$, and Lemma 8.62(iv) applied to $f(x)g(x)/m$ shows that $\mathfrak{a}\mathfrak{b} \subset m\mathcal{O}_K$. Hence $\mathfrak{a}\mathfrak{b} = m\mathcal{O}_K$ and $\mathfrak{a}$ is an invertible ideal. ☐

Definition 8.64. Let R be a subring of a field K.

(i) A finitely generated R-submodule of K is called a *fractional ideal* of R. A fractional ideal generated by a single element is called a *principal fractional ideal*.

(ii) Two fractional ideals $\mathfrak{a}, \mathfrak{b} \subset K$ are *equivalent* if $\mathfrak{a} = c\mathfrak{b}$ for some $c \in K - \{0\}$. ☐

By Proposition 8.63, every nonzero fractional ideal $\mathfrak{a}$ of the ring of algebraic integers $\mathcal{O}_K$ is an invertible $\mathcal{O}_K$-module. Conversely, by choosing a basis for the total quotient module $Q(\mathfrak{a})$ (as a 1-dimensional vector space over $Q(\mathcal{O}_K)$) we see that every invertible $\mathcal{O}_K$-module is isomorphic to some fractional ideal $\mathfrak{a} \subset K$, and the equivalence class of $\mathfrak{a}$ is independent of the choice of basis. We arrive at:

Proposition 8.65. *The Picard group of the ring of integers $\mathcal{O}_K$ in an algebraic number field K is isomorphic to its divisor class group*

$$\mathrm{Cl}\,(\mathcal{O}_K) := \frac{\text{nonzero fractional ideals}}{\text{principal fractional ideals}}.$$

☐

(b) Two quadratic examples

We will compute the Picard group Pic R for two examples: for the ring of integers $R = \mathcal{O}_K$ of an imaginary quadratic number field K, and for the coordinate ring $R = k[C]$ of an affine hyperelliptic curve C. Useful references for this section are Taussky [59], or Borevich and Shafarevich [50].

We shall call a matrix with entries in a ring R an R-matrix for short; and we shall say that two R-matrices A, B of the same size $n \times n$ are *R-similar* if $A = XBX^{-1}$ for some invertible matrix $X \in GL(n, R)$.

Theorem 8.66. *Suppose that the ring of algebraic integers $\mathcal{O}_K$ is generated by a single element $\alpha \in \mathcal{O}_K$, that is, $\mathcal{O}_K = \mathbb{Z}[\alpha]$, and suppose that α is a root of a polynomial of degree $n = [K : \mathbb{Q}]$, irreducible over $\mathbb{Q}$,*

$$f(x) = x^n + a_1 x^{n-1} + \cdots + a_{n-1}x + a_n \in \mathbb{Z}[x].$$

Then there is a natural bijection between the following two sets:

(1) Pic $\mathcal{O}_K$;
(2) $\mathbb{Z}$-similarity classes of $n \times n$ $\mathbb{Z}$-matrices with characteristic polynomial equal to $f(X)$.

Proof. Let $\mathfrak{a}$ be an invertible $\mathcal{O}_K$-module. This is torsion free, and so by Proposition 8.37 it is free of rank n as a $\mathbb{Z}$-module. Let $\alpha_1, \ldots, \alpha_n \in \mathfrak{a}$ be a free basis over $\mathbb{Z}$. Now $\mathfrak{a}$ is a $\mathcal{O}_K = \mathbb{Z}[\alpha]$-module, and so

$$\alpha \begin{pmatrix} \alpha_1 \\ \vdots \\ \alpha_n \end{pmatrix} = M \begin{pmatrix} \alpha_1 \\ \vdots \\ \alpha_n \end{pmatrix}$$

for some $n \times n$ $\mathbb{Z}$-matrix M. Moreover, $f(M) = 0$ since $f(\alpha) = 0$. Since $f(x)$ is irreducible over $\mathbb{Q}$, it is precisely the minimal polynomial of M, and the characteristic polynomial since it has degree n. While M depends on the choice of $\mathbb{Z}$-basis $\alpha_1, \ldots, \alpha_n \in \mathfrak{a}$, its $\mathbb{Z}$-similarity class does not. And if we start with two isomorphic $\mathcal{O}_K$-modules $\mathfrak{a}$, $\mathfrak{a}'$, then the matrices M, M' that we obtain are similar. We have therefore constructed a map from (1) to (2).

Conversely, suppose we are given a $n \times n$ $\mathbb{Z}$-matrix M whose characteristic polynomial is equal to $f(x)$. Then by the Cayley-Hamilton Theorem the mapping $\alpha \mapsto M$ determines a ring homomorphism

$$\mathbb{Z}[\alpha] \to \operatorname{End} \mathbb{Z}^n.$$

This homomorphism makes $\mathbb{Z}^n$ into an $\mathcal{O}_K = \mathbb{Z}[\alpha]$-module; let us denote it by $\mathfrak{a}_M$. Extending the coefficients to $\mathbb{Q}$ makes this module $\mathfrak{a}_M$ into a rational vector space naturally isomorphic to K. In other words, $\mathfrak{a}_M \hookrightarrow K$ as a fractional ideal of $\mathcal{O}_K$. We have therefore constructed a map from (2) to (1) which is precisely the inverse of that above. $\square$

We want to consider, in particular, the case of a quadratic number field $\mathbb{Q}(\sqrt{d})$. We suppose that $d \neq 0, 1$ and that d is squarefree. The ring of integers is then

$$\mathcal{O}_{\sqrt{d}} = \begin{cases} \mathbb{Z}\left[\frac{1+\sqrt{d}}{2}\right] & \text{if } d \equiv 1 \bmod 4, \\[2mm] \mathbb{Z}\left[\sqrt{d}\right] & \text{if } d \not\equiv 1 \bmod 4. \end{cases}$$

Corollary 8.67. *There are a natural bijections among the following three sets:*

(1) $\mathrm{Pic}\,\mathcal{O}_{\sqrt{d}}$;
(2) $\mathbb{Z}$-*similarity classes of* 2×2 $\mathbb{Z}$-*matrices satisfying the condition*

$$\begin{cases} dI_2 = (2M - I_2)^2 & \text{if } d \equiv 1 \bmod 4 \\ dI_2 = M^2 & \text{if } d \not\equiv 1 \bmod 4; \end{cases}$$

(3) $GL(2, \mathbb{Z})$-*orbits of integral quadratic forms* $ax^2 + bxy + cy^2$ *satisfying*

$$b^2 - 4ac = D := \begin{cases} d & \text{if } d \equiv 1 \bmod 4 \\ 4d & \text{if } d \not\equiv 1 \bmod 4, \end{cases}$$

where the $GL(2, \mathbb{Z})$ *action on quadratic forms is the usual one twisted by the determinant:*

$$\begin{pmatrix} a & b/2 \\ b/2 & c \end{pmatrix} \mapsto (\det A)A \begin{pmatrix} a & b/2 \\ b/2 & c \end{pmatrix} A^t.$$

Remark 8.68. If $d < 0$, then the set of quadratic forms with discriminant D is a union of positive definite and negative definite forms. These subsets are transposed by the action of $GL(2, \mathbb{Z})$ but are preserved by $SL(2, \mathbb{Z})$. The set (3) is therefore equivalent to:

(3') $SL(2, \mathbb{Z})$-orbits of positive definite integral quadratic forms $ax^2 + bxy + cy^2$ with discriminant D. $\square$

Proof. The bijection between (1) and (2) follows directly from Theorem 8.66. To map from (2) to (3), suppose first that $dI_2 = M^2$. This is equivalent to saying that M has trace zero and determinant $-d$, so M can be written as

$$M = \begin{pmatrix} b/2 & -a \\ c & -b/2 \end{pmatrix}, \qquad ac - \frac{b^2}{4} = -d.$$

Then

$$M' := M \begin{pmatrix} 0 & 1 \\ -1 & 0 \end{pmatrix} = \begin{pmatrix} a & b/2 \\ b/2 & c \end{pmatrix}$$

is a symmetric matrix defining a quadratic form $ax^2 + bxy + cy^2$ with discriminant $4d$. Moreover, under a similarity $M \mapsto AMA^{-1}$ the matrix M' transforms to

$$AMA^{-1} \begin{pmatrix} 0 & 1 \\ -1 & 0 \end{pmatrix} = AM' \begin{pmatrix} 0 & 1 \\ -1 & 0 \end{pmatrix} A^{-1} \begin{pmatrix} 0 & 1 \\ -1 & 0 \end{pmatrix} = (\det A) A M' A^t.$$

This construction determines a bijection between (2) and (3) in the case $d \not\equiv 1$ mod 4. The case $d \equiv 1$ mod 4 is similar. $\square$

For an imaginary quadratic field, where $d < 0$, the Picard group Pic $\mathcal{O}_{\sqrt{d}}$ is now completely determined by the following fact.

Lemma 8.69 (Gauss). *Each $SL(2, \mathbb{Z})$-orbit of positive definite integral quadratic forms $ax^2 + bxy + cy^2$ has a unique representative satisfying $-a < b \leq a < c$ or $0 \leq b \leq a = c$.* $\square$

The set of complex numbers $(-b + \sqrt{D})/2a$ for which these inequalities are satisfied lie in the region shown in Figure 8.1.

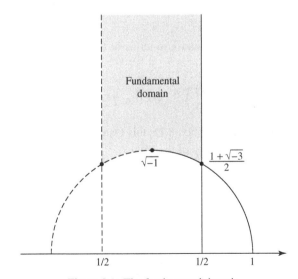

Figure 8.1: The fundamental domain

Example 8.70. Take $d = -41$. Using Lemma 8.69, a complete set of 2×2 $\mathbb{Z}$-matrices M satisfying $M^2 + 41I_2 = 0$, up to $\mathbb{Z}$-similarity, is given by

$$\begin{pmatrix} 0 & 1 \\ -41 & 0 \end{pmatrix}, \begin{pmatrix} 1 & 2 \\ -21 & -1 \end{pmatrix}, \begin{pmatrix} \pm 1 & 3 \\ -14 & \mp 1 \end{pmatrix}, \begin{pmatrix} \pm 1 & 6 \\ -7 & \mp 1 \end{pmatrix}, \begin{pmatrix} \pm 2 & 5 \\ -9 & \mp 2 \end{pmatrix}.$$

Hence the imaginary quadratic field $\mathbb{Q}(\sqrt{-41})$ has class number 8 (that is, $|\text{Pic } \mathcal{O}_{\sqrt{-41}}| = 8$), and its ideal classes are represented by:

$$(1, \sqrt{-41}), \quad (2, 1 + \sqrt{-41}), \quad (3, \mp 1 + \sqrt{-41}),$$

$$(6, \mp 1 + \sqrt{-41}), \quad (5, \mp 2 + \sqrt{-41}).$$

(For an example where $d \equiv 1 \mod 4$, see Exercise 8.9.)   □

We can apply the same reasoning as for $\mathbb{Z}$ to the polynomial ring $k[x]$ (see Proposition 8.37), and we will consider next the Picard group of the quadratic extension $k[x, \sqrt{d(x)}]$, where $d(x) \in k[x]$ is a nonconstant polynomial. We assume that $d(x)$ has no square factors in $k[x]$. Then $k[x, \sqrt{d(x)}]$ is the integral closure of $k[x]$ in the field $k(x, \sqrt{d(x)})$.

Proposition 8.71. *There is a natural bijection between the following two sets:*

(1) Pic $k[x, \sqrt{d(x)}]$,
(2) $k[x]$-similarity classes of 2×2 $k[x]$-matrices M satisfying $M^2 = d(x)I_2$.   □

We can write such a $k[x]$-matrix in terms of three polynomials $f(x)$, $g(x)$, $h(x) \in k[x]$ as

$$M = \begin{pmatrix} g(x) & -f(x) \\ h(x) & -g(x) \end{pmatrix}, \quad g(x)^2 - f(x)h(x) = d(x). \quad (8.22)$$

Under the bijection of Proposition 8.71, this matrix corresponds to the isomorphism class in the Picard group of the ideal

$$(f(x), g(x) - \sqrt{d(x)}) \subset k[x, \sqrt{d(x)}].$$

Within the $k[x]$-similarity class of the matrix (8.22) we can choose $f(x)$ to have minimal degree and replace $g(x)$ by its remainder on division by $f(x)$. By this means, we find a representative for which

$$\deg g(x) < \deg f(x) \le h(x).$$

This is the analogue of Gauss's Lemma 8.69. If $d(x)$ has odd degree, then equality cannot hold on the right, and we conclude:

Lemma 8.72. *If* $\deg d(x) = 2p + 1$*, then the* $k[x]$*-similarity class of the matrix* *(8.22) has a uniquely representative satisfying* $\deg g(x) < \deg f(x) \leq p < h(x)$ *and* $f(x)$ *monic.* □

Example 8.73. Suppose $\deg d(x) = 3$. Then $f(x)$ is either constant or linear. The constant case corresponds to a principal ideal in $k[x, \sqrt{d(x)}]$; otherwise, $f(x)$ is linear and $g(x)$ is constant. If $g(x) = b$, then $f(x) = x - a$, where a is a root of $d(x) - b^2 = 0$. What we have shown is that there is a bijection, when $\deg d(x) = 3$, between the Picard group of $k[x, \sqrt{d(x)}]$ and the elliptic curve

$$C : \{y^2 = d(x)\} \cup \{\infty\},$$

given by:

C	Pic $k[x, \sqrt{d(x)}]$
point (a, b)	ideal class $(x - a, b - \sqrt{x^3 - 1})$
point at infinity ∞	principal ideals

Via this correspondence, in fact, the group structure of Pic $k[x, \sqrt{d(x)}]$ coincides with the well-known group law $\oplus$ on the plane cubic curve $C \subset \mathbb{P}^2$ which is uniquely determined by the rules:

$$p, q, r \in C \text{ collinear} \iff p \oplus q \oplus r = 0,$$

$$\text{point at infinity } \infty \quad = \quad \text{group identity } 0.$$

□

Example 8.74. Suppose $\deg d(x) = 5$. Then $\deg f(x) \leq 2$. If $\deg f(x) = 2$, then the matrix (8.22) takes the form

$$\begin{pmatrix} cx + e & -(x - a_1)(x - a_2) \\ h(x) & -cx - e \end{pmatrix},$$

where, moreover, the line $y = cx + e$ is that passing through the two points $p_1 = (a_1, b_1)$, $p_2 = (a_2, b_2)$ of the affine hyperelliptic curve $C = \{y^2 = d(x)\}$. (If the two points coincide, then $y = cx + e$ is the tangent line to the curve.)

The corresponding ideal is

$$((x - a_1)(x - a_2), cx + e - \sqrt{d(x)}) \subset k[x, \sqrt{d(x)}].$$

It follows that Pic $k[x, \sqrt{d(x)}]$ corresponds birationally to the symmetric product $\mathrm{Sym}^2 C$ of the curve.   ☐

In general, the story is this: the Picard group of $k[x, \sqrt{d(x)}]$ can be given the structure of a p-dimensional algebraic variety, where $\deg d(x) = 2p + 1$, and this variety is birationally equivalent to the symmetric product $\mathrm{Sym}^p C$ of the *hyperelliptic curve of genus p*

$$C : \{y^2 = d(x)\} \cup \{\infty\}. \tag{8.23}$$

(See Example 9.7 in the next chapter.)

8.5 Vector bundles

Let X be an irreducible topological space and F an elementary sheaf on X. If K is the total set in which F takes values (Definition 3.2), it may happen that F actually takes values in some smaller subset $K' \subset K$. It is convenient, and should not lead to any confusion, to agree always to take the smallest such set. This smallest total set of the sheaf is the inductive limit taken over nonempty open sets in X with respect to the inclusion relation:

$$F_{\mathrm{gen}} := \lim_{U \neq \emptyset} F(U) = \bigcup_{U \neq \emptyset} F(U).$$

The set F_{gen} will be called the *(minimal) total set* of the sheaf, or the *stalk at the generic point*.

Example 8.75. If R is an integral domain and $X = \mathrm{Spm}\, R$ with its Zariski topology, then we have defined the structure sheaf as an elementary sheaf of rings in the total set $Q(R)$, the field of fractions of R. (See Section 3.1.) In this case $Q(R)$ is also the minimal total set.   ☐

If we fix a point $p \in X$, we can put the same partial ordering by inclusion on the collection of open sets containing p by

$$U \geq V \quad \Longleftrightarrow \quad U \subset V.$$

Recall that $U \subset V \implies F(U) \supset F(V)$. The limit over all open sets containing p

$$F_p := \lim_{p \in U} F(U) = \bigcup_{p \in U} F(U)$$

is called the *stalk* of F at the point $p \in X$.

Example 8.76. If $X = \mathrm{Spm}\ R$, then $p \in X$ corresponds to a maximal ideal $\mathfrak{m} \subset R$. In this case the stalk of the structure sheaf $F = \mathcal{O}_X$ is the localisation of R at $\mathfrak{m}$,

$$\mathcal{O}_{X,p} = R_{\mathfrak{m}}.$$

$\square$

In general, if $Y \subset X$ is an irreducible closed subset, then the limit

$$\lim_{Y \cap U \neq \emptyset} F(U) = \bigcup_{Y \cap U \neq \emptyset} F(U)$$

is called the stalk of F at (the generic point of) Y. The stalks F_{gen} and F_p are special cases of this.

(a) Elementary sheaves of modules

Let $\mathcal{O}$ be any elementary sheaf of rings on the topological space X.

Definition 8.77. An *elementary sheaf of $\mathcal{O}$-modules* on X is an elementary sheaf $\mathcal{M}$ satisfying the following conditions.

(i) The total set $\mathcal{M}_{\mathrm{gen}}$ is an $\mathcal{O}_{\mathrm{gen}}$-module. Denote the corresponding action by
$\phi : \mathcal{O}_{\mathrm{gen}} \times \mathcal{M}_{\mathrm{gen}} \to \mathcal{M}_{\mathrm{gen}}$.
(ii) For every open set $U \subset X$ we have $\phi(\mathcal{O}(U) \times \mathcal{M}(U)) \subset \mathcal{M}(U)$. $\square$

Condition (ii) says that every $\mathcal{M}(U)$ is an $\mathcal{O}(U)$-module, and it also follows that at every point $p \in X$ the stalk $\mathcal{M}_p$ is an $\mathcal{O}_p$-module.

Definition 8.78. Let R be an integral domain, and let M be a torsion free R-module (that is, $M \hookrightarrow Q(M)$). Then the total fraction module $Q(M)$ is a vector space over the field of fractions $Q(R)$, and on the affine variety $X = \mathrm{Spm}\ R$ we define an elementary sheaf of $\mathcal{O}_X$-modules $\underline{M}$ with total set $\underline{M}_{\mathrm{gen}} = Q(M)$

by assigning to each basic open set $D(a) \subset X$, $a \in R - \{0\}$, the localisation

$$\underline{M}(D(a)) := M_a = \left\{ \frac{m}{a^n} \mid m \in M, \ n \geq 0 \right\} \subset Q(M).$$

$\square$

Note that this generalises the construction of the structure sheaf $\mathcal{O}_X$ itself (Section 3.1). Note also that the stalk of $\underline{M}$ at a point $p \in X$ corresponding to a maximal ideal $\mathfrak{m} \subset R$ is the localisation $\underline{M}_p = M_{\mathfrak{m}}$.

Remark 8.79. We have assumed here that R is an integral domain; but if R is a primary ring (Definition 3.24), a torsion free R-module determines in the same way an elementary sheaf of $\mathcal{O}_X$-modules. $\square$

Example 8.80. Let R be an integral domain, let $\mathfrak{p} \subset R$ be a prime ideal and let M be a torsion free module over $R/\mathfrak{p}$. (In particular, M is also an R-module, though not necessarily torsion free.) Define on $X = \mathrm{Spm}\, R$ an elementary sheaf of $\mathcal{O}_X$-modules with total set $Q(M)$ by assigning to a basic open set $D(a) \subset X$, $a \in R - \{0\}$, the R_a-module

$$\begin{cases} M_{\overline{a}} & \text{if } a \notin \mathfrak{p}, \\ 0 & \text{if } a \in \mathfrak{p}, \end{cases}$$

where $\overline{a} \in R/\mathfrak{p}$ is the residue class of a. This construction gives an extension of the elementary sheaf $\underline{M}$ on the closed subset $Y = \mathrm{Spm}\, R/\mathfrak{p} \subset X$ to the whole of X, which is zero on the complement $X - Y$.

A special case occurs when $\mathfrak{p}$ is a maximal ideal $\mathfrak{m}$ corresponding to $Y = \{p\} \subset X$, a single point, and $M = R/\mathfrak{m}$. In this case we obtain an elementary sheaf on X which assigns to an open set $U \subset X$ the module

$$\begin{cases} R/\mathfrak{m} & \text{if } p \in U, \\ 0 & \text{if } p \notin U. \end{cases}$$

This is called the *skyscraper sheaf* supported at the point p. $\square$

Suppose that $\mathcal{M}, \mathcal{N}$ are elementary sheaves of abelian groups on X. A *sheaf homomorphism* $f : \mathcal{M} \to \mathcal{N}$ consists of a group homomorphism of the total sets

$$f_{\mathrm{gen}} : \mathcal{M}_{\mathrm{gen}} \to \mathcal{N}_{\mathrm{gen}},$$

which for every open set $U \subset X$ satisfies

$$f_{\mathrm{gen}}(\mathcal{M}(U)) \subset \mathcal{N}(U).$$

If $\mathcal{M}, \mathcal{N}$ are elementary sheaves of $\mathcal{O}$-modules and f_{gen} is a homomorphism of $\mathcal{O}_{\text{gen}}$-modules, then $f : \mathcal{M} \to \mathcal{N}$ is called an $\mathcal{O}$-homomorphism.

Definition 8.81. A homomorphism $f : \mathcal{M} \to \mathcal{N}$ of elementary sheaves of modules is said to be *injective, surjective* or an *isomorphism* if at every point $p \in X$ the induced homomorphism on the stalks $f_p : \mathcal{M}_p \to \mathcal{N}_p$ has the respective property. □

If f is injective, then on every open set $U \subset X$ the induced homomorphism $\mathcal{M}(U) \to \mathcal{N}(U)$ is injective. However, the same is not true if we replace injective by surjective. It is for this reason that sheaf cohomology theory is important. (See Section 10.1.)

(b) Line bundles and vector bundles

Definition 8.82. On an algebraic variety X an elementary sheaf L of $\mathcal{O}_X$-modules is called an *invertible sheaf*, or a *line bundle*, if it has the following two properties:

(i) The stalk L_{gen} at the generic point is a 1-dimensional vector space over the function field $k(X)$.

(ii) L is locally isomorphic to the structure sheaf $\mathcal{O}_X$. In other words, there exists an open cover $\{U_i\}_{i \in I}$ of X such that each restriction $L|_{U_i}$ is isomorphic to $\mathcal{O}_{U_i}$. □

We include condition (i) for clarity, although it is not hard to see that it follows from (ii). This definition is the special case $r = 1$ of the following.

Definition 8.83. An elementary sheaf on an algebraic variety X whose stalk at the generic point is an r-dimensional vector space over $k(X)$ and which is locally isomorphic to the direct sum $\mathcal{O}_X^{\oplus r}$ is called a *locally free sheaf*, or a *vector bundle of rank r*. □

Intuitively, a vector bundle of rank r is a family of r-dimensional vector spaces parametrised by the variety X. (This follows from Proposition 8.49.)

Example 8.84. The structure sheaf $\mathcal{O}_X$ is itself a line bundle, called the *trivial line bundle*. The direct sum $\mathcal{O}_X^{\oplus r}$ is a vector bundle called the *trivial vector bundle* of rank r. □

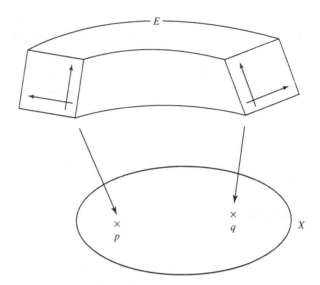

Figure 8.2: A vector bundle E

Just as for modules, the tensor product of two vector bundles of ranks r, s is a vector bundle of rank rs (Proposition 8.51(iii)). In particular, the tensor product of two line bundles is again a line bundle, and the set of isomorphism classes of line bundles on X becomes a group under $\otimes$, called the *Picard group* Pic X.

The following lemma follows from Proposition 8.34.

Lemma 8.85. *Every vector bundle on an affine variety $X = $ Spm R is a sheaf of the form $\underline{M}$* (as in Definition 8.78) *for some locally free R-module M.*   □

We next define the pull-back of a vector bundle E on X under a morphism of varieties $f : Y \to X$. We suppose that X has an affine open cover $\{U_i = $ Spm $R_i\}$ and Y has an affine open cover $\{V_j = $ Spm $S_j\}$, and that f is obtained by gluing affine morphisms $f_j : V_j \to U_j$ corresponding to ring homomorphisms $R_j \to S_j$. Then both S_j and the restriction E_{U_j} are R_j-modules, and we form the tensor product $E_{U_j} \otimes_{R_j} S_j$, which is also an S_j-module. Gluing these S_j-modules we obtain a sheaf of $\mathcal{O}_Y$-modules, which by Lemma 8.53 is locally free. This is called the *pull-back* of the vector bundle E to Y and is denoted by f^*E. Even when S is not an integral domain, the pull-back f^*E under a morphism $f :$ Spm $S \to X$ still makes sense as a locally free S-module, by Proposition 8.34.

On every projective variety Proj R there exists a distinguished line bundle. Here

$$R = \bigoplus_{e \geq 0} R_e$$

is a graded integral domain generated by $R_0 = k$ and R_1. Let $K = Q(R)$ be its field of fractions, and for $i \in \mathbb{Z}$ let

$$K_i = \left\{ \frac{g}{h} \;\middle|\; \begin{array}{l} g, h \in R \text{ are homogeneous elements} \\ \text{and } \deg g - \deg h = i \end{array} \right\} \subset K.$$

Each K_i is a 1-dimensional vector space over the field K_0.

Definition 8.86. For each $i \in \mathbb{Z}$ we define an elementary sheaf of $\mathcal{O}_X$-modules on $X = \text{Proj } R$, with total set K_i, by assigning, on basic open sets U_f where f is homogeneous,

$$U_f \mapsto \left\{ \frac{g}{f^m} \;\middle|\; \begin{array}{l} g \in R \text{ is homogeneous and} \\ \deg g - m \deg f = i \text{ for some } m \geq 0 \end{array} \right\} \subset K_i.$$

This sheaf is a line bundle and is denoted by $\mathcal{O}_X(i)$. In particular, $\mathcal{O}_X(1)$ is called the *tautological line bundle* on Proj R. on Proj R □

We observe that as 1-dimensional vector spaces over K_0 there are canonical isomorphisms

$$K_i \otimes K_j \cong K_{i+j}, \qquad K_i^\vee \cong K_{-i}.$$

These translate into canonical isomorphisms between line bundles on $X = \text{Proj } R$:

$$\mathcal{O}_X(i) \otimes \mathcal{O}_X(j) \cong \mathcal{O}_X(i + j), \qquad \mathcal{O}_X(i)^{-1} \cong \mathcal{O}_X(-i).$$

Example 8.87. Consider projective space

$$\mathbb{P}^n = \text{Proj } R = \text{Proj } k[x_0, x_1, \ldots, x_n].$$

Here

$$K_i = \left\{ \frac{g}{h} \;\middle|\; \begin{array}{l} g, h \in R \text{ are homogeneous polyno-} \\ \text{mials and } \deg g - \deg h = i \end{array} \right\} \subset k(x_0, x_1, \ldots, x_n).$$

The dual $\mathcal{O}_{\mathbb{P}}(-1)$ of the tautological line bundle is in this case called the *tautological line subbundle* and can be viewed as the line subbundle of the trivial vector bundle $\mathcal{O}_{\mathbb{P}}^{\oplus(n+1)}$ spanned by the element $(x_0, x_1, \ldots, x_n)$. More precisely, this can be described in terms of the affine open sets $U_i = \{x_i \neq 0\} \subset \mathbb{P}^n$.

On each $U_i = \text{Spm } R_{x_i,0}$ we can consider the $R_{x_i,0}$-submodule $L_i \subset R_{x_i,0}^{\oplus(n+1)}$ generated by

$$\left(\frac{x_0}{x_i}, \frac{x_1}{x_i}, \ldots, \frac{x_n}{x_i}\right) \in R_{x_i,0}^{\oplus(n+1)}.$$

Since the i-th component is 1, it follows that L_i is a direct summand isomorphic to $R_{x_i,0}$. On the overlaps $U_i \cap U_j$ the submodules L_i and L_j coincide, and hence they glue to a line bundle on $\mathbb{P}^n$. This line bundle is $\mathcal{O}_\mathbb{P}(-1)$. □

This example generalises to the Grassmannian $\mathbb{G} = \mathbb{G}(r, n)$. Namely, the r rows of the matrix (8.1) determine a vector subbundle $\mathcal{F}$ of rank r of the trivial vector bundle $\mathcal{O}_\mathbb{G}^{\oplus n}$. In terms of the affine open cover by sets $D(\det X_I) = \{\det X_I \neq 0\}/GL(r)$, the rows of $(X_I)^{-1}X$ generate a submodule $\mathcal{F}_I \subset \mathcal{O}_{D(\det X_I)}^{\oplus n}$. This is a rank r vector bundle on each affine open set, and glues to a vector bundle $\mathcal{F} \subset \mathcal{O}_\mathbb{G}^{\oplus n}$ on the Grassmannian, called the *universal subbundle* on $\mathbb{G}(r, n)$.

(c) The Grassmann functor

We define an equivalence relation on data consisting of an R-module M and an ordered set of n elements $m_1, \ldots, m_n \in M$ by

$$(M; m_1, \ldots, m_n) \sim (M'; m_1', \ldots, m_n')$$

if and only if there exists an isomorphism $f : M \to M'$ taking each $m_i \mapsto m_i'$.

Definition 8.88. The *Grassmann functor* $\mathcal{G}r(r, n)$ is the functor from the category of rings to the category of sets which assigns to a ring R the set

$$\left\{ \begin{array}{l} (M; m_1, \ldots, m_n) \text{ where } M \text{ is a locally free } R\text{-} \\ \text{module of rank } r \text{ generated by } m_1, \ldots, m_n \end{array} \right\} /\text{isomorphism}$$

and assigns to a ring homomorphism $f : R \to S$ the set mapping

$$[M; m_1, \ldots, m_n] \mapsto [M \otimes_R S; m_1 \otimes 1, \ldots, m_n \otimes 1].$$

(See Lemma 8.53.) □

Although this functor is defined on arbitrary rings R, we will only be concerned, in what follows, with rings containing the field k. What we want to show next anticipates (and serves as a model for) the discussion of Chapter 11, and we refer the reader to Section 11.1(a) for the necessary definitions.

Proposition 8.89. *The functor $\mathcal{G}r(r, n)$ is isomorphic to the functor $\underline{\mathbb{G}(r, n)}$ associated to the Grassmannian variety $\mathbb{G}(r, n)$.*

In the language of Chapter 11 one says that the Grassmannian $\mathbb{G}(r, n)$ is a fine moduli space for the Grassmann functor (Definition 11.5). Before proving the proposition, let us first look at the relation between the functor $\mathcal{G}r(1, n)$ and the projective space $\mathbb{P}^{n-1}$. This functor assigns to a ring R an equivalence class of invertible R-modules M equipped with n generators $m_1, \ldots, m_n \in M$.

(1) Suppose that M is free, and choose an isomorphism $M \xrightarrow{\sim} R$. This identifies $m_1, \ldots, m_n$ with ring elements $a_1, \ldots, a_n \in R$, which define a partition of unity. So the ring homomorphisms for $i = 1, \ldots, n$

$$k\left[\frac{x_1}{x_i}, \ldots, \frac{x_n}{x_i}\right] \to R_{a_i}, \qquad \frac{x_j}{x_i} \mapsto \frac{m_j}{m_i},$$

define, by passing to the spectra and gluing, a morphism

$$\varphi : \operatorname{Spm} R \to \mathbb{P}^{n-1}.$$

If we choose a different isomorphism $M \xrightarrow{\sim} R$, then the ring elements $a_1, \ldots, a_n \in R$ are multiplied by some invertible element of R, and we obtain the same morphism $\operatorname{Spm} R \to \mathbb{P}^{n-1}$.

(2) More generally, M may not be isomorphic to R but is locally isomorphic, so we can choose a covering $\{M_{a_i}\}$ with each $M_{a_i} \cong R_{a_i}$. By (1) we obtain a morphism

$$\varphi_i : \operatorname{Spm} R_{a_i} \to \mathbb{P}^{n-1},$$

and the maps φ_i, φ_j coincide on $\operatorname{Spm} R_{a_i a_j}$. Hence by Proposition 8.33 we get a morphism $\varphi : \operatorname{Spm} R \to \mathbb{P}^{n-1}$.

What we have shown is that the functor $\mathcal{G}r(1, n)$ assigns to a ring R a set which can be viewed as the set of morphisms $\varphi : \operatorname{Spm} R \to \mathbb{P}^{n-1}$. Moreover, this has the property that the sheaf $\underline{M}$ on Spm is the pull-back of the tautological bundle:

$$\varphi^* \mathcal{O}_{\mathbb{P}}(1) \cong \underline{M},$$

and the inclusion $\mathcal{O}_{\mathbb{P}}(-1) \subset \mathcal{O}_{\mathbb{P}}^{\oplus n}$ pulls back to the inclusion of R-modules

$$(m_1, \ldots, m_n) : M^{\vee} \to R^{\oplus n}.$$

Proof of Proposition 8.89. First of all we note that if there is a morphism $\varphi : \operatorname{Spm} R \to \mathbb{G}(r, n)$, then this will determine a vector bundle on $\operatorname{Spm} R$ which is the pull-back of the universal subbundle $\varphi^* \mathcal{F} \subset R^{\oplus n}$. By Lemma 8.85, the dual vector bundle $\varphi^* \mathcal{F}^\vee$ comes from a locally free R-module of rank r and the dual of the inclusion in $R^\oplus$ determines n generators $m_1, \ldots, m_n \in M$. This shows that there is a map of functors

$$\underline{\mathbb{G}}(r, n) \to \mathcal{G}r(r, n), \qquad \varphi \mapsto \varphi^* \mathcal{F}^\vee. \tag{8.24}$$

We will construct the inverse of this map.

Let M be a locally free R-module of rank r. If M is free, then a chosen free basis $m_1, \ldots, m_n \in M \cong R^{\oplus n}$ can be represented as an $r \times n$ R-matrix, and so we obtain a map

$$\text{Spm } R \to \text{Mat}(r, n).$$

Moreover, the composition of this map with the quotient by $GL(r)$

$$\text{Spm } R \to \text{Mat}(r, n) --\to \text{Mat}(r, n) /\!/ GL(r) = \mathbb{G}(r, n)$$

is a morphism independent of the choice of free basis. We denote this morphism by $\varphi_{M,\mathbf{m}}$.

In general, if M is not free, then by gluing affine open sets on which the localisations of M are free an in the discussion (2) above, we obtain a morphism

$$\varphi_{M,\mathbf{m}} : \text{Spm } R \to \text{Mat}(r, n) \to \mathbb{G}(r, n).$$

This is called the *classification morphism* of $(M; m_1, \ldots, m_n)$ and is characterised by the property that it pulls back the universal subbundle $\mathcal{F} \subset \mathcal{O}_{\mathbb{G}}^{\oplus n}$ to the inclusion of R-modules $M^\vee \hookrightarrow R^{\oplus n}$. Thus the map

$$\mathcal{G}r(r, n) \to \underline{\mathbb{G}}(r, n), \qquad (M; m_1, \ldots, m_n) \mapsto \varphi_{M,\mathbf{m}}$$

is the inverse of (8.24). $\qquad\qquad\qquad\qquad\qquad\qquad\qquad\qquad\qquad\qquad\quad\Box$

(d) The tangent space of the functor

Given a ring R and a maximal ideal $\mathfrak{m} \subset R$, the quotient $\mathfrak{m}/\mathfrak{m}^2$ is a vector space ove the field $R/\mathfrak{m}$. Recall that the dual space $(\mathfrak{m}/\mathfrak{m}^2)^\vee$ is called the *(Zariski) tangent space* of R (or of Spm R) at $\mathfrak{m}$ (Definition 4.19). If $R/\mathfrak{m} = k$, then *tangent vectors* of R at $\mathfrak{m}$ are ring homomorphisms

$$f : R \to k[t]/(t^2)$$

for which the composition $R \xrightarrow{f} k[t]/(t^2) \to k[t]/(t) = k$ coincides with the map $R \to R/\mathfrak{m}$.

In what follows we shall write, as usual, ϵ for the residue class t mod t^2 and $k[\epsilon] = k[t]/(t^2)$. (See Examples 3.25 and 4.31.)

Definition 8.90. Given a functor

$$F : \{\text{rings over } k\} \to \{\text{sets}\}$$

and an element $x \in F(k)$, we define the *tangent space* of F at x to be

$$T_x F := \begin{pmatrix} \text{inverse image of } x \in F(k) \text{ under} \\ F_{k[\epsilon] \to k} : F(k[\epsilon]) \longrightarrow F(k) \end{pmatrix} \subset F(k[\epsilon]).$$

This has the structure of a vector space over k. (It is a straightforward exercise to prove this, and the reader may consult Schlessinger [48]. However, for readers meeting these notions for the first time it is probably not a very useful exercise, as in each application the vector space structure will be obvious from the context.) □

An algebraic variety (or, more generally, a scheme) X determines a functor

$$\underline{X} : \{\text{rings}\} \to \{\text{sets}\}.$$

(See Section 3.3(a).) Then, after taking an affine open cover of X, at each k-valued point $x \in \underline{X}(k)$ the tangent space $T_x \underline{X} \subset \underline{X}(k[\epsilon])$ in the sense of Definition 8.90 coincides with the Zariski tangent space.

Proposition 8.91. *The tangent space to the Grassmannian $\mathbb{G}(r, n)$ at a point $[U] \in \mathbb{G}(r, n)$ corresponding to an r-dimensional subspace $U \subset k^n$ is canonically isomorphic to $\operatorname{Hom}_k(U, k^n/U)$.*

Proof. We will use the Grassmann functor $\mathcal{G}r(r, n)$, though alternatively we could use an affine open cover of $\mathbb{G}(r, n)$. To the point $[U] \in \mathbb{G}(r, n)$ there corresponds an exact sequence of vector spaces

$$0 \to U \to k^n \to V \to 0, \tag{8.25}$$

and a tangent vector at $[U]$ is then an exact sequence of free $k[\epsilon]$-modules

$$0 \to \widetilde{U} \to k[\epsilon]^n \to \widetilde{V} \to 0 \tag{8.26}$$

whose reduction modulo (ϵ) coincides with (8.25). Let $u_1, \ldots, u_r \in k^n$ be a basis of U, and let

$$u_1 + \epsilon v_1, \ldots, u_r + \epsilon v_r \in k[\epsilon]^n$$

be a free basis of $\widetilde{U}$ as a $k[\epsilon]$-module. Since $\epsilon^2 = 0$, it follows that $\epsilon u_1, \ldots, \epsilon u_r \in \widetilde{U}$, and this is a basis of $U\epsilon$. This shows that the given tangent vector determines, via (8.26), a well-defined linear map

$$U \to V = k^n/U, \qquad u_i \mapsto v_i \mod U,$$

and this correspondence defines an isomorphism $T_{[U]}\mathbb{G} \xrightarrow{\sim} \mathrm{Hom}_k(U, k^n/U)$.

$\square$

Exercises

1. Show that for $n \geq 3$ the Hilbert series $P_n(t)$ of the semiinvariant ring $k[\mathrm{Mat}(2, n)]^{SL(2)}$ satisfies

$$P_n(t^{-1}) + t^n P_n(t) = 0.$$

(See Proposition 8.4.)

2. Prove Proposition 8.37 by induction on the dimension of the total fraction module $Q(M)$ as a vector space over the field of fractions $Q(R)$.

3. Prove Corollary 8.38 by showing that a minimal system of generators is a free basis.

4. For R-modules L, M, N show that

$$\mathrm{Hom}_R(M \otimes_R N, L) = \mathrm{Hom}_R(M, \mathrm{Hom}_R(N, L)).$$

(This says that the functors $\otimes_R N$ and $\mathrm{Hom}_R(N, \cdot)$ are *adjoint*.) Use this to give another proof of Lemma 8.55.

5. Show that the $\mathbb{Z}$-module $M \subset \mathbb{Q}$ consisting of rational numbers with square free denominator satisfies condition (ii) of Proposition 8.47 but is not locally free. (The author learnt this counterexample from M. Hashimoto.)

6. If Spm R is a union of disjoint open sets U_1, U_2, show that $U_1 = D(e)$, $U_2 = D(1 - e)$ for some idempotent $e^2 = e$. (Note that an idempotent e decomposes the ring as $R = Re \oplus R(1 - e)$.)

7. Let M be a finitely generated module over a local integral domain $(R, \mathfrak{m})$. Show that if

$$\dim_{R/\mathfrak{m}} M/\mathfrak{m}M = \dim_{Q(R)} Q(M),$$

then M is a free module.

8. Show that an invertible ideal $\mathfrak{a} \subset R$ (see Example 8.61) is an invertible R-module.

9. By considering the $\mathbb{Z}$-similarity classes of 2×2 $\mathbb{Z}$-matrices N satisfying

$$N^2 + 119I_2 = 0, \qquad N \equiv I_2 \bmod 2,$$

show that the imaginary quadratic field $\mathbb{Q}(\sqrt{-119})$ has class number 10. (That is, the group Pic $\mathcal{O}_{\sqrt{-119}}$ has order 10.)

9

Curves and their Jacobians

Every curve of genus g has associated to it a g-dimensional algebraic variety called its Jacobian. Analytically, over the field of complex numbers, this is a complex torus $\mathbb{C}^g / \Gamma_C$ where Γ_C is a lattice. Given a basis $\omega_1, \ldots, \omega_g$ of holomorphic 1-forms on the curve, Γ_C is the lattice of periods

$$\Gamma_C = \left\{ \left(\int_\alpha \omega_1, \ldots, \int_\alpha \omega_g \right) \text{ for } \alpha \in H_1(C, \mathbb{Z}) \right\} \subset \mathbb{C}^g. \qquad (9.1)$$

There is a natural map from the curve C to the Jacobian, and this extends to a map, called the Abel-Jacobi map, from its group of divisors to the Jacobian. Classically one uses theta functions to show that the Jacobian is a projective variety. However, in this chapter we will adopt a different approach, using invariants to construct a projective variety whose underlying set is the Picard group of C (Section 9.4), and then showing that over $k = \mathbb{C}$ this variety agrees with the complex torus just described (Section 9.6).

The first three sections and Section 9.5 prepare the way for this construction. A nonsingular algebraic curve is a variety whose local rings are all discrete valuation rings. This leads to the notion of order of pole of a rational function at a point, gap values at a point and the vector space $\Lambda(D)$ of rational functions with poles bounded by a positive divisor D. The (arithmetic) genus is the number of gap values at any point. In Section 9.1 we discuss these notions and prove Riemann's inequality, which gives a lower bound for $\dim \Lambda(D)$ in terms of the genus.

The inequality itself is just a formal consequence of the definitions; however, its depth lies in the underlying facts that the genus is finite and equal to the dimension of the cohomology space $H^1(\mathcal{O}_C)$ (Section 9.2). We show that line bundles on C correspond bijectively to linear equivalence classes of divisors, and in this language $\Lambda(D) = H^0(\mathcal{O}_C(D))$ while the index of speciality $i(D)$ is the dimension of $H^1(\mathcal{O}_C(D))$. Riemann's inequality becomes the

Riemann-Roch formula (9.15) for $L \in \text{Pic } C$,

$$\dim H^0(L) - \dim H^1(L) = \deg L - 1 + g.$$

A variety $X = \text{Spm } R$ is nonsingular at a point $x \in X$ with maximal ideal $\mathfrak{m} \subset R$ if the graded ring $\text{gr}_{\mathfrak{m}} R = \bigoplus \mathfrak{m}^l/\mathfrak{m}^{l+1}$ is isomorphic to a polynomial ring. In Section 9.3, after explaining nonsingularity and differential modules, we extend Theorem 5.3 on the separation of closed orbits to deal with infinitely close orbits. An orbit $G \cdot x \subset X$ is a *free closed orbit* if it is stable with trivial stabiliser; and we show that, if X is nonsingular at all points of a free closed orbit, then the quotient X/G is nonsingular at the image point of this orbit, with dimension $= \dim X - \dim G$.

In Section 9.5 we review duality and de Rham cohomology, and in the final section we show that over the complex numbers our quotient variety is isomorphic to $\mathbb{C}^g/\Gamma_C$. The key to this is Abel's theorem.

9.1 Riemann's inequality for an algebraic curve

Among algebraic varieties, the simplest are the curves, and we begin this section with some facts about affine curves. Let R be a Noetherian integral domain, finitely generated over a field k, and we assume that $\text{Spm } R$ has more than one point. Then a maximal ideal $\mathfrak{m} \subset R$ is nonzero, and so by Nakayama's Lemma 8.23 applied to the localisation $R_{\mathfrak{m}}$ the quotient $\mathfrak{m}/\mathfrak{m}^2$ is also nonzero.

Lemma 9.1. *The following conditions on a maximal ideal* $\mathfrak{m} \subset R$ *are equivalent.*

(i) $\dim_{R/\mathfrak{m}}(\mathfrak{m}/\mathfrak{m}^2) = 1$.
(ii) The localisation $R_{\mathfrak{m}}$ *is a discrete valuation ring.*

Before proving this we note that on passing to the localisation at $\mathfrak{m}$ condition (i) is preserved, in the sense that it implies $\mathfrak{m}R_{\mathfrak{m}}/\mathfrak{m}^2 R_{\mathfrak{m}}$ is 1-dimensional over the field $R_{\mathfrak{m}}/\mathfrak{m}R_{\mathfrak{m}}$. Nakayama's Lemma, in the form of Corollary 8.24, says that if $\overline{\pi}$ spans this space, where $\pi \in \mathfrak{m}R_{\mathfrak{m}}$, $\pi \notin \mathfrak{m}^2 R_{\mathfrak{m}}$, then π generates the maximal ideal $\mathfrak{m}R_{\mathfrak{m}} \subset R_{\mathfrak{m}}$. Such an element $\pi \in \mathfrak{m}R_{\mathfrak{m}}$ is called a *regular parameter* at the maximal ideal $\mathfrak{m}$.

Proof. We shall prove (i) $\implies$ (ii) and leave the converse to the reader. Given property (i), let $\pi \in \mathfrak{m}R_{\mathfrak{m}}$ be a regular parameter.

Claim: The descending chain of ideals $R_{\mathfrak{m}} \supset \mathfrak{m}R_{\mathfrak{m}} = (\pi) \supset (\pi)^2 \supset (\pi)^3 \supset \cdots$ has intersection zero.

To prove this, suppose $a \in \bigcap_{i \geq 0} (\pi)^i$. Then for each $n \geq 0$ there is an element $a_n \in R_\mathrm{m}$ such that $a = a_n \pi^n$. Each $a_n = \pi a_{n+1}$, and so we have an ascending chain of principal ideals

$$(a_0) \subset (a_1) \subset (a_2) \subset (a_3) \subset \cdots.$$

Since R_m is Noetherian, $(a_N) = (a_{N+1})$ for some $N \in \mathbb{N}$, and in particular $a_{N+1} = a_N b$ for some $b \in R_\mathrm{m}$. This implies that $(1 - b\pi)a_N = 0$. But $1 - b\pi$ is invertible (it is outside the unique maximal ideal), and so $a_N = 0$, which implies that $a = 0$, proving the claim.

It follows from this that every nonzero element $a \in R_\mathrm{m}$ has a unique expression $a = u\pi^n$ for $n \geq 0$ and some invertible element $u \in R_\mathrm{m}$. Hence the field of fractions $Q(R) = Q(R_\mathrm{m})$ has the corresponding property that every nonzero element $a \in Q(R)$ has a unique expression $a = u\pi^n$ for some integer $n \in \mathbb{Z}$ and element $u \in Q(R)$, $u \notin \mathrm{m}R_\mathrm{m}$. The map

$$v_\mathrm{m} : Q(R) \to \mathbb{Z} \cup \{\infty\}, \qquad a \mapsto \begin{cases} n & \text{if } a = u\pi^n \neq 0 \\ \infty & \text{if } a = 0 \end{cases} \tag{9.2}$$

is then the required discrete valuation. $\qquad\square$

Since every variety X is constructed by gluing affine varieties, at each point $p \in X$ the stalk of the structure sheaf $\mathcal{O}_{X,p}$ is a local ring by Example 8.76.

Definition 9.2. A *nonsingular algebraic curve* over a field k is an algebraic variety over k such that at every point $p \in C$ the local ring $\mathcal{O}_{C,p}$ is a discrete valuat, thenion ring. We denote by

$$v_p : k(C) \to \mathbb{Z} \cup \{\infty\}$$

the valuation (9.2) at the point $p \in C$. A rational function $f \in k(C)$ is *regular* at p if $v_p(f) \geq 0$ and has a *pole of order* n if $v_p(f) = -n < 0$. $\qquad\square$

Example 9.3. Let $C \subset \mathbb{A}^2$ be an affine plane curve $f(x, y) = 0$ having no singular points in the sense of Definition 1.31. Then C is a nonsingular algebraic curve.

To see this, we consider the maximal ideal of a point $p = (a, b) \in C \subset \mathbb{A}^2$. This is $\widetilde{\mathrm{m}} = (x - a, y - b) \subset k[x, y]$. By hypothesis, the partial derivatives $\partial f/\partial x$, $\partial f/\partial y$ do not both vanish at (a, b), and so $f \notin \widetilde{\mathrm{m}}^2$. Thus at each point $p \in C$, with maximal ideal $\mathrm{m} \subset \mathcal{O}_{C,p}$, the residue class of f spans a 1-dimensional kernel of the restriction $\widetilde{\mathrm{m}}/\widetilde{\mathrm{m}}^2 \to \mathrm{m}/\mathrm{m}^2$. Hence

$$\dim_k \mathrm{m}/\mathrm{m}^2 = \dim_k \widetilde{\mathrm{m}}/\widetilde{\mathrm{m}}^2 - 1 = 1.$$

Hence $\mathcal{O}_{C,p}$ is a discrete valuation ring by Lemma 9.1. $\qquad\square$

In what follows we shall be interested in nonsingular algebraic curves C which are projective. This means that C is embedded as a closed subvariety in projective space and is therefore complete by Corollary 3.60 (or by Corollary 3.73).

Lemma 9.4. *Let C be a projective nonsingular algebraic curve. If a rational function $f \in k(C)$ is regular everywhere, then it is constant. In other words,*

$$\{f \in k(C) \mid v_p(f) \geq 0 \text{ for all } p \in C\} = k.$$

Proof. If $f \in k(C)$ is regular everywhere, then it determines a morphism $C \to \mathbb{A}^1$. By Proposition 3.61, the image is a closed set and therefore a complete affine variety. By Proposition 3.57, this image is a single point, and so f is a constant function. $\square$

From now on, unless indicated otherwise, a *curve* will always mean a projective nonsingular algebraic curve. The simplest examples are projective plane curves

$$C : \{f(x, y, z) = 0\} \subset \mathbb{P}^2,$$

defined by the vanishing of a homogeneous polynomial $f(x, y, z)$ whose partial derivatives $\partial f / \partial x$, $\partial f / \partial y$, $\partial f / \partial z$ have no common zeros in $\mathbb{P}^2$. (See Proposition 1.36.) As an exercise, the reader should note that nonsingularity of a projective plane curve implies irreducibility.

(a) Prologue: gap values and the genus

Fix a point p on a curve C. We say that a natural number $n \in \mathbb{N}$ is a *gap value* at $p \in C$ if there does not exist any rational function $f \in k(C)$ which is regular away from p and has a pole of order n at p.

Example 9.5. Let $C = \mathbb{P}^1$ and $p = \infty \in \mathbb{P}^1$. The rational function field $k(\mathbb{P}^1)$ is the field of Laurent polynomials $f(x)$ in one variable, and a function regular away from ∞ is just a polynomial. The order of pole at ∞ is $n = \deg f(x)$, and such polynomials exist for every $n \geq 1$. Hence there are no gap values. $\square$

Example 9.6. Let C be a 1-dimensional complex torus $\mathbb{C}/(\mathbb{Z} + \mathbb{Z}\tau)$, and let $p = O \in C$ be the origin. By Liouville's Theorem 1.44(ii) there are no meromorphic functions on C holomorphic away from O and with only a simple pole at O. On the other hand, the Weierstrass $\wp$-function is holomorphic away from O with exactly a double pole at O, and by taking successive derivatives $\wp'$, $\wp''$, ... there is a function with exactly one pole of order n at O for arbitrary $n \geq 2$. Hence the set of gap values in this case is $\{1\}$. ☐

Example 9.7. Let $f(x)$ be a polynomial of degree $2g + 1$ without any repeated roots. Then the nonsingular algebraic curve obtained as the 1-point compactification

$$C : \{y^2 = f(x)\} \cup \{\infty\}$$

of the affine plane curve $y^2 = f(x)$ is called a *hyperelliptic curve*. This is a 2-sheeted cover of the projective line $\mathbb{P}^1$ branched over the roots of $f(x) = 0$ and the point $\infty \in \mathbb{P}^1$. (When $k = \mathbb{C}$, this curve is the Riemann surface of the 2-valued function $\sqrt{f(x)}$.) The rational function $x \in k(C)$ has a double pole at $\infty \in C$, while $y \in k(C)$ has a pole of order $2g + 1$ here. Both functions are regular at all other points. Since the function field $k(C)$ is generated by x, y, the set of gap values at ∞ is exactly $\{1, 3, 5, \ldots, 2g - 3, 2g - 1\}$. ☐

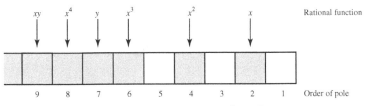

Figure 9.1: The gap values of $y^2 = x^7 - 1$

At a point $p \in C$ the gap values themselves depend on the particular point. (Though Examples 9.5 and 9.6 are exceptional in this respect. In both cases the curve is acted upon transitively by automorphisms, and so the gap values happen, in these examples, to be the same at all points.) However, what turns out to be the case is that the *number of gap values* is independent of the point, and that this number is equal to the genus of the curve C (Corollary 9.21). Thus in the three examples above the genus is $0, 1, g$, respectively.

We can interpret this number in the following way. Choosing a regular parameter t at the point $p \in C$, each rational function can be expanded about p as a Laurent series in t. Taking the principal part of this Laurent series gives a linear map of vector spaces over k

$$\text{pp}_p : \begin{Bmatrix} f \in k(C) \text{ such that} \\ v_q(f) \geq 0 \\ \text{for all } q \in C - \{p\} \end{Bmatrix} \longrightarrow k((t))/k[[t]] = t^{-1}k[t^{-1}].$$

The number of gap values at p is the dimension of the cokernel of this map. Now, not only is the number of gap values independent of the point $p \in C$, but in a certain sense the vector space coker pp_p is also independent of the point. This will become clear when we introduce the cohomology space $H^1(\mathcal{O}_C)$ (see Section 9.2(a)).

(b) Divisors and the genus

Our goal at this point is the Riemann-Roch Formula 9.29 for a line bundle on a curve, and we will approach this in the way it was approached historically, using the language of divisors.

Definition 9.8. The *divisor group* of a curve C is the free abelian group

$$\text{Div } C = \bigoplus_{p \in C} \mathbb{Z}p$$

generated by all the points of C. An element of this group

$$D = \sum_{p \in C} n_p p \qquad \text{(finite sum)}$$

is called a *divisor* on C.   □

The *degree* of a divisor D is defined to be the sum of its coefficients $\deg D = \sum_{p \in C} n_p$, and D is a *positive divisor*, written $D \geq 0$, if $n_p \geq 0$ for all $p \in C$. If the difference of two divisors $D - D'$ is positive, we write $D \geq D'$.

Definition 9.9.

(i) Given a discrete valuation ring R, $v : Q(R) \to \mathbb{Z} \cup \{\infty\}$ and integer $n \in \mathbb{Z}$ we set

$$R(nv) := \{x \in Q(R) \mid v(x) + n \geq 0\}.$$

For example, $R = R(0) \supset R(-v) \supset R(-2v) \supset \cdots$ and $R(-v)$ is a maximal ideal in R.

(ii) For the valuation v_p of the local ring $\mathcal{O}_{C,p}$ at a point on a curve C we write simply $\mathcal{O}_{C,p}(np) := \mathcal{O}_{C,p}(nv_p)$. Then for a divisor $D = \sum_{p \in C} n_p p$ we define

$$\Lambda(D) = \bigcap_{p \in C} \mathcal{O}_{C,p}(n_p p) = \{f \in k(C) \mid v_p(f) + n_p \geq 0 \text{ for all } p \in C\}.$$

In other words, $\Lambda(D)$ is the set of rational functions on C with orders of poles bounded by the coefficients of the divisor. □

The following facts are clear.

(I) $\Lambda(D)$ is a vector subspace of $k(C)$ over k.
(II) If $D \geq D'$, then $\Lambda(D) \supset \Lambda(D')$.
(III) $\Lambda(0) = k$ (by Lemma 9.4).
(IV) For any point $p \in C$ we have $\dim \Lambda(D + p)/\Lambda(D) \leq 1$.
(V) For any positive divisor D,

$$\dim \Lambda(D) \leq \deg D + 1. \tag{9.3}$$

(This follows inductively from (IV), starting from (III).)

Note that for any divisor $D \in \operatorname{Div} C$ there exists some positive divisor D' such that $D' \geq D$. From properties (II) and (V), therefore, we deduce:

Lemma 9.10. *For all $D \in \operatorname{Div} C$ the vector space $\Lambda(D)$ is finite-dimensional.* □

The difference between the two sides of the inequality (9.3) is an important quantity, which we will denote by

$$j(D) := \deg D + 1 - \dim \Lambda(D). \tag{9.4}$$

The following is one of various ways to define the genus of a curve.

Definition 9.11. For any curve C the supremum taken over positive divisors:

$$g := \sup_{D \geq 0} j(D) \quad \in \{0, 1, 2, \ldots, \infty\}$$

is called the *(arithmetic) genus* of C. □

In this language, a gap value at a point $p \in C$ is a natural number $n \in \mathbb{N}$ for which $\Lambda(np) = \Lambda((n-1)p)$. Hence the number of gap values $p \in C$ is given by:

$$\begin{pmatrix} \text{number of gap} \\ \text{values at } p \in C \end{pmatrix} = \sup_{n \geq 0} j(np) \leq g. \tag{9.5}$$

Next, by the same reasoning as for (9.3), we note that if $D \geq D'$, then

$$\dim \Lambda(D) \leq \deg(D - D') + \dim \Lambda(D')$$

and hence:

$$D \geq D' \implies j(D) \geq j(D'). \tag{9.6}$$

Since every divisor is bounded above by a positive divisor, this shows that the positivity condition in Definition 9.11 can be dropped and the supremum taken over arbitrary divisors. That is:

$$g = \sup_{D \in \mathrm{Div}\, C} j(D). \tag{9.7}$$

In particular, this implies

$$\dim \Lambda(D) \geq \deg D + 1 - g \quad \text{for all } D \in \mathrm{Div}\, C. \tag{9.8}$$

This is called *Riemann's inequality*, once we have proved the following:

Theorem 9.12. *The genus g of a curve is finite.*

(c) Divisor classes and vanishing index of speciality

We are going to prove Theorem 9.12 in the next section. For the moment we will assume its validity and examine the divisors on a curve a little more closely. The set of divisors is a slightly artificial object, but it contains a distinguished subset which reflects very closely the world of rational functions on the curve.

Definition 9.13.

(i) For each nonzero rational function $f \in k(C)$ we define a divisor

$$(f) := \sum_{p \in C} v_p(f)p \in \mathrm{Div}\, C.$$

This is called a *principal divisor*.

(ii) The set of all principal divisors $\{(f) \mid f \in k(C) - 0\}$ is a subgroup of $\mathrm{Div}\, C$, and the equivalence relation modulo this subgroup is called *linear equivalence*.

(iii) The quotient group

$$\text{Cl } C := \text{Div } C/\{\text{principal divisors}\}$$

is called the *divisor class group* of the curve. □

Remark 9.14. This is the analogue for the function field $k(C)$ of the divisor class group of an algebraic number field (Section 8.4(a)). The analogue of Proposition 8.65 will be Proposition 9.34. □

If two divisors D, D' are linearly equivalent, then they differ by a principal divisor $D - D' = (h)$, $h \in k(C)$, and the map $f \mapsto fh$ defines a linear isomorphism $\Lambda(D') \overset{\sim}{\to} \Lambda(D)$. In particular, dim $\Lambda(D) = $ dim $\Lambda(D')$, and so:

Lemma 9.15. *The dimension of $\Lambda(D)$ depends only on the divisor class of D.* □

Definition 9.16. For a divisor $D \in \text{Div } C$ the number

$$i(D) := g - j(D) = \text{dim } \Lambda(D) - \text{deg } D - 1 + g \geq 0$$

is called the *index of speciality* of D. □

Note that by (9.6),

$$D \geq D' \quad \Longrightarrow \quad i(D) \leq i(D'). \tag{9.9}$$

Lemma 9.17. *If divisors D, D' are linearly equivalent, then $i(D) = i(D')$.*

Proof. From (9.7) and (9.4),

$$i(D) = \text{dim } \Lambda(D) - \text{deg } D + \sup_{F \in \text{Div } C} \{\text{deg } F - \Lambda(F)\}$$

$$= \text{dim } \Lambda(D) + \sup_{F \in \text{Div } C} \{\text{deg}(F - D) - \Lambda(F)\}.$$

As F ranges through all divisors, so does $F - D$, and so

$$i(D) = \text{dim } \Lambda(D) + \sup_{F \in \text{Div } C} \{\text{deg } F - \Lambda(D + F)\}. \tag{9.10}$$

If D, D' are linearly equivalent, then $D + F$, $D' + F$ are linearly equivalent, and hence Lemma 9.15 implies $i(D) = i(D')$. □

Corollary 9.18. *If divisors D, D' are linearly equivalent, then* $\deg D = \deg D'$. $\qquad\Box$

We will give two sufficient conditions for $i(D) = 0$. The first is in terms of the degree of D. Note that by Definition 9.16, if $\deg D < g - 1$, then $i(D) > 0$. This condition is sharp in the following sense:

Lemma 9.19. *There exists a divisor D_{van} with*

$$\deg D_{\text{van}} = g - 1, \qquad i(D_{\text{van}}) = 0.$$

Proof. Theorem 9.12 implies the existence of some divisor D with $i(D) = 0$. If $\Lambda(D) = 0$, then, by Definition 9.16, $\deg D = g - 1$, and so it suffices to take $D_{\text{van}} = D$. So assume that $\deg D > g - 1$ and $\Lambda(D) \neq 0$. This means we can find a nonzero rational function f for which $(f) + D \geq 0$. Choosing a point $p \in C$ at which $f(p) \neq 0$ we have $f \notin \Lambda(D - p)$, and so $\Lambda(D)/\Lambda(D-p) = k$. This implies $j(D-p) = j(D)$, and so $i(D-p) = i(D) = 0$. We now repeat the argument, subtracting points $n = \deg D - g + 1$ times to obtain

$$\Lambda(D - p_1 - \cdots - p_n) = i(D - p_1 - \cdots - p_n) = 0.$$

We then take $D_{\text{van}} = D - p_1 - \cdots - p_n$. $\qquad\Box$

This has the following important application, complementary to Riemann's inequality (9.8). In fact, its proof works for singular curves as well.

Vanishing Theorem 9.20. *If $\deg D \geq 2g - 1$, then $i(D) = 0$ and*

$$\dim \Lambda(D) = \deg D - g + 1.$$

Proof. We apply Riemann's inequality (9.8) to the difference $D - D_{\text{van}}$, where D_{van} is the divisor constructed in Lemma 9.19. This says, since $\deg(D - D_{\text{van}}) \geq g$, that $\Lambda(D - D_{\text{van}}) \neq 0$. Thus D is linearly equivalent to $D_{\text{van}} + F$ for some positive divisor $F \geq 0$. Hence by (9.9) and Lemma 9.17 we see that $i(D) = 0$. $\qquad\Box$

We have seen in (9.5) that the number of gap values at a point $p \in C$ is the supremum of $j(np)$ for $n \geq 0$, and that this supremum is at most g. But if

$n \geq 2g - 1$, then the vanishing theorem implies that $j(np) = g - i(np) = g$, and so we arrive at:

Corollary 9.21. *At any point $p \in C$ the number of gap values is equal to the genus of C. In particular, this number does not depend on the point.* □

The second sufficient condition for the vanishing of $i(D)$ is that $\Lambda(D)$ has big enough dimension:

Lemma 9.22. *If $\dim \Lambda(D) > g$, then $i(D) = 0$.*

Proof. Suppose, for a contradiction, that $i(D) \neq 0$, and fix a point $p \in C$. Then $i(D + p)$ is equal either to $i(D)$ or to $i(D) - 1$, and so it must happen that $i(D + np) = 1$ for some $n \geq 0$. The divisor $D' = D + np$ then satisfies $\dim \Lambda(D') \geq \dim \Lambda(D) > g$ and $i(D') = 1$. On the other hand, by Theorem 9.20, we must have $\deg D' \leq 2g - 2$. By Definition 9.16, therefore,

$$\dim \Lambda(D') = \deg D' + 1 - g + i(D') \leq g.$$

□

9.2 Cohomology spaces and the genus

In this section we are going to interpret the genus of a curve C as the dimension of a certain cohomology space and deduce from this its finiteness (Theorem 9.12). In fact we shall do the same also for the index of speciality $i(D)$ of a divisor (see (9.13)).

(a) Cousin's problem

We first consider, at a given point $p \in C$, the quotient module $k(C)/\mathcal{O}_{C,p}$ over the local ring $\mathcal{O}_{C,p}$ (Definition 9.9). Let $t_p \in \mathfrak{m}_p \mathcal{O}_{C,p}$ be a regular parameter. We can identify $k(C) = \mathcal{O}_{C,p}[t_p^{-1}]$ (this is localisation at t_p in the sense of Example 8.21), and the elements

$$t_p^{-1}, t_p^{-2}, t_p^{-3}, \ldots \in k(C)$$

can be viewed as a basis of $k(C)/\mathcal{O}_{C,p}$, as a vector space over k. In other words, every rational function $f \in k(C)$ uniquely determines coefficients $c_{-1}, c_{-2}, \ldots, c_{-N}$ such that

$$f \equiv \sum_{i=1}^{N} c_{-i} t_p^{-i} \mod \mathcal{O}_{C,p}.$$

This residue class of f in $k(C)/\mathcal{O}_{C,p}$ is called the *principal part* (or *singular part*) of the function f at $p \in C$.

Cousin's Problem: *Given* Cousin data *consisting of finitely many points* $p_1, \ldots, p_m \in C$ *and a principal part* $\alpha_i \in k(C)/\mathcal{O}_{C,p_i}$ *at each point, when does there exist a rational function* $f \in k(C)$ *satisfying* $f \equiv \alpha_i \bmod \mathcal{O}_{C,p_i}$ *at all of the points?*

(This is also known as Cousin's first problem, or as Mittag-Leffler's problem.)

For $C = \mathbb{P}^1$ it is easy to see, using a global coordinate, that such a function always exists. The same is true for any affine curve $C \in \mathbb{A}^n$ (Exercise 9.4). And for a single point $p \in C$, Cousin's problem corresponds to the problem of computing the gap values at p (Section 9.1(a)).

A different sort of example is the following.

Example 9.23. Consider the field of meromorphic functions on the complex plane $\mathbb{C}$, doubly periodic with respect to a lattice $\Gamma \cong \mathbb{Z}^2$. (See Section 1.5.) By Liouville's Theorem 1.44(ii), the space of functions having a simple pole at each of two cosets $p+\gamma$ and $q+\Gamma$ and holomorphic elsewhere is 1-dimensional (Exercise 9.3). Thus the Cousin problem for simple poles at $p, q \in \mathbb{C}/\Gamma$ cannot in general have a solution. □

The key tool for solving Cousin's problem is the notion of cohomology. We will think of the Cousin data $(\alpha_1, \ldots, \alpha_n)$ as an element of the infinite direct sum $\bigoplus_{p \in C} k(C)/\mathcal{O}_{C,p}$, called the *principal part space* of C (Figure 9.3).

Definition 9.24. The linear map

$$\mathrm{pp} : k(C) \to \bigoplus_{p \in C} k(C)/\mathcal{O}_{C,p}$$

which assigns to a rational function its principal part at each point of the curve is called the *principal part map* on C. Its cokernel

$$H^1(\mathcal{O}_C) := \operatorname{coker} \mathrm{pp}$$

is called the *cohomology space* of the structure sheaf $\mathcal{O}_C$ (or of C with coefficients in the sheaf $\mathcal{O}_C$). □

Thus Cousin's problem is equivalent to that of computing $H^1(\mathcal{O}_C)$.

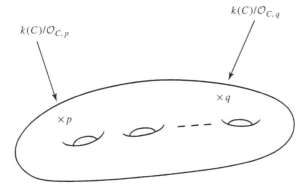

Figure 9.2: The principal part space

Proposition 9.25. *The dimension over k of the cohomology space $H^1(\mathcal{O}_C)$ is equal to the genus g of the curve C.*

Proof. Given a positive divisor $D = \sum_{p \in C} n_p p$, we consider the 'truncated' principal part map

$$\mathrm{pp}_D : \Lambda(D) \to \bigoplus_{p \in C} \mathcal{O}_{C,p}(n_p p)/\mathcal{O}_{C,p} =: \mathcal{O}_D(D). \tag{9.11}$$

The kernel is the field of constant functions k. The dimension of $\mathcal{O}_D(D)$ is $\sum_{p \in C} n_p = \deg D$, and so coker pp_D has dimension equal to $j(D)$. On the other hand, the diagram

$$\Lambda(D) \to \qquad \mathcal{O}_D(D)$$

$$\cap \qquad\qquad \cap$$

$$k(C) \to \bigoplus_{p \in C} k(C)/\mathcal{O}_{C,p}$$

commutes, so that there is an injective linear map

$$\text{coker } \mathrm{pp}_D \hookrightarrow \text{coker } \mathrm{pp} = H^1(\mathcal{O}_C).$$

Now, $H^1(\mathcal{O}_C)$ is the union of the images of coker pp_D as $D \geq 0$ ranges through all positive divisors, and hence

$$\dim H^1(\mathcal{O}_C) = \sup_{D \geq 0} \dim \text{coker } \mathrm{pp}_D = \sup_{D \geq 0} j(D) = g,$$

by Definition 9.11. $\qquad\qquad\qquad\qquad\qquad\qquad\qquad\qquad\qquad\square$

We note from this proof that

$$H^1(\mathcal{O}_C) = \lim_{D \geq 0} \mathrm{coker}\left\{\mathrm{pp}_D : \Lambda(D) \to \mathcal{O}_D(D)\right\}. \tag{9.12}$$

Remark 9.26. One can also consider a multiplicative version of Cousin's problem. We define an elementary sheaf $\mathcal{O}_C^\times$ with the multiplicative group $k(C)^\times$ of nonzero rational functions as its total set, defined on open sets $U \subset C$ by

$$\mathcal{O}_C^\times : U \mapsto \{\text{ regular nowhere vanishing functions on } U \}.$$

At each point $p \in C$ the stalk $\mathcal{O}_{C,p}^\times$ is the group of invertible elements in $\mathcal{O}_{C,p}$. We then have a 'multiplicative principal part map'

$$\delta : k(C)^\times \to \bigoplus_{p \in C} k(C)^\times / \mathcal{O}_{C,p}^\times,$$

and we define $H^1(\mathcal{O}_C^\times)$ to be coker δ. What is the analogue of Cousin's problem in this case? At each point $p \in C$ a valuation gives an isomorphism of $k(C)^\times / \mathcal{O}_{C,p}^\times$ with $\mathbb{Z}$. Thus the direct sum above is nothing other than the divisor group Div C, and the map δ assigns to a function $f \in k(C)^\times$ its principal divisor (f) (Definition 9.13). In particular,

$$H^1(\mathcal{O}_C^\times) = \mathrm{Cl}\, C.$$

(b) Finiteness of the genus

In the expression (9.12) for the cohomology $H^1(\mathcal{O}_C)$ we take a limit over positive divisors as more points are added or, equivalently, as the coefficients

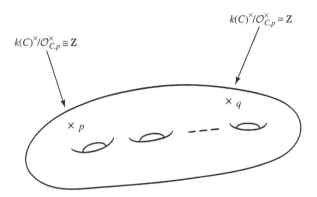

Figure 9.3: Multiplicative principal parts and divisors

over C go to infinity. However, the next proposition shows that the group can be computed by a limit using just one fixed divisor.

Proposition 9.27. *Let D be a positive divisor in C and suppose that the complement of its support $U = C - D$ is affine. Then*

$$H^1(\mathcal{O}_C) = \operatorname{coker} \left\{ \mathcal{O}_C(U) \to \bigoplus_{p \in \operatorname{supp} D} k(C)/\mathcal{O}_{C,p} \right\} = \lim_{n \to \infty} \operatorname{coker} \operatorname{pp}_{nD}.$$

In particular, the genus is given by $g = \sup_{n \geq 0} j(nD)$.

In fact, one can show that the assumption that U is affine is unnecessary, as this is always the case. This is essentially an exercise using Theorem 9.20.

Proof. We decompose the principal part space as

$$\left(\bigoplus_{p \in \operatorname{supp} D} k(C)/\mathcal{O}_{C,p} \right) \oplus \left(\bigoplus_{q \in U} k(C)/\mathcal{O}_{C,q} \right).$$

Since U is affine, the map $k(C) \to \bigoplus_{q \in U} k(C)/\mathcal{O}_{C,q}$ is surjective (that is, Cousin's problem for an affine curve always has a solution) and its kernel is the coordinate ring $k[U] = \mathcal{O}_C(U)$. Hence, by Definition 9.24,

$$H^1(\mathcal{O}_C) = \operatorname{coker} \left\{ \mathcal{O}_C(U) \to \bigoplus_{p \in \operatorname{supp} D} k(C)/\mathcal{O}_{C,p} \right\}.$$

Now since

$$\mathcal{O}_C(U) = \bigcup_{n=1}^{\infty} \Lambda(nD) \quad \text{and, by (9.11),} \quad \bigoplus_{p \in \operatorname{supp} D} k(C)/\mathcal{O}_{C,p} = \bigcup_{n=1}^{\infty} \mathcal{O}_{nD}(nD),$$

we obtain the limit in the propostion. $\qquad \square$

Proof of Theorem 9.12. We consider a fixed embedding of C in some projective space $\mathbb{P}^n$. Let $H \subset \mathbb{P}^n$ be a hyperplane and $C \cap H =: D = p_1 + \cdots + p_d$ its intersection with C. For simplicity we will assume that this intersection is transverse. (There always exists a transverse hyperplane section in this sense (see Exercise 9.5); though in fact this assumption is not essential in the proof that follows.) We take homogeneous coordinates $(x_0 : x_1 : \ldots : x_n)$ for which H has equation $x_0 = 0$. For each intersection point p_i we can find another hyperplane, defined by some linear form $l_i(x)$, passing through p_i but not through the

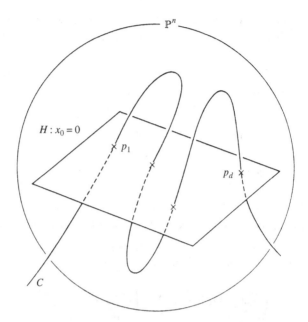

Figure 9.4: A space curve

remaining $d - 1$ points. The ratio $l_i(x)/x_0$ defines a rational function on C
belonging to the space $\Lambda(D)$. We consider the product

$$f_i = \frac{l_1(x) \cdots \widehat{l_i(x)} \cdots l_d(x)}{x_0^{d-1}} \in \Lambda((d-1)D).$$

The function f_i has a pole of order $d - 1$ at p_i and at each p_j, $j \neq i$, a pole
of order at most $d - 2$. Together, therefore, $f_1, \ldots, f_d$ generate $\Lambda((d-1)D)$
modulo $\Lambda((d-2)D)$. Next we introduce a hyperplane not containing any of
the points $p_1, \ldots, p_d$, defined by a linear form $l(x)$, and consider the rational
function $f_0 \in k(C)$ determined by the ratio $l(x)/x_0$. This has a simple pole at
each p_i, and for $a \in \mathbb{N}$ the functions $f_0^a f_1, \ldots, f_0^a f_d$ generate $\Lambda((d-1+a)D)$
modulo $\Lambda((d-2+a)D)$. It follows (denoting the affine open set $C - D$
by U) that

$$\operatorname{coker} \left\{ \mathcal{O}_C(U) \to \bigoplus_{p \in \operatorname{supp} D} k(C)/\mathcal{O}_{C,p} = \bigcup_{n=1}^{\infty} \mathcal{O}_{nD}(nD) \right\}$$

coincides with the cokernel of

$$\operatorname{pp}_{(d-2)D} : \Lambda((d-2)D) \to \mathcal{O}_{(d-2)D}((d-2)D),$$

and in particular it is finite-dimensional. By Proposition 9.27 this shows that $H^1(\mathcal{O}_C)$ is finite-dimensional. □

Example 9.28. A nonsingular plane curve $C \subset \mathbb{P}^2$ of degree d has genus

$$g = \tfrac{1}{2}(d-1)(d-2).$$

To see this, we choose homogeneous coordinates so that the line $x_0 = 0$ intersects C transversally in d points $p_1, \ldots, p_d$. The complementary open set is an affine curve $C_0 : \{f(x, y) = 0\} \subset \mathbb{A}^2$, taking $x_0 = 0$ as the line at infinity. Letting $D = p_1 + \cdots + p_d$, we have

$$\Lambda(nD) = \frac{\{\text{polynomials } h(x, y) \text{ with deg } h \leq n\}}{\{\text{polynomials } f(x, y)h(x, y) \text{ with deg } h \leq n - d\}}.$$

Hence, if $n \geq d$, then

$$\dim \Lambda(nD) = \tfrac{1}{2}(n+1)(n+2) - \tfrac{1}{2}(n+1-d)(n+2-d)$$

$$= nd - \tfrac{1}{2}d(d-3).$$

So $j(nD) = \deg nD + 1 - \dim \Lambda(nD) = (d-1)(d-2)/2$, and the genus formula follows from Proposition 9.27. □

Generalising the cohomology space $H^1(\mathcal{O}_C)$, for any divisor $D = \sum_{p \in C} n_p p \in \operatorname{Div} C$ we define

$$H^1(\mathcal{O}_C(D)) := \operatorname{coker}\left\{ k(C) \to \bigoplus_{p \in C} k(C)/\mathcal{O}_{C,p}(n_p p) \right\}. \tag{9.13}$$

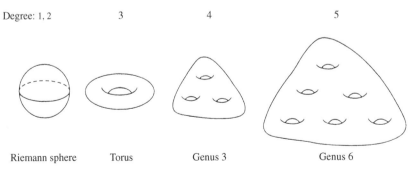

Degree: 1, 2   3   4   5

Riemann sphere   Torus   Genus 3   Genus 6

Figure 9.5: The genus of a plane curve

Note that the kernel is $\Lambda(D)$. The cokernel can also be expressed as the limit, taken over positive divisors $F = \sum_{p \in C} m_p p \in \text{Div } C$,

$$H^1(\mathcal{O}_C(D)) = \varinjlim_{F \geq 0} \text{coker} \left\{ \Lambda(F + D) \to \bigoplus_{p \in C} \mathcal{O}_{C,p}((m_p + n_p)p)/\mathcal{O}_{C,p}(n_p p) \right\}.$$

Its dimension is therefore equal to the index of speciality $i(D)$ (see (9.10)). Substituting into Definition 9.16 we arrive at:

Riemann-Roch Formula 9.29 (Weak form).

$$\dim \Lambda(D) - \dim H^1(\mathcal{O}_C(D)) = \deg D + 1 - g.$$

$\square$

We will write Riemann-Roch in the language of line bundles in (9.15) below. Historically, the formula was written

$$\dim \Lambda(D) - \dim \Lambda(K_C - D) = \deg D + 1 - g$$

(the 'strong form'). We shall discuss the divisor K_C in Section 9.5 and more general Riemann-Roch theorems in Chapter 12.

(c) Line bundles and their cohomology

A divisor on a curve C is an object analogous to a fractional ideal in an algebraic number field K (Definition 8.64); and just as a fractional ideal determines an invertible $\mathcal{O}_K$-module, and hence a line bundle on Spm $\mathcal{O}_K$ (Definition 8.78 and Lemma 8.85), so each divisor on the curve C determines a line bundle on C.

Definition 9.30. Given a divisor $D = \sum_{p \in C} n_p p \in \text{Div } C$, the assignment

$$(\text{open set } U \subset C) \mapsto \bigcap_{p \in U} \mathcal{O}_{C,p}(n_p p)$$

defines an elementary sheaf with total set $k(C)$, which is denoted $\mathcal{O}_C(D)$. $\square$

Note that $D \geq D'$ if and only if $\mathcal{O}_C(D) \supset \mathcal{O}_C(D')$. In particular:

$$D \geq 0 \iff \mathcal{O}_C \subset \mathcal{O}_C(D) \iff \mathcal{O}_C(-D) \subset \mathcal{O}_C.$$

Moreover, $\mathcal{O}_C(D) \cong \mathcal{O}_C(D')$ if and only if D, D' are linearly equivalent; and, in particular, $\mathcal{O}_C(D) \cong \mathcal{O}_C$ if and only if D is a principal divisor.

It is clear that $\mathcal{O}_C(D)$ is a line bundle (Definition 8.82), but in fact the converse is also true:

Lemma 9.31. *Every line bundle with total set $k(C)$ is $\mathcal{O}_C(D)$ for some $D \in$* Div C.

Proof. Let L be a line bundle on C and consider the stalk L_p at a point $p \in C$. Let t_p be a regular parameter at p. Then L_p is an $\mathcal{O}_{C,p}$-module of rank 1 contained in $k(C)$, and so it can be identified with $t_p^n \mathcal{O}_{C,p}$ for some $n =: n_p \in \mathbb{Z}$. Then, on some sufficiently small open set $U \subset C$ containing p, L_U is the same as $t_p^n \mathcal{O}_U$. Consequently, $n_p = 0$ except at finitely many points, and so L determines a divisor $D = -\sum n_p p$. By construction, L and $\mathcal{O}_C(D)$ have the same stalk at all points of the curve, and so they are equal. $\qquad\square$

Of course, any 1-dimensional vector space over $k(C)$ is isomorphic to $k(C)$, so we have proved:

Corollary 9.32. *Every line bundle on a curve C is isomorphic to $\mathcal{O}_C(D)$ for some $D \in$* Div C. $\qquad\square$

Although this corollary appears to reduce the notion of a line bundle on a curve to the simpler idea of a divisor, line bundles remain nevertheless an important tool. Indeed, the key advantage of line bundles is that the total set, as a sheaf, is not restricted to be $k(C)$. The tautological line bundle on projective space is a good example (Definition 8.86), as is the canonical line bundle Ω_C, which will enter the story a little later (in Section 9.5). It is possible to develop the theory of curves without line bundles, using only divisors, but this leads to an unnecessarily constricted view of the subject.

Definition 9.33. The *degree* of a line bundle $L \in$ Pic C is defined to be deg D, where $L \cong \mathcal{O}_C(D)$. $\qquad\square$

It follows from Corollary 9.18 that this definition is independent of the choice of divisor used. Moreover, from the isomorphisms

$$\mathcal{O}_C(D) \otimes \mathcal{O}_C(D') \cong \mathcal{O}_C(D + D'), \qquad \mathcal{O}_C(D)^{-1} \cong \mathcal{O}_C(-D), \qquad (9.14)$$

it follows that the degree satisfies

$$\deg L \otimes M = \deg L + \deg M, \qquad \deg L^{-1} = -\deg L.$$

The isomorphisms (9.14) say that assigning to a divisor on C its associated line bundle defines a group homomorphism

$$\text{Div } C \to \text{Pic } C, \qquad D \mapsto \mathcal{O}_C(D).$$

Since $\mathcal{O}_C(D) \cong \mathcal{O}_C$ if and only if D is a principal divisor, we have the analogue of Proposition 8.65 for a function field:

Proposition 9.34. *The Picard group of a curve is isomorphic to its divisor class group,*

$$\mathrm{Pic}\, C \cong \mathrm{Cl}\, C.$$

□

Given any vector bundle E on C, its space of *global sections* is defined to be

$$H^0(E) := \bigcap_{p \in C} E_p \subset E_{\mathrm{gen}}.$$

For example, in the line bundle case $H^0(\mathcal{O}_C(D)) = \Lambda(D)$. An element of the stalk E_{gen} is a rational section of the bundle, and assigning its principal part at each point of the curve gives a *principal part map* which is a linear map of vector spaces over k,

$$\mathrm{pp}_E : E_{\mathrm{gen}} \to \bigoplus_{p \in C} E_{\mathrm{gen}}/E_p.$$

By definition, $\ker \mathrm{pp}_E = H^0(E)$, and we define

$$H^1(E) := \mathrm{coker}\, \mathrm{pp}_E,$$

called the *cohomology space* of E. In the case $E = \mathcal{O}_C(D)$, this is the same as the cohomology $H^1(\mathcal{O}_C(D))$ already defined. The Riemann-Roch Theorem 9.29 therefore takes the form

$$\dim H^0(L) - \dim H^1(L) = \deg L - 1 + g \qquad (9.15)$$

for any $L \in \mathrm{Pic}\, C$.

The following fact is clear from the definition.

Lemma 9.35. *If a homomorphism of vector bundles $E \to F$ is surjective (Definition 8.81), then the induced linear map $H^1(E) \to H^1(F)$ is surjective.*   □

Given a divisor D and a vector bundle E, the tensor product $E(D) := E \otimes \mathcal{O}_C(D)$ is another vector bundle having the same total set as E. The stalk E_{gen} is the inductive limit (that is, the union) of $H^0(E(D))$ as D ranges over all positive divisors. Suppose that $f \in E_{\mathrm{gen}}$ is a rational section contained in $H^0(E(D))$, where $D = \sum_{p \in C} n_p p \geq 0$. Then its principal part $\text{pp}_E(f)$ is contained in

$$\bigoplus_{p \in C} E_p(n_p p)/E_p, \qquad \text{where } E_p(n_p p) := \mathcal{O}_{C,p}(n_p p) \cdot E_p.$$

Denoting this vector space by $E(D)/E$, the principal part map restricts to

$$\text{pp}_{E,D} : H^0(E(D)) \to E(D)/E,$$

assigning to each rational section of E with poles bounded by the positive divisor D its principal part. Then the linear map $\text{coker pp}_{E,D} \to H^1(E)$ is injective, and $H^1(E)$ is the limit

$$H^1(E) = \lim_{D \geq 0} \text{coker pp}_{E,D} = \bigcup_{D \geq 0} \text{coker pp}_{E,D}. \qquad (9.16)$$

Finally, a linear map of vector bundles $E \times F \to G$ induces a commutative diagram:

$$H^0(E) \times F_{\text{gen}} \qquad \to \qquad G_{\text{gen}}$$

$$\downarrow \qquad\qquad\qquad \downarrow$$

$$H^0(E) \times \bigoplus_{p \in C} F_{\text{gen}}/F_p \to \bigoplus_{p \in C} G_{\text{gen}}/G_p$$

Taking the vertical cokernels gives a linear map

$$H^0(E) \times H^1(F) \to H^1(G), \qquad (9.17)$$

called the *cup product* in cohomology.

(d) Generation by global sections

Let L be a line bundle on C and let $s \in H^0(L)$ be a global section. We choose a neighbourhood $U \subset C$ of p and an isomorphism $\phi : L|_U \overset{\sim}{\to} \mathcal{O}_U$. Via ϕ the restriction $s|_U$ maps to an element of $\mathcal{O}_C(U)$, and, in particular, takes some value $\phi(s)(p) \in k$. This value taken at the point p depends on the choices of U and ϕ made, but whether or not $\phi(s)(p)$ is zero, and indeed the order of zero at p, depend only on p and the section s. The order of vanishing of the function $\phi(s)$ at p is called the *order of s at the point $p \in C$* and denoted $\text{ord}_p(s)$.

Definition 9.36.

(i) Given a nonzero section $s \in H^0(L)$, we define a positive divisor $(s)_0 := \sum_{p \in C}(\text{ord}_p(s))p$, called the *divisor of zeros* of s.

(ii) The set of points $p \in C$ contained in the support of $(s)_0$ for all nonzero sections $s \in H^0(L)$ is called the *base point set* of the line bundle L. If this set is empty, we say that L is *base point free*. □

Let $\{s_1, \ldots, s_n\}$ be a basis of $H^0(L)$ and consider the linear map of vector spaces over $k(C)$,

$$k(C)^{\oplus n} \to L_{\mathrm{gen}}, \qquad (f_1, \ldots, f_n) \mapsto f_1 s_1 + \cdots + f_n s_n.$$

This determines, as the map on the stalks at the generic point, a sheaf homomorphism $\mathcal{O}_C^{\oplus n} \to L$. This can be expressed in basis independent terms as

$$\mathrm{ev}_L : H^0(L) \otimes_k \mathcal{O}_C \to L,$$

called the *evaluation homomorphism* for the line bundle L.

Proposition 9.37. *The following conditions on $L \in \mathrm{Pic}\, C$ are equivalent.*

(i) L is base point free.

(ii) The evaluation homomorphism is surjective. That is, at every point $p \in C$ the induced $\mathcal{O}_{C,p}$-module homomorphism on stalks $H^0(L) \otimes_k \mathcal{O}_{C,p} \to L_p$ is surjective.

Proof. (i) is equivalent to saying that $H^0(L) \to L_p/\mathfrak{m}_p L_p$ is surjective at all points, and this is equivalent to (ii) by Nakayama's lemma. □

A line bundle $L \in \mathrm{Pic}\, C$ enjoying property (ii) is said to be *generated by global sections*.

Proposition 9.38. *If $\deg L \geq 2g$, where g is the genus of the curve, then L is generated by global sections.*

Proof. Given any point $p \in C$, we have $\deg L(-p) = \deg L - 1 \geq 2g - 1$, so that by Theorem 9.20 the line bundle $L(-p)$ has vanishing cohomology. Hence, from (9.16), for every positive divisor $D \geq 0$ the principal part map

$$\mathrm{pp}_{L(-p),D} : H^0(L(D - p)) \to L(D - p)/L(-p)$$

is surjective. In particular, we see by taking $D = p$ that p is not a base point. □

9.3 Nonsingularity of quotient spaces

In this section we introduce differential modules and give a general definition of nonsingularity. We show that the nonsingularity of a free closed orbit under a group action is passed down to the quotient variety. Differential modules on a curve will also play an important part in the discussion of duality in Section 9.5.

(a) Differentials and differential modules

Let k be an algebraically closed field and K a finitely generated extension field of k. A k-linear map $\psi : K \to K$ obeying the Leibniz rule

$$\psi(fg) = f\psi(g) + g\psi(f) \qquad \text{for all } f, g \in K$$

is called a *derivation* of K over k. The set of derivations is a finite-dimensional vector space over K denoted by $\Omega^\vee_{K/k}$, and its dual space $\Omega_{K/k}$ is called the *space of differentials* of K over k. Its dimension is equal to the transcendence degree of the extension:

$$\dim_K \Omega_{K/k} = \text{Tr.deg } K/k.$$

(See, for example, Eisenbud [61] §16.5.) By definition, the space of differentials comes with a k-linear map

$$d_K : K \to \Omega_{K/k},$$

which takes $f \in K$ to the evaluation functional $\psi \mapsto \psi(f)$. This map satisfies

$$d_K(fg) = fd_Kg + gd_Kf \qquad \text{for all } f, g \in K.$$

Every derivation is a composition $K \xrightarrow{d_K} \Omega_{K/k} \to K$ for some linear map $\Omega_{K/k} \to K$.

There are two important variations of this idea. Let R be a finitely generated integral domain over k. Then we can consider derivations of R taking values in the field of fractions of R, and in the residue field at a maximal ideal $\mathfrak{m} \subset R$, respectively:

$$\psi : R \to Q(R) \qquad \text{satisfying } \psi(fg) = f\psi(g) + g\psi(f), \qquad (9.18)$$

$$\psi : R \to R/\mathfrak{m} \cong k \quad \text{satisfying } \psi(fg) = f\psi(g) + g\psi(f). \qquad (9.19)$$

In the first case, derivations (9.18) are in one-to-one correspondence with derivations of the field $Q(R)$ over k. Those of the second case (9.19) are elements of the Zariski tangent space $(\mathfrak{m}/\mathfrak{m}^2)^\vee$ at the maximal ideal (see Definition 4.19 and Section 8.5(d)). This can be seen as follows. The residue ring $R/\mathfrak{m}^2$, as a vector space over $R/\mathfrak{m} \cong k$, has a direct sum decomposition $R/\mathfrak{m}^2 = k\overline{1} \oplus \mathfrak{m}/\mathfrak{m}^2$, and projection on the second summand determines a k-linear map

$$d_{\mathfrak{m}} : R \to R/\mathfrak{m}^2 \to \mathfrak{m}/\mathfrak{m}^2.$$

It is easy to check that this obeys the Leibniz rule, and every derivation (9.19) can be expressed as a composition $R \xrightarrow{d_{\mathfrak{m}}} \mathfrak{m}/\mathfrak{m}^2 \xrightarrow{v} k$ for some linear form (Zariski tangent vector) $v \in (\mathfrak{m}/\mathfrak{m}^2)^{\vee}$.

Definition 9.39. Let R be a finitely generated algebra over k and let I be the kernel of the multiplication map

$$R \otimes_k R \to R, \qquad a \otimes b \mapsto ab.$$

Then the R-module $\Omega_{R/k} := I/I^2$ is called the *(Kähler) differential module* of R. $\qquad\qquad\Box$

If $S \subset R$ is a multiplicative subset, then it is easy to verify that

$$\Omega_{S^{-1}R/k} \cong S^{-1}\Omega_{R/k}. \tag{9.20}$$

Also, the k-linear map

$$d_R : R \to \Omega_{R/k}, \qquad a \mapsto a \otimes 1 - 1 \otimes a \mod I^2$$

satisfies the Leibniz rule, and d_R is universal for linear maps $R \to M$ (where M is an R-module) obeying the Leibniz rule. In particular, we have the following.

Proposition 9.40.

(i) *The total fraction module $Q(\Omega_R)$ (see Section 8.2(a)) is isomorphic to $\Omega_{Q(R)}$.*
(ii) *At a maximal ideal $\mathfrak{m} \subset R$, the quotient $\Omega_R/\mathfrak{m}\Omega_R$ is isomorphic to the Zariski cotangent space $\mathfrak{m}/\mathfrak{m}^2$.* $\qquad\qquad\Box$

When R is an integral domain, $Q(\Omega_R)$ is a vector space over $Q(R)$ of dimension equal to the transcendence degree of $Q(R)/k$. On the other hand, by localisation at $\mathfrak{m}$ we see that $Q(\Omega_R)$ is spanned by $\dim_k(\mathfrak{m}/\mathfrak{m}^2)$

elements, and so:

$$\dim_k(\mathfrak{m}/\mathfrak{m}^2) \geq \text{Tr.deg } Q(R)/k. \tag{9.21}$$

(This is still true if R is not an integral domain, but we will not go into this.)

(b) Nonsingularity

Given a ring R and an ideal $I \subset R$, the direct sum

$$\text{gr}_I R := \bigoplus_{l=0}^{\infty} \frac{I^l}{I^{l+1}} = \frac{R}{I} \oplus \frac{I}{I^2} \oplus \frac{I^2}{I^3} \oplus \frac{I^3}{I^4} \oplus \cdots \tag{9.22}$$

is a graded ring with the natural multiplication.

Definition 9.41. A Noetherian ring R is *regular* at a maximal ideal $\mathfrak{m} \subset R$ if $\text{gr}_\mathfrak{m} R = \bigoplus_{l=0}^{\infty} \mathfrak{m}^l/\mathfrak{m}^{l+1}$ is isomorphic to a polynomial ring over the residue field $R/\mathfrak{m}$. □

We now return to the situation where R is finitely generated over the field k. Let $a_1, \ldots, a_n \in \mathfrak{m}$ be a basis, modulo $\mathfrak{m}^2$, of $\mathfrak{m}/\mathfrak{m}^2$, and consider the ring homomorphism

$$k[x_1, \ldots, x_n] \rightarrow R, \qquad x_i \mapsto a_i.$$

Then, for every $l \geq 0$, the induced map

$$k[x_1, \ldots, x_n]/(x_1, \ldots, x_n)^l \rightarrow R/\mathfrak{m}^l$$

is surjective, and so regularity of R at $\mathfrak{m}$ means that this is an isomorphism for every $l \geq 0$. In particular, the completion of R at $\mathfrak{m}$ is isomorphic to the formal power series ring:

$$k[[x_1, \ldots, x_n]] \stackrel{\sim}{\rightarrow} \widehat{R} := \varprojlim R/\mathfrak{m}^l.$$

By Krull's intersection theorem (see, for example, Eisenbud [61] §5.3) $\bigcap_l \mathfrak{m}^l R_\mathfrak{m} = 0$, so the map $R_\mathfrak{m} \rightarrow \widehat{R}$ is injective. In particular, it follows that $R_\mathfrak{m}$ is an integral domain.

Definition 9.42.

(i) A point $p \in X$ in a variety X is a *nonsingular point* if it is contained in an affine open set $\text{Spm } R \subset X$, where R is an integral domain over k regular at the maximal ideal $\mathfrak{m} \subset R$ corresponding to p.

(ii) A set $\{a_1, \ldots, a_n\} \subset \mathfrak{m}$ whose residue classes define a basis of the Zariski cotangent space $\mathfrak{m}/\mathfrak{m}^2$ is called a *regular system of parameters* at $\mathfrak{m}$. □

This generalises the definition of one regular parameter following Lemma 9.1.

Example 9.43. The polynomial ring $k[x_1, \ldots, x_n]$ is obviously regular at all maximal ideals, and therefore affine space $\mathbb{A}^n$ is nonsingular at all points. Consequently, any variety obtained by gluing affine open sets isomorphic to $\mathbb{A}^n$ is also nonsingular everywhere. Projective space $\mathbb{P}^n$ and Grassmannians $\mathbb{G}(r, n)$ are examples. □

If R is regular at a maximal ideal $\mathfrak{m}$, then a regular system of parameters $a_1, \ldots, a_n \in \mathfrak{m}$ is algebraically independent over k, and hence equality holds in (9.21). In particular, if R is regular at all maximal ideals, then by Proposition 9.40 and Exercise 7.7 the differential module $\Omega_{R/k}$ is locally free.

Definition 9.44. A variety X is *nonsingular* if it is nonsingular at all points. If X is covered by affine open sets Spm R, then by (9.20) the locally free modules $\Omega_{R/k}$ glue together to determine a vector bundle Ω_X, called the *cotangent bundle* of X. □

We will give a functorial characterisation of nonsingular varieties. Consider as an example the case of the affine space $X = \mathbb{A}^n =$ Spm S, where $S = k[x_1, \ldots, x_n]$. We can observe that for any ring homomorphism $\varphi : S \to A$ and any surjective homomorphism $f : A' \to A$ there exists a homomorphism $\varphi' : S \to A'$ such that the following diagram commutes:

$$A'$$

$$\text{lift } \varphi' \ \nearrow \quad \downarrow f \text{ surjective}$$

$$S \ \xrightarrow{\ \varphi\ } A$$

The map φ' is called a *lift* of φ. The existence of this lift means that the map from A'-valued points to A-valued points of X induced by f,

$$\underline{X}(f) : \underline{X}(A') \to \underline{X}(A), \tag{9.23}$$

is surjective (see Section 3.3(a)). A nonsingular variety is one which has this property whenever A, A' are Artin local rings:

Definition 9.45. An *Artin ring* (over k) is a finitely generated ring containing k which satisfies the following equivalent conditions.

(i) R is finite-dimensional as a vector space over k.
(ii) R has only finitely many maximal ideals, and these are all nilpotent. □

Lemma 9.46. *Every Artin ring has a decomposition as a direct sum of Artin local rings.*   □

(See Exercise 9.1.) The next result is the main tool that we will use later for proving nonsingularity of our moduli spaces.

Proposition 9.47. *For a variety X the following properties are equivalent.*

(i) X is nonsingular.

(ii) For any surjective homomorphism of Artin local rings $f : A' \to A$ the map (9.23) is surjective.

Proof. We will prove (ii) $\Longrightarrow$ (i) in the case $X = \mathrm{Spm}\, R$. Choose elements $x_1, \ldots, x_n \in \mathfrak{m}$ giving a basis of $\mathfrak{m}/\mathfrak{m}^2$, and consider a polynomial ring $k[y_1, \ldots, y_n]$. We can construct a ring homomorphism

$$\varphi_2 : R/\mathfrak{m}^2 \to k[y_1, \ldots, y_n]/(y_1, \ldots, y_n)^2$$

by mapping residue classes $\overline{x}_i \mapsto \overline{y}_i$. We now apply condition (ii) to the projection homomorphisms

$$A' := k[y_1, \ldots, y_n]/(y_1, \ldots, y_n)^l \to A := k[y_1, \ldots, y_n]/(y_1, \ldots, y_n)^2.$$

This tells us that for every natural number $l \in \mathbb{N}$ the homomorphism φ_2 extends to a homomorphism

$$\varphi_l : R/\mathfrak{m}^l \to k[y_1, \ldots, y_n]/(y_1, \ldots, y_n)^l,$$

and hence $\overline{x}_1, \ldots, \overline{x}_n$ are algebraically independent in $\mathrm{gr}_\mathfrak{m} R$.   □

(c) Free closed orbits

Suppose that a linearly reductive algebraic group G acts on an affine variety X. In Chapter 5 it was shown that any two distinct closed orbits $O_1 \neq O_2$ are separated by the invariants of the G-action (Theorem 5.3 and Corollary 5.4). To show that the quotient variety is nonsingular, we need to extend this result to the limit as O_1 and O_2 approach infinitely close to each other.

Definition 9.48. A closed orbit $G \cdot x \subset X$ is called a *free closed orbit* if the map $G \to G \cdot x, g \to g \cdot x$ is an isomorphism. Equivalently, the orbit is stable (Definition 5.12) with a trivial stabiliser subgroup.   □

Let $X = \text{Spm } R$ and let $I \subset R$ be the ideal of the orbit $G \cdot x \subset X$. Then I is invariant under the coaction

$$\mu : R \to k[G] \otimes_k R.$$

(That is, $\mu(I) \subset k[G] \otimes_k I$.) The coaction therefore induces a map $R/I \to k[G] \otimes_k R/I$, and for a free closed orbit this is isomorphic to the coproduct $k[G] \otimes_k k[G] \to k[G]$. We want to consider the map

$$I/I^2 \to k[G] \otimes_k I/I^2$$

induced by μ. Via the above isomorphism, this is a homomorphism of $k[G]$-modules. On the other hand, I/I^2 is also a representation of G and is therefore a $k[G]$-module equipped with a G-linearisation (see Definition 6.23). The next lemma shows that I/I^2 is isomorphic to $(I/I^2) \otimes_k k[G]$.

Lemma 9.49. *A $k[G]$-module M having a G-linearisation is free. More precisely, any basis of the space of invariants M^G is a free basis of M as a $k[G]$-module.*

Proof. Let $\nu : M \to k[G] \otimes_k M$ be the coaction of G on M. This is a homomorphism of $k[G]$-modules. Define a linear map

$$\sigma : k[G] \otimes_k M \to k[G] \otimes_k M, \qquad a \otimes m \mapsto \mu_G(a) \cdot (1 \otimes m).$$

If $i : k[G] \to k[G]$ is the coinverse map, then

$$\sigma \circ (i \otimes 1_M) \circ \sigma$$

is an isomorphism, and in particular it follows that σ is an isomorphism. Let $\mu_{G,G} := (\mu_G \otimes 1) \circ \mu_G : k[G] \to k[G] \otimes k[G] \otimes k[G]$. Then by the same reasoning as for σ, the linear map

$$\tau : k[G] \otimes k[G] \otimes M \to k[G] \otimes k[G] \otimes M, \qquad a \otimes b \otimes m \mapsto \mu_{G,G}(a)(1 \otimes b \otimes m)$$

is also an isomorphism. We now define a map f by the commutative diagram

$$
\begin{array}{ccc}
M & \xrightarrow{f} & k[G] \otimes_k M \\
\| & & \downarrow \sigma \\
M & \xrightarrow{\nu} & k[G] \otimes_k M
\end{array}
$$

and apply Lemma 8.39. By the associative law for the (co)action f satisfies the conditions of the lemma, while $\sigma(M_0) = M^G$. Hence

$$M^G \otimes_k k[G] \to M$$

is an isomorphism. □

We now apply the linear reductivity of G to the surjective linear map of representations $I \to I/I^2$. This, together with Lemma 9.49, implies that there exist G-invariant elements $f_1, \ldots, f_r \in I^G$ which form a free basis modulo I^2, where r is the rank of I/I^2 as a $k[G]$-module. By Nakayama's lemma (or, more precisely, by the matrix trick used in its proof) $f_1, \ldots, f_r$ generate I in some neighbourhood of the orbit $G \cdot x$, in the sense that there exists $b \in R$, congruent to 1 modulo I, such that the homomorphism of R_b-modules

$$R_b \oplus \cdots \oplus R_b \to I R_b, \qquad (g_1, \ldots, g_r) \mapsto g_1 f_1 + \cdots + g_r f_r \qquad (9.24)$$

is surjective.

Lemma 9.50. *There exists $b \equiv 1 \bmod I$, as above, which is G-invariant.*

Proof. Let $\mathfrak{a} = \{b \in R \mid bI \subset (f_1, \ldots, f_r)\}$. Then $\mathfrak{a} \subset R$ is an ideal and is also a representation of G. Since $1 \in \mathfrak{a} + I$, it follows from linear reductivity (Proposition 4.37) that $1 \in \mathfrak{a}^G + I^G$. □

Let $\overline{x} \in X /\!/ G$ be the image of $x \in X$ under the affine quotient map $X = \operatorname{Spm} R \to X /\!/ G = \operatorname{Spm} R^G$, corresponding to a maximal ideal $\mathfrak{m} = I \cap R^G = I^G \subset R^G$. Applying linear reductivity to the surjective map of representations (9.24), we see that $\mathfrak{m}$ is generated by $f_1, \ldots, f_r$ in some neighbourhood of $\overline{x}$. The same is true for any power $\mathfrak{m}^l$, and we have shown the following.

Theorem 9.51. *Let $G \cdot x \subset X = \operatorname{Spm} R$ be a free closed orbit with defining ideal $I \subset R$.*

(i) *There exist invariants $f_1, \ldots, f_r \in R^G$ whose residue classes modulo I^2 form a free basis of I/I^2.*
(ii) *There exists a G-invariant affine open set $U \subset X$ containing $G \cdot x$ such that the restrictions of $f_1, \ldots, f_r$ to U generate I.*
(iii) *If $\mathfrak{m} \subset R^G$ is the maximal ideal of the image of x in $X /\!/ G$, then every power $\mathfrak{m}^l$ is generated by $(f_1, \ldots, f_r)^l$.* □

Suppose, further, that X is nonsingular at the point $x \in X$. Then the fibre at x of the graded ring $\mathrm{gr}_I R$ (that is, its quotient by the maximal ideal $\mathfrak{m}$) is a polynomial ring in $r := \dim X - \dim G$ variables. It then follows from part (ii) of the theorem that $\mathrm{gr}_\mathfrak{m} R^G$ is isomorphic to a polynomial ring in r variables, and we conclude:

Corollary 9.52. *If an affine variety X is nonsingular at every point of a free closed orbit $G \cdot x$, then the affine quotient $X /\!/ G$ is nonsingular at the image point $\overline{x} \in X /\!/ G$, with dimension $= \dim X - \dim G$.* $\qquad\Box$

9.4 An algebraic variety with the Picard group as its set of points

Fix a curve C of genus g and an integer $d \in \mathbb{Z}$. We are going to construct in this section a g-dimensional nonsingular projective variety, the *Jacobian* of C, whose underlying set is $\mathrm{Pic}^d C$, the set of isomorphism classes of line bundles on C of degree d.

(a) Some preliminaries

We will assume throughout this section that $d \geq 2g$, and we fix a line bundle $L \in \mathrm{Pic}^{2d} C$. We note that every line bundle $\xi \in \mathrm{Pic}^d C$ has the following properties:

(i) $H^1(\xi) = 0$.
(ii) ξ is generated by global sections.
(iii) $\dim H^0(\xi) = d + 1 - g =: N > g$.

We set $\widehat{\xi} := L \otimes \xi^{-1} \in \mathrm{Pic}^d C$ and note that $\widehat{\xi}$ also has all of the properties (i) to (iii). The key tool in the algebraic construction of the Jacobian is the multiplication map

$$H^0(\xi) \times H^0(\widehat{\xi}) \to H^0(L), \qquad (s, t) \mapsto st.$$

Definition 9.53. Given a line bundle $\xi \in \mathrm{Pic}^d C$, a pair (S, T) consisting of a basis $S = \{s_1, \ldots, s_N\}$ of $H^0(\xi)$ and a basis $T = \{t_1, \ldots, t_N\}$ of $H^0(\widehat{\xi})$ is called a *double marking* of ξ. $\qquad\Box$

Given a line bundle $\xi \in \mathrm{Pic}^d C$ and a double marking (S, T), we introduce the following $N \times N$ matrix with entries in $H^0(L)$:

$$\Psi(\xi, S, T) := \begin{pmatrix} s_1 t_1 & \cdots & s_1 t_N \\ \vdots & & \vdots \\ s_N t_1 & \cdots & s_N t_N \end{pmatrix}.$$

If we fix a rational section of L, then $\Psi(\xi, S, T)$ can be viewed as a matrix of rank 1 over the function field $k(C)$.

Definition 9.54. We denote by $\mathrm{Mat}_N(H^0(L))$ the set of $N \times N$ matrices with entries in $H^0(L)$. The subset of matrices of rank 1 over $k(C)$ or, equivalently, those for which all 2×2 minors vanish, is denoted by $\mathrm{Mat}_{N,1}(H^0(L))$. $\quad\square$

Remark 9.55. A matrix Ψ of rank 1 over a field (or unique factorisation domain) K is expressible as a product:

$$\begin{pmatrix} a_1 \\ \vdots \\ a_n \end{pmatrix} (b_1, \ldots, b_m), \qquad a_i, b_j \in K.$$

Moreover, these vectors are unique up to scalar multiplication in the sense that if

$$\Psi = \begin{pmatrix} a'_1 \\ \vdots \\ a'_n \end{pmatrix} (b'_1, \ldots, b'_m),$$

then $a'_i = c a_i$, $b'_j = c^{-1} b_j$ for some $c \in K$. $\quad\square$

The following proposition says that when Ψ is of the form $\Psi(\xi, S, T)$, the line bundle ξ can be recovered as the image of the linear transformation determined by the matrix.

Proposition 9.56. *Given a matrix* $\Psi \in \mathrm{Mat}_{N,1}(H^0(L))$, *the following two conditions are equivalent.*

(1) The N rows and the N columns of Ψ are linearly independent over k.
(2) $\Psi = \Psi(\xi, S, T)$ for some $\xi \in \mathrm{Pic}^d C$ and double marking (S, T).

Moreover, the line bundle ξ is the image $\xi \subset L^{\oplus N}$ of the sheaf homomorphism determined by Ψ,

$$\langle \Psi \rangle : \mathcal{O}_C^{\oplus N} \to L^{\oplus N}.$$

Proof. (2) $\Longrightarrow$ (1) is clear, and we just have to show (1) $\Longrightarrow$ (2). Let ξ be the image of $\langle \Psi \rangle : \mathcal{O}_C^{\oplus N} \to L^{\oplus N}$. That ξ is a line bundle follows from the fact that Ψ has rank 1 as a matrix over $k(C)$. To see that $\deg \xi = d$, observe that, since Ψ has rank N over k, we have $\dim H^0(\xi) \geq N > g$, and so by Lemma 9.22 we have $H^1(\xi) = 0$. From this it follows that $\deg \xi \geq N + g - 1 = d$.

But the same reasoning $\deg L - \deg \xi = \deg \widehat{\xi} \geq d$ and hence $\deg \xi \leq d$. So $\xi \in \operatorname{Pic}^d C$. That $\Psi = \Psi(\xi, S, T)$ follows from Remark 9.55. $\qquad \square$

Remark 9.57. This construction can also be explained in terms of divisors. The matrix $\Psi(\xi, S, T)$ has rank 1 over the function field $k(C)$ and is therefore a product of $k(C)$-valued vectors $\overrightarrow{f} = (f_1, \ldots, f_N)$ and $\overrightarrow{g} = (g_1, \ldots, g_N)$. Then $\xi = \mathcal{O}_C(D)$, where D is the greatest common (positive) divisor of the polar divisors $(f_1)_\infty, \ldots, (f_N)_\infty$. $\qquad \square$

The space of matrices $\operatorname{Mat}_N(H^0(L))$ is a vector space over k isomorphic to the direct sum of N^2 copies of $H^0(L)$, and the general linear group $GL(N)$ acts on this space by left and right multiplication. In particular, this gives an action of the direct product $GL(N) \times GL(N)$, under which the image of the group homomorphism

$$\mathbb{G}_m \to GL(N) \times GL(N), \qquad t \mapsto (t I_N, t^{-1} I_N) \qquad (9.25)$$

acts trivially. We therefore consider the cokernel

$$GL(N, N) := GL(N) \times GL(N)/\mathbb{G}_m. \qquad (9.26)$$

The coordinate ring of $GL(N)$ is the localisation of the polynomial ring $k[x_{ij}]$ at $\det x$ and is graded by homogeneous degree:

$$k[GL(N)] = k[x_{ij}, (\det x)^{-1}] = \bigoplus_{e \in \mathbb{Z}} k[GL(N)]_e, \qquad \deg x_{ij} = 1.$$

Hence

$$GL(N, N) = \operatorname{Spm} \left(\bigoplus_{e \in \mathbb{Z}} k[GL(N)]_e \otimes_k k[GL(N)]_e \right).$$

Note that, since $GL(N)$ is linearly reductive, so is $GL(N, N)$.

As a representation of $GL(N, N)$ the space $\operatorname{Mat}_N(H^0(L))$ is isomorphic to a direct sum of $\dim H^0(L)$ copies of the space $\operatorname{Mat}_N(k)$ of square matrices over k. This can be viewed as an affine space $\mathbb{A}^n$, where $n = N^2 \dim H^0(L)$, and $\operatorname{Mat}_{N,1}(H^0(L))$ as a closed subvariety. In particular, $\operatorname{Mat}_{N,1}(H^0(L))$ is an affine variety (or, more precisely, each irreducible component is an affine variety, and the discussion below applies to each irreducible component) and is preserved by the action of $GL(N, N)$. This action is of ray type.

The set of matrices Ψ satisfying the linear independence condition (1) in Proposition 9.56 forms an open set

$$\mathcal{U}(L) \subset \operatorname{Mat}_{N,1}(H^0(L)),$$

which is therefore a parameter space for double-marked line bundles (ξ, S, T) of degree d. Moreover, the open set $\mathcal{U}(L)$ is preserved by the action of $GL(N, N)$.

Proposition 9.58. *Matrices* $\Psi, \Psi' \in \mathcal{U}(L)$ *give isomorphic line bundles* ξ, ξ' *if and only if they belong to the same* $GL(N, N)$-*orbit.* $\qquad \square$

We have therefore identified the set $\operatorname{Pic}^d C$ with the space of $GL(N, N)$-orbits in $\mathcal{U}(L) \subset \operatorname{Mat}_{N,1}(H^0(L))$.

(b) The construction

We are going to study the Proj quotient of the action $GL(N, N) \curvearrowright \operatorname{Mat}_{N,1}(H^0(L))$, using the character

$$\delta : GL(N, N) \to \mathbb{G}_m, \qquad (A, B) \mapsto \det A \det B. \qquad (9.27)$$

We denote the kernel of δ by $SL(N, N)$. Semiinvariants and semistability will always be taken with respect to this character. We will show in this section that stability and semistablity coincide and that the open set of stable points is precisely $\mathcal{U}(L) \subset \operatorname{Mat}_{N,1}(H^0(L))$, defined above.

Any linear form $f : H^0(L) \to k$ induces a map, which we will denote by the same symbol, $f : \operatorname{Mat}_N(H^0(L)) \to \operatorname{Mat}_N(k)$. Then the function

$$\operatorname{Mat}_N(H^0(L)) \to k, \qquad \Psi \mapsto \det f(\Psi)$$

is a homogeneous polynomial of degree N and is a semiinvariant of weight 1 for the action of $GL(N, N)$. From this we see (see Definition 6.13):

Lemma 9.59. *Let* Ψ *be a matrix in* $\operatorname{Mat}_N(H^0(L))$. *If there exists a linear form* $f : H^0(L) \to k$ *such that* $\det f(\Psi) \neq 0$, *then* Ψ *is semistable.* $\qquad \square$

An important case arises when f is the *evaluation map* at $p \in C$,

$$\operatorname{ev}_p : H^0(L) \to L/\mathfrak{m}_p L \cong k,$$

or a sum of evaluation maps at points of some divisor. Since the diagram

$$H^0(\xi) \times H^0(\widehat{\xi}) \;\to\; H^0(L)$$

$$\operatorname{ev}_p \downarrow \qquad\qquad \downarrow \operatorname{ev}_p$$

$$\xi/\mathfrak{m}_p\xi \times \widehat{\xi}/\mathfrak{m}_p\widehat{\xi} \to L/\mathfrak{m}_p L$$

commutes, the matrix $\mathrm{ev}_p(\Psi(\xi, S, T)) \in \mathrm{Mat}_N(k)$ is the product of the column and row vectors obtained from the entries $s_1, \ldots, s_N \in H^0(\xi)$ and $t_1, \ldots, t_N \in H^0(\widehat{\xi})$ via the evaluation maps

$$\mathrm{ev}_p : H^0(\xi) \to \xi/\mathfrak{m}_p \xi \cong k, \quad \mathrm{ev}_p : H^0(\widehat{\xi}) \to \widehat{\xi}/\mathfrak{m}_p \widehat{\xi} \cong k.$$

Lemma 9.60. *Let $f_1, \ldots, f_N$ be the nonzero evaluation maps of L at points $p_1, \ldots, p_N \in C$, and let $f = f_1 + \cdots + f_N : H^0(L) \to k$. Then $\det f(\Psi(\xi, S, T))$ is equal to the product of the determinants*

$$\det \left\{ (\mathrm{ev}_{p_1}, \ldots, \mathrm{ev}_{p_N}) : H^0(\xi) \to k^{\oplus N} \right\},$$

$$\det \left\{ (\mathrm{ev}_{p_1}, \ldots, \mathrm{ev}_{p_N}) : H^0(\widehat{\xi}) \to k^{\oplus N} \right\}.$$

Proof. Just observe that

$$f(\Psi(\xi, S, T)) = \begin{pmatrix} f(s_1 t_1) & \cdots & f(s_N t_1) \\ & \cdots & \\ f(s_1 t_N) & \cdots & f(s_N t_N) \end{pmatrix}$$

$$= \begin{pmatrix} s_1(p_1) & \cdots & s_1(p_N) \\ & \cdots & \\ s_N(p_1) & \cdots & s_1(p_N) \end{pmatrix} \begin{pmatrix} t_1(p_1) & \cdots & t_N(p_1) \\ & \cdots & \\ t_1(p_N) & \cdots & t_N(p_N) \end{pmatrix},$$

where $s_i(p_j) := \mathrm{ev}_{p_j}(s_i)$ and so on. On the right-hand side, the first matrix is the matrix representing the map $(\mathrm{ev}_{p_1}, \ldots, \mathrm{ev}_{p_N}) : H^0(\xi) \to k^{\oplus N}$ and the second is the matrix representing $(\mathrm{ev}_{p_1}, \ldots, \mathrm{ev}_{p_N}) : H^0(\widehat{\xi}) \to k^{\oplus N}$. □

Note that the kernel of $(\mathrm{ev}_{p_1}, \ldots, \mathrm{ev}_{p_N}) : H^0(\xi) \to k^{\oplus N}$ is the space $H^0(\xi(-p_1 - \cdots - p_N))$, and similarly for $\widehat{\xi}$. If the points $p_1, \ldots, p_N$ can be chosen so that these kernels vanish, then by the lemma it will follow that $\det f(\Psi(\xi, S, T)) \neq 0$, so that $\Psi(\xi, S, T)$ is semistable. On the other hand, for a general point $p_1 \in C$ we have

$$\dim H^0(\xi(-p_1)) = \dim H^0(\widehat{\xi}(-p_1)) = N - 1.$$

Then, for a general choice of a second point $p_2 \in C$,

$$\dim H^0(\xi(-p_1 - p_2)) = \dim H^0(\widehat{\xi}(-p_1 - p_2)) = N - 2.$$

Repeating this N times we see that there exist points $p_1, \ldots, p_N \in C$ such that

$$\dim H^0(\xi(-p_1 - \cdots - p_N)) = \dim H^0(\widehat{\xi}(-p_1 - \cdots - p_N)) = 0.$$

So indeed, if we take $f : H^0(L) \to k$ to be the sum of the evaluation maps at these points, then Lemma 9.60 guarantees that $\det f(\Psi(\xi, S, T)) \neq 0$, and hence that $\Psi(\xi, S, T)$ is semistable.

Even better, stability follows from the next lemma.

Lemma 9.61. *If $A, B \in GL(N)$ are matrices satisfying $A\Psi(\xi, S, T)B = \Psi(\xi, S, T)$, then A and B are both scalar matrices and $AB = I_N$.*

Proof. By hypothesis, the following diagram commutes:

$$
\begin{array}{ccc}
\mathcal{O}_C^{\oplus N} & \xrightarrow{\langle\Psi\rangle} & L^{\oplus N} \\
A \downarrow & & \uparrow B \\
\mathcal{O}_C^{\oplus N} & \xrightarrow{\langle\Psi\rangle} & L^{\oplus N}
\end{array}
$$

This defines an automorphism of the line bundle $\xi = \operatorname{Im}\langle\Psi\rangle$, and this is just multiplication by a scalar $c \in k$. Thus $A = cI_N$ and $B = c^{-1}I_N$. $\square$

Summarising, we have shown:

Proposition 9.62. *Every matrix $\Psi(\xi, S, T)$ is stable with respect to the action of $GL(N, N)$ and the character δ. That is, $\mathcal{U}(L) \subset \operatorname{Mat}_{N,1}^s(H^0(L))$.* $\square$

Conversely, the matrices $\Psi(\xi, S, T)$ exhaust all the semistable elements of $\operatorname{Mat}_{N,1}(H^0(L))$.

Proposition 9.63. *If $\Psi \in \operatorname{Mat}_{N,1}(H^0(L))$ is semistable, then the N rows and the N columns are linearly independent over k. In other words, $\operatorname{Mat}_{N,1}^{ss}(H^0(L)) \subset \mathcal{U}(L)$.*

Proof. Suppose the rows are linearly dependent. Then we can choose a basis of $H^0(L)$ with respect to which

$$
\Psi = \begin{pmatrix}
0 & 0 & \cdots & 0 \\
* & * & \cdots & * \\
& & \cdots & \\
* & * & \cdots & *
\end{pmatrix}.
$$

Multiplying on the left by the 1-parameter subgroup $A : \mathbb{G}_m \to SL(N)$, $t \mapsto \mathrm{diag}(t^{-N+1}, t, \ldots, t)$, gives

$$
A(t)\Psi = \begin{pmatrix} 0 & 0 & \cdots & 0 \\ t* & t* & \cdots & t* \\ & \cdots & \\ t* & t* & \cdots & t* \end{pmatrix},
$$

which tends to the origin as $t \to 0$. If the columns are linearly independent, we argue similarly, multiplying on the right by $A(t)$. In both cases we see that the closure of the orbit $SL(N, N) \cdot \Psi$ contains the origin, and so Ψ is unstable. $\square$

We have arrived at an action of ray type $GL(N, N) \curvearrowright \mathrm{Mat}_{N,1}(H^0(L))$ for which, by Propositions 9.62 and 9.63, $\mathrm{Mat}^s_{N,1}(H^0(L)) = \mathrm{Mat}^{ss}_{N,1}(H^0(L)) = \mathcal{U}(L)$.

The Proj quotient

$$
\mathrm{Mat}^s_{N,1}(H^0(L))/GL(N, N) = \mathrm{Proj}\, k[\mathrm{Mat}_{N,1}(H^0(L))]^{SL(N,N)} \tag{9.28}
$$

is therefore a good quotient in the sense that its points correspond one-to-one to the orbits of the group action.

(Notice that this quotient construction is valid once we know that the affine variety $\mathrm{Mat}^s_{N,1}(H^0(L))$ is smooth, by Remark 6.14(vi). This will be proved in Proposition 9.68 below.)

We conclude:

Theorem 9.64. *The quotient variety (9.28) is a projective variety whose underlying set is* $\mathrm{Pic}^d C$. $\square$

By the proof of Proposition 9.62 and by construction of the projective quotient, given N distinct points $p_1, \ldots, p_N \in C$, the subset

$$
\left\{ \xi \in \mathrm{Pic}^d C \mid H^0\left(\xi(-p_1 - \cdots - p_N)\right) = H^0\left(\widehat{\xi}(-p_1 - \cdots - p_N)\right) = 0 \right\}
$$

is an affine open set. This is the complement of two translates, in the abelian group $\mathrm{Pic}\, C$, of the *theta divisor*

$$
\Theta := \{\xi \mid H^0(\xi) \neq 0\} \subset \mathrm{Pic}^{g-1} C. \tag{9.29}
$$

The quotient variety (9.28) is covered by affine open sets of this type.

(c) Tangent spaces and smoothness

Let (ξ, S, T) be a line bundle on C with a double marking. What is the tangent space to the affine variety $\mathrm{Mat}_{N,1}(H^0(L))$ at the point $\Psi(\xi, S, T)$? As in Section 8.5(d) we let $k[\epsilon] = k[t]/(t^2)$ with $\epsilon^2 = 0$, $\epsilon \neq 0$. Then a tangent vector to the affine space $\mathrm{Mat}_N(H^0(L))$ at $\Psi(\xi, S, T)$ can be written

$$\Psi(\xi, S, T) + A\epsilon = \begin{pmatrix} s_1 t_1 + a_{11}\epsilon & \cdots & s_1 t_N + a_{1N}\epsilon \\ \vdots & & \vdots \\ s_N t_1 + a_{N1}\epsilon & \cdots & s_N t_N + a_{NN}\epsilon \end{pmatrix}, \quad a_{ij} \in H^0(L).$$

$$(9.30)$$

For this to be a tangent vector to the subvariety $\mathrm{Mat}_{N,1}(H^0(L))$ it is necessary and sufficient to have rank 1 as a matrix over $k(C) \otimes_k k[\epsilon]$. This implies the following.

Lemma 9.65. *The matrix (9.30) is a tangent vector to* $\mathrm{Mat}^s_{N,1}(H^0(L))$ *if and only if there exist rational sections* $s'_1, \ldots, s'_N \in \xi_{\mathrm{gen}}$ *and* $t'_1, \ldots, t'_N \in \widehat{\xi}_{\mathrm{gen}}$ *such that*

$$A = \begin{pmatrix} s'_1 \\ \vdots \\ s'_N \end{pmatrix} (t_1, \ldots, t_N) + \begin{pmatrix} s_1 \\ \vdots \\ s_N \end{pmatrix} (t'_1, \ldots, t'_N).$$

Proof. The condition is equivalent to

$$\Psi(\xi, S, T) + A\epsilon = \begin{pmatrix} s_1 + s'_1\epsilon \\ \vdots \\ s_N + s'_N\epsilon \end{pmatrix} (t_1 + t'_1\epsilon, \ldots, t_N + t'_N\epsilon),$$

which in turn is equivalent to the matrix having rank 1. □

We can rewrite the condition in the lemma as

$$\begin{pmatrix} s'_1 \\ \vdots \\ s'_N \end{pmatrix} (t_1, \ldots, t_N) = - \begin{pmatrix} s_1 \\ \vdots \\ s_N \end{pmatrix} (t'_1, \ldots, t'_N) + A.$$

and interpret this in terms of its principal parts. Namely, at each point $p \in C$ it gives a congruence modulo the stalk L_p,

$$
\begin{pmatrix} s_1' \\ \vdots \\ s_N' \end{pmatrix} (t_1, \ldots, t_N) \equiv - \begin{pmatrix} s_1 \\ \vdots \\ s_N \end{pmatrix} (t_1', \ldots, t_N') \quad \mod L_p.
$$

The local ring $\mathcal{O}_{C,p}$ is a unique factorisation domain, and so (using Remark 9.55) there exists a rational function $h_p \in k(C)$ such that

$$
\begin{pmatrix} s_1' \\ \vdots \\ s_N' \end{pmatrix} \equiv h_p \begin{pmatrix} s_1 \\ \vdots \\ s_N \end{pmatrix} \quad \mod \xi_p, \tag{9.31}
$$

$$
(t_1', \ldots, t_N') \equiv -h_p(t_1, \ldots, t_N) \quad \mod \widehat{\xi}_p.
$$

Although the function h_p itself is not uniquely defined, its principal part h_p mod $\mathcal{O}_{C,p}$ is.

Definition 9.66. Let $T_\Psi \mathrm{Mat}_{N,1}^s(H^0(L))$ be the tangent space to $\mathrm{Mat}_{N,1}^s(H^0(L))$ at Ψ. We define a linear map

$$
\pi_\Psi : T_\Psi \mathrm{Mat}_{N,1}^s(H^0(L)) \to H^1(\mathcal{O}_C)
$$

by assigning to $A \in T_\Psi \mathrm{Mat}_{N,1}^s(H^0(L))$ the cohomology class of the principal part $(h_p)_{p \in C} \in \bigoplus_{p \in C} k(C)/\mathcal{O}_{C,p}$, where h_p is the rational function defined by (9.31). □

Proposition 9.67.

(i) *The kernel of the linear map π_Ψ is equal to the tangent space at $\Psi \in \mathrm{Mat}_{N,1}^s(H^0(L))$ to the orbit of the action $GL(N, N) \curvearrowright \mathrm{Mat}_{N,1}^s(H^0(L))$.*
(ii) *The linear map π_Ψ is surjective.*

In other words, denoting the Lie space of $GL(N, N)$ by $\mathfrak{gl}(N, N)$, we have an exact sequence

$$
\mathfrak{gl}(N, N) \longrightarrow T_\Psi \mathrm{Mat}_{N,1}^s(H^0(L)) \xrightarrow{\pi_\Psi} H^1(\mathcal{O}_C) \longrightarrow 0.
$$

Proof.

(i) If $\pi(A) = 0$, then there exists a rational function $h \in k(C)$ such that, at every point $p \in C$, the components of the vectors

$$
\begin{pmatrix} s_1'' \\ \vdots \\ s_N'' \end{pmatrix} := \begin{pmatrix} s_1' \\ \vdots \\ s_N' \end{pmatrix} - h \begin{pmatrix} s_1 \\ \vdots \\ s_N \end{pmatrix} \quad \mathrm{mod}\ \xi_p
$$

and

$$
(t_1'', \ldots, t_N'') := (t_1', \ldots, t_N') + h(t_1, \ldots, t_N) \quad \mathrm{mod}\ \widehat{\xi}_p
$$

belong to $H^0(\xi)$ and $H^0(\widehat{\xi})$, respectively. Then

$$
A = \begin{pmatrix} s_1'' \\ \vdots \\ s_N'' \end{pmatrix} (t_1, \ldots, t_N) + \begin{pmatrix} s_1 \\ \vdots \\ s_N \end{pmatrix} (t_1'', \ldots, t_N''),
$$

and it is therefore tangent to the $GL(N, N)$-orbit at Ψ.

(ii) Let

$$
\alpha \in \bigoplus_{p \in C} k(C)/\mathcal{O}_{C,p}
$$

be an element of the principal part space. Multiplying by s_i, t_j we obtain elements

$$
\alpha s_i \in \bigoplus_{p \in C} \xi_{\mathrm{gen}}/\xi_p, \qquad \alpha t_j \in \bigoplus_{p \in C} \widehat{\xi}_{\mathrm{gen}}/\widehat{\xi}_p
$$

whose cohomology classes are zero by the hypotheses made at the beginning of Section 9.4(a). We can therefore find rational sections $s_i' \in \xi_{\mathrm{gen}}$ and $t_j' \in \widehat{\xi}_{\mathrm{gen}}$ having $\alpha s_i, \alpha t_j$ as their principal parts. Then for each i, j we have $s_i' t_j - s_i t_j' \in H^0(L)$, and if we take this as (i, j)-th entry of a matrix A, then $\pi(A)$ is precisely the cohomology class of α. $\qquad \square$

Proposition 9.68. *The open set* $\mathrm{Mat}_{N,1}^{\mathrm{s}}(H^0(L)) \subset \mathrm{Mat}_{N,1}(H^0(L))$ *of stable points for the action of* $GL(N, N)$ *is nonsingular.*

Proof. We will use Proposition 9.47. Let $f : A' \to A$ be a surjective homomorphism of Artinian local rings over k with maximal ideals $\mathfrak{n} \subset A$, $\mathfrak{n}' \subset A'$. As a vector space $A = k \cdot 1 \oplus \mathfrak{n}$, and so an A-valued point of $\mathrm{Mat}_{N,1}^{\mathrm{s}}(H^0(L))$

can be written

$$\Psi(\xi, S, T) + P = \begin{pmatrix} s_1 t_1 + a_{11} & \cdots & s_1 t_N + a_{1N} \\ \vdots & & \vdots \\ s_N t_1 + a_{N1} & \cdots & s_N t_N + a_{NN} \end{pmatrix}, \tag{9.32}$$

where $a_{ij} \in H^0(L) \otimes_k A$, and the matrix (9.32) has rank 1 over $k(C) \otimes_k A$. This can therefore be expressed as

$$\Psi(\xi, S, T) + P = \begin{pmatrix} s_1 + q_1 \\ \vdots \\ s_N + q_N \end{pmatrix} (t_1 + r_1, \ldots, t_N + r_N)$$

for some rational sections q_i of the vector bundle $\xi \otimes_k \mathfrak{n}$ on C and r_j of the vector bundle $\widehat{\xi} \otimes_k \mathfrak{n}$. The sections s_i, t_j are nonzero, and so the functions

$$1 + \frac{q_i}{s_i}, \quad 1 + \frac{r_j}{t_j} \in k(C) \otimes_k A$$

have well-defined logarithms. We define:

$$s_i' = s_i \log\left(1 + \frac{q_i}{s_i}\right) \in \xi_{\text{gen}} \otimes_k \mathfrak{n}, \qquad t_j' = t_j \log\left(1 + \frac{r_j}{t_j}\right) \in \widehat{\xi}_{\text{gen}} \otimes_k \mathfrak{n}.$$

These satisfy

$$\begin{pmatrix} s_1' \\ \vdots \\ s_N' \end{pmatrix} (t_1, \ldots, t_N) \equiv - \begin{pmatrix} s_1 \\ \vdots \\ s_N \end{pmatrix} (t_1', \ldots, t_N') \quad \mod L \otimes_k \mathfrak{n}$$

and therefore determine, by taking principal parts as in Definition 9.66, an element of $H^1(\mathcal{O}_C \otimes_k \mathfrak{n})$ which we will denote by $\pi(\log P)$.

Now by Lemma 9.35 there exists an element $\alpha' \in H^1(\mathcal{O}_C \otimes_k \mathfrak{n}')$ which maps under f to $\pi(\log P)$. By the same argument as in the proof of Proposition 9.67(ii) we can construct an A'-valued point $\Psi(\xi, S, T) + P' \in \text{Mat}_{N,1}^s(H^0(L))$ for which $\pi(\log P') = \alpha'$, and by using the exponential function for matrices we see that this maps under f to the A-valued point $\Psi(\xi, S, T) + P$, as required. $\qquad\square$

Remark 9.69. This proof makes implicit use of the series expansion

$$\log(1 + x) = -\sum_{n=1}^{\infty} \frac{(-x)^n}{n}$$

and is therefore only valid in characteristic zero. In fact, one could give an alternative proof which also works in positive characteristic, using the methods of Section 10.4(a) later on. □

By Lemma 9.61, every closed orbit of the action $GL(N, N)$ $\curvearrowright$ $\mathrm{Mat}_{N,1}(H^0(L))$ is isomorphic to $GL(N, N)$; and hence from Proposition 9.68 and Corollary 9.52 we conclude:

Theorem 9.70. *The projective variety* $\mathrm{Mat}^s_{N,1}(H^0(L))/GL(N, N)$ *is nonsingular, and at every point its tangent space is isomorphic to* $H^1(\mathcal{O}_C)$. □

9.5 Duality

Let C be a (nonsingular projective) curve. By using a distinguished line bundle, the dualising line bundle on C, it is possible to express cohomology spaces as the duals of spaces of global sections, a tool which is endlessly useful. Although this duality has a very powerful abstract formulation, we will give a quite concrete account using differentials and residues. Later in the section we will define algebraic de Rham cohomology for use in Section 9.6.

(a) Dualising line bundles

We begin by re-examining the definition of the cohomology space $H^1(L)$. The stalk L_{gen} at the generic point is a 1-dimensional vector space over $k(C)$. We consider the diagonal linear map

$$\Delta : L_{\mathrm{gen}} \to \bigoplus_{p \in C} L_{\mathrm{gen}}, \qquad f \mapsto (\dots, f, f, f, \dots).$$

Definition 9.71. We define $\mathcal{H}^\vee(L)$ to be the vector space of linear forms

$$\alpha : \bigoplus_{p \in C} L_{\mathrm{gen}} \to k$$

which vanish on the diagonal $\Delta(L_{\mathrm{gen}})$. □

Clearly $\mathcal{H}^\vee(L)$ is a vector space over $k(C)$, and the dual of the cohomology space $H^1(L)$ can be identified with the k-linear subspace

$$H^1(L)^\vee = \{\alpha \mid \alpha \text{ vanishes on } \bigoplus_{p \in C} L_p\} \subset \mathcal{H}^\vee(L).$$

More generally, for any divisor $D = \sum_p n_p p$ we can identify

$$H^1(L(-D))^\vee = \{\alpha \mid \alpha \text{ vanishes on } \bigoplus_{p \in C} L_p(-n_p p)\} \subset \mathcal{H}^\vee(L). \qquad (9.33)$$

The next definition gives a natural choice for the line bundle L and the linear map α.

Definition 9.72.

(i) A linear form $\alpha : H^1(L) \to k$, viewed as an element of $\mathcal{H}^\vee(L)$

$$\alpha : \bigoplus_{p \in C} L_{\text{gen}} \to k, \qquad \alpha \circ \Delta = 0,$$

is called *nowhere vanishing* if for all $q \in C$ it satisfies

$$\alpha \left(L_q(q) \oplus \bigoplus_{p \in C - \{q\}} L_p \right) \neq 0.$$

(ii) If there exists a nowhere vanishing linear form on $H^1(L)$, then L is called a *dualising line bundle* on C.   □

One can see the existence of a dualising line bundle in the following way. Let L be a line bundle of maximal degree such that $H^1(L) \neq 0$ – such a line bundle exists by Theorem 9.20 and turns out to be a dualising line bundle. This is because at every point $q \in C$ the cohomology of $L(q) = L \otimes \mathcal{O}_C(q)$ vanishes, so that

$$\left(L_q(q) \oplus \bigoplus_{p \in C - \{q\}} L_p \right) + \Delta(L_{\text{gen}}) = \bigoplus_{p \in C} L_{\text{gen}}.$$

Hence any nonzero $\alpha \in H^1(L)^\vee \subset \mathcal{H}^\vee(L)$ is nowhere vanishing.

Next, fixing a nonzero linear form $\alpha \in H^1(L)^\vee$, consider the $k(C)$-linear map

$$k(C) \to \mathcal{H}^\vee(L), \qquad f \mapsto f\alpha.$$

This map is clearly injective, and its restriction to functions f such that $(f) + D \geq 0$ determines a linear map $H^0(\mathcal{O}_C(D)) \to H^1(L(-D))^\vee$.

Theorem 9.73. *If $\alpha \in H^1(L)^\vee$ is a nowhere vanishing linear form, then for any divisor $D \in \mathrm{Div}\, C$ the map*

$$H^0(\mathcal{O}_C(D)) \to H^1(L(-D))^\vee, \qquad f \mapsto f\alpha$$

is an isomorphism. Equivalently, the composition

$$H^0(\mathcal{O}_C(D)) \times H^1(L(-D)) \xrightarrow{\text{cup}} H^1(L) \xrightarrow{\alpha} k,$$

where the cup product is (9.17), is a nondegenerate pairing.

To prove this we first need:

Lemma 9.74. *If $\alpha \in H^1(L)^\vee$ is nowhere vanishing and $\beta \in H^1(L(-D))^\vee$ satisfies $\beta = f\alpha$, then $f \in H^0(\mathcal{O}_C(D))$.*

Proof. Let $(f) = \sum_p a_p p$ and $D = \sum_p n_p p$. The hypothesis that $\beta \in H^1(L(-D))^\vee$ means that $\bigoplus_{p \in C} L_p(-n_p p) \subset \ker \beta$, and so $\beta = f\alpha$ implies that

$$\bigoplus_{p \in C} L_p(-(a_p + n_p)p) \subset \ker \alpha.$$

But $\bigoplus_{p \in C} L_p \subset \ker \alpha$, and hence the assumption that α is nowhere vanishing forces $a_p + n_p \geq 0$ for all $p \in C$. Hence $(f) + D \geq 0$. □

Proof of Theorem 9.73. (Weil [68].) We just have to show surjectivity; pick $\beta \in H^1(L(-D))^\vee$. Then, for any positive divisor $F \geq 0$ we have injective linear maps

$$H^0(\mathcal{O}_C(F)) \hookrightarrow H^1(L(-D-F))^\vee, \qquad h \mapsto h\beta$$

and

$$H^0(\mathcal{O}_C(D+F)) \hookrightarrow H^1(L(-D-F))^\vee, \qquad h' \mapsto h'\alpha.$$

If the degree of F is sufficiently large, then the dimension of $H^1(L(-D-F))^\vee$ is just $\deg F$ plus a constant, and the same is true of the dimensions of $H^0(\mathcal{O}_C(F))$ and $H^0(\mathcal{O}_C(D+F))$. For $F \geq 0$ of sufficiently high degree, therefore, these two subspaces have nonzero intersection; that is, there exist $h \in H^0(\mathcal{O}_C(F))$ and $h' \in H^0(\mathcal{O}_C(D+F))$ satisfying $h\beta = h'\alpha$. Hence $\beta = (h'/h)\alpha$, and we just need to check that $h'/h \in H^0(\mathcal{O}_C(D))$. But this follows from Lemma 9.74. □

A divisor $K \in \operatorname{Div} C$ such that $L \cong \mathcal{O}_C(K)$ is a dualising line bundle is called a *canonical divisor*. (Such divisors exist by Corollary 9.32.) Applying Theorem 9.73 to K and $0 \in \operatorname{Div} C$ we obtain the following.

Corollary 9.75. *If L is a dualising line bundle on C, then the vector spaces $H^1(\mathcal{O}_C)$ and $H^0(L)$ are dual and $\dim H^1(L) = 1$.* □

Combining this corollary with the Riemann-Roch formula (9.15) yields $g - 1 = \dim H^0(L) - \dim H^1(L) = \deg L + 1 - g$, and hence:

Corollary 9.76. *If $L = \mathcal{O}_C(K)$ is a dualising line bundle, then* $\deg K = 2g - 2$, *where g is the genus of C.*   □

(b) The canonical line bundle

The cotangent bundle Ω_C of a curve C is a line bundle whose total set $\Omega_{k(C)}$ is a 1-dimensional vector space over $k(C)$. This is called the *canonical line bundle* on C. At each point $p \in C$ the stalk $\Omega_{C,p}$ of Ω_C is generated by dt_p, where $t_p \in \mathfrak{m}_p \mathcal{O}_{C,p}$ is a regular parameter. (See Section 9.3(a).)

Now, the principal parts of Ω_C have a very special property. Call an element $\omega \in \Omega_{k(C)}$ a *rational differential*. At each point $p \in C$ this has a Laurent expansion

$$\sum_{n=-N}^{\infty} a_n t_p^n dt_p$$

with respect to a regular local parameter, and the coefficient a_{-1} is independent of the choice of this parameter. This coefficient is called the *residue* of ω at p and denoted

$$\operatorname*{Res}_p \omega := a_{-1}.$$

The sum of residues over all points of the curve vanishes:

Residue Theorem 9.77. *Every rational differential $\omega \in \Omega_{k(C)}$ satisfies* $\sum_{p \in C} \operatorname*{Res}_p \omega = 0$.

Outline of the proof. For any finite sheeted cover $C \to C'$ we can define a trace map on differentials,

$$\operatorname{tr} : \Omega_{k(C)} \to \Omega_{k(C')},$$

with the property that for any $\omega \in \Omega_{k(C)}$ one has

$$\sum_{p \in C} \operatorname*{Res}_p \omega = \sum_{q \in C'} \operatorname*{Res}_q \operatorname{tr}(\omega).$$

(See, for example, Iwasawa [63].) To prove the theorem for a general curve C one can use this fact to reduce to the case of $C' = \mathbb{P}^1$, for which the proof is a simple computation.

Alternatively, when the ground field is $k = \mathbb{C}$, the curve C can be viewed as a compact Riemann surface. In this case we can take a triangulation containing all the poles of ω in the interiors of faces and apply Cauchy's residue formula to the faces. The total integral vanishes since C is orientable, so that all the contour integrals along the edges cancel out. $\square$

It follows from this result that

$$\alpha : \bigoplus_{p \in C} \Omega_{k(C)} / \Omega_{C,p} \to k, \qquad (\omega_p) \mapsto \sum_{p \in C} \operatorname*{Res}_{p} \omega_p$$

defines a nonzero linear form on $H^1(\Omega_C)$, and hence that $H^1(\Omega_C) \neq 0$. More than this, if we view α as an element of $\mathcal{H}^\vee(C)$, then it is nowhere vanishing in the sense of Definition 9.72 – hence Ω_C is a dualising line bundle. From Theorem 9.73 and Corollary 9.75 we get the following.

Theorem 9.78.

(i) For any divisor $D \in \operatorname{Div} C$ the vector spaces $H^0(\mathcal{O}_C(D))$ and $H^1(\Omega_C(-D))$ are canonically dual.

(ii) The cohomology space $H^1(\mathcal{O}_C)$ is canonically dual to the space $H^0(\Omega_C)$ of regular differentials (also called differentials of the first kind). In particular, the genus g of C is equal to the number of linearly independent regular differentials. (This is called the geometric genus.)

(iii) $\dim H^1(\Omega_C) = 1$ and $\deg \Omega_C = 2g - 2$. $\square$

It follows from the Residue Theorem 9.77 that $H^0(\Omega_C(p)) = H^0(\Omega_C)$ for every point $p \in C$. By Theorem 9.78(i), on the other hand, $H^1(\Omega_C(p)) = 0$. It follows that for every positive divisor $D \geq 0$,

$$\dim H^0(\Omega_C(D)) = \dim H^0(\Omega_C) + \deg D - 1.$$

Applying this to $D = np$ for $n \geq 2$ and to $D = p + q$ for $p, q \in C$, we obtain:

Theorem 9.79.

(i) (Existence of differentials of the second kind.) For every point $p \in C$ and integer $n \geq 2$ there exists a rational differential with a pole of order n at p and regular elsewhere.

(ii) (Existence of differentials of the third kind.) For every pair of distinct points $p, q \in C$ there exists a rational differential with a simple pole at each of p, q and regular elsewhere. Moreover, there exists such a differential with residues $1, -1$ at p, q, respectively. $\square$

(c) De Rham cohomology

A rational differential $\varphi \in \Omega_{k(C)}$ with zero residue at every point is called a *differential of the second kind*.

Example 9.80. The differential df of a rational function $f \in k(C)$, called an *exact differential*, is a differential of the second kind. ☐

Although this example is an extreme case, it is nevertheless the case that, given a differential φ of the second kind, there exists for every $p \in C$ a rational function $f_p \in k(C)$ satisfying

$$df_p \equiv \varphi \mod \Omega_{C,p}.$$

(This requires that the ground field k has characteristic zero, an assumption that we have not needed up to now.) The function f_p is uniquely determined modulo $\mathcal{O}_{C,p}$, and taking the principal part at all points of the curves defines an element

$$\int_{\text{prin}} \varphi := (f_p)_{p \in C} \in \bigoplus_{p \in C} k(C)/\mathcal{O}_{C,p}.$$

Theorem 9.79(i) says that there is an exact sequence:

$$0 \longrightarrow H^0(\Omega_C) \longrightarrow \left\{\begin{matrix}\text{differentials of the}\\ \text{second kind}\end{matrix}\right\} \xrightarrow{\int_{\text{prin}}} \bigoplus_{p \in C} k(C)/\mathcal{O}_{C,p} \longrightarrow 0.$$

$$(9.34)$$

Definition 9.81. The quotient of the vector space of differentials of the second kind modulo exact differentials,

$$H^1_{\text{dR}}(C) := \frac{\{\text{differentials of the second kind }\}}{\{df \mid f \in k(C)\}},$$

is called the *algebraic de Rham cohomology space* of the curve C. ☐

Factoring out the sequence (9.34) by the exact differentials yields an exact sequence

$$0 \longrightarrow H^0(\Omega_C) \longrightarrow H^1_{\text{dR}}(C) \longrightarrow H^1(\mathcal{O}_C) \longrightarrow 0, \qquad (9.35)$$

called the *Hodge filtration* of $H^1_{\text{dR}}(C)$. We are now going to construct a distinguished bilinear form on the de Rham cohomology.

Definition 9.82. Let $\varphi, \psi \in \Omega_{k(C)}$ be differentials of the second kind.

(i) Given a point $p \in C$, define

$$\langle \varphi \mid \psi \rangle_p := \operatorname*{Res}_p f_p \psi \in k, \quad \text{where } f_p \in k(C), df_p \equiv \varphi \bmod \Omega_{C,p}.$$

(ii) Set $\langle \varphi \mid \psi \rangle = \sum_{p \in C} \langle \varphi \mid \psi \rangle_p \in k$. □

Clearly, $\langle \mid \rangle_p$ and $\langle \mid \rangle$ are both k-valued bilinear forms on the vector space of differentials of the second kind.

Lemma 9.83.

(i) $\langle \mid \rangle$ is skew-symmetric, that is, $\langle \varphi \mid \psi \rangle + \langle \psi \mid \varphi \rangle = 0$.
(ii) $\langle \varphi \mid df \rangle = 0$ for any rational function $f \in k(C)$.
(iii) $\langle \varphi \mid \psi \rangle = 0$ if both $\varphi, \psi \in H^0(\Omega_C)$. □

Proof.

(i) Pick $p \in C$ and suppose that $df = \varphi, dg = \psi$ at p. Then $f\psi + g\varphi = d(fg)$ at p and has has no residue at this point. Hence $\langle \mid \rangle_p$ is skew-symmetric.

(ii) $\langle \varphi \mid df \rangle = \sum_p \operatorname*{Res}_p (f\varphi) = 0$ by the Residue Theorem 9.77.

(iii) is immediately clear from the definition. □

It follows from part (ii) of the lemma that $\langle \mid \rangle$ induces a skew-symmetric bilinear form on algebraic de Rham cohomology, for which we will use the same notation:

$$\langle \mid \rangle : H^1_{\mathrm{dR}}(C) \times H^1_{\mathrm{dR}}(C) \to k. \tag{9.36}$$

This is called the *de Rham cup product*.

Proposition 9.84. *The de Rham cup product is a nondegenerate pairing.*

Proof. Pick a basis of $H^1_{\mathrm{dR}}(C)$ extending a basis of $H^0(\Omega_C)$. With respect to such a basis, according to Lemma 9.83, the cup product is represented by a (skew-symmetric) matrix

$$\begin{pmatrix} 0 & A \\ -A & * \end{pmatrix}.$$

The matrix A represents the cup product $H^0(\Omega_C) \times H^1(\mathcal{O}_C) \to k$, which we have seen in Theorem 9.78 is degenerate. It follows that the above matrix has nonzero determinant, and so the de Rham cup product is nondegenerate. □

Corollary 9.85. *If* $\psi \in H^1_{\mathrm{dR}}(C)$ *annihilates* $H^0(\Omega_C)$ *in the cup product, that is,* $\langle \psi \mid \omega \rangle = 0$ *for all* $\omega \in H^0(\Omega_C)$*, then* $\psi \in H^0(\Omega_C)$*.* □

9.6 The Jacobian as a complex manifold

In this section we will let k be the field $\mathbb{C}$ of complex numbers and consider a nonsingular projective algebraic curve, which we shall view as a compact Riemann surface. In particular, the topology on C will be the usual complex topology. Fixing a point $p_0 \in C$ and choosing a basis of holomorphic differentials $\omega_1, \ldots, \omega_g \in H^0(\Omega_C)$, we define a holomorphic map

$$AJ : C \to \mathbb{C}/\Gamma_C, \qquad p \mapsto \int_{p_0}^{p} \vec{\omega} := \left(\int_{p_0}^{p} \omega_1, \ldots, \int_{p_0}^{p} \omega_g \right)^t, \qquad (9.37)$$

where $\Gamma_C \subset \mathbb{C}$ is the period lattice (9.1). This is called the *Abel-Jacobi map*. This map expends additively to the abelian group of divisors $\mathrm{Div}\, C$, and we use the same symbol AJ to denote the Abel-Jacobi map on divisors. Note that its restriction to divisors of degree zero,

$$AJ : \mathrm{Div}^0 C \to \mathbb{C}/\Gamma_C, \qquad D = \sum n_p p \mapsto \sum n_p AJ(p),$$

does not depend on the choice of base point $p_0 \in C$. Abel's Theorem says that $\ker AJ \subset \mathrm{Div}^0 C$ is precisely the subgroup of principal divisors.

Abel's Theorem 9.86. *If* $D = \sum n_p p \in \mathrm{Div}^0 C$ *is a divisor of degree zero, then* D *is a principal divisor if and only if the abelian integral*

$$\sum_{p \in C} n_p \int_{p_0}^{p} \vec{\omega}$$

is contained in the period lattice $\Gamma_C \subset \mathbb{C}^g$*. In other words,* D *is principal if and only if* $AJ(D) = 0$*.* □

Note that the case $g = 1$ is Liouville's Theorem 1.44(iv) and its converse (Exercise 9.12). We will give a proof of Abel's Theorem below, but first we use it to show the following.

Theorem 9.87. *When* $k = \mathbb{C}$, *the quotient variety* $\mathrm{Mat}_{N,1}(H^0(L))/GL(N,N)$ *constructed in Section 9.4 is isomorphic to the complex torus* $\mathbb{C}^g / \Gamma_C$. *In particular, it is irreducible.*

Proof. Given a point $\Psi = (a_{ij})_{1 \le i,j \le N} \in \mathrm{Mat}^s_{N,1}(H^0(L))$, where $a_{ij} \in H^0(L)$, let $D_{\Psi,1} \mathrm{Div}\, C$ be the greatest common divisor of the zero-sets $(a_{1j})_0$, $1 \le j \le N$, along the first row (see Definition 9.36). Then we consider the holomorphic map

$$\mathrm{Mat}^s_{N,1}(H^0(L)) \to \mathbb{C}^g / \Gamma_C, \qquad \Psi \mapsto AJ(D_{\Psi,1}).$$

Each $GL(N,N)$ maps down to a single point, and so this induces a map

$$\phi : \mathrm{Mat}^s_{N,1}(H^0(L))/GL(N,N) \to \mathbb{C}^g / \Gamma_C.$$

Then ϕ is holomorphic and, by Abel's Theorem, is injective. Since it is a map between compact complex manifolds of the same dimension, it is an isomorphism. □

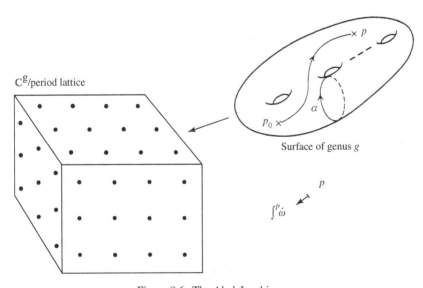

Figure 9.6: The Abel-Jacobi map

(a) Compact Riemann surfaces

If we regard the complex curve C as a Riemann surface, then we have a notion of *holomorphic functions* and *meromorphic functions* on C, the latter locally a ratio of holomorphic functions. If $f(z)$ is a holomorphic function (where z is a complex coordinate on the Riemann surface), then, for small $r > 0$,

$$f(z) = \frac{1}{2\pi} \int_0^{2\pi} f(z + re^{i\theta})d\theta, \qquad (9.38)$$

and from this it follows that, if $f(z)$ is nonconstant, then it cannot attain its maximum modulus at an interior point of its domain. (This is the *maximum modulus principle*.) Since C is compact, this implies that there are no nonconstant functions holomorphic everywhere on C.

A rational function is meromorphic on its domain, and in fact the converse is also true:

Proposition 9.88. *Every meromorphic function on C is rational, that is, it belongs to $\mathbb{C}(C)$.*

Proof. For any divisor $D \in \operatorname{Div} C$ denote by $\Lambda^{\mathrm{an}}(D)$ the set of meromorphic functions satisfying $(f) + D \geq 0$. We have just shown, for example, that $\Lambda^{\mathrm{an}}(0) = \mathbb{C}$. Note that by the same reasoning as for (9.3) this space satisfies

$$\dim \Lambda^{\mathrm{an}}(D) \leq \deg D + 1.$$

Now, given a meromorphic function f, we fix a divisor D for which $f \in \Lambda^{\mathrm{an}}(D)$, and for some positive divisor $F \geq 0$ we consider the two subspaces $\Lambda(D + F)$ and $f\Lambda(F) \subset \Lambda^{\mathrm{an}}(D + F)$. By Riemann's inequality (9.8), both subspaces have dimension bounded below by $\deg F +$ constant, while we have just shown that $\dim \Lambda^{\mathrm{an}}(D + F) \leq \deg F +$ constant. Hence, when $\deg F$ is sufficiently large,

$$\Lambda(D + F) \cap f\Lambda(F) \neq \emptyset.$$

In other words, we can find rational functions g, h such that $g = fh \neq 0$, and so f is itself rational. $\qquad\square$

Next we will view C as a real 2-dimensional manifold and consider its homology groups $H_i(C, \mathbb{Z})$. The alternating sum of its Betti numbers

$$e(C) := \sum_i (-1)^i b_i = 2 - b_1, \qquad b_i = \operatorname{rank}_{\mathbb{Z}} H_i(C, \mathbb{Z}),$$

is called the *Euler number* of C, and it is well known that this can be computed from a triangulation as

$$e(C) = \begin{pmatrix} \text{number} \\ \text{of vertices} \end{pmatrix} - \begin{pmatrix} \text{number} \\ \text{of edges} \end{pmatrix} + \begin{pmatrix} \text{number} \\ \text{of faces} \end{pmatrix}.$$

Proposition 9.89. *If C has arithmetic genus g (Definition 9.11), then $e(C) = 2 - 2g$.*

Sketch Proof. For the Riemann sphere $\mathbb{P}^1_{\mathbb{C}}$ this is clear (Example 9.5 and Corollary 9.21). For a general curve C it is enough, by Theorem 9.78(iii), to show that $e(C) = -\deg \Omega_C$, and to do this we use a finite cover $\pi : C \to \mathbb{P} = \mathbb{P}^1_{\mathbb{C}}$. Let $p_1, \ldots, p_h \in C$ be the ramification points of π, with ramification indices $e_1, \ldots, e_h \in \mathbb{N}$. Then the canonical line bundle on C is given by

$$\Omega_C \cong \pi^* \Omega_{\mathbb{P}} \otimes \mathcal{O}_C \left(\sum_{i=1}^{h} (e_i - 1) p_i \right).$$

Taking degrees on both sides,

$$\deg \Omega_C = d \cdot \deg \Omega_{\mathbb{P}} + \sum_{i=1}^{h} (e_i - 1),$$

where $d = \deg \pi$ is the degree of the cover. On the other hand, by taking suitable triangulations (so that on C lifts the triangulation on $\mathbb{P}^1_{\mathbb{C}}$ and all of the branch points in $\mathbb{P}^1_{\mathbb{C}}$ are vertices) the Euler numbers upstairs and downstairs are related by

$$e(C) = d \cdot e(\mathbb{P}^1_{\mathbb{C}}) - \sum_{i=1}^{h} (e_i - 1),$$

and from these two identities it follows that $e(C) = -\deg \Omega_C$. $\qquad\square$

Corollary 9.90. *A complex curve of genus g has Betti number $b_1 = 2g$.* $\qquad\square$

In other words, the genus coincides with the *topological genus* of C as a Riemann surface – the number of 'holes' in C, in its well-known guise as a rubber tube.

(b) The comparison theorem and the Jacobian

When $k = \mathbb{C}$, a rational differential $\varphi \in \Omega_{\mathbb{C}(C)}$ will be called an *abelian differential*. One can consider the contour integrals of φ; in particular, if α is a closed contour on C avoiding the poles of φ, then

$$\int_\alpha \varphi \in \mathbb{C}$$

is called a *period* of the abelian differential φ.

Lemma 9.91. *An abelian differential φ is a closed 1-form, meaning that $d\varphi = 0$ away from the poles of φ, where d is the exterior derivative on the real surface C.*

Proof. Locally (away from its poles) φ can be written as $f(z)dz$, where $z = x + \sqrt{-1}y$ is a local complex coordinate and $f(z)$ is a holomorphic function. Let $u(x, y)$, $v(x, y)$ be the real and imaginary parts of $f(z)$ so that

$$d\varphi = (u + \sqrt{-1}v)(dx + \sqrt{-1}dy) = (udx - vdy) + \sqrt{-1}(udy + vdx).$$

The exterior derivative is then

$$d\varphi = (du \wedge dx - dv \wedge dy) + \sqrt{-1}(du \wedge dy + dv \wedge dx)$$

$$= \left\{ -\left(\frac{\partial u}{\partial y} + \frac{\partial v}{\partial x}\right) + \sqrt{-1}\left(\frac{\partial u}{\partial x} - \frac{\partial v}{\partial y}\right) \right\} dx \wedge dy,$$

and this vanishes by the Cauchy-Riemann equations. $\square$

From this lemma and Stokes' Theorem,

$$\int_{\partial\beta} \varphi = \int_\beta d\varphi,$$

it follows that the period φ around a closed contour α in $C - \{\text{poles of } \varphi\}$ depends only on the homology class of α. If, in addition, φ is a differential of the second kind, that is, φ has no residues, then it follows that the period is also defined independently of how the contour winds around the poles, and hence depends only on the homology class of α in C. On the other hand, the periods of exact differentials df and logarithmic exact differentials df/f all vanish, and it follows that contour integration induces a bilinear pairing of abelian groups

$$H_1(C, \mathbb{Z}) \times H^1_{dR}(C) \to \mathbb{C}, \qquad (\alpha, \varphi) \mapsto \int_\alpha \varphi. \tag{9.39}$$

This can be re-expressed as

$$H_1(C, \mathbb{Z}) \to H^1_{dR}(C)^\vee$$

or as

$$H^1_{dR}(C) \to \mathrm{Hom}(H_1(C, \mathbb{Z}), \mathbb{C}) = H^1(C, \mathbb{C}). \tag{9.40}$$

Let us call (9.40) the *comparison map*.

Proposition 9.92. *The comparison map is an isomorphism* $H^1_{\mathrm{dR}}(C) \xrightarrow{\sim} H^1(C, \mathbb{C})$.

Proof. By Corollary 9.90 we only need to show that it is injective. Fix a base point $p_0 \in C$ and consider the path integral

$$f(p) = \int_{p_0}^{p} \varphi$$

as a function of a moving endpoint $p \in C$, where φ is a differential of the second kind. To suppose that (the de Rham class) φ maps to zero under the comparison map means that f is a single-valued function on the Riemann surface C. But then f is meromorphic and $\varphi = df$, and so, using Proposition 9.88, φ is an exact differential. □

A Riemann surface carries a natural orientation, and so any two closed paths α, β have a well-defined intersection number $(\alpha \cdot \beta) \in \mathbb{Z}$. This depends only on the homology classes of the paths, and so it determines a bilinear form

$$H_1(C, \mathbb{Z}) \times H_1(C, \mathbb{Z}) \to \mathbb{Z}. \tag{9.41}$$

Poincaré duality says that this pairing is *unimodular*: that is, there exist homology bases

$$\{\alpha_1, \ldots, \alpha_{2g}\}, \quad \{\alpha_1^*, \ldots, \alpha_{2g}^*\}$$

with respect to which the intersection pairing is given by the identity matrix, $(\alpha_i \cdot \alpha_j^*) = \delta_{ij}$. The *cup product* of two cohomology classes $f_1, f_2 \in H^1(C, \mathbb{C}) \cong \mathrm{Hom}(H_1(C, \mathbb{Z}), \mathbb{C})$ is then defined by

$$f_1 \cup f_2 = \sum_{i,j=1}^{2g} f_1(\alpha_i) f_2(\alpha_j^*).$$

This does not depend on the choice of the bases.

Proposition 9.93 (Bilinear relations for differentials of the second kind).
The following diagram commutes, relating the comparison map and the cup products in cohomology.

$$
\begin{array}{ccc}
H^1_{\mathrm{dR}}(C) \times H^1_{\mathrm{dR}}(C) & \xrightarrow{\langle|\rangle} & \mathbb{C} \\
\downarrow & & \downarrow \times 2\pi\sqrt{-1} \\
H^1(C, \mathbb{C}) \times H^1(C, \mathbb{C}) & \xrightarrow{\cup} & \mathbb{C}
\end{array}
$$

Proof. Taking a symplectic basis of homology

$$\{\alpha_1, \ldots, \alpha_g, \beta_1, \ldots, \beta_g\} \tag{9.42}$$

we can cut the Riemann surface open as in Figure 9.7.
Since, for all $1 \le i, j \le g$,

$$(\alpha_i \cdot \alpha_j) = (\beta_i \cdot \beta_j) = 0, \quad (\alpha_i \cdot \beta_j) = \delta_{ij},$$

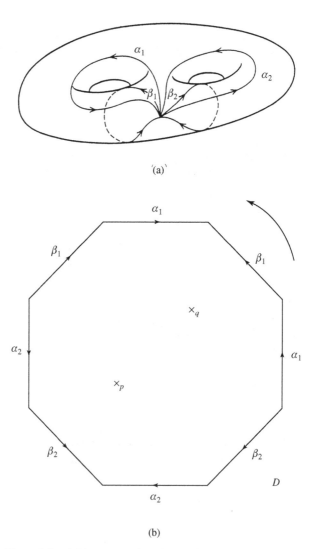

Figure 9.7: (a) Riemann surface C, and (b) the surface cut open.

the dual basis is

$$\{\beta_1, \ldots, \beta_g, -\alpha_1, \ldots, -\alpha_g\}.$$

What we have to show, therefore, given differentials of the second kind φ, ψ, is the relation

$$\sum_{i=1}^{g} \left(\int_{\alpha_i} \varphi \int_{\beta_i} \psi - \int_{\alpha_i} \psi \int_{\beta_i} \varphi \right) = 2\pi \sqrt{-1} \langle \varphi \mid \psi \rangle.$$

Consider, in the domain D obtained by cutting C open, the path integral

$$f(p) = \int^{p} \psi.$$

This is a meromorphic function of $p \in D$ and can integrate the product $f\varphi$ around the boundary of D. By Cauchy's residue theorem this is

$$\int_{\partial D} f\varphi = 2\pi \sqrt{-1} \sum_{p \in D} \operatorname*{Res}_{p} f\varphi,$$

and the bilinear relation follows from this. □

From now on we will identify $H^1_{\mathrm{dR}}(C)$ and $H^1(C, \mathbb{C})$ and simply write $H^1(C)$.

The bilinear relations and the identification of Proposition 9.93 allow us to define the Jacobian by an analytic construction. In the first instance it is the real $2g$-dimensional torus

$$(S^1)^{2g} = \frac{\mathbb{R}}{2\pi \sqrt{-1} \mathbb{Z}^{2g}} = \frac{H^1(C, \mathbb{R})}{2\pi \sqrt{-1} H^1(C, \mathbb{Z})}. \tag{9.43}$$

From the Hodge filtration (9.35),

$$0 \to H^0(\Omega_C) \to H^1(C) \to H^1(\mathcal{O}_C) \to 0$$

$$\cup$$

$$H^1(C, \mathbb{R})$$

and Lemma 9.95 below we obtain an isomorphism of real vector spaces $H^1(C, \mathbb{R}) \xrightarrow{\sim} H^1(\mathcal{O}_C)$.

Definition 9.94. The *analytic Jacobian* of C is the torus (9.43) equipped with the complex structure coming from the above isomorphism:

$$\operatorname{Jac}^{\mathrm{an}} C := \frac{H^1(\mathcal{O}_C)}{2\pi \sqrt{-1} H^1(C, \mathbb{Z})}.$$

□

By Poincaré duality we obtain an isomorphism

$$\mathrm{Jac}^{\mathrm{an}}C \xrightarrow{\sim} \frac{H^0(\Omega_C)^\vee}{H_1(C, \mathbb{Z})} = \mathbb{C}^g / \Gamma_C,$$

where $\Gamma_C \subset \mathbb{C}$ is the period lattice (9.1) of C. An alternative expression again is

$$\mathrm{Jac}^{\mathrm{an}}C = \frac{H^1(C)}{H^0(\Omega_C) + 2\pi\sqrt{-1}H^1(C, \mathbb{Z})} = \frac{H^1(C, \mathbb{C}^*)}{H^0(\Omega_C)}.$$

Lemma 9.95. $H^0(\Omega_C) \cap H^1(C, \mathbb{R}) = 0$ *in* $H^1(C)$.

Proof. Suppose $\omega \in H^0(\Omega_C) \cap H^1(C, \mathbb{R})$. Then the periods of ω are all real numbers, so the imaginary part of the integral

$$f(p) = \int^p \omega$$

is a (single-valued) function on C and has the harmonicity property (9.38). By the maximum principle the imaginary part is therefore constant, and hence by the Cauchy-Riemann equations the function f itself is constant, and hence $\omega = df = 0$. $\square$

(c) Abel's Theorem

Given an abelian differential ψ, we can represent its residues as a divisor with complex coefficients

$$\mathrm{Res}\,\psi := \sum_{p \in C} (\mathop{\mathrm{Res}}_p \psi)p \in \mathrm{Div}\,C \otimes_{\mathbb{Z}} \mathbb{C}.$$

Definition 9.96. An abelian differential with only simple poles is said to be of *logarithmic type*, and if, in addition, the residue at every point is an integer, then a differential of logarithmic type is said to be of *divisor type*. $\square$

Example 9.97. For any rational function $f \in \mathbb{C}(C)$ the logarithmic derivative df/f is an abelian differential of divisor type, called a *logarithmic exact differential*. Its residue $\mathrm{Res}\,df/f$ is just the principal divisor $(f) \in \mathrm{Div}\,C$ determined by f (Definition 9.13). $\square$

By the residue theorem, the divisor Res ψ has degree zero. By Theorem 9.79, therefore, there is an exact sequence

$$0 \longrightarrow H^0(\Omega_C) \longrightarrow \left\{ \begin{matrix} \text{abelian} & \text{differentials} \\ \text{of divisor type} \end{matrix} \right\} \xrightarrow{\text{Res}} \text{Div}^0 C \longrightarrow 0. \qquad (9.44)$$

Remark 9.98. The quotient

$$H^1_{\text{Del}}(C) := \frac{\text{abelian differentials of divisor type}}{\text{logarithmic exact differentials } df/f}$$

is called the *Deligne cohomology* of C. Dividing (9.44) through by the group $\mathbb{C}(C)^\times$ then gives an exact sequence

$$0 \to H^0(\Omega_C) \to H^1_{\text{Del}}(C) \to \text{Pic}^0 C \to 0.$$

Figure 9.8 summarises the geography of abelian differentials on a curve. $\qquad \square$

Consider the integral of an abelian differential ψ along a path α in C avoiding the poles of ψ. Replacing α by a path making one additional circuit of a pole $p \in C$ changes the integral by the addition of $2\pi\sqrt{-1}\,\underset{p}{\text{Res}}\,\psi$; and so if ψ is of

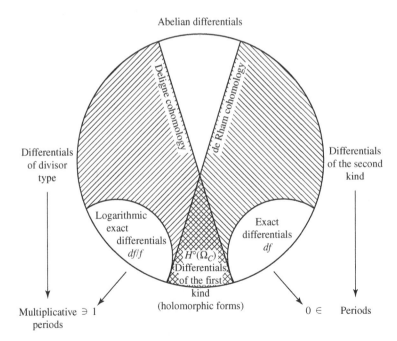

Figure 9.8: Abelian differentials

divisor type, then the quantities

$$\int_\alpha \psi \quad \mathrm{mod}\ 2\pi\sqrt{-1}\mathbb{Z}$$

and

$$\exp \int_\alpha \psi \in \mathbb{C}^*$$

depend only on the homology class of α in the curve. The exponential $\exp \int_\alpha \psi$ is called a *multiplicative period* and defines multiplicative versions of (9.39) and (9.40),

$$\left\{ \begin{array}{l} \text{abelian}\quad\text{differentials} \\ \text{of divisor type} \end{array} \right\} \times H_1(C, \mathbb{Z}) \to \mathbb{C}^*$$

and

$$\rho : \left\{ \begin{array}{l} \text{abelian}\quad\text{differentials} \\ \text{of divisor type} \end{array} \right\} \to \mathrm{Hom}(H_1(C, \mathbb{Z}), \mathbb{C}^*) = H^1(C, \mathbb{C}^*). \qquad (9.45)$$

By the same reasoning as in Proposition 9.92 we get the following.

Proposition 9.99. *The group homomorphism (9.45) has kernel* $\ker \rho = \{$logarithmic exact differentials$\}$, *and so induces an isomorphism*

$$H^1_{\mathrm{Del}}(C) \overset{\sim}{\to} H^1(C, \mathbb{C}^*).$$

$\square$

The bilinear relation for pairs of differentials of the second kind (Proposition 9.93) can be extended to pairs consisting of a holomorphic differential and an abelian differential of divisor type:

Proposition 9.100 (Bilinear relations for differentials of divisor type). *The following diagram commutes.*

$$
\begin{array}{ccc}
\left\{ \begin{array}{l} \text{abelian}\quad\text{differentials} \\ \text{of divisor type} \end{array} \right\} & \overset{\rho}{\longrightarrow} & H^1(C, \mathbb{C}^*) \\[2em]
\mathrm{Res} \downarrow & & \downarrow \\[2em]
\mathrm{Div}^0 C & \overset{AJ}{\longrightarrow} & \mathrm{Jac}^{\mathrm{an}} C
\end{array}
$$

Proof. Let ψ be an abelian differential of divisor type and let $\vec{\omega} = (\omega_1, \ldots, \omega_g)$, where $\{\omega_1, \ldots, \omega_g\}$ is a basis of $H^0(\Omega_C)$. Using the symplectic basis (9.42) of $H_1(C, \mathbb{Z})$, what we have to show is the relation

$$\sum_{i=1}^{g}\left(\int_{\alpha_i}\psi\int_{\beta_i}\vec{\omega}-\int_{\alpha_i}\vec{\omega}\int_{\beta_i}\psi\right)\equiv 2\pi\sqrt{-1}AJ(\,\mathrm{Res}\,\psi)$$

modulo the period lattice. We use the same proof as for Proposition 9.93: define a holomorphic vector-valued function on the domain D

$$\vec{f}(p)=\int^{p}\vec{\omega}$$

and integrate the product $\psi\vec{f}$ around the boundary of D. By Cauchy's residue theorem

$$\int_{\partial D}\psi\vec{f}=2\pi\sqrt{-1}\sum_{p\in D}\mathrm{Res}_{p}\psi\vec{f},$$

and the bilinear relation follows. $\qquad\square$

Proof of Abel's Theorem 9.86. First, we apply the above bilinear relation to a logarithmic exact differential df/f. By Proposition 9.99, this maps under ρ to the trivial element, and hence $\mathrm{Res}\,df/f=(f)\in\mathrm{Div}^0 C$ lies in the kernel of the Abel-Jacobi map. So we have proved half of Abel's Theorem. For the converse, suppose that $D\in\mathrm{Div}^0 C$ and $AJ(D)=0$. Since Res is surjective (Theorem 9.79), we can write $D=\mathrm{Res}\,\psi$ for some abelian differential of divisor type ψ. By Proposition 9.100 again, $\rho(\psi)$ is in the kernel of $H^1(C,\mathbb{C}^*)\to\mathrm{Jac}^{\mathrm{an}}C$, that is, $\rho(\psi)$ is in the image of $H^0(\Omega_C)$. In other words, there exists $\omega\in H^0(\Omega_C)$ such that all the multiplicative periods of $\psi-\omega$ are trivial,

$$\int_{\alpha}(\psi-\omega)\in 2\pi\sqrt{-1}\mathbb{Z}\quad\text{for all }\alpha\in H_1(C,\mathbb{Z}).$$

So by Proposition 9.99, $\psi-\omega=df/f$ for some $f\in\mathbb{C}(C)$, and hence $D=\mathrm{Res}\,\psi=\mathrm{Res}\,(\psi-\omega)=(f)$ is a principal divisor. $\qquad\square$

Exercises

1. Show that the two conditions of Definition 9.45 are equivalent, and prove Lemma 9.46.
2. Show that a curve of genus 0 is isomorphic to $\mathbb{P}^1$.
3. In the situation of Example 9.23, use the Weierstrass ζ-function

$$\zeta(z)=\frac{1}{z}+\sum_{\gamma\in\Gamma-\{0\}}\left(\frac{1}{z-\gamma}+\frac{1}{\gamma}+\frac{z}{\gamma^2}\right)$$

to show that there exists a meromorphic function on $\mathbb{C}/\Gamma$ with a simple
pole at each of $p+\Gamma, q+\Gamma$, and holomorphic elsewhere. (Consider linear
combinations of $\zeta(z-p)$ and $\zeta(z-q)$.)

4. If C is a nonsingular affine curve, show that the principal part map

$$\mathrm{pp} : k(C) \to \bigoplus_{p\in C} k(C)/\mathcal{O}_{C,p}$$

is surjective.

5. If $C \subset \mathbb{P}^n$ is a nonsingular projective curve, show that there exists a hyperplane $H \subset \mathbb{P}^n$ intersecting C transversally. (This means that at each
intersection point the tangent spaces of C and H intersect transversally in
the tangent space of $\mathbb{P}^n$.) *Hint:* Use the discussion of Section 5.2(a) to show
that the parameter space for hyperplanes not intersecting C transversally
has dimension at most $n-1$.

6. Show that if a line bundle $L \in \mathrm{Pic}\, C$ is generated by global sections, then
it defines a morphism to projective space

$$C \to \mathbb{P}^{n-1}, \qquad n = \dim H^0(L).$$

(See Section 8.5(c).) Prove that this is an embedding if $\deg L \geq 2g+1$.

7. If R is a ring generated over k by $a_1, \ldots, a_n$, show that the differential
module $\Omega_{R/k}$ is a generated as an R-module by $da_1, \ldots, da_n$.

8. Suppose that a ring R over k is nonsingular at every maximal ideal. Show
that R is a direct sum of integral domains.

9. Show that any two dualising line bundles on C are isomorphic.

10. Prove that if the genus of C is positive, then the canonical line bundle Ω_C
is generated by global sections.

11. Let E be a vector bundle on a curve C.
    (i) If $U \subset C$ is an affine open set, show that there is an exact sequence

$$0 \longrightarrow H^0(E) \longrightarrow E(U) \longrightarrow \bigoplus_{p\notin U} E_p \longrightarrow H^1(E) \longrightarrow 0.$$

    (Use the proof of Proposition 9.27.)
    (ii) If $U, V \subset C$ are two open sets, show that there is an exact sequence

$$0 \longrightarrow H^0(E) \longrightarrow E(U) \oplus E(V) \longrightarrow E(U \cap V) \longrightarrow H^1(E) \longrightarrow 0.$$

12. In the setting of Exercise 9.3, let $D = \sum_p n_p p$ be a divisor of degree zero
on the complex torus $\mathbb{C}/\Gamma$. For the function

$$f(z) = \exp\left(2\pi\sqrt{-1}\sum_p n_p \int^z \zeta(z-p)dz\right)$$

show the following.

(i) $f(z)$ is a meromorphic function on the complex plane $\mathbb{C}$.

(ii) If the sum $\sum_p n_p p \in \mathbb{C}$ belongs to the lattice $\Gamma \subset \mathbb{C}$, then $f(z)$ is doubly periodic with period lattice Γ.

(iii) In case (ii), the principal divisor determined by $f(z)$, viewed as a meromorphic function on $\mathbb{C}/\Gamma$, is equal to D. (This is a special case of Abel's Theorem.)

10

Stable vector bundles on curves

As in the last chapter, C will be a nonsingular projective algebraic curve of genus g, which we just call a *curve* for short. In this chapter we are going to study vector bundles on C. The key point in the construction of a moduli space for vector bundles is the notion of *stability* introduced by Mumford [31] (see Definition 10.20). The goal of this chapter is to show that, fixing any line bundle L on C, the set of isomorphism classes of stable vector bundles with determinant line bundle isomorphic to L

$$SU_C(2, L) := \left\{ \begin{array}{l} \text{rank 2 stable vector bundles } E \\ \text{with det } E \cong L \end{array} \right\} \Big/ \text{isomorphism}$$

can be given the structure of an algebraic variety. Briefly, the idea of the construction is the following. Just as for the construction of the Jacobian in the previous chapter, we assume that L has sufficiently high degree to guarantee that E is generated by global sections, and we consider the skew-symmetric bilinear map

$$H^0(E) \times H^0(E) \to H^0(L), \qquad (s, s') \mapsto s \wedge s'.$$

This form has rank 2, and we denote by $\text{Alt}_{N,2}(H^0(L))$ the affine variety which parametrises such skew-symmetric forms of rank ≤ 2 in dimension $N = \dim H^0(E)$. (See Section 10.3(b) for notation.) We will use this wedge product to reduce our moduli problem to the quotient problem for the action of $GL(N)$ on $\text{Alt}_{N,2}(H^0(L))$. One encounters various difficulties that do not appear in the line bundle case of the last chapter, but it turns out that the notion of stability is the correct way to resolve these problems, and one proves the following.

Theorem 10.1. *Suppose that the line bundle L has degree $\geq 4g - 1$.*

(i) There exists a Proj quotient

$$\mathrm{Alt}^{ss}_{N,2}(H^0(L)) /\!/ GL(N)$$

which is a projective variety of dimension $3g - 3$.
(ii) The open set

$$\mathrm{Alt}^{s}_{N,2}(H^0(L)) / GL(N)$$

has an underlying set $SU_C(2, L)$. Moreover, it is nonsingular and at each point $E \in SU_C(2, L)$ its tangent space is isomorphic to $H^1(\mathfrak{sl}\, E)$.
(iii) If $\deg L$ is odd, then

$$\mathrm{Alt}^{ss}_{N,2}(H^0(L)) /\!/ GL(N) = \mathrm{Alt}^{s}_{N,2}(H^0(L)) / GL(N) = SU_C(2, L)$$

is a smooth projective variety. □

One basic technique for working with vector bundles is, by passing to sub-bundles and quotients, to reduce to the case of line bundles. In Section 10.1 we illustrate this method by proving various basic results that will be needed later. We prove the Riemann-Roch Theorem for vector bundles, the unique decomposition of a bundle into indecomposable subbundles and the classification of extensions by their cohomology classes. In Section 10.2 we restrict our attention to rank 2 vector bundles and investigate some of their properties. In Section 10.3 we introduce the notion of a Gieseker point of a rank 2 vector bundle, and we show that semistability under the action of the general linear group is equivalent to the semistability of the vector bundle. We prove this by direct construction of semiinvariants, using Pfaffians, and without recourse to the Hilbert-Mumford numerical criterion of Chapter 7. (The same statement is true with semistability replaced by stablility, but for that we do need the numerical criterion, and this is discussed briefly in Section 10.4(c).) In Section 10.4 we put all of these ideas together to prove Theorem 10.1.

10.1 Some general theory

Let E be a vector bundle on C. Thus E is an elementary sheaf with total set E_{gen}, a finite-dimensional vector space over $k(C)$, and for each nonempty open set $U \subset C$ defines a $\mathcal{O}_C(U)$-module $E(U)$. We denote the stalk at a point $p \in C$ by $E_p \subset E_{\mathrm{gen}}$. For each positive integer i we can define in a natural way a vector bundle $\bigwedge^i E$ whose total set is the exterior power $\bigwedge^i E_{\text{gen}}$. In particular, if $r = r(E)$ is the rank of E, then

$$\det E := \bigwedge^r E$$

is a line bundle, called the *determinant line bundle* of E. An $\mathcal{O}_C$-module ho-momorphism $f : E \to F$ between vector bundles E, F will simply be called a homomorphism, and we denote the set of these by $\text{Hom}(E, F)$.

(a) Subbundles and quotient bundles

Note that the following two definitions are not the same!

Definition 10.2. A vector bundle F on C is a *subsheaf* of E if:

(i) $F_{\text{gen}} \subset E_{\text{gen}}$ is a vector subspace over $k(C)$; and
(ii) $F(U) \subset E(U)$ is a submodule over $\mathcal{O}_C(U)$, for every open set $U \subset C$.

In this case we write $E \subset F$. □

Definition 10.3. A *subbundle* $F \subset E$ is a subsheaf which satisfies, in addition:
(iii) $F(U) = E(U) \cap F_{\text{gen}}$ for every open set $U \subset C$. □

If $D \geq 0$ is a positive divisor, then $\mathcal{O}_C(-D) \subset \mathcal{O}_C$ is a subsheaf but not a subbundle. Indeed, a line bundle has no nonzero subbundles. More generally, if $F \subset E$ is a subbundle, then $F(-D) \subset E$ is a subsheaf, but not a subbundle if $D \neq 0$. But in the other direction, starting with any subsheaf $F \subset E$ we can construct a subbundle, called the *saturation* of F in E, as the elementary sheaf

$$U \mapsto E(U) \cap F_{\text{gen}}.$$

If $F \subset E$ is a subbundle, then, although for each open set $U \subset C$ the quotient $E(U)/F(U)$ is a submodule of the quotient vector space $E_{\text{gen}}/F_{\text{gen}}$, in general the mapping $U \mapsto E(U)/F(U)$ does not define an elementary sheaf. However, using the stalks, the mapping

$$U \mapsto \{f \in E_{\text{gen}}/F_{\text{gen}} \mid f \in E_p/F_p \text{ for all } p \in U\}$$

does define an elementary sheaf, and this is what we call the *quotient E/F*. Note that the natural projection maps

$$E(U) \to (E/F)(U)$$

are not in general surjective, although they are surjective on the stalks, and so $E \to E/F$ is called a surjective map of sheaves (see Definition 8.81). This is a delicate point in the theory of sheaves, and it is important to treat it with care.

We now take an affine open cover $\{U_i\}$, $U_i = \text{Spm } R_i$, of our curve. The restrictions $E|_{U_i}$ and $F|_{U_i}$ come from some R_i-module M_i and submodule $N_i \subset M_i$. The condition that $F \subset E$ is a subbundle is equivalent to requiring that each quotient R_i-module be torsion free, and the quotient bundle E/F is the elementary sheaf obtained by gluing the modules M_i/N_i. Since the local ring $\mathcal{O}_{C,p}$ at each point $p \in C$ is a discrete valuation ring, it follows that each M_i/N_i is a locally free R_i-module (Corollary 8.38). From this we conclude:

Lemma 10.4. *If E is a vector bundle on a curve C and $F \subset E$ is a subbundle, then the quotient E/F is a vector bundle.* $\qquad\square$

A sequence of vector bundle homomorphisms

$$0 \to F \to E \to G \to 0 \qquad (10.1)$$

is *exact* if at each point $p \in C$ the stalk maps

$$0 \to F_p \to E_p \to G_p \to 0$$

form an exact sequence of $\mathcal{O}_{C,p}$-modules. In this case we will view F as a subbundle of E and G as its quotient bundle E/F.

Definition 10.5. The exact sequence (10.1) is *split* if any of the following equivalent conditions are satisfied.

(i) There exists a subbundle (or subsheaf) $G' \subset E$ for which the composition $G' \hookrightarrow E \to G$ is an isomorphism.
(ii) There exists a homomorphism $f : G \to E$ for which the composition $G \xrightarrow{f} E \to G$ is an isomorphism.
(iii) There exists a homomorphism $g : E \to F$ for which the composition $F \hookrightarrow E \xrightarrow{g} F$ is an isomorphism.

In this case, either of the maps f or g is called a *splitting* of the sequence. $\quad\square$

Let $f : E \to F$ be a vector bundle homomorphism. Locally this is a homomorphism of $\mathcal{O}_C$-modules, and the *image* Im f is the sheaf obtained by gluing the image modules. This is a vector bundle with total set equal to Im $\{f_{\text{gen}} : E_{\text{gen}} \to F_{\text{gen}}\}$ and is a subsheaf of F. It can also be defined as follows. First note that

$$U \mapsto \ker \{E(U) \to F(U)\}$$

defines an elementary sheaf with total set $\ker f_{\text{gen}}$. This is a subbundle $\ker f \subset E$, and we can define the image sheaf to be the quotient

$$\operatorname{Im} f := E/\ker f \hookrightarrow F.$$

A homomorphism $f : E \to F$ induces homomorphisms of exterior powers

$$\textstyle\bigwedge^i f : \bigwedge^i E \to \bigwedge^i F, \qquad 0 \le i \le \min\{r(E), r(F)\}.$$

If $r(E) = r(F) = r$, then we write $\det f = \bigwedge^r f$ for the homomorphism of determinant line bundles.

Proposition 10.6. *Let $f : E \to F$ be a homomorphism between vector bundles of the same rank. Then f is an isomorphism if and only if $\det f : \det E \to \det F$ is an isomorphism of line bundles.* $\qquad\square$

(b) The Riemann-Roch formula

Let E be a vector bundle of rank r. If $N \subset E_{\text{gen}}$ is a vector subspace over $k(C)$, then

$$U \mapsto E(U) \cap N$$

defines a subbundle of E. In this way it is always possible to construct exact sequences of the form (10.1); and by using this one can deduce properties of E from properties of bundles of lower rank, and finally from line bundles. In this section we are going to use this method to derive a Riemann-Roch formula for vector bundles.

Definition 10.7. The *degree* $\deg E \in \mathbb{Z}$ of a vector bundle is the degree of its determinant line bundle $\det E$. $\qquad\square$

If E lies in an exact sequence (10.1), then there is an isomorphism

$$\det F \otimes \det G \cong \det E,$$

and hence $\deg E = \deg F + \deg G$. In other words, degree is additive on exact sequences.

Recall from Section 9.2 that the spaces $H^0(E)$ and $H^1(E)$ are defined to be the kernel and cokernel of the principal part map:

$$0 \longrightarrow H^0(E) \longrightarrow E_{\text{gen}} \longrightarrow \bigoplus_{p \in C} E_{\text{gen}}/E_p \longrightarrow H^1(E) \longrightarrow 0.$$

We will write $h^i(E) = \dim H^i(E)$ for $i = 0, 1$.

The exact sequence (10.1) gives rise to a long exact sequence of cohomology spaces as follows. First, it induces a commutative diagram, in which each row is exact:

$$
\begin{array}{ccccccc}
0 \longrightarrow & F_{\text{gen}} & \longrightarrow & E_{\text{gen}} & \longrightarrow & G_{\text{gen}} & \longrightarrow 0 \\
& \downarrow & & \downarrow & & \downarrow & \\
0 \longrightarrow & \bigoplus_{p \in C} F_{\text{gen}}/F_p & \longrightarrow & \bigoplus_{p \in C} E_{\text{gen}}/E_p & \longrightarrow & \bigoplus_{p \in C} G_{\text{gen}}/G_p & \longrightarrow 0
\end{array}
$$

(10.2)

Applying the Snake Lemma 8.57 to (10.2) yields an exact sequence

$$0 \to H^0(F) \to H^0(E) \to H^0(G) \xrightarrow{\delta} H^1(F) \to H^1(E) \to H^1(G) \to 0.$$

(10.3)

The connecting map $\delta : H^0(G) \to H^1(F)$ is in this case called the *coboundary map*.

Proposition 10.8. *For any vector bundle E on C the vector spaces $H^0(E)$ and $H^1(E)$ are both finite-dimensional over k.*

Proof. When E is a line bundle this has been proved in the last chapter (see Section 9.2(c)). For higher rank we can use induction on the rank of E. Using any proper $k(C)$-vector subspace of E_{gen} we have an exact sequence (10.1), and hence an exact sequence (10.3). By the inductive hypothesis, the spaces $H^0(F), H^0(G), H^1(F), H^1(G)$ are all finite-dimensional, and hence so are $H^0(E), H^1(E)$. □

Corollary 10.9. *If E is a vector bundle, then the degree of its line subbundles $L \subset E$ is bounded above.* □

Proof. Since $H^0(L)$ is a subspace of $H^0(E)$, we have $h^0(L) \le h^0(E)$. Then by Theorem 9.20 either $\deg L \le 2g - 2$ or $\deg L = h^0(L) + g - 1 \le h^0(E) + g - 1$. □

Since the sequence (10.3) is exact, it follows that the alternating sum of the dimensions of its terms is zero (Exercise 10.3). Thus

$$h^0(E) - h^1(E) = \left(h^0(F) - h^1(F)\right) + \left(h^0(G) - h^1(G)\right).$$

In other words, $h^0 - h^1$ is additive on exact sequences.

Riemann-Roch Formula 10.10 (Weak form). *If E is a vector bundle of rank r on a curve C of genus g, then*

$$h^0(E) - h^1(E) = \deg E - r(g - 1).$$

Proof. Let $RR(E) = h^0(E) - h^1(E) - \deg E + r(g - 1)$. We shall show that $RR(E) = 0$ by induction on r; in the line bundle case $r = 1$ we have already seen this in (9.15). In general, we can construct an exact sequence (10.1), and this sequence satisfies $RR(E) = RR(F) + RR(G)$. But $RR(F) = RR(G) = 0$ by the inductive hypothesis.   □

In the same spirit, we can derive duality for vector bundles. This says that the cup product defines an isomorphism $H^1(E) \xrightarrow{\sim} H^0(E^\vee \otimes \Omega_C)^\vee$, proved again by induction on the rank (see Exercise 10.4):

Theorem 10.11. *For any vector bundle E on a curve C the cup product*

$$H^1(E) \times H^0(E^\vee \otimes \Omega_C) \to H^1(\Omega_C) \cong k,$$

where Ω_C is the canonical line bundle, is a nondegenerate pairing.   □

Finally, given two vector bundles E, F, we can define an elementary sheaf $\mathcal{H}om(E, F)$ of local homomorphisms from E to F by

$$U \mapsto \operatorname{Hom}_{\mathcal{O}(U)}(E(U), F(U)).$$

This has total set $\operatorname{Hom}_{k(C)}(E_{\text{gen}}, F_{\text{gen}})$ and is isomorphic to $E^\vee \otimes F$. It has stalk $\operatorname{Hom}_{\mathcal{O}_{C,p}}(E_p, F_p)$ at $p \in C$, and its space of global sections is the vector space $\operatorname{Hom}(E, F)$ of (global) homomorphisms from E to F. In the case $E = F$ we write

$$\mathcal{E}nd\ E = \mathcal{H}om(E, E), \quad \mathfrak{sl}\ E := \ker\ \{\text{tr}\ : \mathcal{E}nd\ E \to \mathcal{O}_C\}.$$

Both of these bundles have degree zero, and so by Theorem 10.10 they satisfy

$$\begin{aligned}
h^0(\mathcal{E}nd\ E) - h^1(\mathcal{E}nd\ E) &= r^2(g - 1), \\
h^0(\mathfrak{sl}\ E) - h^1(\mathfrak{sl}\ E) &= (r^2 - 1)(g - 1).
\end{aligned} \tag{10.4}$$

(c) Indecomposable bundles and stable bundles

Definition 10.12. A vector bundle E is *decomposable* if it is isomorphic to the direct sum $E_1 \oplus E_2$ of two nonzero vector bundles; otherwise, E is *indecomposable*. □

Example 10.13. Every line bundle is indecomposable. □

If E admits an *idempotent*, that is, an endomorphism $f \in \text{End } E$ satisfying $f^2 = f$, then $E = \ker f \oplus \ker (1 - f)$; and conversely, if $E = E_1 \oplus E_2$, then $f = (0, 1)$ is such an idempotent. Thus decomposability is equivalent to the existence of an idempotent not equal to 0 or 1 (the identity endomorphism) on E. An arbitrary vector bundle is isomorphic to a direct sum of indecomposable bundles, and this decomposition is unique in the following sense.

Theorem 10.14 (Atiyah [70]). *If a vector bundle E has two direct sum decompositions into indecomposable bundles, $E = E_1 \oplus \cdots \oplus E_m = F_1 \oplus \cdots \oplus F_n$, then $m = n$ and $E_1, \ldots, E_m$ are isomorphic to $F_1, \ldots, F_m$ after reordering suitably.* □

The completeness of the curve C is essential in this theorem, as Exercise 10.5 shows. We prove the theorem first for the case of a rank 2 bundle. Suppose that

$$E = L_1 \oplus L_2 = M_1 \oplus M_2,$$

where L_1, L_2 and M_1, M_2 are line bundles. We have to show that either $L_1 \cong M_1$, $L_2 \cong M_2$ or $L_1 \cong M_2$, $L_2 \cong M_1$. For each $M = M_1$ or M_2 we have homomorphisms

$$i : M \to E, \qquad j : E \to M,$$

where i is injective, j is surjective and $j \circ i = \text{id}_M$. In terms of the decomposition $E = L_1 \oplus L_2$, we can write $i = (i_1, i_2)$, where $i_1 : M \to L_1$, $i_2 : M \to L_2$ and $j = j_1 + j_2$, where $j_1 : L_1 \to M$, $j_2 : L_2 \to M$. These maps then satisfy

$$j_1 \circ i_1 + j_2 \circ i_2 = \text{id}_M.$$

An endomorphism of the line bundle M is just multiplication by a scalar (i.e. an element of k) and so at least one of $j_1 \circ i_1$ or $j_2 \circ i_2$ must be an isomorphism – suppose $j_1 \circ i_1$. But then M is a direct summand of L_1, and since L_1 is irreducible, we conclude that $M = L_1$. □

The general case of Theorem 10.14 follows from:

Proposition 10.15. *The following conditions on a vector bundle E are equivalent.*

(1) E is indecomposable.
(2) If f_1, $f_2 \in$ End E and $f_1 + f_2$ is an isomorphism, then one of f_1 or f_2 is an isomorphism.   □

Remark 10.16. Condition (2) is equivalent to saying that End E is a local ring: that the set of noninvertible elements is the maximal ideal.   □

Given an endomorphism $f : E \to E$, consider the determinant $\det f$: $\det E \to \det E$. This is just multiplication by a scalar because $\det E$ is a line bundle, and this scalar is nonzero if and only if f is an isomorphism. Now, for an arbitrary scalar λ consider $\det(f - \lambda\text{id})$. This is a polynomial of degree $r(E)$ in λ and is the characteristic polynomial of the endomorphism f. In particular, if α is an eigenvalue, then $f - \alpha \cdot \text{id}$ fails to be an isomorphism.

Lemma 10.17. *If E is indecomposable, then $f \in$ End E has only one eigenvalue.*   □

Proof. Suppose f has distinct eigenvalues α, β. Then its characteristic polynomial can be expressed as a product of two polynomials without common factors:

$$\det(f - \lambda \cdot \text{id}) = p(\lambda)q(\lambda), \qquad p(\alpha) = q(\beta) = 0.$$

There exist polynomials $a(\lambda)$, $b(\lambda)$ satisfying

$$p(\lambda)a(\lambda) + q(\lambda)b(\lambda) = 1,$$

so that the endomorphism $h = p(f)a(f) \in$ End E satisfies $h(1 - h) = 0$ by the Cayley-Hamilton Theorem. This implies that E is the direct sum $\ker h \oplus \ker(1 - h)$, and since h and $1 - h$ are both nonzero, we conclude that E is decomposable.   □

Proof of Proposition 10.15. (1) $\implies$ (2) Suppose that E is indecomposable and that $f = f_1 + f_2$ is an isomorphism but f_1 is not an isomorphism. Then $f_1 f^{-1}$ fails to be an isomorphism. By the lemma, its only eigenvalue is 0, so it is nilpotent. But then $f_2 f^{-1} = 1 - f_1 f^{-1}$ is an isomorphism, and this implies that f_2 is an isomorphism.

(2) $\implies$ (1) If E is decomposable, then there exists a nonzero, nonidentity idempotent $f = f^2$. Then neither of $f, 1 - f$ is an isomorphism, although the sum is an isomorphism. $\qquad \square$

Definition 10.18. A vector bundle E is *simple* if its only endomorphisms are scalars, End $E = k$. $\qquad \square$

A simple vector bundle is necessarily indecomposable, though the converse is not true, as we will see in the next section (Proposition 10.45). But (10.4) implies the following, which is relevant to the moduli theory for vector bundles (see Exercise 10.1).

Lemma 10.19. *If E is indecomposable, then*

$$h^1(\mathcal{E}nd \ E) = r^2(g-1) + 1,$$
$$h^1(\mathfrak{sl} \ E) = (r^2 - 1)(g - 1).$$

$\qquad \square$

Definition 10.20. A vector bundle E is *stable* (or *semistable*, respectively) if every vector subbundle $F \subset E$ satisfies

$$\frac{\deg F}{\operatorname{rank} F} < \frac{\deg E}{\operatorname{rank} E} \qquad (\text{or} \leq \text{respectively}).$$

The ratio $\mu(E) := \deg E/\operatorname{rank} E$ is called the *slope* of E, and the stability of E can be expressed as $\mu(F) < \mu(E)$ for all subbundles $F \subset E$. To avoid confusion with other notions of stability we shall sometimes refer to this property as the *slope stability*. $\qquad \square$

Remark 10.21. Note that when $\deg E$ and $\operatorname{rank} E$ are coprime, stability and semistability are equivalent. $\qquad \square$

The following lemma follows from Exercise 10.1.

Lemma 10.22. *Let L be a line bundle on C. Then a vector bundle E is (semi)stable if and only if $L \otimes E$ is (semi)stable.* $\qquad \square$

The stability of E can also be expressed by saying that $\mu(G) > \mu(E)$ for every quotient bundle $G = E/F$ of E. Passing to the dual bundle, quotients become subbundles and we see:

Lemma 10.23. *A vector bundle E is (semi)stable if and only if its dual bundle $E^\vee$ is (semi)stable.* $\square$

The next fact will be important in the moduli theory.

Proposition 10.24. *Let E, E' be semistable vector bundles of the same rank and degree, and suppose that one of them is stable. Then every nonzero homomorphism between E and E' is an isomorphism.*

Proof. Let r and d be the common rank and degree of the two bundles, and let $f : E \to E'$ be a homomorphism with image $F \subset E'$. Since E, E' are semistable, we have

$$\frac{d}{r} \le \mu(F) \le \frac{d}{r},$$

and so $\mu(F) = d/r$. If rank $F < r$, then this contradicts the stability of E or E', and hence rank $F = r$. In particular, this means that $f_{\text{gen}} : E_{\text{gen}} \to E'_{\text{gen}}$ is an isomorphism of vector spaces over $k(C)$, and so $\det f_{\text{gen}}$ is also an isomorphism. This implies that $\det f : \det E \to \det E'$ is injective, and since $\deg E = \deg E'$, it follows that $\det f$ is an isomorphism. Hence f is an isomorphism by Proposition 10.6. $\square$

Corollary 10.25. *Every stable vector bundle is simple.*

Proof. An endomorphism $f \in \text{End } E$ induces, at each point $p \in C$, an endomorphism of the fibre $E/E(-p) \cong k^{\oplus r}$. Let $\alpha \in k$ be an eigenvalue of this map, and consider $f - \alpha \cdot \text{id} \in \text{End } E$. This is not an isomorphism, so by Proposition 10.24 it is zero. $\square$

Recall that if L is a line bundle with $\deg L > 2g - 2$, then $H^1(L) = 0$ (Theorem 9.20). For general vector bundles this kind of vanishing condition on cohomology does not hold; however, for semistable bundles one can show something similar.

Proposition 10.26. *If E is a semistable vector bundle with $\mu(E) > 2g - 2$, or if E is stable and $\mu(E) \geq 2g - 2$, then $H^1(E) = 0$.*

Proof. By hypothesis, every quotient line bundle of E has degree greater than $2g - 2$. On the other hand, the canonical line bundle Ω_C has degree equal to $2g - 2$, and so there is no nonzero homomorphism $E \to \Omega_C$. Hence $H^1(E) = 0$ by Theorem 10.11. $\square$

Proposition 9.38 also generalises to semistable vector bundles:

Proposition 10.27. *If E is semistable and $\mu(E) > 2g - 1$, or if E is stable and $\mu(E) \geq 2g - 1$, then E is generated by global sections.*

Proof. By the previous proposition, $H^1(E(-p)) = 0$ for every point $p \in C$. It follows that, for every positive divisor $D \geq 0$, the restricted principal part map

$$H^0(E(D - p)) \to E(D - p)/E(-p)$$

is surjective. In particular, taking $D = p$ shows that the evaluation map $H^0(E) \to E/E(-p)$ is surjective at every point $p \in C$. $\square$

(d) Grothendieck's Theorem

Grothendieck's Theorem gives a complete classification of vector bundles on the projective line $\mathbb{P}^1$. First consider, on any curve, a short exact sequence of vector bundles

$$\mathbb{E}: \quad 0 \to M \to E \to L \to 0. \tag{10.5}$$

Tensoring with the dual bundle $L^\vee$ gives a short exact sequence

$$0 \to \mathcal{H}om(L, M) \to \mathcal{H}om(L, E) \to \mathcal{E}nd(L) \to 0, \tag{10.6}$$

and its associated long exact sequence of vector spaces

$$\begin{aligned} 0 &\longrightarrow \mathrm{Hom}(L, M) \longrightarrow \mathrm{Hom}(L, E) \longrightarrow \mathrm{End}(L) \\ &\xrightarrow{\delta} H^1(\mathcal{H}om(L, M)) \longrightarrow H^1(\mathcal{H}om(E, M)) \longrightarrow H^1(\mathcal{E}nd(L)) \longrightarrow 0. \end{aligned}$$

Definition 10.28. The image under the coboundary map δ of $\mathrm{id} \in \mathrm{End}(L)$, which we will denote by

$$\delta(\mathbb{E}) \in H^1(\mathcal{H}om(L, M)),$$

is called the *extension class* of the exact sequence (10.5). $\square$

By construction, if $\delta(\mathbb{E}) = 0$, then there exists a homomorphism $f : L \to E$ for which the composition $L \xrightarrow{f} E \to L$ (where the second map is the surjection of (10.5)) is the identity endomorphism of L. In other words, the sequence (10.5) splits. In particular, we see:

Proposition 10.29. *If* $H^1(\mathcal{H}om(L, M)) = 0$, *then every exact sequence (10.5) splits.*

We will apply this fact to $\mathbb{P}^1$. Since the genus is zero, the cohomology of a line bundle L is particularly simple. Namely,

$$H^0(L) = 0 \text{ if } \deg L \leq -1,$$
$$H^1(L) = 0 \text{ if } \deg L \geq -1,$$

while, by Riemann-Roch,

$$h^0(L) - h^1(L) = \deg L + 1.$$

Lemma 10.30. *Every rank 2 vector bundle on* $\mathbb{P}^1$ *is isomorphic to a direct sum of two line bundles.*

Proof. Tensoring with a line bundle if necessary, it is enough to assume that $\deg E = 0$ or -1. First, by the Riemann-Roch Theorem 10.10 we note that $H^0(E) \neq 0$, and so E contains $\mathcal{O}_C$ as a subsheaf. This saturates to a line subbundle $M \subset E$, and $M \cong \mathcal{O}_C(D)$ for some positive divisor $D \geq 0$. In particular, $\deg M \geq 0$, and denoting the quotient by $L = E/M$ we have an exact sequence

$$0 \to M \to E \to L \to 0.$$

However, $\deg L^{-1} \otimes M = -\deg E + 2 \deg M \geq -1$, so that $H^1(L^{-1} \otimes M) = 0$. By Proposition 10.29, therefore, the sequence splits.   □

Grothendieck's Theorem 10.31. *Every vector bundle on* $\mathbb{P}^1$ *is isomorphic to a direct sum of line bundles.*

Proof. We prove this by induction on the rank $r \geq 2$ of E, starting with the previous lemma. By Corollary 10.9 there exists a line subbundle $M \subset E$ whose degree $m = \deg M$ is maximal among line subbundles of E.

Claim: Every line subbundle $L \subset F := E/M$ has $\deg L \leq m$.

Consider the preimage $\widetilde{L} \subset E$ of L under the projection $E \to F$. This is a rank 2 vector bundle, and $\deg \widetilde{L} = m + \deg L$. By Lemma 10.30, it contains a line subbundle of degree at least $\deg \widetilde{L}/2$, so that, by the way M was chosen, we have $(m + \deg L)/2 \leq \deg L$. The claim follows from this.

By the inductive hypothesis, and the claim, the quotient bundle F is isomorphic to a direct sum $L_1 \oplus \cdots \oplus L_{r-1}$ of line bundles of degrees $\deg L_i \leq m$. Since $H^1(L_i^{-1} \otimes M) = 0$ for each i, it follows that the exact sequence

$$0 \to M \to E \to F = \bigoplus_{i=1}^{r-1} L_i \to 0$$

splits. $\qquad\qquad\qquad\qquad\qquad\qquad\qquad\qquad\qquad\qquad\qquad\qquad\qquad\qquad$ □

(e) Extensions of vector bundles

Given vector bundles L and M, we are going to classify bundles E having M as a subbundle with quotient L. First of all, let us make precise the meaning of the classification problem.

Definition 10.32.

(i) A short exact sequence (10.5)

$$\mathbb{E}: \quad 0 \to M \to E \to L \to 0$$

is called an *extension of L by M*.

(ii) Two extensions $\mathbb{E}$ and $\mathbb{E}'$ are *equivalent* if there exists an isomorphism of vector bundles $f : E \overset{\sim}{\to} E'$ and a commutative diagram:

$$\begin{array}{ccccccccc}
0 & \longrightarrow & M & \longrightarrow & E & \longrightarrow & L & \longrightarrow & 0 \\
& & \| & & f \downarrow & & \| & & \\
0 & \longrightarrow & M & \longrightarrow & E' & \longrightarrow & L & \longrightarrow & 0
\end{array}$$

$\qquad\qquad\qquad\qquad\qquad\qquad\qquad\qquad\qquad\qquad\qquad\qquad\qquad\qquad$ □

Note that the extension class $\delta(\mathbb{E})$ defined in Definition 10.28 depends only on the equivalence class of the extension in this sense. The main result of this section is the following.

Theorem 10.33. *The assignment*

$$\left\{\begin{matrix} \text{extensions} \\ \text{of } L \text{ by } M \end{matrix}\right\} /\text{equivalence} \longrightarrow H^1(\mathcal{H}om(L, M))$$

given by $\mathbb{E} \mapsto \delta(\mathbb{E})$ (Definition 10.28) *is a bijection.*

We will prove this in the case when L is a line bundle. Tensoring with L^{-1} transforms the exact sequence of Definition 10.32(i) to

$$0 \to L^{-1} \otimes M \to E \to \mathcal{O}_C \to 0,$$

and it is therefore enough to consider the case $L = \mathcal{O}_C$:

$$\mathbb{E}: \quad 0 \to M \to E \to \mathcal{O}_C \to 0. \tag{10.7}$$

The coboundary map in the induced long exact cohomology sequence is

$$\delta : H^0(\mathcal{O}_C) \to H^1(M),$$

and the extension class $\delta(\mathbb{E}) \in H^1(M)$ is the image under this map of the constant section $1 \in H^0(\mathcal{O}_C)$.

First of all, let us follow carefully the construction of Lemma 8.57 which defines the coboundary map. We choose a rational section $s \in E_{\text{gen}}$ mapping to the constant section $1 \in H^0(\mathcal{O}_C)$. The principal part $(s \bmod E_p)_{p \in C}$ can then be viewed as belonging to the principal part space of M; let us denote this by $\sigma \in \bigoplus_{p \in C} M_{\text{gen}}/M_p$. The extension class $\delta(\mathbb{E})$ is then $\sigma \bmod M_{\text{gen}} \in H^1(M)$. We are going to prove Theorem 10.33 by actually giving a finer classification using σ and not just its cohomology class.

Definition 10.34.

(i) A *framed extension* is a pair $(\mathbb{E}, s)$ consisting of an extension $\mathbb{E}$ as in Definition 10.32(i) and a splitting $s : L_{\text{gen}} \to E_{\text{gen}}$ of the exact sequence of vector spaces

$$0 \to M_{\text{gen}} \to E_{\text{gen}} \to L_{\text{gen}} \to 0.$$

(ii) Two framed extensions $(\mathbb{E}, s)$ and $(\mathbb{E}', s')$ are *equivalent* if there exists an isomorphism of extensions $f : E \to E'$ such that the diagram

$$E_{\mathrm{gen}} \xleftarrow{\ s\ } L_{\mathrm{gen}}$$

$$f_{\mathrm{gen}} \downarrow \qquad \|$$

$$E'_{\mathrm{gen}} \xleftarrow{\ s'\ } L_{\mathrm{gen}}$$

commutes with the diagram in Definition 10.32(ii). □

In the case when $L = \mathcal{O}_C$, a framed extension determines naturally an element $\sigma = \sigma(\mathbb{E}, s) \in \bigoplus_{p \in C} M_{\mathrm{gen}}/M_p$, and again this depends only on the equivalence class of the framed extension.

Proposition 10.35. *The map*

$$\left\{ \begin{array}{l} \text{framed extensions} \\ \text{of } \mathcal{O}_C \text{ by } M \end{array} \right\} \Big/ \text{equivalence} \longrightarrow \bigoplus_{p \in C} M_{\mathrm{gen}}/M_p$$

given by $(\mathbb{E}, s) \mapsto \sigma(\mathbb{E}, s)$ *is a bijection.*

If we fix an extension of $\mathcal{O}_C$ by the M, then the choice of a framing is up to an element of M_{gen}. Replacing s by $s + m$ for some rational section $m \in M_{\mathrm{gen}}$ has the effect of adding the principal part of m (at each point of C) to $\sigma(\mathbb{E}, s)$. The cohomology class of $\sigma(\mathbb{E}, s)$ does not, therefore, depend on s, and this cohomology class is precisely $\delta(\mathbb{E}) \in H^1(M)$. It follows that in the case $L = \mathcal{O}_C$, Theorem 10.33 follows from Proposition 10.35.

Proof of Proposition 10.35. We will construct an inverse map. As a simplest case let us construct an extension (10.7) starting from a point $p \in C$ and a rational section $s \in M_{\mathrm{gen}}$. To do this we first take an affine open neighbourhood $U \subset C$ of the point, chosen small enough that s is regular on $U - \{p\}$. Here and below, when working on an affine variety we will always denote a module and its corresponding sheaf of modules by the same symbol. We let

$$M|_U \rtimes_{(p,s)} \mathcal{O}_U \subset M_{\mathrm{gen}} \oplus k(U) = M_{\mathrm{gen}} \oplus k(C)$$

be the $\mathcal{O}_U$-submodule generated by $M|_U \oplus 0$ and $(s, 1)$:

$$M|_U \rtimes_{(p,s)} \mathcal{O}_U = \{(m + fs, f) \mid m \in M|_U, \ f \in \mathcal{O}_U\}.$$

This submodule depends only on the principal part $(s \bmod M_p) \in M_{\text{gen}}/M_p$ of s at p. Clearly:

(1) $M|_U \rtimes_{(p,s)} \mathcal{O}_U$ contains as a submodule $M|_U (= M|_U \oplus 0)$, with quotient module isomorphic to $\mathcal{O}_U$; and

(2) the two submodules

$$M|_U \rtimes_{(p,s)} \mathcal{O}_U, \quad M|_U \oplus \mathcal{O}_U \subset M_{\text{gen}} \oplus k(C)$$

are equal on $U - \{p\}$.

Consequently, we obtain a vector bundle on C by gluing along $U - \{p\}$ the two bundles:

(a) $M|_U \rtimes_{(p,s)} \mathcal{O}_U$ on U;
(b) $M \oplus \mathcal{O}_C$ on $C - \{p\}$.

We can denote the resulting bundle by $M \rtimes_{(p,s)} \mathcal{O}_C$.

More generally, given a collection of points $p_1, \ldots, p_n \in C$ and rational sections $s_1, \ldots, s_n \in M_{\text{gen}}$, choose affine neighbourhoods $p_i \in U_i \subset C$ so that each s_i is regular on $U_i - \{p_i\}$. Then, just as above, we can construct a vector bundle on C which restricts to:

(a) $M|_{U_i} \rtimes_{(p_i, s_i)} \mathcal{O}_{U_i}$ on each U_i;
(b) $M \oplus \mathcal{O}_C$ on $C - \{p_1, \ldots, p_n\}$.

This vector bundle depends only on the principal part

$$\sigma = (s_i)_{1 \leq i \leq n} \in \bigoplus_{i=1}^{n} M_{\text{gen}}/M_{p_i},$$

and we will denote it by $M \rtimes_\sigma \mathcal{O}_C$. By construction there is an exact sequence

$$0 \to M \to M \rtimes_\sigma \mathcal{O}_C \to \mathcal{O}_C \to 0,$$

and the assignment $\sigma \mapsto M \rtimes_\sigma \mathcal{O}_C$ is inverse to that of Proposition 10.35. $\qquad \square$

It remains to prove Theorem 10.33 for a general cokernel line bundle L. The exact sequence (10.6) in Section 10.1(d) determines at the generic point an exact sequence of vector spaces over $k(C)$,

$$0 \to \text{Hom}(L_{\text{gen}}, M_{\text{gen}}) \to \text{Hom}(L_{\text{gen}}, E_{\text{gen}}) \to \text{End } L_{\text{gen}} \to 0.$$

Choosing a lift of the identity $1 \in \text{End } L_{\text{gen}}$ to $\text{Hom}(L_{\text{gen}}, E_{\text{gen}})$ is equivalent to giving a framing of the extension $\mathbb{E}$, and in just the same way as for

Proposition 10.35 one can prove:

Proposition 10.36. *The map*

$$\begin{Bmatrix} \text{framed extensions} \\ \text{of } L \text{ by } M \end{Bmatrix} /\text{equivalence} \longrightarrow \bigoplus_{p \in C} \frac{\text{Hom}(L_{\text{gen}}, M_{\text{gen}})}{\text{Hom}(L_p, M_p)}$$

given by $(\mathbb{E}, s) \mapsto \delta(\mathbb{E}, s)$ *is a bijection.* $\qquad\square$

Dividing out this bijection by the action of $\text{Hom}(L_{\text{gen}}, M_{\text{gen}})$ now yields Theorem 10.33.

10.2 Rank 2 vector bundles

We are now going to look at vector bundles of rank 2 in more detail.

(a) Maximal line subbundles

Given a vector bundle E, we have seen (Corollary 10.9) that the degrees of its line subbundles are bounded above. Let us look more closely at this in the case when E has rank 2. We suppose that E is an extension

$$0 \to L \to E \to M \to 0,$$

where L, M are line bundles, and that $N \subset E$ is a line subbundle. If N is a subsheaf of L, then $N = L$; if not, then the composition

$$N \hookrightarrow E \to M$$

is nonzero, and so N is isomorphic to a subsheaf of M. This shows:

Lemma 10.37. *If E is an extension of rank 2 as above and $N \subset E$ is a line subbundle, then either $N = L$ or $\deg N \leq \deg M$. In particular, every line subbundle satisfies $\deg N \leq \max\{\deg L, \deg M\}$.* $\qquad\square$

If E is not semistable, then it has a line subbundle of degree strictly greater than $\deg E/2$. Such a line bundle is called a *destabilising line subbundle*, and the lemma implies the following.

Proposition 10.38. *A rank 2 vector bundle E has at most one destabilising line subbundle.* $\qquad\square$

If E is indecomposable or simple, then the next two lemmas give upper bounds for the degree of its line subbundles in terms of the degree of E.

Lemma 10.39. *If E is a simple vector bundle of rank 2, then every line subbundle $L \subset E$ satisfies*

$$2 \deg L \le \deg E + g - 2.$$

Proof. Let M be the quotient line bundle E/L. The simplicity of E implies that $H^0(M^{-1} \otimes L) = \mathrm{Hom}(M, L) = 0$. On the other hand, indecomposability and Proposition 10.29 imply that $H^1(M^{-1} \otimes L) \ne 0$. Applying Riemann-Roch to the line bundle $M^{-1} \otimes L$, therefore, we get

$$-1 \ge h^0(M^{-1} \otimes L) - h^1(M^{-1} \otimes L) = -\deg M + \deg L + 1 - g.$$

The inequality in the lemma follows from this and the relation $\deg E = \deg L + \deg M$. $\qquad\square$

The following is proved in a similar manner, and we omit the details.

Lemma 10.40. *If E is an indecomposable vector bundle of rank 2, then every line subbundle $L \subset E$ satisfies*

$$2 \deg L \le \deg E + 2g - 2.$$

$\qquad\square$

(b) Nonstable vector bundles

Let us summarise some of the conclusions of the previous sections:

$$\text{indecomposable} \Longleftarrow \text{simple} \Longleftarrow \text{stable} \Longrightarrow \text{semistable}.$$

The reverse implications, however, do not hold in general. Propositions 10.38 and 10.45 below show that simple $\not\Longrightarrow$ stable and indecomposable $\not\Longrightarrow$ simple, respectively.

Proposition 10.41. *Let L, M be line bundles satisfying $\deg L > \deg M$ and $\mathrm{Hom}(M, L) = 0$. Then every nonsplit extension of M by L is simple.*

Proof. Suppose that E is such an extension

$$0 \to L \to E \to M \to 0$$

and that f is an endomorphism of E. By hypothesis, L is a destabilising subbundle of E, so that by Proposition 10.38 the image $f(L)$ is either L or 0. If it is 0, then f factors through the quotient M:

$$f : E \to M \to E.$$

But since the extension is nonsplit, $\mathrm{Hom}(M, E) = \mathrm{Hom}(M, L) = 0$, and so $f = 0$.

Suppose, on the other hand, that $f(L) = L$. Then the restriction $f|_L : L \to L$ is multiplication by some constant $a \in k$. This implies that the endomorphism $f - a \cdot \mathrm{id}_E$ is zero on L, so, by the first part of the proof, it follows that $f = a \cdot \mathrm{id}_E$. □

Example 10.42. Let C be a curve of genus $g \geq 3$. For each $1 \leq d \leq g - 2$ there exists a line bundle $\xi \in \mathrm{Pic}^d C$ with $H^0(\xi) = 0$. By the Riemann-Roch Theorem, $H^1(\xi) \neq 0$, and so there exists a nontrivial extension

$$0 \to \xi \to E \to \mathcal{O}_C \to 0,$$

which is unstable but simple. □

We next consider the case when E contains a line subbundle of degree exactly $\deg E/2$. Letting $M = E/L$, we have an extension

$$0 \to L \to E \to M \to 0, \qquad \deg L = \deg M. \tag{10.8}$$

In this case E is semistable but not stable.

Definition 10.43. Given a bundle E as in (10.8) which is semistable but not stable, we let $\mathrm{gr}(E) = L \oplus M$. If E is stable, then $\mathrm{gr}(E) = E$. □

Proposition 10.44. *The direct sum* $\mathrm{gr}(E)$ *depends only on the vector bundle E and not on the choice of extension (10.8).*

Proof. Suppose that $L' \subset E$ is another line subbundle of degree $\deg E/2$ and consider the composition $L' \hookrightarrow E \to M$. If this is zero, then $L' = L$; otherwise, $L' \cong M$. On the other hand, $E/L' \cong \det E \otimes L'^{-1}$, and so the bundle $\mathrm{gr}(E)$ obtained from $L' \subset E$ is isomorphic to $L \oplus M$. □

Next, we construct indecomposable bundles which are not simple.

Proposition 10.45. *Let* $0 \to L \to E \to M \to 0$ *be a nonsplit extension with* $\mathrm{Hom}(M, L) \neq 0$. *Then E is indecomposable but not simple.*

Proof. It is enough to assume that $L = \mathcal{O}_C$. The exact sequnce can then be written as

$$0 \to \mathcal{O}_C \to E \to \mathcal{O}_C(-D) \to 0, \qquad (10.9)$$

where D is some positive divisor. First note that $H^0(E)$ is 1-dimensional. If ϕ is an endomorphism of E, then we will denote by $H^0(\phi)$ the induced linear automorphism of $H^0(E)$.

Suppose that $H^0(\phi) = 0$. Then ϕ maps the line subbundle $\mathcal{O}_C$ to zero and therefore factors through the quotient $\mathcal{O}_C(-D)$:

$$\phi : E \to \mathcal{O}_C(-D) \to E.$$

Since the sequence (10.9) is nonsplit, it follows that the composition of ϕ with the surjection $E \to \mathcal{O}_C(-D)$ is zero. In other words, the image of ϕ is contained in the line subbundle $\mathcal{O}_C \subset E$, so that ϕ s induced by an element of $\mathrm{Hom}(\mathcal{O}_C(-D), \mathcal{O}_C)$:

$$\phi : E \to \mathcal{O}_C(-D) \to \mathcal{O}_C \to E.$$

If, on the other hand, $H^0(\phi) \neq 0$, then it is multiplication by a constant $a \in k$. Then, by considering $\phi - a \cdot \mathrm{id}_E$ we reduce to the previous case, and this shows that

$$\mathrm{End}\, E = k \oplus \mathrm{Hom}(\mathcal{O}_C(-D), \mathcal{O}_C).$$

Hence E is not simple, and E is indecomposable by Proposition 10.15   □

Remarks 10.46.

(i) Every rank 2 vector bundle which is indecomposable but not simple is described by the construction of this proposition.

(ii) There exist exact sequences

$$0 \to L \to E \to M \to 0$$

which are nonsplit but in which the vector bundle E is decomposable. For example, on the projective line $\mathbb{P} = \mathbb{P}^1$ with homogeneous coordinates $(x : y)$:

$$0 \longrightarrow \mathcal{O}_{\mathbb{P}}(-1) \longrightarrow \mathcal{O}_{\mathbb{P}} \oplus \mathcal{O}_{\mathbb{P}} \longrightarrow \mathcal{O}_{\mathbb{P}}(1) \longrightarrow 0$$
$$c \quad \mapsto \quad (cy, -cx)$$
$$(a, b \quad \mapsto \quad ax + by$$

This sequence is nonsplit by Theorem 10.14.   □

(c) Vector bundles on an elliptic curve

We now suppose that C has genus 1. By the Riemann-Roch Theorem every line bundle L on C satisfies

$$h^0(L) - h^1(L) = \deg L.$$

Moreover:

(i) if $\deg L > 0$, then $H^1(L) = 0$,
(ii) if $\deg L < 0$, then $H^0(L) = 0$,
(iii) if $\deg L = 0$ and $L \not\cong \mathcal{O}_C$, then $H^0(L) = H^1(L) = 0$.

We will give a complete classification of all indecomposable rank 2 vector bundles on the elliptic cure C. We consider first the case of odd degree.

Proposition 10.47. *On a curve of genus 1, given a line bundle L of odd degree, there exists, up to isomorphism, a unique indecomposable rank 2 vector bundle E with $\det E \cong L$.*

Proof. It is enough to consider the case $\deg L = 1$. Since $H^1(L^{-1})$ is 1-dimensional, there is, up to isomorphism, just one nonsplit exact sequence

$$0 \to \mathcal{O}_C \to E \to L \to 0.$$

Claim: $h^0(E) = 1$.

Since $H^0(L) \neq 0$, it follows that L contains $\mathcal{O}_C$ as a subsheaf. Let $E' \subset E$ be the inverse image of this subsheaf. The dual $L^{-1} \hookrightarrow \mathcal{O}_C$ of the inclusion $\mathcal{O}_C \hookrightarrow L$ induces an injective map

$$H^1(L^{-1}) \hookrightarrow H^1(\mathcal{O}_C),$$

and it follows from this that E' is a nonsplit extension

$$0 \to \mathcal{O}_C \to E' \to \mathcal{O}_C \to 0.$$

Hence $h^0(E') = 1$ and, since $h^0(L) = 1$, this shows that $h^0(E) = 1$ too.

The indecomposability of the vector bundle E can now be proved by the same reasoning as in the proof of Proposition 10.45. We have therefore proved the existence part of the proposition, and it remains to show uniqueness. Fixing L of degree 1 and E with $\det E \cong L$, we have $H^0(E) \neq 0$ by Riemann-Roch, so that E contains $\mathcal{O}_C$ as a subsheaf. But applying Lemma 10.40 to the saturation of this subsheaf shows that $\mathcal{O}_C$ must be a line subbundle, and so E is precisely the bundle constructed above.   □

In fact, the vector bundle constructed in Proposition 10.47 is also stable, and therefore simple.

We next consider the case of even degree.

Proposition 10.48. *On a curve of genus 1 every indecomposable rank 2 vector bundle of even degree is an extension of the form*

$$0 \to M \to E \to M \to 0$$

for some line bundle M on C.

Proof. It will be sufficient to consider the case $\deg E = 2$. By Riemann-Roch, $h^0(E) \geq 2$, so that we can use two linearly independent sections $s, t \in H^0(E)$ to construct a homomorphism

$$f : \mathcal{O}_C \oplus \mathcal{O}_C \to E, \qquad (a, b) \mapsto as + bt.$$

Claim: f is injective.

Suppose not. Then the image is a subsheaf of rank 1, and we denote its saturation by $L \subset E$. Then $\deg L \geq 2$ because $h^0(L) \geq 2$. But this contradicts Lemma 10.40.

Since $\deg E = 2$, the homomorphism f cannot be surjective. In other words, there exists some point $p \in C$ at which the induced map $k \oplus k \to E/E(-p)$ fails to be an isomorphism. This means that E contains $\mathcal{O}_C(p)$ as a subsheaf and, by Lemma 10.40, as a line subbundle. The quotient line bundle $M = E/\mathcal{O}_C(p)$ then has degree 1. But indecomposability of E implies that $H^1(M^{-1}(p)) \neq 0$, and hence M is isomorphic to $\mathcal{O}_C(p)$, proving the proposition.   □

Putting these results together we obtain:

Proposition 10.49 (Atiyah [71])**.** *Let E be a rank 2 vector bundle over a curve C of genus 1, with determinant line bundle $\det E = L$.*

(i) *If $\deg L$ is odd, then E indecomposable $\Longleftrightarrow$ E simple $\Longleftrightarrow$ E stable $\Longleftrightarrow$ E semistable. Moreover, $SU_C(2, L)$ is a single point.*

(ii) *If $\deg L$ is even, then are no simple bundles (and therefore no stable bundles), but E indecomposable $\Longrightarrow$ E semistable.*   □

10.3 Stable bundles and Pfaffian semiinvariants

Under the action of the algebraic group $GL(N, N)$ on the affine space of square matrices $\text{Mat}_N(k)$ the determinant function is a semiinvariant, and in the last chapter we used this fact to show stability under the group action on matrices $\Psi(\xi, S, T)$ (in $\text{Mat}_N(H^0(L))$ for a fixed line bundle L) representing line bundles $\xi \in \text{Pic}^d C$. In this section, in the same spirit, we are going to study the semistability of *Gieseker points* associated to rank 2 vector bundles. For this, the central notion, with which we will build our semiinvariants, is that of the Pfaffian of a skew-symmetric matrix, and we begin with a discussion of Pfaffians.

(a) Skew-symmetric matrices and Pfaffians

Let $\text{Alt}_N(k)$ be the vector space of skew-symmetric $N \times N$ matrices over k. A matrix $A = (a_{ij})_{1 \leq i,j \leq N}$ in $\text{Alt}_N(k)$ has zeros on the main diagonal and is determined by the entries above the diagonal, and we will adopt the following notation:

$$A = \begin{bmatrix} a_{12} & a_{13} & \cdots & a_{1,N-1} & a_{1,N} \\ & a_{23} & \cdots & a_{2,N-1} & a_{2,N} \\ & & \ddots & \vdots & \vdots \\ & & & a_{N-2,N-1} & a_{N-2,N} \\ & & & & a_{N-1,N} \end{bmatrix}.$$

Thus, for example, an element of $\text{Alt}_2(k)$ is written $[a]$.

Such a matrix determines an element of degree 2 in the exterior algebra $\bigwedge\langle e_1, \ldots, e_N \rangle$, which we will denote by

$$\sigma_A := \sum_{1 \leq i < j \leq N} a_{ij} e_i \wedge e_j \in \overset{2}{\bigwedge}\langle e_1, \ldots, e_N \rangle,$$

and this correspondence defines an isomorphism $\text{Alt}_N(k) \overset{\sim}{\to} \bigwedge^2 \langle e_1, \ldots, e_N \rangle$.

The general linear group $GL(N)$ acts on $\text{Alt}_N(k)$ by

$$A \mapsto XAX^t, \qquad A \in \text{Alt}_N(k), \quad X \in GL(N).$$

Under this action the rank of A is invariant; more precisely, the matrices $A \in \text{Alt}_N(k)$ of any constant rank make up a single $GL(N)$-orbit. The following fact is fundamental here.

Proposition 10.50. *Every skew-symmetric matrix has even rank.* $\qquad\square$

Because of this, the properties of the group action $GL(N) \curvearrowright \mathrm{Alt}_N(k)$ depend in an essential way on whether N is even or odd.

Even skew-symmetric matrices Let $N \in \mathbb{N}$ be an even number.

Definition 10.51. The *Pfaffian* of a skew-symmetric matrix $A \in \mathrm{Alt}_N(k)$ is the number Pfaff $A \in k$ defined by

$$\sigma_A^{N/2} = \sigma_A \wedge \ldots \wedge \sigma_A = (N/2)!(\mathrm{Pfaff}\ A)e_1 \wedge \ldots \wedge e_N \in \bigwedge^N \langle e_1, \ldots, e_N \rangle \cong k.$$

$\square$

The Pfaffian is a homogeneous polynomial of degree $N/2$ in the entries a_{ij} of A and can be written

$$\mathrm{Pfaff}\ A = \frac{1}{N!!} \sum_{f \in \Sigma_N} \mathrm{sgn}(f) a_{f(1)f(2)} a_{f(3)f(4)} \cdots a_{f(N-1)f(N)},$$

where Σ_N is the symmetric group of permutations of N letters and $N!!$ denotes the 'subfactorial'

$$N!! = N(N-2)(N-4)\cdots 6 \cdot 4 \cdot 2, \qquad (N-1)!! = (N-1)(N-3)\cdots 5 \cdot 3 \cdot 1.$$

Note that in the case $N = 2$ this is simply

$$\mathrm{Pfaff}\ [a] = a.$$

For $N > 2$, the Pfaffian can be evaluated by expansion in a similar manner to the determinant: for $i < j$, let A_{ij} denote the $(N-2) \times (N-2)$ skew-symmetric submatrix obtained from A by deleting the i-th and j-th rows and columns. Then

$$\mathrm{Pfaff}\ A = \sum_{j=2}^{N} (-1)^j a_{1j} \mathrm{Pfaff}\ A_{1j}. \qquad (10.10)$$

As a polynomial, Pfaff A is a sum of $(N-1)!!$ monomials with coefficients ± 1.

Example 10.52. In the case $N = 4$,

$$\mathrm{Pfaff} \begin{bmatrix} a_{12} & a_{13} & a_{14} \\ & a_{23} & a_{24} \\ & & a_{34} \end{bmatrix} = a_{12}a_{34} - a_{13}a_{24} + a_{14}a_{23}.$$

In the case $N = 6$, one sees:

$$\text{Pfaff} \begin{bmatrix} a_{12} & a_{13} & a_{14} & a_{15} & a_{16} \\ & a_{23} & a_{24} & a_{25} & a_{26} \\ & & a_{34} & a_{35} & a_{36} \\ & & & a_{45} & a_{46} \\ & & & & a_{56} \end{bmatrix}$$

$$= a_{12}\text{Pfaff} \begin{bmatrix} a_{34} & a_{35} & a_{36} \\ & a_{45} & a_{46} \\ & & a_{56} \end{bmatrix}$$

$$-a_{13}\text{Pfaff} \begin{bmatrix} a_{24} & a_{25} & a_{26} \\ & a_{45} & a_{46} \\ & & a_{56} \end{bmatrix} + a_{14}\text{Pfaff} \begin{bmatrix} a_{23} & a_{25} & a_{26} \\ & a_{35} & a_{36} \\ & & a_{56} \end{bmatrix}$$

$$-a_{15}\text{Pfaff} \begin{bmatrix} a_{23} & a_{24} & a_{26} \\ & a_{34} & a_{36} \\ & & a_{46} \end{bmatrix} + a_{16}\text{Pfaff} \begin{bmatrix} a_{23} & a_{24} & a_{25} \\ & a_{34} & a_{35} \\ & & a_{45} \end{bmatrix}$$

$$= -\det \begin{vmatrix} a_{14} & a_{15} & a_{16} \\ a_{24} & a_{25} & a_{26} \\ a_{34} & a_{35} & a_{36} \end{vmatrix} + (a_{23}, -a_{13}, a_{12}) \begin{pmatrix} a_{14} & a_{15} & a_{16} \\ a_{24} & a_{25} & a_{26} \\ a_{34} & a_{35} & a_{36} \end{pmatrix} \begin{pmatrix} a_{56} \\ -a_{46} \\ a_{45} \end{pmatrix}.$$

$\square$

The following facts are easy to check.

Proposition 10.53.

(i) The Pfaffian is a square root of the determinant:

$$\det A = (\text{Pfaff } A)^2.$$

In particular, Pfaff $A \neq 0$ if and only if A has rank N.

(ii) For any $N \times N$ matrix X we have

$$\text{Pfaff } (X A X^t) = (\det X)(\text{Pfaff } A).$$

Thus the Pfaffian is a semiinvariant of weight 1 (with respect to the character $\chi = \det$) for the action $GL(N) \curvearrowright \text{Alt}_N(k)$.

(iii) For any $B \in \mathrm{Mat}_{N/2}(k)$ and $C \in \mathrm{Alt}_{N/2}(k)$ we have

$$\mathrm{Pfaff} \begin{pmatrix} 0 & B \\ -B^t & C \end{pmatrix} = (-1)^{\frac{N}{2}+1} \det B.$$

$\square$

Odd skew-symmetric matrices Now let N be an odd number. The Pfaffian of $A \in \mathrm{Alt}_N(k)$ is no longer defined, but instead we can consider the Pfaffians of its diagonal $(N-1) \times (N-1)$ minors.

Definition 10.54. The *radical vector* of an odd skew-symmetric matrix $A \in \mathrm{Alt}_N(k)$ is the N-vector

$$\mathrm{rad}\, A = \left((-1)^{i-1} \mathrm{Pfaff}\, A_i \right)_{1 \leq i \leq N} = \begin{pmatrix} \mathrm{Pfaff}\, A_1 \\ -\mathrm{Pfaff}\, A_2 \\ \mathrm{Pfaff}\, A_3 \\ \vdots \\ (-1)^{N-1} \mathrm{Pfaff}\, A_N \end{pmatrix},$$

where A_i denotes the submatrix obtained from A by deleting the i-th row and column. $\square$

We can begin by noting the following formula, which will be needed in Chapter 12. (The case $N = 3$ follows from Example 10.52.)

Example 10.55. If A, A' are skew-symmetric $N \times N$ matrices, where N is odd, and B is any $N \times N$ matrix of rank ≤ 2, then

$$\mathrm{Pfaff} \begin{pmatrix} A & B \\ -B^t & A' \end{pmatrix} = (\mathrm{rad}\, A)^t B (\mathrm{rad}\, A').$$

$\square$

The radical vector is essentially the power

$$\sigma_A^{(N-1)/2} \in \bigwedge^{N-1} \langle e_1, \ldots, e_N \rangle$$

of $\sigma_A \in \bigwedge^2 \langle e_1, \ldots, e_N \rangle$. The following properties are easily verified.

Proposition 10.56. *Consider $A \in \mathrm{Alt}_N(k)$.*

(i) rank $A \leq n - 1$, *and* rank $A < n - 1$ *if and only if* rad $A = 0$.
(ii) $A \cdot \mathrm{rad}\, A = 0$.

(iii) If X is an N × N matrix and X* is its matrix of cofactors, then

$$\text{rad}\,(XAX^t) = X^{*,t}\text{rad}\,A.$$

$\square$

Although in this case the only semiinvariants of the action $GL(N) \curvearrowright \text{Alt}_N(k)$ are the constants, we can nevertheless do the following. Given three skew-symmetric matrices $A, B, C \in \text{Alt}_N(k)$, we can consider the scalar product

$$(\text{rad}\,A)^t B\,\text{rad}\,C. \tag{10.11}$$

Under the action of $X \in GL(N)$, this product transforms to

$$(\text{rad}\,A)^t X^*(XBX^t)(X^{*,t}\text{rad}\,C) = (\det X)^2(\text{rad}\,A)^t B\,\text{rad}\,C. \tag{10.12}$$

It follows that the expression (10.11) is a semiinvariant of weight 2 for the diagonal action of $GL(N)$ on the direct sum

$$\text{Alt}_N(k) \oplus \text{Alt}_N(k) \oplus \text{Alt}_N(k)$$

(or direct product, if we view $\text{Alt}_N(k)$ as an affine space).

Skew-symmetric matrices of rank 2   A skew-symmetric matrix $A \in \text{Alt}_N(k)$ has rank $A \leq 2$ if and only if all of its 4×4 minor Pfaffians vanish. Matrices with this property will play an important role in what follows; the following is the skew-symmetric analogue of Lemma 9.55.

Proposition 10.57. *Let K be a field* (or a local Artinian ring – this possibility will be needed in Section 10.4(a)) *and let A ∈* $\text{Alt}_N(K)$*. If A has rank 2, then there exists a 2 × N matrix*

$$W = \begin{pmatrix} a_1 & a_2 & \cdots & a_N \\ b_1 & b_2 & \cdots & b_N \end{pmatrix}$$

such that, for all $1 \leq i, j \leq N$*, the (i, j)-th entry of A is the (i, j)-th 2 × 2 minor of W. In other words,*

$$A = W^t \begin{pmatrix} 0 & 1 \\ -1 & 0 \end{pmatrix} W.$$

Moreover, the matrix W is unique up to the action of SL(2, K) on the left. $\square$

We omit the proof of this result. The matrix W represents a point of the Grassmannian $\mathbb{G}(2, N)$, and the entries of A are the Plücker coordinates of this point.

(b) Gieseker points

We are now going to establish a one-to-one correspondence between isomorphism classes of rank 2 vector bundles satisfying some appropriate conditions and $GL(N)$-orbits in an affine space $\mathrm{Alt}_N(V)$ of skew-symmetric matrices with entries in a suitable vector space V. The precise statement is Corollary 10.62, and this is the analogue of Proposition 9.58 in the line bundle case. This construction is the key to proving Theorem 10.1.

Notation 10.58. For the rest of this chapter we fix a line bundle $L \in \mathrm{Pic}\, C$ and consider rank 2 vector bundles E with $\det E = L$. By Riemann-Roch we have

$$h^0(E) - h^1(E) = \deg L + 2 - 2g =: N.$$

The natural number N will always take this value.

A set $S = \{s_1, \ldots, s_N\} \subset H^0(E)$ of N linearly independent global sections is called a *marking* of the vector bundle E, and the pair (E, S) is called a *marked vector bundle*.   □

In the line bundle case of the previous chapter (see Section 9.4(a)) we needed the key properties that:

(i) $H^1(E) = 0$.
(ii) E is generated by global sections.

These properties were guaranteed by taking large enough degree. For rank $E \geq 2$, this is no longer quite the case, and we will usually impose conditions (i) and (ii) as additional hypotheses – though note that they are satisfied by *semistable* bundles of sufficiently high degree, by Propositions 10.26 and 10.27. When they are satisfied we have $N = h^0(E)$, and a marking S is a basis of $H^0(E)$. Moreover, generation by global sections means that the homomorphism

$$(s_1, \ldots, s_N) : \mathcal{O}_C^{\oplus N} \to E, \qquad (f_1, \ldots, f_N) \mapsto \sum_{i=1}^{N} f_i s_i, \qquad (10.13)$$

is surjective. At the same time, there is a homomorphism

$$(s_1\wedge, \ldots, s_N\wedge) : E \to (\det E)^{\oplus N}, \qquad t \mapsto (s_1 \wedge t, \ldots, s_N \wedge t) \qquad (10.14)$$

which, if E is generated by global sections, is injective. To explain this, recall that the stalk at the generic point E_{gen} is a 2-dimensional vector space over the function field $k(C)$, so there is a skew-symmmetric bilinear form

$$\wedge : E_{\mathrm{gen}} \times E_{\mathrm{gen}} \to \det E_{\mathrm{gen}} \cong k(C). \qquad (10.15)$$

Thus $s \wedge s = 0$ and $s \wedge s' + s' \wedge s = 0$ for $s, s' \in E_{\text{gen}}$. Moreover, if s, s' are global sections of E, then $s \wedge s'$ is a global section of det E, and so restriction of (10.15) defines a skew-symmetric k-bilinear map

$$H^0(E) \times H^0(E) \to H^0(\det E), \qquad (s, s') \mapsto s \wedge s'.$$

The bilinear form (10.15) induces an isomorphism $E_{\text{gen}} \xrightarrow{\sim} \text{Hom}(E_{\text{gen}}, \det E_{\text{gen}})$. In particular, each global section $s \in H^0(E)$ determines a homomorphism

$$s \wedge : E \to \det E, \qquad t \mapsto s \wedge t.$$

Definition 10.59 (Gieseker [44]). Given a vector space V, we will denote by $\text{Alt}_N(V)$ the set of skew-symmetric $N \times N$ matrices whose entries belong to V. Given a marked vector bundle (E, S) with det $E = L$, the skew-symmetric matrix

$$T_{E,S} = \begin{pmatrix} s_1 \\ \vdots \\ s_N \end{pmatrix} \wedge (s_1, \ldots, s_N)$$

$$= \begin{bmatrix} s_1 \wedge s_2 & s_1 \wedge s_3 & \cdots & s_1 \wedge s_N \\ & s_2 \wedge s_3 & \cdots & s_2 \wedge s_N \\ & & \ddots & \vdots \\ & & & s_{N-1} \wedge s_N \end{bmatrix} \in \text{Alt}_N(H^0(L))$$

will be called the *Gieseker matrix*, or *Gieseker point*, of E corresponding to the marking S. □

Proposition 10.60. *Given $S = \{s_1, \ldots, s_N\} \subset H^0(E)$, the composition of (10.13) and (10.14)*

$$\mathcal{O}_C^{\oplus N} \xrightarrow{(s_1, \ldots, s_N)} E \xrightarrow{(s_1 \wedge, \ldots, s_N \wedge)} L^{\oplus N}$$

is given by the matrix $T_{E,S} \in \text{Alt}_N(H^0(L))$. □

Note that any matrix $T \in \text{Alt}_N(H^0(L))$ determines a vector bundle map

$$\langle T \rangle : \mathcal{O}_C^{\oplus N} \to L^{\oplus N},$$

and this is skew-symmetric in the sense that the dual map $\langle T \rangle^t : (L^{-1})^{\oplus N} \to \mathcal{O}_C^{\oplus N}$, after tensoring with L, is equal to $-\langle T \rangle$. Moreover, because of Proposition 10.60, when T is a Gieseker matrix of a bundle E the image sheaf of $\langle T \rangle$

is nothing other than E itself:

Proposition 10.61. *Suppose that $H^1(E) = 0$ and that E is generated by global sections. Then, for any marking S, E is isomorphic to the image of the homomorphism*

$$\langle T_{E,S} \rangle : \mathcal{O}_C^{\oplus N} \to L^{\oplus N}$$

defined by its Gieseker point. □

We now consider the action $GL(N) \curvearrowright \mathrm{Alt}_N(H^0(L))$ given by

$$T \mapsto XTX^t, \qquad T \in \mathrm{Alt}_N(H^0(L)), \quad X \in GL(N),$$

where we view $\mathrm{Alt}_N(H^0(L))$ as an affine space $\mathbb{A}^n$, where $n = h^0(L) \times N(N - 1)/2$. This action is of ray type. If we assume $H^1(E) = 0$, so that the marking S is a basis of $H^0(E)$, then the $GL(N)$-orbit of its Gieseker points depends only on the isomorphism class of E and not on the choice of S. Conversely, Proposition 10.61 guarantees that the vector bundle E can be recovered from any Gieseker point, and hence:

Corollary 10.62. *The mapping* (in the setting of Notation 10.58)

$$\left\{ \begin{array}{l} \text{isomorphism classes of} \\ \text{vector bundles } E \text{ with } H^1(E) = 0 \\ \text{and generated by global sections} \end{array} \right\} \longrightarrow \left\{ \begin{array}{l} GL(N)\text{-orbits} \\ \text{in } \mathrm{Alt}_N(H^0(L)) \end{array} \right\}$$

sending E to the orbit of its Gieseker points $T_{E,S}$ is injective. □

Example 10.63. Suppose that $E = \xi \oplus \widehat{\xi}$ is a direct sum of line bundles, and that $S \subset H^0(E) = H^0(\xi) \oplus H^0(\widehat{\xi})$ is the union of sets $\{s_1, \ldots, s_m\} \subset H^0(\xi)$ and $\{t_1, \ldots, t_n\} \subset H^0(\widehat{\xi})$. Then $L = \xi \otimes \widehat{\xi}$ and

$$T_{E,S} = \begin{pmatrix} 0 & \Psi \\ -\Psi^t & 0 \end{pmatrix}, \qquad \text{where } \Psi = \begin{pmatrix} s_1 t_1 & \cdots & s_1 t_m \\ & \cdots & \\ s_n t_1 & \cdots & s_n t_m \end{pmatrix}.$$

In this case, everything we are going to do for rank 2 bundles reduces to the constructions of the preceding chapter for the action $GL(N, N) \curvearrowright \mathrm{Mat}_{N,1}(H^0(L))$. □

We next ask for the stabilisers of these points.

Lemma 10.64. *Suppose that $H^1(E) = 0$, that E is generated by global sections and that E is simple. Given a marking S and a matrix $X \in GL(N)$,*

$$XT_{E,S}X^t = T_{E,S} \quad \text{if and only if} \quad X = \pm I_N.$$

Proof. The hypothesis $XT_{E,S}X^t = T_{E,S}$ is equivalent to the commutativity of the diagram:

$$\mathcal{O}_C^{\oplus N} \xrightarrow{\langle T_{E,S} \rangle} L^{\oplus N}$$

$$X^t \downarrow \qquad \uparrow X$$

$$\mathcal{O}_C^{\oplus N} \xrightarrow{\langle T_{E,S} \rangle} L^{\oplus N}$$

This diagram determines an endomorphism ϕ of E, and the assumption that E is simple implies that $\phi = c\,\mathrm{id}_E$ for some $c \in k$. But then $X = X^t = c \cdot I_N$, and in particular, $c^2 = 1$. □

Remark 10.65. Note that $-I_N \in GL(N)$ acts trivially on the whole space $\mathrm{Alt}_N(H^0(L))$. □

(c) Semistability of Gieseker points

We now need to consider the question of (semi)stability of a point $T \in \mathrm{Alt}_N(H^0(L))$ under the action of $GL(N)$, with respect to the determinant character $g \mapsto \det g$. We will show that if E is a rank 2 vector bundle with $H^1(E) = 0$ and $\deg E \geq 4g - 2$, then the Gieseker points $T_{E,S}$ are semistable if and only if E is slope-semistable as a vector bundle. (Conversely, we will see that if $\deg L \geq 4g - 2$, then every semistable $T \in \mathrm{Alt}_N(H^0(L))$ is a Gieseker point of a semistable vector bundle – this is Proposition 10.81 below.)

A semiinvariant of weight w is a polynomial function $F = F(T) \in k[\mathrm{Alt}_N(H^0(L))]$ with the property

$$F(g \cdot T) = (\det g)^w F(T), \quad \text{for all } g \in GL(N),$$

and the unstable set in $\mathrm{Alt}_N(H^0(L))$ is the common zero-set of all semiinvariants of positive weight. Recall, moreover, that a point T is unstable if and only if the closure of its $SL(N)$-orbit contains the origin. A 'Gieseker point' $\Psi(\xi, S, T)$ of a line bundle ξ is always stable (Proposition 9.62). However, for vector bundles this is no longer the case. For rank greater than 1 the following phenomenon appears.

Proposition 10.66. *Let S be a marking and $M \subset E$ a line subbundle of the vector bundle E, and consider the vector subspaces $\langle S \rangle \subset H^0(E)$ (of dimension N) and $H^0(M) \subset H^0(E)$.*

(i) If there exists $M \subset E$ such that

$$\dim \left(H^0(M) \cap \langle S \rangle \right) > \frac{N}{2},$$

then the Gieseker point $T_{E,S} \in \mathrm{Alt}_N(H^0(L))$ is unstable.
(ii) If there exists $M \subset E$ such that

$$\dim \left(H^0(M) \cap \langle S \rangle \right) \geq \frac{N}{2},$$

then $T_{E,S} \in \mathrm{Alt}_N(H^0(L))$ fails to be stable.

Proof. Let $a = \dim H^0(M) \cap \langle S \rangle$ and $b = N - a$. Since the question is independent of the choice of Gieseker point within its $GL(N)$-orbit, it likewise depends only on the linear span $\langle S \rangle$ and not on S itself. We may therefore assume S chosen so that its first a vectors belong to $H^0(M) \cap \langle S \rangle$. The skew-symmetric matrix $T_{E,S}$ will then have a block decomposition in which the top left-hand $a \times a$ block contains only zeros:

$$T_{E,S} = \left(\begin{array}{c|c} 0 & B \\ \hline -B^t & C \end{array} \right).$$

We now consider the 1-parameter subgroup

$$t \mapsto g(t) = \left(\begin{array}{ccc|ccc} t^{-b} & & & & & \\ & \ddots & & & 0 & \\ & & t^{-b} & & & \\ \hline & & & t^a & & \\ & 0 & & & \ddots & \\ & & & & & t^a \end{array} \right) \in SL(N).$$

This acts by

$$T_{E,S} \mapsto g(t)T_{E,S}g(t)^t = \left(\begin{array}{c|c} 0 & t^{a-b}B \\ \hline -t^{a-b}B^t & t^{2a}C \end{array} \right).$$

In case (i) we are assuming that $a > b$. So letting $t \to 0$ shows that $0 \in \overline{SL(N) \cdot T_{E,S}}$, so the Gieseker point is unstable.

For case (ii), assume that $a = b$. In this case the limit as $t \to 0$ is the matrix

$$T_0 = \left(\begin{array}{c|c} 0 & B \\ \hline -B^t & 0 \end{array}\right).$$

Either $C = 0$, so that $T_{E,S} = T_0$ already has a positive-dimensional stabiliser, or else $C \neq 0$ but the orbit of $T_{E,S}$ contains the nonstable point T_0 in its closure and therefore fails to be closed. In either case, $T_{E,S}$ fails to be stable. □

On account of this phenomenon we make the following definition.

Definition 10.67. Let E be a rank 2 vector bundle. If

$$h^0(M) \leq \tfrac{1}{2} h^0(E) \quad (\text{resp. } <) \quad \text{for every line subbundle } M \subset E,$$

then we say that E is H^0-*semistable* (resp. H^0-*stable*). □

If $H^1(E) = 0$, then in Proposition 10.66 we have $N = h^0(E)$ and $\langle S \rangle = H^0(E)$. The proposition therefore says:

Corollary 10.68. *Suppose that $H^1(E) = 0$ and let $T = T_{E,S}$ be any Gieseker point of E. Then:*

(i) if T is $GL(N)$-semistable, then E is H^0-semistable;
(ii) if T is $GL(N)$-stable, then E is H^0-stable. □

The relationship with slope-semistability is given by the following (but see also Exercise 9.8):

Proposition 10.69. *Suppose that $H^1(E) = 0$ and $\deg E \geq 4g - 2$. Then E is H^0-semistable if and only if it is slope-semistable.*

Proof. First observe that by Riemann-Roch any line bundle M satisfies

$$h^0(M) - h^1(M) - \frac{h^0(E) - h^1(E)}{2} = \deg M - \frac{\deg E}{2}.$$

Since $H^1(E) = 0$, this implies

$$\frac{h^0(E)}{2} - h^0(M) \leq \left(\frac{h^0(E)}{2} - h^0(M)\right) + h^1(M) = \frac{\deg E}{2} - \deg M. \quad (10.16)$$

Letting M run through the line subbundles of E, this shows at once that H^0-semistability of E implies slope-semistability.

For the converse, suppose that there exists a line subbundle $M \subset E$ for which the left-hand side of (10.16) is negative. Note that, by hypothesis,

$$h^0(E) = \deg E + 2 - 2g \geq 2g,$$

and therefore $h^0(M) > h^0(E)/2 \geq g$. By Lemma 9.22 this implies that $H^1(M) = 0$, so equality holds in (10.16). Hence the right-hand side of (10.16) is negative. $\square$

We next show the converse of Corollary 10.68(i). (We can also show the converse of part (ii) if we use the Hilbert-Mumford numerical criterion. For this, see Section 10.4(c).)

Proposition 10.70. *Suppose that $H^1(E) = 0$. Then, if the vector bundle E is H^0-semistable, then its Gieseker points $T_{E,s} \in \mathrm{Alt}_N(H^0(L))$ are semistable for the action of $GL(N)$.*

The proof of this will occupy the remainder of this section. As preparation, we investigate some elementary properties of H^0-semistability.

A quotient line bundle $Q = E/M$ of an H^0-semistable vector bundle E satisfies

$$h^0(Q) \geq h^0(E) - h^0(M) \geq \tfrac{1}{2}h^0(E).$$

Lemma 10.71. *If E is H^0-semistable and $h^0(E) \geq 2$, then E is generated by global sections at a general point $p \in C$. In particular, $h^0(E(-p)) = h^0(E) - 2$ at the general point.*

Proof. Consider the evaluation homomorphism

$$H^0(E) \otimes \mathcal{O}_C \to E.$$

The image sheaf has $h^0(E)$ linearly independent sections; if it had rank 1, then its saturation would be a line bundle violating H^0-semistability. So the image has rank 2. $\square$

The following is the technical key to proving Proposition 10.70.

Lemma 10.72. *If E is H^0-semistable and $h^0(E) \geq 4$, then there exists a point $p \in C$ for which the bundle $E(-p)$ is H^0-semistable.*

Proof (Raynaud [75]). Let $h^0(E) = n$. At a general point $p \in C$ we have $h^0(E(-p)) = n - 2$ by Lemma 10.71. We suppose that at every point the bundle $E(-p)$ is H^0-unstable and therefore contains some line subbundle, which we will denote by $M^p(-p) \subset E(-p)$, with

$$h^0(M^p(-p)) > \frac{n}{2} - 1.$$

Claim: The line subbundle $M^p \subset E$ is independent of the choice of the general point $p \in C$.

Granted the claim, we have a line subbundle $M (= M^p) \subset E$ which satisfies

$$h^0(M(-p)) > \frac{n}{2} - 1$$

at a general point of the curve. But this implies $h^0(M) > n/2$, contradicting the H^0-semistability of E, and we are done.

To prove the claim, let $q \in C$ be another, distinct, point. We first consider the case $n \geq 5$. Then

$$h^0(E(-p)) = n - 2 > \frac{n}{2},$$

and this implies that $E(-p)$ is generically generated by global sections (otherwise we get a line subbundle of $E(-p) \subset E$ violating the H^0-semistability of E). Hence $h^0(E(-p-q)) = n - 4$. On the other hand,

$$h^0(M^p(-p-q)) + h^0(M^q(-p-q)) = h^0(M^p(-p)) - 1 + h^0(M^q(-q))$$
$$-1 > n - 4.$$

This implies that

$$0 \neq H^0(M^p(-p-q)) \cap H^0(M^q(-p-q)) \subset H^0(E(-p-q)),$$

and hence the line subbundles $M^p(-p-q)), M^q(-p-q) \subset E(-p-q)$ coincide. Hence $M^p = M^q$.

Now consider the case $n = 4$. We now have $h^0(E(-p)) = 2$ and $h^0(M^p(-p)) \geq 2$, and so $H^0(E(-p)) = H^0(M^p(-p))$. In particular,

$$H^0(E(-p-q)) = H^0(M^p(-p-q)) \cong k.$$

Similarly

$$H^0(E(-p-q)) = H^0(M^q(-p-q)) \cong k.$$

So again the two line subbundles $M^p(-p-q))$, $M^q(-p-q) \subset E(-p-q)$ have a common global section, and they therefore coincide.   □

Proof of Proposition 10.70. To show semistability of a Gieseker point $T_{E,S}$ we have to exhibit a semiinvariant of positive weight which is nonzero at $T_{E,S}$. We consider separately the cases when N is even or odd. (Note that $N \equiv \deg L$ mod 2.)

When N is even we can construct semiinvariants as follows. For any linear form $f : H^0(L) \rightarrow k$ we can evaluate f on the entries of a matrix $T \in \mathrm{Alt}_N(H^0(L))$ to obtain a skew-symmetric matrix $f(T) \in \mathrm{Alt}_N(k)$. Then, by Proposition 10.53(ii), the function

$$\mathrm{Alt}_N(H^0(L)) \rightarrow k, \qquad T \mapsto \mathrm{Pfaff}\, f(T)$$

is a semiinvariant of weight 1.

By repeated use of Lemmas 10.71 and 10.72 we can find points $p_1, \ldots, p_{N/2} \in C$ such that

$$H^0(E(-p_1 - \cdots - p_{N/2})) = 0. \tag{10.17}$$

If we let $\mathrm{ev}_i = \mathrm{ev}_{p_i} : H^0(L) \rightarrow k$ be the evaluation map at the i-th point, then (10.17) says that the linear map of N-dimensional vector spaces

$$g := (\mathrm{ev}_1, \ldots, \mathrm{ev}_{N/2}) : H^0(E) \rightarrow \bigoplus_{i=1}^{N/2} E/E(-p_i)$$

is an isomorphism. Now consider the skew-symmetric pairing

$$H^0(E) \times H^0(E) \xrightarrow{\wedge} H^0(L) \xrightarrow{f} k,$$

where $f := \mathrm{ev}_1 + \cdots + \mathrm{ev}_{N/2} : H^0(L) \rightarrow k$. This pairing has matrix $f(T_{E,S})$ and transforms via the isomorphism g to a skew-pairing $k^N \times k^N \rightarrow k$ with matrix $\begin{pmatrix} 0 & I_{N/2} \\ -I_{N/2} & 0 \end{pmatrix}$. In other words, there is a commutative diagram:

$$H^0(E) \times H^0(E) \xrightarrow{\wedge} H^0(L)$$

$$g \times g \downarrow \qquad\qquad \downarrow f \tag{10.18}$$

$$k^N \times k^N \longrightarrow k$$

Using Proposition 10.53(ii), it follows that $\mathrm{Pfaff}\, f(T_{E,S})$ is equal to $\det g \neq 0$. Hence the Gieseker point $T_{E,S} \in \mathrm{Alt}_N(H^0(L))$ is semistable.   □

We turn now to the case when N is odd. In this case the strategy for producing semiinvariants, using Proposition 10.56 and the remarks following, is to use triples of linear forms $f, f', h : H^0(L) \to k$. From these and from $T \in \mathrm{Alt}_N(H^0(L))$ we get vectors rad $f(T)$, rad $f'(T) \in k^N$ and a skew-symmetric matrix $h(T) \in \mathrm{Alt}_N(k)$. We then form the scalar product

$$\mathrm{Alt}_N(H^0(L)) \to k, \qquad T \mapsto (\mathrm{rad}\ f(T))^t h(T)\mathrm{rad}\ f'(T).$$

By (10.12), this is a semiinvariant of weight 2.

Remark 10.73. Since minus the identity $-I_N \in GL(N)$ acts trivially on $\mathrm{Alt}_N(H^0(L))$, it follows that if F is a semiinvariant of weight w, then $F = \det(-I_N)^w F$. When N is odd this implies that there are no nonzero semiinvariants of odd weight. $\qquad\square$

We now choose N points $p_1, \dots, p_N \in C$, and we let $\mathrm{ev}_i : H^0(L) \to k$ be the i-th evaluation map, as before. We set

$$f = \mathrm{ev}_1 + \cdots + \mathrm{ev}_{\frac{N-1}{2}}, \quad f' = \mathrm{ev}_{\frac{N+1}{2}} + \cdots + \mathrm{ev}_{N-1}, \quad h = \mathrm{ev}_N.$$

The proof of Proposition 10.70 is now completed by the following:

Lemma 10.74. *If E is H^0-semistable, then there exist points $p_1, \dots, p_N \in C$ such that, for any marking $S \subset H^0(E)$,*

$$(\mathrm{rad}\ f(T_{E,S}))^t h(T_{E,S})\mathrm{rad}\ f'(T_{E,S}) \neq 0,$$

where f, f', h are defined as above.

Proof. Let $n := (N-1)/2$. The function $f : H^0(L) \to k$ is the sum of the evaluation maps at the points $p_1, \dots, p_n \in C$, and by Proposition 10.56(i)

$$\mathrm{rad}\ f(T_{S,E}) \neq 0 \quad\Longleftrightarrow\quad h^0(E(-p_1 - \cdots - p_n)) = 1.$$

Moreover, via the diagram (10.18) one sees that if these equivalent conditions hold, then the vector rad $f(T_{S,E})$ spans the 1-dimensional space

$$H^0(E(-p_1 - \cdots - p_n)) = \ker\{(\mathrm{ev}_1, \dots, \mathrm{ev}_n) : H^0(E) \to k^{2n}\},$$

relative to the basis $S \subset H^0(E)$.

Now by repeated use of Lemma 10.72 we can find points $p_1, \dots, p_{n-1} \in C$ such that $E(-p_1 - \cdots - p_{n-1})$ is H^0-semistable and $h^0(E(-p_1 - \cdots - p_{n-1})) = 3$. We then pick two general points $p_n, p_{n+1} \in C$ and two global sections $s, t \in H^0(E(-p_1 - \cdots - p_{n-1}))$ such that $s(p_n) = t(p_{n+1}) = 0$. These sections are necessarily linearly independent and, by H^0-semistability, generate a subsheaf of rank 2. Thus if p_n is general, then the fibre at this point will be generated by global sections. Hence, with respect to the N points

$$p_1, \ldots, p_{n-1}, p_n; p_{n+1}, p_1, \ldots, p_{n-1}; p_n,$$

the scalar product of the lemma is nonzero.   □

10.4 An algebraic variety with $SU_C(2, L)$ as its set of points

Our aim is now to prove Theorem 10.1. For this we need to study the $GL(N)$-orbits in the affine space $\mathrm{Alt}_N(H^0(L))$ coming from vector bundles via Corollary 10.62.

By identifying $L \cong \mathcal{O}_C(D)$ for some divisor $D \in \mathrm{Div}\,C$ we can view elements $T \in \mathrm{Alt}_N(H^0(L))$ as skew-symmetric matrices with entries in the function field $k(C)$; we then observe that the Gieseker points $T_{E,S}$, as matrices over $k(C)$, have rank 2 (Proposition 10.60).

Definition 10.75. The set of matrices $T \in \mathrm{Alt}_N(H^0(L))$ of rank ≤ 2 over $k(C)$ is a closed subvariety which we denote by $\mathrm{Alt}_{N,2}(H^0(L)) \subset \mathrm{Alt}_N(H^0(L))$.   □

Let $x_{ij}^{(\alpha)}$, for $1 \leq i, j \leq N$ and $1 \leq \alpha \leq h^0(L)$, be coordinates in the affine space $\mathrm{Alt}_N(H^0(L))$. Then $\mathrm{Alt}_{N,2}(H^0(L)) \subset \mathrm{Alt}_N(H^0(L))$ is defined by $\binom{N}{4}h^0(L^2)$ equations determined by the vanishing of global sections,

$$\mathrm{Pfaff} \begin{bmatrix} x_{ij} & x_{ik} & x_{il} \\ & x_{jk} & x_{jl} \\ & & x_{kl} \end{bmatrix} = x_{ij} \circ x_{kl} - x_{ik} \circ x_{jl} + x_{il} \circ x_{jk} \in H^0(L^2),$$

for $1 \leq i < j < k < l \leq N$, and where $x_{ij} := (x_{ij}^{(\alpha)})_{1 \leq \alpha \leq h^0(L)} \in H^0(L)$ and $\circ : H^0(L) \times H^0(L) \to H^0(L^2)$ is the natural multiplication map.

If $T \neq 0$, then the rank condition is equivalent to saying that the image E of the sheaf homomorphism

$$\langle T \rangle : \mathcal{O}_C^{\oplus N} \to L^{\oplus N}$$

is a rank 2 vector bundle.

Remarks 10.76.

(i) E can also be described in the following way. The rank condition of Definition 10.75 together with Proposition 10.57 says that

$$T = W^t \begin{pmatrix} 0 & 1 \\ -1 & 0 \end{pmatrix} W$$

for some $2 \times N$ matrix

$$W = \begin{pmatrix} f_1 & f_2 & \cdots & f_N \\ g_1 & g_2 & \cdots & g_N \end{pmatrix}$$

of functions $f_i, g_j \in k(C)$. The N columns of W span a subsheaf of the constant sheaf $k(C) \oplus k(C)$, and this subsheaf is precisely E.

(ii) Alternatively, E is then the $\mathcal{O}_C$-module spanned by the two rows of W in $k(C)^{\oplus N}$. In other words, we are viewing E as a point in the Grassmannian $\mathbb{G}(2, N)$ over the function field $k(C)$, S as a choice of homogeneous coordinates in $\mathbb{P}_{k(C)}^{N-1}$, and the Gieseker point $T_{E,S}$ as the corresponding matrix of *Plücker coordinates* of E. $\square$

(a) Tangent vectors and smoothness

In this section we will prove the following.

Proposition 10.77. *Let E be a rank 2 vector bundle with $\det E = L$ and $H^1(E) = 0$. Then:*

(i) $\mathrm{Alt}_{N,2}(H^0(L))$ is smooth at each Gieseker point $T_{E,S}$.

(ii) If E is simple, then the quotient of the tangent space to $\mathrm{Alt}_{N,2}(H^0(L))$ at a Gieseker point $T_{E,S}$ by the Lie space $\mathfrak{gl}(N)$ is isomorphic to $H^1(\mathfrak{sl}\, E)$:

$$T_{T_{E,S}}\mathrm{Alt}_{N,2}(H^0(L))/\mathfrak{gl}(N) \cong H^1(\mathfrak{sl}\, E).$$

Given vector spaces U, V, the space $\mathrm{Hom}(U, V)$ of linear maps $f : U \to V$ can be viewed as an affine space. There is then, for each natural number r, a subset $\mathrm{Hom}_r(U, V) \subset \mathrm{Hom}(U, V)$ consisting of linear maps of rank $\leq r$ and defined as a closed subvariety by the vanishing of all the $(r + 1) \times (r + 1)$ minors.

Lemma 10.78. *Suppose that $f \in \mathrm{Hom}_r(U, V)$ has rank exactly equal to r. Then the tangent space to $\mathrm{Hom}_r(U, V)$ at f is equal to*

$$S_f := \{h \mid h(\ker f) \subset \mathrm{Im}\, f\} \subset \mathrm{Hom}(U, V).$$

Proof. Choose bases of U and V so that the matrix representing $f : U \to V$ is in canonical form $\mathrm{diag}(1, \ldots, 1, 0, \ldots, 0)$. If $h : U \to V$ is another linear map, then $f + \epsilon h$, where $\epsilon^2 = 0$, $\epsilon \neq 0$, is represented by a matrix

$$\left(\begin{array}{c|c} I_r & 0 \\ \hline 0 & 0 \end{array}\right) + \epsilon \left(\begin{array}{c|c} A & B \\ \hline C & D \end{array}\right).$$

Since $\epsilon^2 = 0$, the only possible nonzero $(r + 1) \times (r + 1)$ minors in this matrix are the entries of D (concatenated with the block I_r). Hence the condition that all $(r + 1) \times (r + 1)$ minors be zero is equivalent to $D = 0$. But this is the case if and only if $h(\ker f) \subset \operatorname{Im} f$. ☐

In the tangent vector space S_f there are two vector subspaces to consider. One consists of h satisfying $h(\ker f) = 0$, which is equivalent to factoring through an element of $\operatorname{Hom}(\operatorname{Im} f, V)$. The other consists of h satisfying $h(U) \subset \operatorname{Im} f$ or, in other words, h comes from an element of $\operatorname{Hom}(U, \operatorname{Im} f)$. The intersection consists of endomorphisms of $\operatorname{Im} f$, and in this way we obtain an exact sequence of vector spaces:

$$0 \to \operatorname{End}(\operatorname{Im} f) \to \operatorname{Hom}(\operatorname{Im} f, V) \oplus \operatorname{Hom}(U, \operatorname{Im} f) \to S_f \to 0. \qquad (10.19)$$

Now suppose that $V = U^\vee$, and consider the subset $\operatorname{Hom}^-(U, U^\vee)$ of skew-symmetric linear maps: those $f : U \to U^\vee$, that is, equal to minus their transpose (dual) map. Suppose that $f \in \operatorname{Hom}^-(U, U^\vee)$ has rank $\leq r$. This means that all its $(r + 2) \times (r + 2)$ Pfaffian minors vanish, and these Pfaffians define a closed subvariety $\operatorname{Hom}_r^-(U, U^\vee) \subset \operatorname{Hom}^-(U, U^\vee)$.

Lemma 10.79. *Suppose that $f : U \to U^\vee$ is skew-symmetric and has rank equal to r. Then the tangent space to $\operatorname{Hom}_r^-(U, U^\vee)$ at f is equal to*

$$S_f^- := \{h \mid h(\ker f) \subset \operatorname{Im} f\} \subset \operatorname{Hom}^-(U, U^\vee).$$

☐

We will skip the proof as it is exactly the same as that of the previous lemma, replacing determinants with Pfaffians.

The two subspaces $\{h \mid h(\ker f) = 0\}$ and $\{h \mid h(U) \subset \operatorname{Im} f\}$, when the maps f, h are skew-symmetric, are exchanged by taking the transpose; moreover, the intersection

$$\{h \mid h(\ker f) = 0\} \cap \{h \mid h(U) \subset \operatorname{Im} f\} \cap \operatorname{Hom}^-(U, U^\vee)$$

is exactly the space of endomorphisms of $\operatorname{Im} f$ which preserve a skew-symmetric form. We will denote this space by $\operatorname{End}^-(\operatorname{Im} f)$. In the case $r = 2$, for example, this is just the subspace $\mathfrak{sl}(\operatorname{Im} f) \subset \operatorname{End}(\operatorname{Im} f)$ of linear endomorphisms with trace zero. From (10.19) we obtain an exact sequence:

$$0 \to \operatorname{End}^-(\operatorname{Im} f) \to \operatorname{Hom}(U, \operatorname{Im} f) \to S_f^- \to 0. \qquad (10.20)$$

In order to prove Proposition 10.77, we are going to apply Lemma 10.79 over the function field $k(C)$. Before doing that we will use Lemma 10.78 to show again that the Picard variety constructed in the last chapter has tangent space $H^1(\mathcal{O}_C)$.

Second proof of Proposition 9.67. From a double marked line bundle (ξ, S, T) we have constructed a matrix

$$\Psi = \Psi(\xi, S, T) = \begin{pmatrix} s_1 t_N & \cdots & s_1 t_N \\ \vdots & & \vdots \\ s_N t_1 & \cdots & s_N t_N \end{pmatrix}.$$

This determines a sheaf homomorphism $\langle \Psi \rangle : \mathcal{O}_C^{\oplus N} \to L^{\oplus N}$ whose image is isomorphic to ξ, and we consider the subsheaf

$$\mathcal{S}_\Psi := \{h \mid h(\ker \langle \Psi \rangle) \subset \xi\} \subset \mathcal{H}om(\mathcal{O}_C^{\oplus N}, L^{\oplus N}) \cong L^{\oplus N^2}.$$

This is a subbundle, and we will apply Lemma 10.78 to the map on stalks at the generic point

$$\langle \Psi \rangle_{\text{gen}} : k(C)^{\oplus N} \to L_{\text{gen}}^{\oplus N}.$$

This says that the tangent space to $\text{Mat}_{N,1}(L_{\text{gen}})$ at $\langle \Psi \rangle_{\text{gen}}$ is the space of global sections $H^0(\mathcal{S}_\Psi)$. Corresponding to (10.19), there is an exact sequence of vector bundles on C

$$0 \to \mathcal{E}nd\, \xi \to \mathcal{H}om(\xi, L^{\oplus N}) \oplus \mathcal{H}om(\mathcal{O}_C^{\oplus N}, \xi) \to \mathcal{S}_\Psi \to 0,$$

and hence

$$0 \to \mathcal{O}_C \to \widehat{\xi}^{\oplus N} \oplus \xi^{\oplus N} \to \mathcal{S}_\Psi \to 0.$$

Now $H^1(\xi) = H^1(\widehat{\xi}) = 0$ by the hypotheses made at the beginning of Section 9.4(a), so that taking global sections gives an exact sequence:

$$0 \longrightarrow H^0(\mathcal{O}_C) \longrightarrow H^0(\xi) \oplus H^0(\widehat{\xi}) \longrightarrow H^0(\mathcal{S}_\Psi) \longrightarrow H^1(\mathcal{O}_C) \longrightarrow 0$$
$$\|$$
$$\mathfrak{gl}(N) \oplus \mathfrak{gl}(N)$$

This is exactly the sequence asserted in Proposition 9.67. $\qquad\square$

The vector bundle case is entirely similar.

Proof of Proposition 10.77(ii). Let E be a simple rank 2 vector bundle with Gieseker point $T = T_{E,S} \in \text{Alt}_{N,2}(H^0(L))$. We apply Lemma 10.79 to the map on stalks at the generic point coming from $\langle T \rangle : \mathcal{O}_C^{\oplus N} \to L^{\oplus N}$ (whose image is E). This determines a subbundle

$$\mathcal{S}_T^- := \{h \mid h(\ker \langle T \rangle) \subset E\} \subset \mathcal{H}om^-(\mathcal{O}_C^{\oplus N}, L^{\oplus N}) \cong L^{\oplus N(N-1)/2}.$$

The tangent space to $\mathrm{Alt}_{N,2}(L_{\mathrm{gen}})$ at $\langle T \rangle_{\mathrm{gen}}$ is then the space of rational sections of $\mathcal{S}_T^-$, and that of $\mathrm{Alt}_{N,2}(H^0(L))$ is $H^0(\mathcal{S}_T^-)$. Corresponding to (10.20), we have an exact sequence of vector bundles on C:

$$0 \to \mathfrak{sl}\ E \to \mathcal{H}om(\mathcal{O}_C^{\oplus N}, E) \to \mathcal{S}_T^- \to 0.$$

But $\mathcal{H}om(\mathcal{O}_C^{\oplus N}, E) \cong E^{\oplus N}$ while $H^1(E)$ by hypothesis, and so taking global sections gives:

$$0 \longrightarrow H^0(\mathfrak{sl}\ E) \longrightarrow \mathrm{Hom}(\mathcal{O}_C^{\oplus N}, E) \longrightarrow H^0(\mathcal{S}_T^-) \longrightarrow H^1(\mathfrak{sl}\ E) \longrightarrow 0$$
$$\|$$
$$\mathfrak{gl}(N)$$

Since E is simple, we have $\mathfrak{sl}\ E = 0$, while $\mathrm{Hom}(\mathcal{O}_C^{\oplus N}, E)$ is the tangent space at the Gieseker point of the $GL(N)$-orbit. $\qquad\square$

Proof of Proposition 10.77(i). We will use Proposition 9.47. Let $f : A' \to A$ be a surjective homomorphism of local Artinian rings, with maximal ideals $\mathfrak{n} \subset A$, $\mathfrak{n}' \subset A'$. Let $\mathcal{T} = (a_{ij})_{1 \le i,j \le N}$ be an A-valued point of $\mathrm{Alt}_{N,2}(H^0(L))$ whose reduction modulo $\mathfrak{n}$ is the Gieseker matrix T. We have to show that this lifts to an A'-valued point. It is enough to prove this for the case $\dim \ker f = 1$.

Let ϵ be a vector spanning $\ker f$. By Proposition 10.57, the matrix $\mathcal{T}$ can be expressed as

$$\mathcal{T} = \begin{bmatrix} s_1 \wedge s_2 & s_1 \wedge s_3 & \cdots & s_1 \wedge s_N \\ & s_2 \wedge s_3 & \cdots & s_2 \wedge s_N \\ & & \ddots & \vdots \\ & & & s_{N-1} \wedge s_N \end{bmatrix}$$

for some rational sections $s_i \in E_{\mathrm{gen}} \otimes_k A$. Since this is an A-valued point of $\mathrm{Alt}_{N,2}(H^0(L))$, the entries $a_{ij} := s_i \wedge s_j$ belong to $H^0(L \otimes_k A)$. Since f is surjective, we can lift each s_i to an element $s_i' \in E_{\mathrm{gen}} \otimes_k A'$ and each a_{ij} to an element $a_{ij}' \in H^0(L \otimes_k A')$, preserving the skew-symmetry. The matrix

$$\left(s_i' \wedge s_j' - a_{ij}'\right)_{1 \le i,j \le N} \tag{10.21}$$

then determines a rational section of $\mathcal{H}om^-(\mathcal{O}_C^{\oplus N}, L^{\oplus N}) \otimes_k A'$, and since this section vanishes when we apply f, it is in fact a rational section of

$$\mathcal{H}om^-(\mathcal{O}_C^{\oplus N}, L^{\oplus N}) \otimes_k \ker f = \mathcal{H}om^-(\mathcal{O}_C^{\oplus N}, L^{\oplus N})\epsilon.$$

Moreover, its principal part is $(s_i' \wedge s_j')_{1 \le i, j \le N}$ and is contained in $\mathcal{S}_T^- \otimes_k A'$. It follows that at each point $p \in C$ this matrix determines a principal part in the vector bundle $\mathcal{S}_T^- \otimes_k \ker f$. By hypothesis, $H^1(\mathcal{S}_T^-) = 0$, and so these principal parts come from a global rational section. In other words, there exist $s_1'', \ldots, s_N'' \in E_{\text{gen}}$ such that (10.21) is everywhere the principal part of

$$\left((\bar{s}_i + s_i''\epsilon) \wedge (\bar{s}_j + s_j''\epsilon)\right)_{1 \le i, j \le N} - (\bar{s}_i \wedge \bar{s}_j)_{1 \le i, j \le N},$$

where $\bar{s}_i$ is the reduction of s_i modulo $\mathfrak{n}$. Hence, if we set

$$T' = \left((s_i' + s_i''\epsilon) \wedge (s_j' + s_j''\epsilon)\right)_{1 \le i, j \le N},$$

then the entries of T' are everywhere regular and T' is an A'-valued point of $\text{Alt}_{N,2}(H^0(L))$ lifting T. $\qquad\square$

(b) Proof of Theorem 10.1

We now take our fixed line bundle L to have degree $\ge 4g - 1$, and we consider the action $GL(N) \curvearrowright \text{Alt}_{N,2}(H^0(L))$.

Suppose that $E \in SU_C(2, L)$. Then by Proposition 10.26 we have $H^1(E) = 0$, so the orbit $GL(N) \cdot T_{E,S}$ of a Gieseker point depends only on E and not on the marking S. By Proposition 10.27, moreover, E is generated by global sections and is therefore recovered up to isomorphism from its Gieseker points (Lemma 10.61). And by Propositions 10.69 and 10.70, the Gieseker points of E are semistable for the action of $GL(N)$.

Conversely, suppose that $T \in \text{Alt}_{N,2}(H^0(L))$ is a semistable point for the $GL(N)$ action. The columns of T are vectors in $H^0(L)^{\oplus N}$, and as for Proposition 9.63 in the line bundle case we can show the following.

Lemma 10.80. *If $T \in \text{Alt}_{N,2}(H^0(L))$ is semistable, then the N columns of T are linearly independent vectors in $H^0(L)^{\oplus N}$ over k.*

Proof. Suppose not. Then by a suitable change of basis (that is, by moving within the $GL(N)$-orbit) we can assume that the first row and column of

T are zero:

$$T = \begin{pmatrix} 0 & 0 & \cdots & 0 \\ 0 & * & \cdots & * \\ \vdots & & \ddots & \\ 0 & * & \cdots & * \end{pmatrix}.$$

The action of the 1-parameter subgroup

$$t \mapsto g(t) := \begin{pmatrix} t^{-N+1} & & & \\ & t & & \\ & & \ddots & \\ & & & t \end{pmatrix} \in SL(N)$$

maps T to

$$g(t)Tg(t)^t = \begin{pmatrix} 0 & 0 & \cdots & 0 \\ 0 & t^2* & \cdots & t^2* \\ \vdots & & \ddots & \\ 0 & t^2* & \cdots & t^2* \end{pmatrix},$$

and letting $t \to 0$ shows that the origin is in the closure of the $SL(N)$-orbit; so T is unstable. $\qquad\qquad\square$

Given a semistable point $T \in \mathrm{Alt}_{N,2}(H^0(L))$, let $E \subset L^{\oplus N}$ be the image of the homomorphism

$$\langle T \rangle : \mathcal{O}_C^{\oplus N} \to L^{\oplus N}.$$

Proposition 10.81. *Suppose that* $\deg L \geq 4g - 2$ *and that* $T \in \mathrm{Alt}_{N,2}(H^0(L))$ *is semistable for the action of* $GL(N)$. *Then* $E = \mathrm{Im}\,\langle T \rangle \subset L^{\oplus N}$ *satisfies:*

(i) $H^1(E) = 0$;
(ii) $\det E \cong L$;
(iii) E *is semistable.*

Proof.

(i) Let $V \subset H^0(E)$ be the space of global sections coming from the surjection $\mathcal{O}_C^{\oplus N} \to E$. Lemma 10.80 implies that $\dim V = N$, and, in particular, that $h^0(E) \geq N$. By Theorem 10.11, the vanishing of $H^1(E)$ implies that there is a nonzero homomorphism $f : E \to \Omega_C$, and this induces a linear map $V \to H^0(\Omega_C)$. The kernel of this map then has dimension at least

$N - g \geq g$, and so, letting $M := \ker f \subset E$, we have

$$\dim H^0(M) \cap V \geq \frac{N}{2}.$$

From (the proof of) Proposition 10.66, it follows that T is unstable.

(ii) Consider the bilinear pairing

$$\mathcal{O}_C^{\oplus N} \times \mathcal{O}_C^{\oplus N} \to L, \qquad (u, v) \mapsto u^t T v.$$

This is skew-symmetric and vanishes if u or $v \in \ker \langle T \rangle$, and hence defines a sheaf homomorphism

$$\bigwedge^2 E \to L.$$

We have to show that this is an isomorphism, and for this it is enough to check that $\deg L \leq \deg E$ (and hence $\deg L = \deg E$). But by construction

$$\deg L - 2g + 2 = N \leq h^0(E),$$

while by part (i) $H^1(E) = 0$, so that

$$h^0(E) = \deg E - 2g + 2,$$

and we are done.

(iii) By construction T is a Gieseker point of the vector bundle E, and so semistablility follows from Propositions 10.66 and 10.70.   □

Proof of Theorem 10.1. In view of Remark 6.14(vi), we have a Proj quotient once we know that the semistable set $\mathrm{Alt}_{N,2}^{ss}(H^0(L))$ is smooth. But this follows from Proposition 10.81(i), which guarantees the condition $H^1(E) = 0$, together with Proposition 10.77(i).

Now consider the open set $\mathrm{Alt}_{N,2}^s(H^0(L))/GL(N)$ of stable orbits. First note that, for each stable Gieseker point T, the vector bundle $E = \mathrm{Im}\,\langle T \rangle$ is stable. This follows from Corollary 10.68 and the proof of Proposition 10.69. Conversely, if E is stable as a vector bundle, then it is simple, so by Lemma 10.64 its Gieseker points T have a finite stabiliser and hence are stable for the $GL(N)$-action. We therefore arrive at a bijection:

$$SU_C(2, L) \overset{\sim}{\to} \mathrm{Alt}_{N,2}^s(H^0(L))/GL(N).$$

By Lemma 10.64, moreover, under the action $GL(N)/\{\pm I_N\} \curvearrowright \mathrm{Alt}_{N,2}^s(H^0(L))$ all orbits are free closed orbits. Thus, by Corollary 9.52, the open set $\mathrm{Alt}_{N,2}^s(H^0(L))$ is nonsingular.

Moreover, when E is stable,

$$\dim H^1(\mathfrak{sl}\, E) = 3g - 3$$

by Lemma 10.19. This proves parts (i) and (ii). For part (iii) we note that when $\deg L$ is odd, stability and semistability of E are equivalent by Remark 10.21, and so the conclusions of parts (i) and (ii) coincide. □

(c) Remarks on higher rank vector bundles

One can generalise the Gieseker matrices of this chapter to higher rank vector bundles. Let E be a vector bundle on C of rank r, and suppose that $\det E \cong L$. The total set E_{gen} is an r-dimensional vector space over the function field $k(C)$, and one can consider skew-symmetric multilinear maps

$$E_{\text{gen}} \times \cdots \times E_{\text{gen}} \to L_{\text{gen}}.$$

Such a map defines by restriction a skew-symmetric multilinear map over k

$$H^0(E) \times \cdots \times H^0(E) \to H^0(L), \qquad (s_1, \ldots, s_r) \mapsto s_1 \wedge \ldots \wedge s_r.$$

If $N := \dim H^0(E)$, then, after choosing a basis, this defines an element of $\mathrm{Hom}\left(\bigwedge^r k^{\oplus N}, H^0(L)\right)$, called a *Gieseker point* of E. There is a natural group action on the Gieseker points,

$$GL(N) \curvearrowright \mathrm{Hom}\left(\bigwedge^r k^{\oplus N}, H^0(L)\right),$$

coming from the action on k^n or, equivalently, by changing basis in $H^0(E)$.

Definition 10.82. A vector bundle E is called H^0-*semistable* if

$$\frac{h^0(F)}{\mathrm{rank}\, F} \leq \frac{h^0(E)}{\mathrm{rank}\, E} \quad \text{for every subbundle } F \subset E.$$

□

The following extends (and gives an alternative proof of) Proposition 10.70.

Proposition 10.83 (Gieseker [72]). *If a vector bundle E is H^0-semistable and is generated by global sections, except possibly at finitely many points, then its Gieseker points are semistable for the action of $GL(N)$.*

Proof. Notationally this is rather cumbersome in general, so we will just illustrate the proof by giving it in the case of rank $r = 4$. Let us suppose that some Gieseker point $T_E \in \mathrm{Hom}\left(\bigwedge^r k^{\oplus N}, H^0(L)\right)$ is unstable. By the numerical criterion (7.2) this means that some 1-PS $\lambda : \mathbb{G}_m \to SL(N)$ satisfies

$$\lim_{t \to 0} \lambda(t) \cdot T_E = 0.$$

Take a basis $S = \{s_1, s_2, \ldots, s_N\}$ of $H^0(E)$ which diagonalises λ, that is:

$$\lambda : t \mapsto \begin{pmatrix} t^{r_1} & & & \\ & t^{r_2} & & \\ & & \ddots & \\ & & & t^{r_N} \end{pmatrix}, \qquad \begin{array}{l} \text{where } r_1 + \cdots + r_N = 0 \text{ and} \\ r_1 \leq r_2 \leq \cdots \leq r_N. \end{array}$$

Now, each $(s_i, s_j, s_k, s_l) \in H^0(E) \times \cdots \times H^0(E)$ maps, under $\lambda(t) \cdot T_E$, to $t^{r_i + r_j + r_k + r_l} s_i \wedge s_j \wedge s_k \wedge s_l$. By hypothesis, then,

$$s_i \wedge s_j \wedge s_k \wedge s_l = 0$$

whenever $r_i + r_j + r_k + r_l \leq 0$.

We now introduce three conditions on the basis $S \subset H^0(E)$:

$\star_1$ $s_1 \wedge s_{i+1} = 0 \in H^0(\bigwedge^2 E)$ for all $i \leq N/4$.

$\star_2$ $s_1 \wedge s_{i+1} \wedge s_{j+1} = 0 \in H^0(\bigwedge^3 E)$ for all $i \leq N/4$ and $j \leq N/2$.

$\star_3$ $s_1 \wedge s_{i+1} \wedge s_{j+1} \wedge s_{k+1} = 0 \in H^0(\bigwedge^4 E) = H^0(L)$ for all $i \leq N/4$, $j \leq N/2$ and $k \leq 3N/4$.

Claim: $r_1 + r_{i+1} + r_{j+1} + r_{k+1} \leq 0$ whenever $i \leq N/4$, $j \leq N/2$ and $k \leq 3N/4$.

This claim, which we will prove in a moment, implies that $\star_3$ is always satisfied. Suppose that $\star_2$ is not. Then some suitable choice of $s_1, s_{i+1}, s_{j+1} \in S$ spans a subsheaf whose saturation is a rank 3 subbundle $F \subset E$. Condition $\star_3$ tells us that for each $k \leq 3N/4$ the section s_{k+1} is contained in $H^0(F) \subset H^0(E)$, and this implies that E is H^0-unstable.

If $\star_2$ is satisfied but $\star_1$ is not, then similarly we can construct a rank 2 subbundle with enough sections to violate H^0-semistability; while if $\star_1$ is satisfied, then we get a destabilising line subbundle. In each case we have a contradiction. It just remains to prove the claim.

For this, consider the step function $f(x)$ defined on the half-open interval $[0, N)$ by $f(x) = r_{p+1}$ for $x \in [p, p+1)$. By monotonicity of the function we have

$$r_1 + r_{i+1} + r_{j+1} + r_{k+1} = f(0) + f(i) + f(j) + f(k)$$

$$\leq f(0) + f\left(\frac{N}{4}\right) + f\left(\frac{2N}{4}\right) + f\left(\frac{3N}{4}\right)$$

$$\leq \frac{4}{N} \int_0^N f(x)dx = \frac{4}{N} \sum_{i=1}^N r_i = 0. \qquad \square$$

Using Proposition 10.83, one can construct a quasiprojective moduli space $SU_C(r, L)$ for stable vector bundles of rank r and fixed determinant line bundle L, as the quotient by $GL(N)$ of a closed subvariety of $\mathrm{Hom}\left(\bigwedge^r k^{\oplus N}, H^0(L)\right)$. (More precisely, a subscheme, though the subset of semistable points that we quotient is in fact nonsingular.) When the rank r and degree of L are coprime, stability and semistability coincide and the quotient is a nonsingular projective variety. There even exists, in these cases, a universal vector bundle on the product $C \times SU_C(r, L)$.

Exercises

1. Let E be a vector bundle of rank r on a curve C. Show the following.
   (i) If L is a line bundle, then $\deg(E \otimes L) = \deg E + r \deg L$.
   (ii) If F is a vector bundle of rank s, then

$$\deg(E \otimes F) = s \deg E + r \deg F, \qquad \mu(E \otimes F) = \mu(E) + \mu(F).$$

2. For a rank 2 vector bundle E, prove the following isomorphism:

$$E^\vee \cong E \otimes (\det E)^{-1}.$$

3. If

$$0 \to V_1 \longrightarrow V_2 \to \cdots \to V_{n-1} \longrightarrow V_n \to 0$$

is an exact sequence of vector spaces, show that $\sum_{i=1}^n (-1)^i \dim V_i = 0$.

4. (i) **The Five Lemma.** In the commutative diagram of abelian groups

$$
\begin{array}{ccccccccc}
U_1 & \longrightarrow & U_2 & \longrightarrow & U_3 & \longrightarrow & U_4 & \longrightarrow & U_5 \\
\downarrow & & \downarrow & & \downarrow & & \downarrow & & \downarrow \\
V_1 & \longrightarrow & V_2 & \longrightarrow & V_3 & \longrightarrow & V_4 & \longrightarrow & V_5
\end{array}
$$

the rows are exact and the vertical maps $U_i \to V_i$ are isomorphisms for $i = 1, 2, 4, 5$. Show that $U_3 \to V_3$ is an isomorphism.

   (ii) Use the Five Lemma to complete the proof of Theorem 10.11

5. (i) In the ring of integers $R = \mathbb{Z}[\sqrt{-5}]$ of the algebraic number field $\mathbb{Q}[\sqrt{-5}]$, show that the ideal $\mathfrak{a} = (2, 1 + \sqrt{-5})$ is a direct summand of the free module $R \oplus R$.

   (ii) Let $R = k[x, \sqrt{x^3 + 1}]$. Show that the ideal $\mathfrak{a} = (x - 2, \sqrt{x^3 + 1} - 3)$ is a direct summand of the free module $R \oplus R$ and is not isomorphic to R.

6. If E, E' are semistable vector bundles on a curve satisfying $\mu(E) > \mu(E')$, show that $\mathrm{Hom}(E, E') = 0$.

7. Given a vector bundle E on a curve, show that the slopes of all subbundles of E are bounded above.

8. Prove Proposition 10.69 without the hypothesis $H^1(E) = 0$.

11

Moduli functors

'A moduli space is an algebraic variety which parametrises the set of equivalence classes of some objects.' This explanation is reassuring psychologically, but it is not terribly precise. A moduli space of vector bundles, for example, ought to carry a family of bundles which 'controls' all equivalence classes. (Similarly for a moduli space of varieties, although we do not treat this case in this book.) In Chapter 9 we constructed a projective variety, as a quotient of $\mathrm{Mat}_{N,1}(H^0(L))$ (for some fixed line bundle L), whose underlying set was $\mathrm{Pic}^d C$, the set of line bundles of degree d. But this raises some obvious questions:

(1) The *set* $\mathrm{Pic}^d C$ is uniquely determined by the curve C, but is the same true of the *algebraic variety* with $\mathrm{Pic}^d C$ as its set of points? In particular, our construction depended on the choice of a line bundle L. Is the isomorphism class of the quotient variety independent of this choice?

(2) By tensoring with line bundles we get surjective maps $\mathrm{Pic}^d C \to \mathrm{Pic}^e C$. With respect to the algebraic variety structures that we have constructed on these sets, do these maps become morphisms (that is, polynomial maps)?

In order to answer these questions, the fundamental notion is that of a family and, following from this, the notions of fine and coarse moduli space, which we explain in Section 11.1(a). The variety $\mathrm{Pic}^d C$ not only has the set of isomorphism classes of line bundles as its set of points – it actually supports a family of line bundles in which each isomorphism class is uniquely represented. This is the first main result in this chapter:

Theorem 11.1. *The quotient variety* $\mathrm{Mat}^s_{N,1}/GL(N,N)$ *represents the Picard functor* $\mathcal{P}ic^d_C$ *for families of line bundles of degee d on C.* □

398

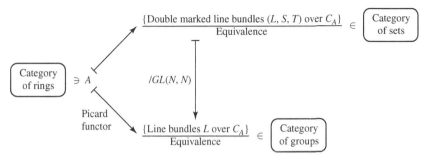

Figure 11.1: The Picard functor

In particular, it will follow from this that the projective variety $\text{Pic}^d C$ does not depend on the choice of auxiliary line bundle L. It is called the *algebraic Jacobian* (or *Picard variety*) of the curve.

The answer to question (2) is also affirmative:

Corollary 11.2. $\text{Pic}^0 C$ *is an algebraic group.*

The linear algebraic groups that we have considered so far in this book, such as $GL(N)$ and $GL(N, N)$, are affine algebraic groups. $\text{Pic}^0 C$ is not, and it has very different properties. It is a projective algebraic group, and in particular it is complete. (A complete algebraic group is called an *abelian variety*.)

In the second part of this chapter we study the analogue of the Picard functor for rank 2 vector bundles. Although the definition is very simple, it turns out that we lose many good properties enjoyed by the Picard functor. To begin with, the functor no longer takes values in the category of groups, but only sets. Second, because of the jumping phenomenon, the moduli functor is not representable by an algebraic variety and does not even admit a coarse moduli space (Definition 11.6). It admits a best approximation by an algebraic variety only after we restrict the class of vector bundles which we study: namely, the (semi)stable bundles. We therefore restrict our attention mainly to the moduli functor $\mathcal{SU}_C(2, L)$ for stable rank 2 vector bundles with fixed determinant line bundle L. In the case when $\deg L$ is odd we can obtain results similar to those for the Picard functor. (When $\deg L$ is odd, $N := \deg L + 2 - 2g$ is also odd. See Section 10.3(b).)

Theorem 11.3. *Assume that* $\deg L$ *is odd. Then the projective quotient* $\text{Alt}^s_{N,2}(H^0(L))/GL(N)$ *is a fine moduli space for stable rank 2 vector bundles* on C with determinant line bundle L. In other words, it represents the moduli functor $\mathcal{SU}_C(2, L)$. $\square$

The main tools used to prove these two theorems are direct images and co-homology modules, and we explain these in Section 11.1. In particular, the key idea here is the Base Change Theorem 11.15. We then introduce the Picard functor, and prove Theorem 11.1 at the end of this section. For this we have to construct the Poincaré line bundle. This really comes out of the quotient construction, using rings of invariants, of the Jacobian: the affine variety $\text{Mat}_{N,1}(H^0(L))$ carries a line bundle, or more fundamentally a module over the coordinate ring, and the Poincaré line bundle is constructed as the submodule of invariant elements.

In Section 11.2, as well as proving Theorem 11.3, we consider the moduli problem for vector bundles of even degree $\deg L$. In this case, the moduli space that we obtain by considering only stable vector bundles is not complete but is contained as an open set in the projective quotient variety $\text{Alt}_{N,2}(H^0(L))/\!/GL(N)$. This open set parametrises isomorphism classes of stable vector bundles; its complement is the quotient of the Jacobian of C by ± 1 (the *Kummer variety*) and parametrises S-equivalence classes (Proposition 11.37) of semistable bundles.

Finally, in Section 11.3 we present various explicit examples. We also touch a little on the question of moduli over nonalgebraically closed ground fields.

11.1 The Picard functor

(a) Fine moduli and coarse moduli

Typically, a 'moduli problem' for some class of objects in algebraic geometry consists of a notion of *family* parametrised by affine varieties Spm R, and the problem is thought of as solved if there is a universal family parametrised by a variety X with the property that every family over Spm R is uniquely induced by pulling back via a morphism Spm $R \to X$.

Formally, the moduli problem is a functor:

$$F : \{\text{algebras over } k\} \to \{\text{sets}\}, \qquad R \mapsto (\text{set of familes over Spm } R)$$

On the other hand, recall from Section 3.3(a) the interpretation of a variety X as a functor $\underline{X}$ from (finitely generated) k-algebras to sets, by assigning to an algebra R the set of solutions over R to the equations defining X. More precisely, it follows from (3.10) that, given a k-algebra R, the set $\underline{X}(R)$ can be identified with the set of morphisms Spm $R \to X$.

Definition 11.4. Suppose that F, G are two functors from the category of k-algebras to the category of sets. A *natural transformation* (or *functorial morphism*) $\rho : F \to G$ is a family

$$\{\rho(R) : F(R) \to G(R)\}_R$$

assigning a set mapping $\rho(R)$ to every k-algebra R and such that, for every k-algebra homomorphism $f : R \to S$, the following diagram commutes:

$$F(R) \xrightarrow{\rho(R)} G(R)$$

$$F(f) \downarrow \qquad \downarrow G(f)$$

$$F(S) \xrightarrow{\rho(S)} G(S)$$

If there exist natural transformations $\rho : F \to G$ and $\tau : G \to F$ satisfying $\tau\rho = \mathrm{id}_F$ and $\rho\tau = \mathrm{id}_G$ (where id_F means the natural transformation $F \to F$, which is the identity mapping on every set), then the functors F, G are said to be *isomorphic*. □

Now, every morphism of varieties $X \to Y$ induces a natural transformation of functors $\underline{X} \to \underline{Y}$; and conversely it can be shown that every natural transformation $\underline{X} \to \underline{Y}$ arises from a morphism of varieties in this way. (This is easy in the case when X and Y are both affine varieties.)

Returning to our moduli problem, a solution, or 'moduli space', is an isomorphism of the functor F with the functor

$$\underline{X} : \{\text{algebras over } k\} \to \{\text{sets}\}, \qquad R \mapsto (\text{set of morphisms } \mathrm{Spm}\, R \to X).$$

Definition 11.5. A functor F from k-algebras to sets is said to be *representable*, and to be *represented by a variety* X (or more generally a scheme X), if it is isomorphic to the functor $\underline{X}$. X is called a *fine moduli space* for the functor F. □

Unfortunately, in many moduli problems one cannot expect to have a representable functor. (See Example 11.32 below.) For this reason the following notion of 'coarse moduli' was proposed by Mumford.

Definition 11.6. Given a functor

$$F : \{\text{algebras over } k\} \to \{\text{sets}\},$$

a variety (or scheme) X is said to be a *best approximation* to F if it satisfies the following condition.

(i) There exists a natural transformation $\rho : F \to \underline{X}$ which is universal among natural transformations from F to variety functors (or, more precisely, scheme functors). In other words, given any $\tau : F \to \underline{Y}$ there exists a unique morphism $f : X \to Y$ making the following diagram commute:

$$F \xrightarrow{\ \rho\ } \underline{X}$$

$$\tau \searrow \quad \downarrow \underline{f}$$

$$\underline{Y}.$$

A best approximation X is called a *coarse moduli space* for the functor F if it satisfies, in addition:

(ii) for every algebraically closed field $k' \supset k$, the (set) map $\rho(k') : F(k') \to \underline{X}(k')$ is bijective. $\square$

There are two immediate remarks to make: first, that fine implies coarse, and second, that coarse (or best approximation) implies unique. That is, if Y is also a coarse moduli space for the functor F, then by the universal property there are natural transformations between $\underline{X}$ and $\underline{Y}$ in both directions, and these are inverse to each other and hence come from an isomorphism $X \cong Y$.

Example 11.7. By Proposition 8.89, the variety $\mathbb{G}(r, n)$ is a fine moduli space for the Grassmann functor $\mathcal{G}r(r, n)$. $\square$

Example 11.8. Approximation of the quotient functor. We can explain the meaning of the affine quotient map of Chapter 5 from this point of view. Recall that for an algebraic group G the functor $\underline{G}$ takes values in the category of groups. Moreover, if G acts on a variety X, then the functor $\underline{G}$ acts on the functor $\underline{X}$ and so determines a *quotient functor*

$$\underline{X}/\underline{G} : \{\text{algebras over } k\} \to \{\text{sets}\}, \qquad R \mapsto \underline{X}(R)/\underline{G}(R).$$

The affine quotient map $\Phi : X \to X/\!/G$ is characterised as being a best approximation of the quotient functor $\underline{X}/\underline{G}$. In other words, Φ has the following universal property: given any morphism $\phi : X \to Y$ which is constant on G-orbits in X, there exists a unique morphism $f : X/\!/G \to Y$ making the following diagram commute:

$$X \xrightarrow{\Phi} X /\!/ G$$

$$\phi \searrow \quad \downarrow f$$

$$Y.$$

By Corollary 5.17, moreover, the open subset X^s/G is a coarse moduli space for the functor $\underline{X^s}/\underline{G}$. □

(b) Cohomology modules and direct images

Let A be a finitely generated algebra over k. We shall consider the extension $A \otimes_k k(C)$ of A, extending coefficients from k to the function field $k(C)$. Also, we denote by $A \otimes_k \mathcal{O}_C$ the elementary sheaf on C defined by the ring extensions

$$U \mapsto A \otimes_k \mathcal{O}_C(U).$$

The pair consisting of the topological space C and the elementary sheaf $A \otimes_k \mathcal{O}_C$ we denote by C_A.

Definition 11.9. A *vector bundle on C_A* is an elementary sheaf $\mathcal{E}$ of $A \otimes_k \mathcal{O}_C$ modules satisfying the following conditions.

(i) The total set (denoted $\mathcal{E}_{\text{gen}}$) is a locally free $A \otimes_k k(C)$-module.
(ii) If $U \subset C$ is an affine open set, then $\mathcal{E}(U)$ is a locally free $A \otimes_k \mathcal{O}_C(U)$-module. □

In the case $A = k$, of course, $\mathcal{E}$ is nothing but a vector bundle on the curve C.

In what follows we will only consider A-modules of finite rank. Since the rank of a locally free module is locally constant, this number depends only on the connected component of Spm A.

Let $\mathcal{E}$ be a vector bundle on C_A and let $f : A \to A'$ be a ring homomorphism. Then we can define a vector bundle on $C_{A'}$, denoted $\mathcal{E} \otimes_A A'$, by taking the total set $\mathcal{E}_{\text{gen}} \otimes_A A'$ and assigning to each open set $U \subset C$ the extension $\mathcal{E}(U) \otimes_A A'$. The vector bundle $\mathcal{E} \otimes_A A'$ is called the *pullback* under the morphism Spm $A' \to$ Spm A. In the case $A \hookrightarrow A'$, we also refer to *extension of coefficients* from A to A' and, when $A' = A/I$, to *reduction modulo an ideal* $I \subset A$. In particular, if $\mathfrak{m} \subset A$ is a maximal ideal, then we get a vector bundle $\mathcal{E} \otimes_A (A/\mathfrak{m}) = \mathcal{E}/\mathfrak{m}\mathcal{E}$ on C. If this maximal ideal corresponds to a point $t \in$ Spm A, then this vector bundle is denoted by $\mathcal{E}|_{C \times t}$, or simply by $\mathcal{E}_t$ when this is not likely to lead to confusion.

For vector bundles on C_A, just as for vector bundles on C, we can define a space of global sections H^0 and cohomology space H^1. Namely, the stalk of $\mathcal{E}$ at a point,

$$\mathcal{E}_p = \bigcup_{p \in U} \mathcal{E}(U),$$

is a module over $A \otimes_k \mathcal{O}_{C,p}$. We then define H^0 and H^1 to be the kernel and cokernel, respectively, of the principal part map:

$$0 \longrightarrow H^0(\mathcal{E}) \longrightarrow \mathcal{E}_{\text{gen}} \xrightarrow{\ \text{pp}\ } \bigoplus_{p \in C} \mathcal{E}_{\text{gen}}/\mathcal{E}_p \longrightarrow H^1(\mathcal{E}) \longrightarrow 0.$$

This is an exact sequence of A-modules.

Definition 11.10. The A-module $H^0(\mathcal{E})$ is called the *(zeroth) direct image* of $\mathcal{E}$ on Spm A. The A-module $H^1(\mathcal{E})$ is called the *cohomology module*, or the *first direct image* of $\mathcal{E}$ on Spm A. $\qquad\square$

Example 11.11. If E is a vector bundle on C and $\mathcal{E} = E \otimes_k A$, then $H^0(\mathcal{E}) = H^0(E) \otimes A$ and $H^1(\mathcal{E}) = H^1(E) \otimes A$. For example, if $\mathcal{E} = \mathcal{O}_C \otimes_k A$, then $H^0(\mathcal{E}) = A$ and $H^1(\mathcal{E}) = H^1(\mathcal{O}_C) \otimes A \cong A^{\oplus g}$. $\qquad\square$

In this example we see that each $H^i(\mathcal{E})$ is a finitely generated free A-module. More generally:

Theorem 11.12. *If $\mathcal{E}$ is a vector bundle on C_A, then $H^0(\mathcal{E})$, $H^1(\mathcal{E})$ are both finitely generated A-modules.*

Corollary 11.13. *If $\mathcal{E}$ is a vector bundle on C_A, then $H^1(\mathcal{E} \otimes \mathcal{O}_C(D)) = 0$ for some positive divisor $D \geq 0$.*

Proof. For each point $t \in$ Spm A there exists some positive divisor D_t on C satisfying $H^1(\mathcal{E}|_{C \times t}(D_t)) = 0$. By Nakayama's Lemma 8.23 (and Theorem 11.12) t has an open neighbourhood $U_t \subset$ Spm A such that $H^1(\mathcal{E}|_{C \times t'}(D_t)) = 0$ for every $t' \in U_t$. One can cover Spm A with finitely many such neighbourhoods and then take D to be a divisor bounding the corresponding divisors D_t. $\qquad\square$

In order to prove the theorem we need:

Lemma 11.14. *For any vector bundle $\mathcal{E}$ on C_A there exists a finite set of line bundles $L_1, \ldots, L_N$ on C together with a surjective sheaf homomorphism $(L_1 \oplus \cdots \oplus L_N) \otimes_k A \to \mathcal{E}$.*

Proof. The stalk at the generic point $\mathcal{E}_{\mathrm{gen}}$ is generated by finitely many elements as a module over $k(C) \otimes_k A$. This means that there is a positive divisor $D \geq 0$ and a sheaf homomorphism $\mathcal{O}_C(-D)^{\oplus N(0)} \to \mathcal{E}$ which is surjective away from finitely many points $p_1, \ldots, p_n \in C$. We can then find rational sections $s_1, \ldots, s_{N(1)} \in \mathcal{E}_{\mathrm{gen}}$ which are regular away from p_1 and generate the stalk $\mathcal{E}_{p_1}$. Hence there exists some positive divisor $D_1 \geq 0$ and a sheaf homomorphism $\mathcal{O}_C(-D)^{\oplus N(0)} \oplus \mathcal{O}_C(-D_1)^{\oplus N(1)} \to \mathcal{E}$ surjective away from the points $p_2, \ldots, p_n$. Repeating this construction at the remaining points, we get the lemma. $\square$

Proof of Theorem 11.12. Let $\mathcal{K}$ be the kernel of the homomorphism given in the lemma, and consider the exact sequence

$$0 \to \mathcal{K} \to (L_1 \oplus \cdots \oplus L_N) \otimes_k A \to \mathcal{E} \to 0.$$

From the part of the exact cohomology sequence

$$\left(H^1(L_1) \oplus \cdots \oplus H^1(L_N) \right) \otimes_k A \longrightarrow H^1(\mathcal{E}) \longrightarrow 0,$$

where each $H^1(L_i)$ is a finite-dimensional vector space, it follows that $H^1(\mathcal{E})$ is a finitely generated A-module. In particular, $H^1(\mathcal{K})$ is also finitely generated, and so the part of the cohomology sequence

$$\left(H^0(L_1) \oplus \cdots \oplus H^0(L_N) \right) \otimes_k A \longrightarrow H^0(\mathcal{E}) \longrightarrow H^1(\mathcal{K})$$

shows that $H^0(\mathcal{E})$ is also finitely generated. $\square$

Base Change Theorem 11.15. *Let $\mathcal{E}$ be a vector bundle on C_A. Then for any ring homomorphism $A \to A'$ there exists, for each $i \geq 0$, a natural homomorphism of A'-modules*

$$H^i(\mathcal{E}) \otimes_A A' \to H^i(\mathcal{E} \otimes_A A'). \tag{11.1}$$

In particular, it will be important to understand the case

$$H^i(\mathcal{E}) \otimes_A A/\mathfrak{m} \to H^i(\mathcal{E} \otimes_A A/\mathfrak{m}), \tag{11.2}$$

where $A' = A/\mathfrak{m}$ at some maximal ideal $\mathfrak{m} \subset A$.

By right-exactness of the tensor product (Lemma 8.55) together with Nakayama's Lemma and Lemma 8.31 we obtain the following facts.

Lemma 11.16.

(i) For $i = 1$ the base change homomorphisms (11.1) and (11.2) are surjective.
(ii) If $H^1(\mathcal{E}_t) = 0$ for every $t \in \mathrm{Spm}\,A$, then $H^1(\mathcal{E}) = 0$. $\qquad\square$

If we express $H^1(\mathcal{E})$ as the cokernel of a homomorphism of free A-modules

$$f : A^{\oplus M} \to A^{\oplus N},$$

then we have

$$\dim H^1(\mathcal{E} \otimes_A A/\mathfrak{m}) = N - \mathrm{rank}\,(f \otimes_A A/\mathfrak{m}).$$

The rank of a linear map is a lower semicontinuous function, and so we obtain:

Lemma 11.17. *For any vector bundle $\mathcal{E}$ on C_A the function*

$$\mathrm{Spm}\,A \to \mathbb{Z}, \qquad t \mapsto \dim H^1(\mathcal{E}_t)$$

is upper semicontinuous with respect to the Zariski topology. In other words, for each $a \in \mathbb{Z}$,

$$\{t \mid \dim H^1(\mathcal{E}_t) \geq a\} \subset \mathrm{Spm}\,A$$

is a closed subset. $\qquad\square$

For $H^0(\mathcal{E})$ the following facts are fundamental.

Theorem 11.18. *Suppose $\mathcal{E}$ is a vector bundle on C_A with $H^1(\mathcal{E}) = 0$. Then the following hold:*

(i) $H^0(\mathcal{E})$ is locally free as an A-module; and
(ii) for every ring homomorphism $A \to A'$ the base change homomorphism

$$H^0(\mathcal{E}) \otimes_A A' \to H^0(\mathcal{E} \otimes_A A')$$

is an isomorphism.

Proof.

(i) Let $\{U, V\}$ be an affine open cover of C. By the same reasoning as for Exercise 8.11 there is an exact sequence

$$0 \longrightarrow H^0(\mathcal{E}) \longrightarrow \mathcal{E}(U) \oplus \mathcal{E}(V) \longrightarrow \mathcal{E}(U \cap V) \longrightarrow H^1(\mathcal{E}) \longrightarrow 0.$$

Note that $\mathcal{E}(U)$, $\mathcal{E}(V)$, $\mathcal{E}(U \cap V)$ are all flat as A-modules. Since $H^1(\mathcal{E}) = 0$, therefore, it follows from Proposition 8.54 that $H^0(\mathcal{E})$ is also flat. But $H^0(\mathcal{E})$ is finitely generated by Theorem 11.12, so by Proposition 8.48 it is locally free.

(ii) By Lemma 8.58, tensoring with any A-algebra A' preserves exactness of the above sequence, and this implies that $H^0(\mathcal{E}) \otimes_A A' \cong H^0(\mathcal{E} \otimes_A A')$. $\square$

Remark 11.19. In fact, a little more than this is true when Spm A is reduced (that is, A contains no nilpotents). Namely, the A-module $H^0(\mathcal{E})$ is flat, and hence the conclusions (i) and (ii) of the theorem hold, provided the dimension $h^0(\mathcal{E}_t)$ is constant over Spm A. (See Mumford [81], pp. 50–51.) $\square$

Intuitively, $H^0(\mathcal{E})$ can be thought of as the vector bundle on Spm A whose fibre at $t \in$ Spm A is the space of global sections $H^0(\mathcal{E}_t) = H^0(\mathcal{E}|_{C \times t})$.

Corollary 11.20. *If $\mathcal{E}$ is a vector bundle on C_A, then the function* Spm $A \to \mathbb{Z}$, $t \mapsto \deg \mathcal{E}_t$ *is constant on connected components of* Spm A.

Proof. We apply Theorem 11.18 to the vector bundle $\mathcal{E}(D)$, where D is the divisor constructed in Corollary 11.13. This satisfies the requirement that H^1 vanishes, so the function $t \mapsto \dim H^0(\mathcal{E}_t(D))$ is locally constant (using Proposition 8.49). By Riemann-Roch, therefore, $\deg \mathcal{E}_t(D)$ is locally constant. $\square$

It follows from Corollary 11.20 and Riemann-Roch that the integer $h^0(\mathcal{E}_t) - h^1(\mathcal{E}_t)$ is locally constant on Spm A. Combining this with Lemma 11.17 we see:

Proposition 11.21. *For any vector bundle $\mathcal{E}$ on C_A the function*

$$\text{Spm } A \to \mathbb{Z}, \qquad t \mapsto \dim H^0(\mathcal{E}_t)$$

is upper semicontinuous: for each $a \in \mathbb{Z}$ the subset $\{t \mid \dim H^0(\mathcal{E}_t) \geq a\} \subset$ Spm A *is closed.* $\square$

(c) Families of line bundles and the Picard functor

A vector bundle $\mathcal{E}$ on C_A associates to each point $t \in$ Spm A (corresponding to a maximal ideal $\mathfrak{m} \subset A$) a vector bundle $\mathcal{E}_t$ on C (by pull-back via Spm $A/\mathfrak{m} \hookrightarrow$ Spm A), and we can observe that this correspondence does not change if $\mathcal{E}$ is replaced with $\mathcal{E} \otimes_A M$ for any invertible A-module M.

Definition 11.22.

(i) Two vector bundles $\mathcal{E}$, $\mathcal{E}'$ on C_A are *equivalent* if $\mathcal{E}' \cong \mathcal{E} \otimes_A M$ for some invertible A-module M.

(ii) By an *algebraic family of vector bundles on C parametrised by* Spm A we mean an equivalence class (in the sense of (i)) of vector bundles on C_A.

The set of families of line bundles parametrised by Spm A becomes a group under a tensor product, and this group is just Pic C_A/Pic A. Furthermore, given a ring homomorphism $f : A \to A'$, the pullback of a family via Spm $A' \to$ Spm A is well defined (if $\mathcal{E}$ and $\mathcal{E}'$ are equivalent, then so are $\mathcal{E} \otimes_A A'$ and $\mathcal{E}' \otimes_A A'$), and the pullback of families of line bundles is a group homomorphism

$$\otimes f : \text{Pic } C_A/\text{Pic } A \to \text{Pic } C_{A'}/\text{Pic } A', \qquad \mathcal{L} \mapsto \mathcal{L} \otimes_A A'.$$

If $g : A' \to A''$ is another ring homomorphism, then this operation satisfies

$$(\otimes g)(\otimes f) = \otimes gf.$$

Definition 11.23. The covariant functor

$$\mathcal{P}ic_C : \left\{ \begin{array}{l} \text{finitely generated} \\ \text{rings over } k \end{array} \right\} \longrightarrow \left\{ \text{groups} \right\}$$

which assigns $A \mapsto \text{Pic } C_A/\text{Pic } A$ is called the *Picard functor* for the curve C. $\square$

Given a family of line bundles $\mathcal{L} \in \text{Pic } C_A/\text{Pic } A$, the degree of $\mathcal{L}|_{C \times t}$ is constant on connected components of Spm A (Corollary 11.20). We will denote by $\mathcal{P}ic_C^d \subset \mathcal{P}ic_C$ the subfunctor which assigns families of line bundles of degree d. The following proposition is then half of Theorem 11.1; the remaining part, that the moduli space is fine, will be proved in Section 11.1(d).

Proposition 11.24. *Let* $L \in \text{Pic}^{2d} C$, *where* $d \geq 2g$, *and let* $N = d + 1 - g$, *as in Section 9.4(a). Then the projective quotient* $\text{Mat}_{N,1}^s(H^0(L))/GL(N, N)$ *is a coarse moduli space for the Picard functor* $\mathcal{P}ic_C^d$. $\square$

Corollary 11.25. *The isomorphism class of the variety* $J_d := \text{Mat}_{N,1}^s$ $(H^0(L))/GL(N, N)$ *depends only on* C *and* d, *and not on the line bundle* $L \in \text{Pic}^{2d} C$.

Proof of Proposition 11.24. The idea is similar to that of Proposition 8.89 for the Grassmann functor. For each finitely generated k-algebra A, our aim is to find a natural bijection between line bundles Ξ on C_A such that $\deg \Xi_t = d$ at every $t \in \operatorname{Spm} A$, up to equivalence, and morphisms $\operatorname{Spm} A \to J_d$. Let $\widehat{\Xi} := L_A \otimes_\mathcal{O} \Xi$. Then by Theorem 11.18 both of $H^0(\Xi)$, $H^0(\widehat{\Xi})$ are localy free A-modules of rank N, and their fibres at a point $t \in \operatorname{Spm} A$ are the spaces $H^0(C, \Xi_t)$, $H^0(C, \widehat{\Xi}_t)$. There then exists a bilinear homomorphism of A-modules:

$$H^0(\Xi) \times H^0(\widehat{\Xi}) \to H^0(L) \otimes_k A. \qquad (11.3)$$

Step 1. We first consider the case when both of $H^0(\Xi)$, $H^0(\widehat{\Xi})$ are free A-modules. Let S, $\widehat{S}$ be free bases. Via (11.3), these determine an $N \times N$ matrix with entries in $H^0(L) \otimes_k A$, and so we get a morphism to an (affine) space of matrices, $\operatorname{Spm} A \to \operatorname{Mat}_N(H^0(L))$. This maps into the closed subvariety

$$\operatorname{Mat}_{N,1}(H^0(L)) \subset \operatorname{Mat}_N(H^0(L))$$

defined by the vanishing of the 2×2 minors and, moreover, into the open set $\operatorname{Mat}^s_{N,1}(H^0(L))$, since for all $t \in \operatorname{Spm} A$, the line bundles Ξ_t, $\widehat{\Xi}_t$ are generated by global sections. We will denote this map by

$$\widetilde{\varphi} : \operatorname{Spm} A \to \operatorname{Mat}^s_{N,1}(H^0(L))$$

and the composition of $\widetilde{\varphi}$ with the quotient map by

$$\varphi : \operatorname{Spm} A \to J_d = \operatorname{Mat}^s_{N,1}(H^0(L))/GL(N,N)$$

The map φ depends only on the equivalence class of Ξ (in the sense of Definition 11.22) and not on the choice of S, $\widehat{S}$.

Step 2. We now take an affine open cover

$$\operatorname{Spm} A = U_1 \cup \ldots \cup U_n$$

such that the A-modules $H^0(\Xi)$ and $H^0(\widehat{\Xi})$ restrict to free modules on each U_i. For each i, by choosing free bases of $H^0(\Xi)|_{U_i}$ and $H^0(\widehat{\Xi})|_{U_i}$ we obtain a map $\widetilde{\varphi}_i : U_i \to \operatorname{Mat}^s_{N,1}(H^0(L))$ as in Step 1, and on intersections $U_i \cap U_j$ the maps $\widetilde{\varphi}_i$ and $\widetilde{\varphi}_j$ differ only by the choice of free bases of $H^0(\Xi)|_{U_i \cap U_j}$ and $H^0(\widehat{\Xi})|_{U_i \cap U_j}$. It follows that the corresponding maps $\varphi_i : U_i \to J_d$ and $\varphi_j : U_j \to J_d$ agree on the intersection $U_i \cap U_j$, and by gluing we therefore obtain a morphism

$$\varphi : \operatorname{Spm} A \to J_d.$$

This is called the *classifying map* for the family of line bundles Ξ.

Let Ξ, Ξ' be two line bundles on C_A which are locally equivalent as families of line bundles on C. By this we mean that there is an open cover of $\operatorname{Spm} A$ as above, such that on each open set the restrictions $\Xi|_{U_i}$ and $\Xi'|_{U_i}$ are equivalent.

For these line bundles the classifying maps φ, φ' : Spm $A \to J_d$ are the same, and in particular we see that φ depends only on the equivalence class of Ξ (Definition 11.22). This verifies the first requirement for J_d to be a coarse moduli space: we have constructed a natural transformation of functors $\mathcal{P}ic_C^d \to \underline{J_d}$. Moreover, it follows from the results of Chapter 9 that over an algebraically closed field this is bijective in the sense of Definition 11.6(ii), as required.

Step 3. Finally, we have to show universality (Definition 11.6(i)). Suppose that we have a natural transformation $\psi : \mathcal{P}ic_C^d \to \underline{Y}$ for some variety Y. Over the product $C \times \text{Mat}_{N,1}(H^0(L))$ there is a tautological homomorphism of vector bundles

$$\mathcal{O}_{C \times \text{Mat}}^{\oplus N} \to (L \boxtimes \mathcal{O}_{\text{Mat}})^{\oplus N}.$$

Restricted to the open set $C \times \text{Mat}_{N,1}^s(H^0(L))$, this map has rank 1 at each point, and its image is a line bundle. We denote this line bundle on $C \times \text{Mat}_{N,1}^s(H^0(L))$ by $\mathcal{Q}$, called the *universal line bundle*. Composing the natural transformation $\overline{\text{Mat}_{N,1}^s(H^0(L))} \to \mathcal{P}ic_C^d$, defined in the obvious way by the pullback of $\mathcal{Q}$, with ψ gives a natural transformation of functors

$$\overline{\text{Mat}_{N,1}^s(H^0(L))} \longrightarrow \mathcal{P}ic_C^d \xrightarrow{\psi} \underline{Y}.$$

Since the line bundle $\mathcal{Q}$ is trivial on $GL(N, N)$-orbits, it follows that the corresponding morphism $\text{Mat}_{N,1}^s(H^0(L)) \to Y$ descends to the quotient, and so we obtain a morphism $J_d \to Y$ with the required properties.   □

(d) Poincaré line bundles

We begin by returning to Section 9.3(c) and extending the discussion of that section from ideals (that is, orbits) to modules (vector bundles). Let G be a linearly reductive algebraic group acting on an affine variety $X = \text{Spm } R$, and let M be an R-module with a G-linearisation (Definition 6.23). Thus M is also a representation of G and has a subset of invariants $M^G \subset M$ which, by definition of a linearisation, is an R^G-module.

Lemma 11.26. *If M is a finitely generated module over a Noetherian ring R, then M^G is finitely generated as an R^G-module.*

Proof. The idea of the proof is the same as that of Hilbert's Theorem 4.51. Denote by $M' \subset M$ the submodule generated by M^G. Since R (and M) are Noetherian, it follows that M' is finitely generated. Letting $m_1, \dots, m_n \in M'$

be generators, the map

$$R \oplus \cdots \oplus R \to M', \qquad (a_1, \ldots, a_n) \mapsto a_1 m_1 + \cdots + a_n m_n$$

is surjective, and hence by linear reductivity the induced map

$$R^G \oplus \cdots \oplus R^G \to (M')^G = M^G$$

is surjective, so that M^G is generated as an R^G-module by $m_1, \ldots, m_n$. $\qquad \square$

Proposition 11.27. *Suppose that all orbits of $G \curvearrowright \mathrm{Spm}\, R$ are free closed orbits and that M is a locally free R-module. Then M^G is a locally free R^G-module and $M \cong M^G \otimes_{R^G} R$.*

Proof. Let $I \subset R$ be the ideal of an orbit, with corrsponding maximal ideal $\mathfrak{m} = I \cap R^G \subset R^G$. Then M/IM is a $k[G]$-module with a G-linearisation and so is a free $k[G]$-module by Lemma 9.49. By linear reductivity we can find $m_1, \ldots, m_r \in M^G$ whose residue classes $\overline{m}_1, \ldots, \overline{m}_r \in M/IM$ are a free basis, and hence the natural homomorphism of R-modules

$$M^G \otimes_{R^G} R \to M$$

is an isomorphism along each orbit. Since, by Lemma 11.26, M^G is finitely generated, it follows from Nakayama's Lemma that this homomorphism is surjective. But it is also injective because M is locally free. $\qquad \square$

The projective quotient map in the direction of some character $\chi \in \mathrm{Hom}(G, \mathbb{G}_m)$,

$$\Phi = \Phi_\chi : X^{ss} \to X /_\chi G,$$

is locally an affine quotient map, and so we obtain:

Corollary 11.28. *Suppose that all semistable points are stable and that every orbit of the action $G \curvearrowright X^{ss} = X^s$ is a free closed orbit. Then, given a vector bundle E on X with a G-linearisation, there exists a vector bundle E_0 on X^s/G such that $E \cong \Phi^* E_0$.* $\qquad \square$

We now return to the proof of Proposition 11.24, and we will apply Corollary 11.28 to the universal line bundle $\mathcal{Q}$. This had the property that

$$(1 \times \widetilde{\varphi})^* \mathcal{Q} \cong \Xi \quad \text{under } 1 \times \widetilde{\varphi} : C \times \mathrm{Spm}\, A \to C \times \mathrm{Mat}^s_{N,1}(H^0(L)).$$

Let R be the coordinate ring of the affine variety $\text{Mat}_{N,1}(H^0(L))$. There is a tautological homomorphism of R-modules $\tau : R^{\oplus N} \to R^{\oplus N} \otimes_k H^0(L)$ given in the obvious way by matrix multiplication. Given a linear map $f : H^0(L) \to k$, we then get a homomorphism of R-modules as the composition

$$\phi_f : R^{\oplus N} \xrightarrow{\tau} R^{\oplus N} \otimes_k H^0(L) \xrightarrow{1 \otimes f} R^{\oplus N} \otimes_k k = R^{\oplus N}.$$

When f is the evaluation map at a point $p \in C$ this homomorphism has rank ≤ 1 everywhere, and on the open set $\text{Mat}^s_{N,1}(H^0(L))$ its image is precisely the line bundle

$$\mathcal{Q}_p := \mathcal{Q}|_{p \times \text{Mat}}.$$

The group $GL(N) \times GL(N)$ acts on R, and using this action we let it act on the source and target $R^{\oplus N}$ of the homomorphism ϕ_f, respectively, by

$$\begin{pmatrix} f_1 \\ \vdots \\ f_N \end{pmatrix} \mapsto A \begin{pmatrix} g \cdot f_1 \\ \vdots \\ g \cdot f_N \end{pmatrix}, \qquad \begin{pmatrix} f_1 \\ \vdots \\ f_N \end{pmatrix} \mapsto \begin{pmatrix} g \cdot f_1 \\ \vdots \\ g \cdot f_N \end{pmatrix} B^{t,-1}.$$

The map ϕ_f is then a $GL(N) \times GL(N)$-homomorphism, and in particular $GL(N) \times GL(N)$ acts on the R-module $\mathcal{Q}_p$. Similarly, the universal line bundle $\mathcal{Q}$ carries a $GL(N) \times GL(N)$-linearisation, under which $(t, t^{-1}) \in GL(N) \times GL(N)$ acts nontrivially. However, this element acts trivially on the line bundle $\mathcal{Q} \otimes \mathcal{Q}_p^{-1}$ with its induced linearisation, and so this line bundle possesses a $GL(N, N)$-linearisation. According to Corollary 11.28, therefore, $\mathcal{Q} \otimes \mathcal{Q}_p^{-1}$ is the pullback of some line bundle

$$\mathcal{P}_d \in \text{Pic } C \times J_d.$$

This is called the *Poincaré line bundle*.

Remark 11.29. More generally, let $D = \sum_i m_i p_i \in \text{Div} C$ be a divisor of degree 1. Then the line bundle $\mathcal{Q} \otimes_R \prod_i \mathcal{Q}_{p_i}^{-m_i}$ descends to $C \times J_d$. $\square$

Lemma 11.30. *Let Ξ be a line bundle on C_A with classifying map $\varphi : \text{Spm } A \to J_d$. Then Ξ is equivalent to the pullback $(1 \times \varphi)^* \mathcal{P}_d$ via $1 \times \varphi : C_A \to C \times J_d$.*

Proof. Let $\mathcal{L} = (1 \times \varphi)^* \mathcal{P}_d$. By construction of the classifying map, Ξ is already locally isomorphic to $\mathcal{L}$. In other words, $\Xi|_{C \times U_i} \cong \mathcal{L}|_{C \times U_i}$ over some affine open cover $\text{Spm } A = U_1 \cup \ldots \cup U_n$. Thus $M := H^0(\Xi \otimes \mathcal{L}^{-1})$ is an invertible

A-module, and the natural homomorphism

$$\mathcal{L} \otimes_A M \to \Xi$$

is an isomorphism. Hence Ξ and $\mathcal{L}$ are equivalent. $\qquad\square$

This lemma gives the crucial 'universal' property of the Poincaré line bundle, which makes J_d a fine, and not just a coarse, moduli space. Precisely, it says that the correspondence $\varphi \mapsto (1 \times \varphi)^* \mathcal{P}_d$ gives a natural transformation of functors

$$\underline{J_d} \to \mathcal{P}ic_C^d,$$

which is inverse to the natural transformation given in Proposition 11.24. Hence the Picard functor $\mathcal{P}ic_C^d$ is represented by the quotient variety J_d, and we have proved Theorem 11.1.

11.2 The moduli functor for vector bundles

Given a finitely generated ring A over k, a family of vector bundles of rank r on C parametrised by Spm A is an equivalence class of vector bundles of rank r on C_A (Definition 11.22). Denote the set of such families by $\mathcal{VB}_C(r)(A)$. Given a ring homomorphism $f : A \to A'$, let $\mathcal{VB}_C(r)(f) : \mathcal{VB}_C(r)(A) \to \mathcal{VB}_C(r)(A')$ denote the pullback of families via Spm $A' \to$ Spm A. We have then defined a covariant functor

$$\mathcal{VB}_C(r) : \left\{ \begin{array}{c} \text{finitely generated} \\ \text{rings over } k \end{array} \right\} \longrightarrow \left\{ \text{sets} \right\},$$

called the *moduli functor* for vector bundles of rank r on the curve C. Of course, the case $r = 1$ is (after forgetting the group structure) the Picard functor

$$\mathcal{VB}_C(1) = \mathcal{P}ic_C.$$

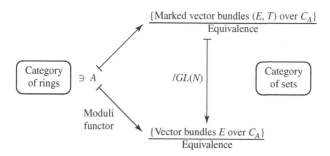

Figure 11.2: The moduli functor for vector bundles

If $\mathcal{E}$ is a vector bundle on C_A, then its determinant $\det \mathcal{E}$ is a line bundle on C_A, and this operation commutes with pullback under morphisms Spm $A' \rightarrow$ Spm A. Moreover, if $\mathcal{E}$ and $\mathcal{E}'$ are equivalent vector bundles, then $\det \mathcal{E}$ and $\det \mathcal{E}'$ are equivalent line bundles, and so the determinant is well defined on families. We therefore have a natural transformation of functors,

$$\det : \mathcal{VB}_C(r) \rightarrow \mathcal{Pic}_C.$$

In particular, if L is any line bundle on C, then we can define (for any A) a family $L_A := L \otimes_k A$ on C_A. This is called a *constant family*, and we consider its preimage under det (the 'fibre functor'):

Definition 11.31. Given $L \in \mathrm{Pic}\, C$, we denote by

$$\mathcal{VB}_C(r, L) : \left\{ \begin{array}{l} \text{finitely generated} \\ \text{rings over } k \end{array} \right\} \longrightarrow \left\{ \text{sets} \right\}$$

the functor which associates to objects A the set of families $\mathcal{E}$ on C_A for which $\det \mathcal{E}$ is equivalent to the constant family $L_A \in \mathcal{Pic}_C(A)$.

A 'moduli space' for vector bundles (of rank r) is an object which in some suitable sense approximates the functor $\mathcal{VB}_C(r)$ (or its 'connected components' of vector bundles with fixed degree) or $\mathcal{VB}_C(r, L)$. However, as soon as $r \geq 2$, a *coarse moduli space*, or a best approximation in the sense of Definition 11.6, cannot exist for the following reason.

Example 11.32. The jumping phenomenon. There exist families of vector bundles of rank ≥ 2, say $\mathcal{E}$ parametrised by $T = $ Spm A, with the following property. For some dense open set $U \subset T$ and all $u \in U, t \in T$,

$$\mathcal{E}_t \begin{cases} \cong \mathcal{E}_u & \text{if } t \in U, \\ \ncong \mathcal{E}_u & \text{if } t \in T - U. \end{cases}$$

For example, let $L \in \mathrm{Pic}\, C$ be a line bundle and fix an extension of $\mathcal{O}_C$ by L with (nonzero) extension class $e \in H^1(L)$. Then each $a \in k$ determines an extension

$$0 \rightarrow L \rightarrow E_a \rightarrow \mathcal{O}_C \rightarrow 0$$

with extension class $ae \in H^1(L)$. This illustrates the jumping phenomenon, with $T = \mathbb{A}^1$, because $E_a \cong E_1$ for all $a \neq 0$, and we can guarantee that this is indecomposable by choosing $H^0(L) \neq 0$ (by Proposition 10.45) while, on the other hand, $E_0 \cong L \oplus \mathcal{O}_C \ncong E_1$.

How do we see that this is an algebraic family of vector bundles? We claim that there exists an exact sequence of vector bundles on C_A, where $A = k[s]$,

$$0 \to L \otimes_k k[s] \to \mathcal{E} \to \mathcal{O}_C \otimes_k k[s] \to 0$$

with the property that for every $a \in k$ the reduction of $\mathcal{E}$ modulo $\mathfrak{m} = (s-a) \subset k[s]$ is isomorphic to E_a, that is,

$$\mathcal{E}|_{C \times a} \cong E_a.$$

To prove this, we let $\{U, V\}$ be an open cover of C and represent the extension class $e \in H^1(L)$ by an element $b \in L(U \cap V)$ (Exercise 8.11). The vector bundle E_1 is then obtained by gluing the bundles $\mathcal{O}_U \oplus L_U$ and $\mathcal{O}_V \oplus L_V$ along $U \cap V$ using the transition matrix $\begin{pmatrix} 1 & b \\ 0 & 1 \end{pmatrix}$. The extension $\mathcal{E}$ is then defined by gluing $(\mathcal{O}_U \oplus L_U) \otimes_k k[s]$ and $(\mathcal{O}_V \oplus L_V) \otimes_k k[s]$ along $U \cap V$ using the transition matrix $\begin{pmatrix} 1 & bs \\ 0 & 1 \end{pmatrix}$. $\qquad\square$

Now suppose that X were a coarse moduli space for the moduli functor $\mathcal{VB}_C(r)$, $r \geq 2$, and $\{\mathcal{E}_t\}_{t \in T}$ some jumping family as above. By the coarse moduli property there is then a morphism $f : T \to X$ with the property that some dense open set $U \subset T$ maps to a single point, but whose image $f(T) \subset X$ contains more than one point, a contradiction.

This phenomenon forces us, if we want to construct a moduli space as an algebraic variety, to restrict to some smaller class among the vector bundles that we are considering.

Definition 11.33. Given a line bundle $L \in \operatorname{Pic} C$, we denote by $\mathcal{SU}_C(r, L) \subset \mathcal{VB}_C(r, L)$ the subfunctor which associates to a ring A the set of families $\mathcal{E}$ on C_A for which $\mathcal{E}_t$ is stable for all points $t \in \operatorname{Spm} A$. $\qquad\square$

Note that stability depends only on the equivalence class of the vector bundle $\mathcal{E}$ on C_A representing the family (Lemma 10.22). For the same reason, tensoring with any line bundle $\xi \in \operatorname{Pic} C$, $\mathcal{E} \mapsto \mathcal{E} \otimes_{\mathcal{O}_C} \xi$ induces an isomorphism of functors

$$\mathcal{SU}_C(r, L) \overset{\sim}{\to} \mathcal{SU}_C(r, L \otimes \xi^r). \qquad (11.4)$$

It follows that the isomorphism class of the functor $\mathcal{SU}_C(r, L)$ depends only on $\deg L \bmod r$, and not on L itself. In the rest of this section we restrict to $r = 2$ and consider separately the cases when $\deg L$ is odd or even.

(a) Rank 2 vector bundles of odd degree

When deg L is odd we have already shown in the last chapter that the quotient variety $\mathrm{Alt}^s_{N,2}(H^0(L))/GL(N)$ has $SU_C(2, L)$ as its underlying set of points. In this section we are going to prove Theorem 11.3.

As in the last chapter (see Theorem 10.1) we assume that deg $L \geq 4g - 1$ and we let $N = \deg L + 2 - 2g$.

Proposition 11.34. *Suppose that* deg $L \geq 4g - 1$ *is odd. Then the quotient variety* $\mathcal{M}_L := \mathrm{Alt}^{ss}_{N,2}(H^0(L))/\!/GL(N)$ *is a coarse moduli space for the functor* $\mathcal{SU}_C(2, L)$.

The proof is similar to that of Proposition 11.24 for the Jacobian case. First we construct a natural transformation of functors

$$\mathcal{SU}_C(2, L) \to \underline{\mathcal{M}_L}, \tag{11.5}$$

and then we show that this satisfies the universal property.

Let A be a finitely generated ring over the field k and $\mathcal{E}$ be a rank 2 vector bundle on C_A. We suppose that $\mathcal{E}_t$ is stable for all $t \in \mathrm{Spm}\, A$ and that $\det \mathcal{E} \cong L_A \otimes M$ for some invertible A-module M (that is, for some line bundle on Spm A). We consider the A-module $H^0(\mathcal{E})$ of global sections (Definition 11.10). By Proposition 10.26 and Theorem 11.18, this is a locally free A-module of rank N. We will consider the skew-symmetric A-bilinear map:

$$H^0(\mathcal{E}) \times H^0(\mathcal{E}) \to H^0(\det \mathcal{E}) = H^0(L) \otimes_k M, \qquad (s, t) \mapsto s \wedge t.$$

Step 1. We first consider the case where $H^0(\mathcal{E})$ and M are both free A-modules. Then, by choosing free bases, the above map determines a skew-symmetric $N \times N$ matrix with entries in $H^0(L) \otimes_k A$. In other words, we get a morphism

$$\widetilde{\varphi} : \mathrm{Spm}\, A \to \mathrm{Alt}_N(H^0(L)).$$

The image of this map is contained in the zero-set of the 4×4 Pfaffian minors

$$\mathrm{Alt}_{N,2}(H^0(L)) \subset \mathrm{Alt}_N(H^0(L)),$$

and by Theorem 11.18 the image of each $t \in \mathrm{Spm}\, A$ is exactly a Gieseker point of the vector bundle $\mathcal{E}_t = \mathcal{E}|_{C \times t}$. As we saw in the last chapter, this is stable for the action of $GL(N)$, and so the morphism $\widetilde{\varphi}$ maps into the open set $\mathrm{Alt}^s_{N,2}(H^0(L))$. We denote the composition of $\widetilde{\varphi}$ with the quotient map

$\text{Alt}_{N,2}^s(H^0(L)) \to \mathcal{M}_L$ by

$$\varphi : \text{Spm } A \to \mathcal{M}_L.$$

This map depends only on $\mathcal{E}$ and not on the choice of basis for $H^0(\mathcal{E})$.

Step 2. Choose an affine open cover

$$\text{Spm } A = U_1 \cup \cdots \cup U_n$$

such that the restriction of $H^0(\mathcal{E})$ and M to each U_i are both free modules. Just as for the Picard functor, this gives, using Step 1, a morphism $\varphi : \text{Spm } A \to \mathcal{M}_L$.

Step 3. We have therefore constructed the natural transformation (11.5), and we will now check that it has the universal property. On the product $C \times \text{Alt}_N(H^0(L))$ there is a natural tautological homomorphism,

$$\mathcal{O}_{C \times \text{Alt}}^{\oplus N} \to (L \boxtimes \mathcal{O}_{\text{Alt}})^{\oplus N},$$

given by matrix multiplication, whose restriction to $C \times \text{Alt}_{N,2}^s(H^0(L))$ has rank 2. The image is then a rank 2 vector bundle whose restriction to $C \times \text{Alt}_{N,2}^s(H^0(L))$ we denote by $\mathcal{Q}$.

Now suppose that we have a natural transformation of functors $\mathcal{SU}_C(2, L) \to \underline{X}$ for some other variety X. Applying it to the vector bundle $\mathcal{Q}$ then determines a morphism $\text{Alt}_{N,2}^s(H^0(L)) \to X$; and since $\mathcal{Q}$ is preserved by the action of $GL(N)$, this descends to a morphism of the quotient $\mathcal{M}_L \to X$. $\qquad\square$

Proof of Theorem 11.3. It is enough to show that the vector bundle $\mathcal{Q}$ descends to the product $C \times \mathcal{M}_L$. However, by construction $\mathcal{Q}$ comes with a natural $GL(N)$-linearisation in which the element $-I_N \in GL(N)$ (which acts trivially on $\text{Alt}_{N,2}^s(H^0(L))$) acts as -1, so that $\mathcal{Q}$ cannot descend as it is. We can solve this problem by 'twisting' $\mathcal{Q}$ as a GR-module (where $G = GL(N)$ and R is the coordinate ring of $\text{Alt}_{N,2}^s(H^0(L))$). That is, we consider $\mathcal{Q}' := \mathcal{Q} \otimes_R \mathcal{D}$, where $\mathcal{D}$ denotes the trivial R-module linearised by $\det g$ for $g \in GL(N)$. Here $-I_N$ acts trivially, and we have a line bundle $\mathcal{Q}'$ which carries a $GL(N)/\pm I_N$-linearisation. (Note that this is only possible when $\deg L$, and hence N, is odd!)

By Lemma 10.64, all the orbits of the action $GL(N)/\pm I_N \curvearrowright \text{Alt}_{N,2}^s(H^0(L))$ are free closed orbits, so that, by Corollary 11.28, $\mathcal{Q}'$ is the pullback of some vector bundle $\mathcal{U}$ on $C \times \mathcal{M}_L$. Pulling back $\mathcal{U}$ via morphisms $\text{Spm } A \to \mathcal{M}_L$ then defines an inverse of the natural transformation (11.5), and hence the functor $\mathcal{S}_C(2, L)$ is represented by the variety $\mathcal{M}_L$. $\qquad\square$

More precisely, the functor $\mathcal{SU}_C(2, L)$ is represented by the pair consisting of $\mathcal{M}_L$ and the vector bundle $\mathcal{U}$ on $C \times \mathcal{M}_L$. This is called the *universal bundle*.

(b) Irreducibility and rationality

It follows from (11.4) that the functor $\mathcal{SU}_C(2, L)$ is represented by an algebraic variety for *every* line bundle L of odd degree, and from now on we will denote this variety by $SU_C(2, L)$. It is independent of L up to isomorphism, and we will show that it is irreducible and rational.

Consider a stable vector bundle $E \in SU_C(2, L)$ when $\deg L = 2g - 1$. By Riemann-Roch we have $h^0(E) - h^1(E) = 1$, so that $h^0(E) > 0$ and E contains $\mathcal{O}_C$ as a subsheaf. The saturation of this subsheaf is a line subbundle isomorphic to $\mathcal{O}_C(D)$ for some positive divisor $D \in \mathrm{Div} C$, and there is an exact sequence:

$$0 \to \mathcal{O}_C(D) \to E \to L(-D) \to 0. \tag{11.6}$$

Note that stability of E implies that $d := \deg D \le g - 1$.

Let us first consider the case $D = 0$. Here, the equivalence class of the extension

$$0 \to \mathcal{O}_C \to E \to L \to 0$$

is parametrised by the cohomology space $H^1(L^{-1})$, and by Riemann-Roch this has dimension $3g - 2$. The isomorphism class of the bundle is parametrised by the projectivisation of this space, and there is a moduli map from the open set U_0 parametrising stable bundles:

$$U_0 \qquad \subset \quad \mathbb{P}H^1(L^{-1}) \cong \mathbb{P}^{3g-3}$$

$$f_0 \downarrow$$

$$SU_C(2, L)$$

What about the case $D > 0$? Positive divisors of degree d on C are parametrised by the d-fold symmetric product $\mathrm{Sym}^d C$. (This is a nonsingular variety of dimension d.) For each $D \in \mathrm{Sym}^d C$, the extensions (11.6) are parametrised by the cohomology space $H^1(L^{-1}(2D))$, which has dimension $3g - 2 - 2d$, and in this case we obtain a moduli map from the open set U_d of stable bundles in a projective bundle over $\mathrm{Sym}^d C$ with fibre $\mathbb{P}H^1(L^{-1}(2D))$:

$$U_d \qquad \subset \quad \mathbb{P}^{3g-3-2d}\text{-bundle} \to \mathrm{Sym}^d C$$

$$f_d \downarrow$$

$$SU_C(2, L)$$

The images of the maps $f_d : U_d \to SU_C(2, L)$ for $0 \le d \le g - 1$ cover the moduli space. On the other hand, each U_d, if nonempty, has dimension $3g-3-d$; since $\dim SU_C(2, L) = 3g - 3$, it follows that U_0 must be nonempty. The map f_0 is therefore dominant; and since U_0 is irreducible, it follows at once that $SU_C(2, L)$ is irreducible.

Proposition 11.35. $SU_C(2, L)$ *is a rational variety when* $\deg L$ *is odd.*

Proof. We have seen that a general stable bundle $E \in SU_C(2, L)$ is a (nonsplit) extension of L by $\mathcal{O}_C$. In particular, $h^0(E) \ge 1$. If $h^0(E) > 1$ for a general stable bundle, then the general fibre of f_0 wold be positive-dimensional, contradicting the fact that $\dim U_0 = \dim SU_C(2, L)$. Hence $h^0(E) = 1$ for general $E \in SU_C(2, L)$, and so f_0 is birational – that is, it is an isomorphism over an open subset of the moduli space. $\qquad\qquad\square$

(c) Rank 2 vector bundles of even degree

We showed in the last chapter that when $\deg L$ is even the Gieseker points $T_{E,S}$ of a semistable vector bundle $E \in SU_C(2, L)$ are semistable for the action of $GL(N)$ on $\mathrm{Alt}_{N,2}(H^0(L))$ (Propositions 10.69 and 10.70). In a moment we will show that if E is a stable vector bundle, then its Gieseker points $T_{E,S}$ are $GL(N)$-stable. This implies, in particular, that the quotient variety $\mathrm{Alt}^s_{N,2}(H^0(L))/GL(N)$ has $SU_C(2, L)$ as its underlying set of points. However, unlike the odd degree case, there are now semistable vector bundles which are not stable, and as a consequence the quotient variety $\mathrm{Alt}^s_{N,2}(H^0(L))/GL(N)$ is not complete. It is contained as an open set in the projective variety

$$\overline{SU}_C(2, L) := \mathrm{Alt}^{ss}_{N,2}(H^0(L))/\!/GL(N),$$

and one can ask what the geometric points of this bigger variety correspond to in terms of vector bundles. This is answered by Proposition 11.37.

We will assume that $\deg L \ge 4g - 2$.

Proposition 11.36. *The Gieseker points of a stable vector bundle are* $GL(N)$-*stable.*

Proof. We have already observed in the proof of Theorem 10.1 (Section 10.4(b)) that a Gieseker point $T_{E,S}$ of a stable bundle E has a finite stabiliser, using Lemma 10.64 and the fact that E is simple. So we just have to show that the orbit of $T_{E,S}$ is closed.

So suppose that $T \in \mathrm{Alt}_{N,2}(H^0(L))$ is in the closure W of the orbit $GL(N) \cdot T_{E,S}$ of E. We have seen in Section 10.4(b) that such T is the Gieseker point of some semistable vector bundle E'. Moreover, there exists a vector bundle $\mathcal{E}$ on $C \times W$ such that $\mathcal{E}|_{C \times t} \cong E$ for t in an open set of W and $\mathcal{E}|_{C \times t} \cong E'$ for t in the boundary. Namely, $\mathcal{E}$ is the restriction of the universal bundle $\mathcal{Q}$ constructed in the proof of Proposition 11.34. If we apply semicontinuity (Proposition 11.21) to the bundle $E^{\vee} \otimes \mathcal{E}$, we see that

$$\dim \mathrm{Hom}(E, E') \geq \dim \mathrm{Hom}(E, E) \geq 1,$$

so that there exists a nonzero homomorphism $E \rightarrow E'$. But this is then an isomorphism by Proposition 10.24. This shows that the orbit $GL(N) \cdot T_{E,S}$ is closed. □

Proposition 11.37. *If E, E' are rank 2 vector bundles with the same determinant line bundle L and Gieseker points $T_{E,S}$, $T_{E',S'}$, then the following are equivalent:*

(i) $\mathrm{gr}\, E \cong \mathrm{gr}\, E'$ *(see Definition 10.43);*
(ii) $T_{E,S}$, $T_{E',S'}$ *are closure-equivalent under the action of $GL(N)$.*

Bundles E, E' satisfying condition (i) are said to be *S-equivalent*.

Proof. By Proposition 11.36, the Gieseker orbit of a stable vector bundle is closed and of maximal dimension; so if either of E, E' is stable, then (by Corollary 5.5 and its proof) condition (ii) is equivalent to $T_{E,S}$, $T_{E',S'}$ being in the same $GL(N)$-orbit. We therefore only need to consider the case where neither of E, E' is stable: in other words, we assume that they are extensions of line bundles:

$$0 \rightarrow L \rightarrow E \rightarrow M \rightarrow 0, \qquad \deg L = \deg M$$
$$0 \rightarrow L' \rightarrow E' \rightarrow M' \rightarrow 0, \qquad \deg L' = \deg M'.$$

Since, by hypothesis, $\deg L = \deg M \geq 2g - 1$, it follows that $h^0(L) = h^0(M) = h^0(E)/2$. So let $S = \{s_1, \dots, s_N\}$ be a basis of $H^0(E)$ in which $s_1, \dots, s_{N/2}$ are a basis of $H^0(L) \subset H^0(E)$. Then $T_{E,S}$ has the form

$$\left(\begin{array}{c|c} 0 & B \\ \hline -B^t & C \end{array} \right).$$

Now let

$$g(t) = \begin{pmatrix} t^{-1} & & & & & \\ & \ddots & & & 0 & \\ & & t^{-1} & & & \\ \hline & & & t & & \\ & 0 & & & \ddots & \\ & & & & & t \end{pmatrix} \in SL(N).$$

Then, as we have already seen in the proof of Proposition 10.66,

$$\lim_{t \to 0} g(t) T_{E,S} g(t)^t = \left(\begin{array}{c|c} 0 & B \\ \hline -B^t & 0 \end{array} \right).$$

But this is the Gieseker point of the decomposable vector bundle $L \oplus M = \mathrm{gr}(E)$, and we see that $T_{\mathrm{gr}(E),S}$ is contained in the closure of the orbit $GL(N) \cdot T_{E,S}$. Hence we have shown that (i) implies (ii).

For the converse, the idea is the same as in the proof of Proposition 11.36. Suppose that the two orbit closures have an intersection point:

$$T \in \overline{GL(N) \cdot T_{E,S}} \cap \overline{GL(N) \cdot T_{E',S'}}.$$

Then T is a Gieseker point of some semistable vector bundle F, and, as in the proof of Proposition 11.36, we can find a family of vector bundles $\mathcal{E}$ which is equal to E on an open set and jumps to F on the boundary. We then apply upper semicontinuity (Proposition 11.21) to the family $L^{-1} \otimes \mathcal{E}$, where $L \subset E$ is the same line subbundle as above. This gives

$$\dim \mathrm{Hom}(L, F) \geq \dim \mathrm{Hom}(L, E) \geq 1,$$

so that L is contained as a line subbundle in F. By the same reasoning L' is also a line subbundle of F. But since F is semistable, this implies that either $L \cong L'$ or $F \cong L \oplus L'$. Either way, we conclude that $\mathrm{gr}(E) \cong \mathrm{gr}(E')$. □

Recall that if ξ is a line bundle, $\widehat{\xi} = L \otimes \xi^{-1}$, and the multiplication map

$$H^0(\xi) \times H^0(\widehat{\xi}) \to H^0(L)$$

is represented by a matrix T, then the vector bundle $E = \xi \oplus \widehat{\xi}$ has as a Gieseker point the matrix

$$\begin{pmatrix} 0 & T \\ -T^t & 0 \end{pmatrix}$$

(Example 10.63). This means that the map

$$\text{Pic}^{d/2}C \to \overline{SU}_C(2, L), \qquad \xi \mapsto \xi \oplus \hat{\xi},$$

is induced in the quotient by the map

$$\text{Mat}_{N/2,1}(H^0(L)) \to \text{Alt}_{N,2}(H^0(L)), \qquad T \mapsto \begin{pmatrix} 0 & T \\ -T^t & 0 \end{pmatrix}.$$

Thus $\overline{SU}_C(2, L)$ contains $SU_C(2, L)$ as a an open set, with complement (the semistable boundary) equal to the image of $\text{Pic}^{d/2}C$ (called the *Kummer variety*).

Remark 11.38. Unlike the odd degree case, $SU_C(2, L)$ is not a fine moduli space. That is, it can be shown that there is no universal vector bundle on the product $C \times SU_C(2, L)$. (See Ramanan [74].)

11.3 Examples

In this section we explain, first, how one can write down explicitly the construction of the Jacobian given in Section 9.4, and then we give some examples of moduli spaces of vector bundles.

(a) The Jacobian of a plane quartic

Let $C \subset \mathbb{P}^2$ be a nonsingular plane curve of degree e defined by a homogeneous polynomial equation $f_e(x, y, z) = 0$. As we saw in Section 9.2, this has genus $g = (e - 1)(e - 2)/2$. For the auxiliary line bundle L of degree $2d$ used in Section 9.4 we will take the restriction to C of $\mathcal{O}_{\mathbb{P}^2}(e - 1)$, with $d = e(e - 1)/2$. When $e = 3, 4$ we have $d \geq 2g$, so that by the methods of Section 9.4 we can construct $\text{Pic}^d C$ as the quotient variety

$$\text{Mat}^s_{e,1}(H^0(L))/GL(e, e).$$

We may identify $H^0(L)$ with the set of homogeneous polynomials of degree $e - 1$ in coordinates x, y, z.

Let us consider the case of a plane quartic curve – that is, $e = 4$. In this case,

$$\text{Pic}^6 C = \text{Mat}^s_{4,1}(H^0(L))/GL(4, 4),$$

where $\text{Mat}_{4,1}(H^0(L))$ consists of 4×4 matrices

$$A = \begin{pmatrix} d_{11}(x, y, z) & d_{12}(x, y, z) & d_{13}(x, y, z) & d_{14}(x, y, z) \\ d_{21}(x, y, z) & d_{22}(x, y, z) & d_{23}(x, y, z) & d_{24}(x, y, z) \\ d_{31}(x, y, z) & d_{32}(x, y, z) & d_{33}(x, y, z) & d_{34}(x, y, z) \\ d_{41}(x, y, z) & d_{42}(x, y, z) & d_{43}(x, y, z) & d_{44}(x, y, z) \end{pmatrix}$$

of cubic forms $d_{ij}(x, y, z)$ all of whose 2×2 minors vanish on C; in other words, these 2×2 minors are 36 sextics which are divisible by the quartic $f_4(x, y, z)$ defining $C \subset \mathbb{P}^2$.

Proposition 11.39. *A matrix* $A \in \text{Mat}_{4,1}(H^0(L))$, *as above, is stable under the* $GL(4, 4)$ *action if and only if it is of one of the following forms.*

(1) A is the matrix of cofactors of a 4×4 matrix of linear forms

$$M = (l_{ij}(x, y, z))_{1 \le i, j \le 4}, \quad \text{where } \det M = f_4(x, y, z).$$

(2) A is of the form, up to the action of $GL(4, 4)$,

$$A = \left(\begin{array}{c|c} d(x, y, z) & q(x, y, z)\mathbf{x}^t \\ \hline q'(x, y, z)\mathbf{x} & l(x, y, z)\mathbf{x}\mathbf{x}^t \end{array} \right), \quad \mathbf{x} = \begin{pmatrix} x \\ y \\ z \end{pmatrix},$$

where $\deg l = 1$, $\deg q = \deg q' = 2$, $\deg d = 3$, *and* $qq' - ld = f_4$.

Proof. By Propositions 9.62 and 9.63, stability is equivalent to the matrix having rank 4 over the field k. We consider the cofactor matrix A^* of A. Recall that for an $N \times N$ matrix this has the properties:

(i) $(A^*)^* = (\det A)^{N-2}A$;   (ii) $\det A^* = (\det A)^{N-1}$.

In the present case, A^* is a 4×4 matrix of forms of degree 9, and these are all divisible by f_4^2. There are therefore two possibilities: either $A^* \equiv 0$, or $A^* = f_4^2 M$, where $M = (l_{ij}(x, y, z))$ is a nonzero matrix of linear forms. Let us first deal with the latter case. Taking determinants, we have $(\det A)^3 = f_4^8 \det M$, and by unique factorisation this implies that $\det M = f_4$, $\det A = f_4^3$. Hence

$$A = \frac{(A^*)^*}{(\det A)^2} = \frac{1}{f_4^6}(A^*)^* = \frac{1}{f_4^6}(f_4^2 M)^* = M^*,$$

and we have case (1) of the proposition.

The second case is when A^* vanishes identically or, equivalently, A has rank 2 over the function field $k(C)$. In this case, up to the $GL(4, 4)$ action A can be written

$$
\left(
\begin{array}{c|ccc}
d(x,y,z) & * & * & * \\
\hline
* & & & \\
* & & B & \\
* & & &
\end{array}
\right),
$$

where B is a 3×3 matrix of cubic forms with rank 1. By Remark 9.55 it is easy to see that, up to the action of $GL(4,4)$, the block B can be written as $l(x,y,z)\mathbf{x}\mathbf{x}^t$ for some linear form l. Since A has rank 1 along the curve $C \subset \mathbb{P}^2$, this forces it to take the form (2) of the proposition. $\qquad\square$

The second case corresponds to an element of $\mathrm{Pic}^6 C$ of the form $\mathcal{O}_C(1) \otimes \mathcal{O}_C(p + p')$ for some points $p, p' \in C$, while all other line bundles correspond to the first case. We can identify the same quotient with $\mathrm{Pic}^2 C$ by the bijection

$$
\mathrm{Pic}^2 C \xrightarrow{\sim} \mathrm{Pic}^6 C, \qquad \xi \mapsto \xi \otimes \mathcal{O}_C(1),
$$

and then the set of matrices (2) is the theta divisor

$$
\Theta = \{\mathcal{O}_C(p + p') \mid p, p' \in C\} \cong \mathrm{Sym}^2 C,
$$

while the set of matrices (1) is its complement.

(b) The affine Jacobian of a spectral curve

For any curve of genus g the set

$$
\Theta := \{\mathcal{O}_C(p_1 + \cdots + p_{g-1}) \mid p_1, \ldots, p_{g-1} \in C\} \subset \mathrm{Pic}^{g-1} C
$$

is called the *theta divisor* in $\mathrm{Pic}^{g-1} C$, and its complement is an affine variety. As we have seen in Section 9.4(b), $\mathrm{Pic}^{g-1} C$ is a projective variety and by construction there exists a semiinvariant whose zero-set is Θ (or, more precisely, is 2Θ). This complement,

$$
\mathrm{Aff.Jac}\ C := \mathrm{Pic}^{g-1} C - \Theta,
$$

is called the *affine Jacobian* of C. The affine Jacobian of a plane curve of degree e can be described explicitly in the following way.

Proposition 11.40. *Given a homogeneous polynomial $f_e(x,y,z)$ of degree e, denote by $\mathrm{Mat}_e(x,y,z;f_e)$ the set of $e \times e$ matrices of linear forms*

$$
M = (l_{ij}(x,y,z))_{1 \le i,j \le e} \quad \text{such that } \det M = f_e(x,y,z).
$$

This is a closed subvariety in an affine space $\mathbb{A}^n$, $n = 3e^2$, *which is acted on by* $SL(e) \times SL(e)$ *via* $M \mapsto gMg'$. *Then the plane curve* $C = \{f_e(x, y, z) = 0\} \subset \mathbb{P}^2$ *has an affine Jacobian equal to the affine quotient*

$$\text{Aff.Jac } C = \text{Mat}_e^s(x, y, z; f_e)/SL(e) \times SL(e).$$

$\square$

This proposition is a special case of Theorem 11.41 below. When $e = 3$ or 4, we obtain a completion (or, in other words, a compactification) of the affine Jacobian by assigning to a matrix $M \in \text{Mat}_e(x, y, z; f_e)$ its matrix of cofactors. The e-sheeted cover $C \to \mathbb{P}^1$ defined by the equation

$$\tau^e + f_m(x, y)\tau^{e-1} + f_{2m}(x, y)\tau^{e-2} + \cdots + f_{(e-1)m}(x, y)\tau$$
$$+ f_{em}(x, y) = 0, \tag{11.7}$$

where each $f_i(x, y)$ is homogeneous of degree i, if it is nonsingular, is called a *spectral curve* of degree e and index m. In particular, every nonsingular plane curve $C \subset \mathbb{P}^2$ is a spectral curve of index 1.

Theorem 11.41. *Let* $C \to \mathbb{P}^1$ *be the spectral curve (11.7) and denote by* $\text{Mat}_e(x, y; C)$ *the set of* $e \times e$ *matrices* $M = (h_{ij}(x, y))_{1 \le i, j \le e}$ *of homogeneous polynomials of degree m, with a fixed characteristic polynomial*

$$\det(M - \tau I_e) = \text{left-hand side of (11.7)}.$$

Then $\text{Mat}_e(x, y; C)$ *is an affine variety on which* $SL(e)$ *acts by conjugation,* $M \mapsto gMg^{-1}$, $g \in SL(e)$, *and the affine quotient is the affine Jacobian of* C,

$$\text{Aff.Jac } C = \text{Mat}_e^s(x, y; C)/SL(e).$$

$\square$

This theorem is the analogue for function fields of Theorem 8.66. It can be proved in a similar manner, making use also of Grothendieck's Theorem 10.31. We omit the details here, but see Beauville et al. [78], [79].

(c) The Jacobian of a curve of genus 1

Let C be a curve of genus 1. Over an algebraically closed field the variety $\text{Pic}^d C$ is always isomorphic to C. Here we shall consider what happens when the field k is not necessarily algebraically closed. (For example, the rational numbers $k = \mathbb{Q}$. For simplicity we shall, nevertheless, continue to assume that k has characteristic zero.)

First of all, note that C need not necessarily possess any rational points over k. If C does have a rational point, then by taking this to be on the line at infinity in the projective plane C can be represented as the plane cubic

$$y^2 = 4x^3 - g_2 x - g_3, \qquad g_2, g_3 \in k.$$

This is called the *Weierstrass canonical form* of a plane cubic.

Definition 11.42. Let C be a curve of genus 1 over k. A curve J over k which possesses a rational point over k and which is isomorphic to C over the algebraic closure $\overline{k}$ is said to be *arithmetically* the Jacobian of C. $\qquad\square$

In the construction of $\mathrm{Pic}^d C$ given in Section 9.4, suppose that the auxiliary line bundle L is defined over k. Then the quotient variety $\mathrm{Pic}^d C$ is also defined over k.

Remark 11.43. We are not assuming here that the Poincaré line bundle is defined over k, and so we view $\mathrm{Pic}^d C$ only as a coarse moduli space. However, it is clear from the construction that there will exist such a Poincaré line bundle *provided* C possesses a k-rational point or, more generally, a divisor of degree 1 defined over k. (See Remark 11.29.) $\qquad\square$

When C is defined over k there exists some positive line bundle L_0 defined over k.

$\underline{\deg L_0 = 1}$. In this case, by taking the zero-set of a global section we get a rational point of C over k, and hence C is arithmetically its own Jacobian.

When $\deg L_0 \geq 2$, we see by taking $L = L_0^2$ in the quotient construction of Section 9.4 that $\mathrm{Pic}^d C$ is arithmetically the Jacobian of C. This can be identified with $\mathrm{Pic}^0 C$ by $\otimes L_0 : \mathrm{Pic}^0 C \xrightarrow{\sim} \mathrm{Pic}^d C$.

$\underline{\deg L_0 = 2}$. In this case $\dim H^0(L_0) = 2$, and the ratio of two linearly independent global sections determines a 2-sheeted cover $C \to \mathbb{P}^1$. This has four branch points, which are the zeros of a binary quartic

$$(a \lozenge x, y) = a_0 x^4 + 4a_1 x^3 y + 6a_2 x^2 y^2 + 4a_3 x y^3 + a_4 y^4,$$

and C can be represented as

$$C: \quad \tau^2 = a_0 x^4 + 4a_1 x^3 y + 6a_2 x^2 y^2 + 4a_3 x y^3 + a_4 y^4. \tag{11.8}$$

Since C is defined over k, the coefficients $a_0, \ldots, a_4$ belong to k.

Theorem 11.44. *If C is the genus 1 curve (11.8), then the arithmetic Jacobian* Pic^0*C is the affine plane curve*

$$y^2 = \det \left| x \begin{pmatrix} & & 2 \\ & -1 & \\ 2 & & \end{pmatrix} - \begin{pmatrix} a_0 & a_1 & a_2 \\ a_1 & a_2 & a_3 \\ a_2 & a_3 & a_4 \end{pmatrix} \right|$$

together with its point at infinity.

Remark 11.45. The right-hand side in this equation is the cubic used in the solution by radicals of the quartic $(a \lozenge x, y) = 0$. We have already used this equation in Section 1.3(b) (See Remark 1.26.) ☐

Theorem 11.44 will follow from the next proposition. If L is any line bundle of degree 4, then dim $H^0(L) = 4$, and four linearly independent sections define an embedding $C \hookrightarrow \mathbb{P}^3$, whose image is a curve of degree 4 and is the intersection of two quadric surfaces $q_0(x_0, x_1, x_2, x_3) = q_1(x_0, x_1, x_2, x_3) = 0$. For each $i = 0, 1$ we can write

$$q_i(x_0, x_1, x_2, x_3) = \mathbf{x}^t Q_i \mathbf{x}, \qquad \mathbf{x} = (x_0, x_1, x_2, x_3)^t$$

for symmetric 4×4 matrices Q_0, Q_1, and we consider the relative characteristic polynomial

$$\det(x Q_0 + y Q_1).$$

This is a binary quartic in x, y.

Proposition 11.46. *If C is the genus 1 curve* $q_0(x_0, x_1, x_2, x_3) = q_1(x_0, x_1, x_2, x_3) = 0$*, then the arithmetic Jacobian* Pic^2*C is the double cover of* $\mathbb{P}^1$ *with equation*

$$\tau^2 = \det(x Q_0 + y Q_1).$$

☐

Proof of Theorem 11.44. Let $L = L_0^2$, and let $x^2, 2xy, y^2, \tau \in H^0(L)$ be a basis for the global sections, where $x, y \in H^0(L_0)$. This basis determines an embedding,

$$C \to \mathbb{P}^3, \qquad (x : y : \tau) \mapsto (x^2 : 2xy : y^2 : \tau) =: (x_0 : x_1 : x_2 : x_3),$$

whose image is contained in the quadric surfaces

$$x_1^2 - 4x_0x_2 = 0,$$
$$a_0x_0^2 + 2a_1x_0x_1 + a_1(2x_0x_2 + x_1^2) + 2a_3x_1x_2 + a_4x_2^2 - x_3^2 = 0.$$

The relative characteristic polynomial of these quadrics is

$$\det \left| -x \left(\begin{array}{ccc|c} & & 2 & \\ & -1 & & \\ 2 & & & \\ \hline & & & 0 \end{array} \right) + \left(\begin{array}{ccc|c} a_0 & a_1 & a_2 & \\ a_1 & a_2 & a_3 & \\ a_2 & a_3 & a_4 & \\ \hline & & & -1 \end{array} \right) \right|,$$

and this is the right-hand side of the equation in the theorem.   □

In order to prove Proposition 11.46 we follow the construction of Section 9.4 when $d = 2$ and the auxiliary line bundle is L as above. $\text{Pic}^2 C$ is the quotient variety

$$\text{Mat}_{2,1}(H^0(L)) /\!/ GL(2,2), \tag{11.9}$$

where $\text{Mat}_{2,1}(H^0(L))$ consists of 2×2 matrices of linear forms in $x = (x_0, x_1, x_2, x_3)$

$$\begin{pmatrix} l_{11}(x) & l_{12}(x) \\ l_{21}(x) & l_{22}(x) \end{pmatrix}$$

whose determinant vanishes on $C \subset \mathbb{P}^3$. In other words, the determinant of such a matrix is a linear combination of $q_0(x)$ and $q_1(x)$. It happens that the ring of semiinvariants determining the quotient variety (11.9) is well known and can be written down explicitly.

Theorem 11.47. *Let* $\text{Mat}_2(x_0, \ldots, x_m)$ *be the set of* 2×2 *matrices of linear forms in* $x_0, \ldots, x_m$. *This space is acted on by* $GL(2,2)$, *and the ring of semiinvariants has the following generators.*

(1) Weight 1. The $(m+1)(m+2)/2$ coefficients of the quadratic form $\det M(x)$, where $M(x) \in \text{Mat}_2(x_0, \ldots, x_m)$. These are quadratic forms on $\text{Mat}_2(x_0, \ldots, x_m)$.

(2) Weight 2. Writing $M(x) \in \text{Mat}_2(x_0, \ldots, x_m)$ as

$$M_0 x_0 + M_1 x_1 + \cdots + M_m x_m,$$

the $\binom{m+1}{4}$ determinants $\det |M_i, M_j, M_k, M_l|$ for $i < j < k < l$, where each M_i is viewed as a vector in k^4. These are quartic forms on Mat_2 $(x_0, \ldots, x_m)$.   □

This follows from the next result, due to Weyl ([60] Theorems (2.9A) and (2.17A)). The 4-dimensional vector space of 2×2 matrices $V = \mathrm{Mat}_2(k)$ carries an inner product,

$$M, M' \mapsto \mathrm{tr}\,(M^*M'),$$

where M^* denotes the matrix of cofactors of M. With respect to this inner product the action of the image of $SL(2) \times SL(2)$ in $GL(2, 2)$ lives in the special orthogonal group $SO(V)$.

Theorem 11.48. *Let* $V, \langle\,,\,\rangle$ *be any n-dimensional inner product space. Under the diagonal action of* $SO(V)$ *on a direct sum* $V \oplus \cdots \oplus V = \bigoplus^r V$, *the ring of invariants has the following generators. Moreover, when* $r \leq n$ *these generators are algebraically independent.*

(1) The $\binom{r+1}{2}$ *inner products* $f_{ij}(v_1, \ldots, v_r) := \langle v_i, v_j \rangle$, *for* $1 \leq i \leq j \leq r$.
(2) The $\binom{r}{n}$ *determinants* $f_I(v_1, \ldots, v_r) := \det |v_{i_1}, \ldots, v_{i_n}|$, *where* $I = \{1 \leq i_1 < \cdots < i_n \leq r\}$. $\qquad\square$

The quotient

$$\mathrm{Mat}_2(x_0, x_1, x_2, x_3)//GL(2, 2)$$

is a variety of dimension $4 \times 4 - 7 = 9$, and by Theorem 11.47 it is embedded in 10-dimensional weighted projective space $\mathbb{P}(1^{10} : 2)$. The square of the semiinvariant (2) can be expressed as a quartic form in the semiinvariants (1) (in fact the determinant), and hence the quotient is a quartic hypersurface in $\mathbb{P}(1^{10} : 2)$. To say this another way, it is a double cover

$$\pi : \mathrm{Mat}_2(x_0, x_1, x_2, x_3)//GL(2, 2) \xrightarrow{2:1} \mathbb{P}^9 = \{\text{quadrics in } \mathbb{P}^3\}$$

branched over a quartic hypersurface $B \subset \mathbb{P}^9$.

Proof of Proposition 11.46. The variety $\mathrm{Pic}^2 C$ for the curve $C \subset \mathbb{P}^3$ is the inverse image $\pi^{-1}(\mathbb{P}^1)$, where $\mathbb{P}^1 \subset \mathbb{P}^9$ is the span of Q_0 and Q_1. The intersection of this line with B is the zero-set of the relative characteristic polynomial, so the proposition follows. $\qquad\square$

$\underline{\deg L_0 = 3.}$ In this case $\dim H^0(L_0) = 3$ and the ratios of the global sections define an embedding of the curve as a plane cubic $C \subset \mathbb{P}^2$. Since L_0 is defined over k, the equation $f_3(x, y, z) = 0$ of the cubic has coefficients in k. By

Proposition 11.40, the affine Jacobian of C is the quotient variety

$$\text{Mat}_3(x, y, z; f_3)//SL(3) \times SL(3),$$

and the arithmetic Jacobian of C is the one-point compactification of this variety. Let us denote a general element of $\text{Mat}_3(x, y, z; f_3)$ by $M = M_0x + M_1y + M_2z$.

Proposition 11.49. *The ring of semiinvariants of the action* $GL(3, 3) \curvearrowright$ $\text{Mat}_3(x, y, z)$ *has Hilbert series*

$$\frac{1 - t^6}{(1 - t)^{10}(1 - t^2)(1 - t^3)}$$

and has the following 12 generators.

(1) Weight 1. The 10 coefficients (that is, mixed determinants) of the cubic form

$$\det(M_0x + M_1y + M_2z).$$

(2) Weight 2. The trace

$$f(M) = \text{tr}\,(M_0^* M_1^* M_2^*),$$

where M_i^* is the matrix of cofactors of M_i.
(3) Weight 3. The trace

$$g(M) = \text{tr}\,(M_0 M_1 M_2 M_0^* M_1^* M_2^*).$$

$\square$

(This proposition is a rephrasing of the results of Teranishi [87] on the ring of invariants of the conjugation action $SL(3) \curvearrowright \text{Mat}_3(x, y, z)$.)

From the Hilbert series and the known weights of the generators it follows that there must be a relation among these generators of the form

$$g^2 + a_3g = a_0 f^3 + a_2 f^2 + a_4 f + a_6, \qquad (11.10)$$

in which each a_i is a homogeneous polynomial of degree i in the 10 generators of weight 1. This shows that the affine Jacobian of the plane cubic C is the plane cubic (11.10) in which the a_i are obtained by specialising the 10 weight 1 semiinvariants to the 10 coefficients of the defining equation $f_3(x, y, z)$ of C.

Remark 11.50. The arithmetic Jacobian of a curve of genus 1 in degree 2 was investigated by Weil in [88]. In a remark on this paper in his collected works he makes the following observation.

The 'covariants' of a plane cubic $f(x, y, z) = 0$ are generated by three forms denoted classically by H, J, Θ (see Salmon [35]). Of these, $H = H(f)$ is the

Hessian (see Example 5.26), of degree 3, and J, Θ are both sextics. This means that every covariant plane curve with respect to the action of the projective group is a polynomial in H, J, Θ with coefficients in $k[\mathbb{V}_{2,3}]^{SL(3)} = k[S, T]$. (For example, the degree 9 covariant consisting of the inflectional lines of the cubic can be shown to have equation $5Sf^2H(f) - H(f)^3 - f\Theta(f) = 0$.)

These generating covariants satisfy a single relation:

$$y^2 = 4x^3 + 108Sxz^2 - 27Tz^3, \qquad \text{where } x = \Theta, \, y = J, z = H^2.$$

Weil's observation is that this is precisely the equation of the arithmetic Jacobian of $f(x, y, z) = 0$. □

(d) Vector bundles on a spectral curve

We will extend Theorem 11.41 in this section to describe the moduli of rank 2 vector bundles on the spectral curve (11.7). First, let us make some remarks about the cofactors of a skew-symmetric $e \times e$ matrix. Let $A = (a_{ij})$ be such a matrix, where e is even, and denote by A_{ij} the skew-symmetric $(e-2) \times (e-2)$ submatrix obtained by deleting the i-th and j-th rows and columns of A.

Definition 11.51. The *cofactor matrix* of a 2×2 skew-symmetric matrix is defined to be

$$\begin{pmatrix} 0 & a \\ -a & 0 \end{pmatrix}^{\mathrm{adj}} := \begin{pmatrix} 0 & -1 \\ 1 & 0 \end{pmatrix}.$$

If A is an $N \times N$ skew-symmetric matrix for even $N \geq 4$, then its cofactor matrix A^{adj} is defined to be minus the $N \times N$ skew-symmetric matrix whose (i, j)-th entry is $(-1)^{i+j}\mathrm{Pfaff}\, A_{ij}$. □

For example, when $N = 4$,

$$\begin{bmatrix} a & b & c \\ & d & e \\ & & f \end{bmatrix}^{\mathrm{adj}} = \begin{bmatrix} f & -e & d \\ & c & -b \\ & & a \end{bmatrix}^{t} = \begin{bmatrix} -f & e & -d \\ & -c & b \\ & & -a \end{bmatrix}.$$

Corresponding to the identity $AA^* = (\det A)I$, where A^* is the matrix of cofactors (transposed) of a general square matrix A, the cofactor matrix of a skew-symmetric matrix satisfies:

$$AA^{\mathrm{adj}} = (\mathrm{Pfaff}\, A)I.$$

Before we come to general spectral curves, let us return to plane quartics $C \subset \mathbb{P}^2$. With respect to the line bundle $L = \mathcal{O}_C(3)$, the compactified moduli space of semistable vector bundles $\overline{SU}_C(2, L)$ is the projective quotient

$$\text{Alt}_{8,2}(H^0(\mathcal{O}_C(3)))//GL(8),$$

where $\text{Alt}_{8,2}(H^0(\mathcal{O}_C(3)))$ is the affine variety of 8×8 skew-symmetric matrices $[d_{ij}(x, y, z)]$ whose entries are cubic forms in the homogenous coordinates x, y, z and whose 70 4×4 Pfaffian minors are all divisible by the quartic $f_4(x, y, z)$ which defines the curve.

Moreover, mapping $A \mapsto A^{\text{adj}}$ defines a morphism

$$\text{Alt}_{8,2}(H^0(\mathcal{O}_C(1)))//SL(8) \to \text{Alt}_{8,2}(H^0(\mathcal{O}_C(3)))//GL(8) \cong \overline{SU}_C(2, L),$$

where the left-hand space is the affine quotient of the affine variety $\text{Alt}_{8,2}(H^0(\mathcal{O}_C(1)))$ (in the affine space $\mathbb{A}^{84}$) of 8×8 skew-symmetric matrices of linear forms in x, y, z with Pfaffian equal to $f_4(x, y, z)$. This map is an open immersion and its image is the set of (S-equivalence classes of) semistable vector bundles E satisfying $H^0(E(-1)) = 0$.

Note that $\det E(-1) \cong \Omega_C$. More generally, suppose that E is a vector bundle with canonical determinant $\det E \cong \Omega_C$ on a curve C. If $H^0(E) = 0$, then E is semistable, and this condition defines an open subset of $\overline{SU}_C(2, L)$. In fact, as we saw in the proof of Proposition 10.70, its complement is the zero-set of a semiinvariant and this open subset is therefore an affine variety,

$$\overline{SU}_C^{\text{aff}}(2, \Omega_C) = \{E \mid H^0(E) = 0\} \subset \overline{SU}_C(2, \Omega_C).$$

Proposition 11.40 and Theorem 11.41 now extend to the following.

Proposition 11.52. *If C is the plane curve $\{f_e(x, y, z) = 0\} \subset \mathbb{P}^2$, then*

$$\overline{SU}_C^{\text{aff}}(2, \Omega_C) \cong \text{Alt}_{2e}(x, y, z; f_e)//SL(2e),$$

where $\text{Alt}_{2e}(x, y, z; f_e)$ is the affine variety of skew-symmetric $2e \times 2e$ matrices with Pfaffian equal to $f_e(x, y, z)$.   □

Theorem 11.53. *Let $C \to \mathbb{P}^1$ be the spectral curve (11.7) of degree e and index m,*

$$\tau^e + f_m(x, y)\tau^{e-1} + f_{2m}(x, y)\tau^{e-2} + \cdots + f_{(e-1)m}(x, y)\tau + f_{em}(x, y) = 0.$$

Let $\text{Alt}_{2e}(x, y, z; C)$ be the set of $2e \times 2e$ skew-symmetric matrices $A = (h_{ij}(x, y))$ of homogeneous polynomials of degree m, with fixed characteristic *polynomial*

$$\mathrm{Pf}\left(A + \tau \begin{pmatrix} 0 & I_e \\ -I_e & 0 \end{pmatrix}\right) = \tau^e + f_m(x, y)\tau^{e-1} + f_{2m}(x, y)\tau^{e-2} + \cdots$$
$$\cdots + f_{(e-1)m}(x, y)\tau + f_{em}(x, y).$$

Then

$$\overline{SU}_C^{\mathrm{aff}}(2, \Omega_C) = \mathrm{Alt}_{2e}(x, y, z; C)//\mathrm{Sp}(2e).$$

$\square$

(e) Vector bundles on a curve of genus 2

A curve that can be expressed as a double cover $C \xrightarrow{2:1} \mathbb{P}^1$ of the projective line is called a *hyperelliptic curve* (Example 9.7). The number of branch points is always even and is $2g + 2$, where g is the genus. We will consider here the case $g = 2$. Then, if we take homogeneous coordinates $(x : y)$ on $\mathbb{P}^1$, the curve can be expressed as

$$C : \quad \tau^2 = f_6(x, y)$$

for some sextic form $f_6(x, y) = (a_0, \ldots, a_6 \emptyset x, y)$. We denote by $\mathcal{O}_C(1)$ the pullback of the tautological line bundle on $\mathbb{P}^1$. (In terms of divisors, this is $\mathcal{O}_C(p + q)$, where $\{p, q\} \subset C$ is the inverse image of a point of $\mathbb{P}^1$.) Since $\deg \mathcal{O}_C(1) = 2$ and $h^0(\mathcal{O}_C(1)) \geq 2$, it follows from Riemann-Roch that $\mathcal{O}_C(1) \cong \Omega_C$, the canonical line bundle, and that $h^0(\mathcal{O}_C(1)) = 2$. By Riemann-Roch and the Vanishing Theorem 9.20 we have $h^0(\mathcal{O}_C(2)) = 3$ and $h^0(\mathcal{O}_C(3)) = 5$. We can therefore take as bases

$$x^2, 2xy, y^2 \in H^0(\mathcal{O}_C(2)), \qquad x^3, 3x^2y, 3xy^2, y^3, \tau \in H^0(\mathcal{O}_C(3)).$$

The latter determines an embedding $C \hookrightarrow \mathbb{P}^4$ whose image is the intersection of four quadrics (using homogeneous coordinates $(x_0 : x_1 : x_2 : x_3 : x_4)$)

$$x_0x_2 - x_1^2 = x_0x_3 - x_1x_2 = x_1x_3 - x_2^2 = 0$$
$$a_0x_0^2 + 2a_1x_0x_1 + a_2(2x_0x_2 + x_1^2) + 2a_3(x_0x_3 + x_1x_2) \qquad (11.11)$$
$$+a_4(2x_1x_3 + x_2^2) + 2a_5x_2x_3 + a_6x_3^2 = x_4^2.$$

The construction of Chapter 10 gave a variety of parametrising rank 2 vector bundles E with $\det E \cong L$, for some line bundle $L \in \mathrm{Pic}\, C$, which satisfy $H^1(E) = 0$ and the condition that E is generated by global sections. By Propositions 10.26 and 10.27, these two conditions are guaranteed for all semistable bundles if $\deg L \geq 4g$ and for all stable bundles if $\deg L \geq 4g - 2$. For $g = 2$, let us take $L = \mathcal{O}_C(3)$ and construct the moduli space $SU_C(2, \mathcal{O}_C(3))$ of stable vector bundles with this determinant.

In this case, a stable bundle E has $h^0(E)$ equal to $N = 4$ (see Notation 10.58), and so $SU_C(2, \mathcal{O}_C(3))$ is the open set of stable points in the projective quotient of the action

$$GL(4) \curvearrowright \mathrm{Alt}_{4,2}(H^0(\mathcal{O}_C(3))).$$

An element of $\mathrm{Alt}_{4,2}(H^0(\mathcal{O}_C(3)))$ is a 4×4 skew-symmetric matrix of linear forms in $x_0, \ldots, x_4$ which has rank 2 along the curve $C \subset \mathbb{P}^4$, or, in other words, whose Pfaffian is a linear combination of the four quadrics (11.11).

First of all, we consider the action of $GL(4)$ on the affine space $\mathrm{Alt}_4(x_0, x_1, x_2, x_3, x_4) = \mathrm{Alt}_4(k) \otimes H^0(\mathcal{O}_C(3))$ of matrices without any rank condition. The 6-dimensional vector space $\mathrm{Alt}_4(k)$ comes with an inner product,

$$(X, Y) \mapsto \mathrm{tr}\, X^{\mathrm{adj}} Y,$$

and this inner product is preserved by the subgroup $SL(4) \subset GL(4)$; that is, $SL(4)$ acts on $\mathrm{Alt}_4(x_0, x_1, x_2, x_3, x_4)$ by $SL(4) \to SO(6)$. We can therefore apply Theorem 11.48 to obtain:

Proposition 11.54. *The ring of semiinvariants of the action* $GL(4) \curvearrowright$ $\mathrm{Alt}_4(x_0, x_1, x_2, x_3, x_4)$ *has 15 algebraically independent generators of weight 1. These are the coefficients* Pf_{ij} *of the quadratic form*

$$\mathrm{Pfaff}\, A = \sum_{i \leq j} \mathrm{Pf}_{ij}(A) x_i x_j, \qquad A \in \mathrm{Alt}_4(x_0, x_1, x_2, x_3, x_4).$$

$\square$

It follows that the projective quotient $\mathrm{Alt}_4(x_0, x_1, x_2, x_3, x_4)/\!/GL(4)$ is isomorphic to $\mathbb{P}^{14}$ and that

$$\mathrm{Alt}_{4,2}(x_0, x_1, x_2, x_3, x_4)/\!/GL(4) = \begin{pmatrix} \text{linear span of 4 points} \\ \text{corresponding to the} \\ \text{quadrics (11.11)} \end{pmatrix} \cong \mathbb{P}^3 \subset \mathbb{P}^{14}.$$

One can be more precise than this. Let $(\lambda : \mu : \nu : \rho)$ be homogeneous coordinates in the $\mathbb{P}^3$ spanned by the quadrics (11.11). Then the general quadric in this space has discriminant

$$\det \left[\begin{pmatrix} & & \lambda & \mu \\ -2\lambda & & -\mu & \nu \\ \lambda & -\mu & -2\nu & \\ \mu & \nu & & \end{pmatrix} + \rho \begin{pmatrix} a_0 & a_1 & a_2 & a_3 \\ a_1 & a_2 & a_3 & a_4 \\ a_2 & a_3 & a_4 & a_5 \\ a_3 & a_4 & a_5 & a_6 \end{pmatrix} \right].$$

The zero-set of this determinant is a quartic surface $\mathcal{K}_4 \subset \mathbb{P}^3$, and the moduli space $SU_C(2, \mathcal{O}_C(3))$ is precisely the complement $\mathbb{P}^3 - \mathcal{K}_4$. In fact, $\mathcal{K}_4$ is a well-known surface called the *Kummer quartic surface* (see, for example, Hudson [80]): it is the quotient of the Jacobian of C by the involution $[-1]$ and has exactly 16 nodes.

Remark 11.55. The four quadrics (11.11) can be used to define a rational map $\mathbb{P}^4 - - \to \mathbb{P}^3$, indeterminate along the curve $C \hookrightarrow \mathbb{P}^4$. This resolves to a morphism of the blow-up along the curve:

$$f : \mathrm{Bl}_C \mathbb{P}^4 \to \mathbb{P}^3.$$

There are three possibilities for the preimage of a point $p \in \mathbb{P}^3$.

 (i) If $p \in \mathbb{P}^3 - \mathcal{K}_4$, then $f^{-1}(p)$ is a nonsingular conic (isomorphic to $\mathbb{P}^1$).
 (ii) If $p \in \mathcal{K}_4 - \{16 \text{ nodes}\}$, then $f^{-1}(p)$ is a reducible conic (a pair of lines).
 (iii) If $p \in \{16 \text{ nodes of } \mathcal{K}_4\}$, then $f^{-1}(p)$ is a cone over a twisted cubic $\mathbb{P}^1$.

When $g \geq 3$ and L has odd degree, the moduli space $\overline{SU}_C(2, L)$ of semistable vector bundles is singular along the boundary. However, it has a good desingularisation $\widetilde{SU}_C(2, L)$ and also a conic bundle:

$$\mathcal{P} \xrightarrow{\varphi} \widetilde{SU}_C(2, L) \xrightarrow{\pi} \overline{SU}_C(2, L).$$

That is, π is an isomorphism over the open set $SU_C(2, L)$ of stable bundles, and the fibres of φ are conics. These conics are nonsingular (irreducible) over $\pi^{-1} SU_C(2, L)$ and are line pairs over the boundary. See Seshadri [86] or Narasimhan and Ramanan [82]. □

Finally, we remark that one can construct moduli spaces not only for stable bundles, but more generally for simple bundles. Let us see what sort of object this is in the case of genus 2 that we have been considering. Consider the set $\mathrm{Sim}_C(2, \mathcal{O}_C)$ of simple rank 2 vector bundles with a trivial determinant. It follows from Lemma 10.39 that every simple vector bundle is semistable, and this makes it quite easy to describe the space $\mathrm{Sim}_C(2, \mathcal{O}_C)$ as there is a surjective map

$$\mathrm{Sim}_C(2, \mathcal{O}_C) \to \mathbb{P}^3 - \{16 \text{ nodes of } \mathcal{K}_4\} \subset \overline{SU}_C(2, \mathcal{O}_C). \tag{11.12}$$

Away from the Kummer surface $\mathcal{K}_4$ this map is an isomorphism, while the simple bundles which are not stable are parametrised by the Jacobian $J_C := \mathrm{Pic}^0 C$ away from its 2-torsion points $J_C[2]$, in the following way. Given

$\alpha \in J_C, \alpha^2 \not\cong \mathcal{O}_C$, we can associate extensions

$$0 \to \alpha^{-1} \to E_1 \to \alpha \to 0,$$

$$0 \to \alpha \to E_2 \to \alpha^{-1} \to 0.$$

Each E_i is simple if and only if the extension is nonsplit, and in this case it is unique up to isomorphism since $h^1(\alpha^2) = 1$. As vector bundles, E_1 and E_2 are not isomorphic, but they represent the same point of $\overline{SU}_C(2, \mathcal{O}_C)$ by Proposition 11.37. Hence the map (11.12) is 2-to-1 over the Kummer surface $\mathcal{K}_4$. Thus $\text{Sim}_C(2, \mathcal{O}_C)$ is a nonseparated algebraic space, or (over $k = \mathbb{C}$) a non-Hausdorff complex space, and cannot be made into a variety or scheme.

Remark 11.56. On the other hand, when $g = 2$ and L has odd degree (so stable and semistable are equivalent) it happens that $\text{Sim}_C(2, L) = SU_C(2, L)$. Moreover, it was shown by Newstead that this variety embeds in $\mathbb{P}^5$ as a complete intersection of two quadrics. (See Newstead [83] or Desale and Ramanan [91].)   □

Exercises

1. Prove the following claims concerning the tangent space of functors (Definition 8.90).
   (i) The tangent space of the Picard functor $\mathcal{P}ic_C$ is isomorphic at every point to $H^1(\mathcal{O}_C)$.
   (ii) The functors $\mathcal{VB}_C(r)$ and $\mathcal{VB}_C(r)$ have tangent spaces, at a vector bundle E, isomorphic to $H^1(\mathcal{E}nd\, E)$ and $H^1(\mathfrak{sl}\, E)$, respectively.

2. Let $C \subset \mathbb{P}^2$ be a nonsingular plane cubic $f_3(x, y, z) = 0$. Let

$$M := \begin{pmatrix} q_{11}(x,y,z) & q_{12}(x,y,z) & q_{13}(x,y,z) \\ q_{21}(x,y,z) & q_{22}(x,y,z) & q_{23}(x,y,z) \\ q_{31}(x,y,z) & q_{32}(x,y,z) & q_{33}(x,y,z) \end{pmatrix}$$

be a 3×3 matrix of quadratic forms, all of whose 2×2 minors are divisible by $f_3(x, y, z)$. Show that M is stable under the action of $GL(3, 3)$ if and only if one of the following holds:
   (i) M is the matrix of cofactors of a 3×3 matrix of linear forms $N = (l_{ij}(x, y, z))$ satisfying $\det N = f_3(x, y, z)$; or
   (ii) $M = \begin{pmatrix} x^2 & xy & xz \\ xy & y^2 & yz \\ xz & yz & z^2 \end{pmatrix}.$

12

Intersection numbers and the Verlinde formula

Let C be a curve of genus g and let L be a line bundle on C. In Chapter 10 we constructed a projective moduli space $\overline{SU}_C(2, L)$ for semistable rank 2 vector bundles on C with determinant line bundle L. As a variety this depends only on the parity of $\deg L$, so there are two cases: when $\deg L$ is odd it is equal to $SU_C(2, L)$ and is nonsingular; when $\deg L$ is even it is the completion of $SU_C(2, L)$ by adding the Kummer variety. Let us denote it by $\mathcal{N}_C^+$ or $\mathcal{N}_C^-$ for the even and or odd cases, respectively. On $\mathcal{N}_C^\pm$ there exists (in each case) a naturally defined *standard line bundle* $\mathcal{L}$ (see Section 12.3), and these satisfy the *Verlinde formulae*:

$$\dim H^0(\mathcal{N}_C^+, \mathcal{L}^{\otimes k}) = \left(\frac{k+2}{2}\right)^{g-1} \sum_{j=1}^{k+1} \frac{1}{\left(\sin\frac{j\pi}{k+2}\right)^{2g-2}} \qquad (12.1)$$

$$\dim H^0(\mathcal{N}_C^-, \mathcal{L}^{\otimes k}) = (k+1)^{g-1} \sum_{j=1}^{2k+1} \frac{(-1)^{j-1}}{\left(\sin\frac{j\pi}{2k+2}\right)^{2g-2}}. \qquad (12.2)$$

On account of their beauty and importance, these formulae have attracted the interest of many mathematicians since Thaddeus [111].

In this chapter we will restrict ourselves to the odd degree case, for which the moduli space is fine, and we will prove the Verlinde formula (12.2) via the intersection numbers among cohomology classes

$$c_1(\mathcal{L}) = \alpha \in H^2(\mathcal{N}_C^-), \quad \beta \in H^4(\mathcal{N}_C^-), \quad \gamma \in H^2(\mathcal{N}_C^-),$$

defined by means of the universal vector bundle. These classes are called the

Newstead classes, and the full intersection formula among them is

$$(\alpha^m \beta^n \gamma^p) = (-1)^n 2^{2g-2-p} \frac{g!\, m!}{(g-p)!} b_{g-1-n-p} \qquad \text{for } m + 2n + 3p = 3g - 3,$$
(12.3)

where $b_k \in \mathbb{Q}$ is a rational number defined by the Taylor expansion $x / \sin x = \sum_k b_k x^{2k}$ when $k \geq 0$ and is zero when $k < 0$. In fact, we only need this for $p = 0$,

$$(\alpha^m \beta^n) = (-1)^n m!\, 4^{g-1} b_{g-1-n}, \qquad m + 2n = 3g - 3,$$
(12.4)

and it this form of the formula that we prove here.

The first two sections of the chapter introduce some background for proving the Verlinde formula (12.2). In Section 12.1 we show that the right-hand side is a polynomial in k; in Section 12.2 we review the Riemann-Roch Theorem in order to show that the left-hand side is a polynomial in k.

The heart of the chapter is Section 12.3, where we begin by defining the standard line bundle $\mathcal{L}$ on $\mathcal{N}_g^-$. This is closely tied up with the invariant theoretic construction of $SU_C(2, L)$ from Chapter 10 – in fact, the space of global sections $H^0(\mathcal{N}_g^-, \mathcal{L}^k)$, for $k \geq 0$, is exactly the space of weight $2k$ semiininvariants under the action $GL(n) \curvearrowright \mathrm{Alt}_{2,N}(H^0(L))$ from which the moduli space was constructed as quotient.

As a differentiable manifold, $\mathcal{N}_C^-$ is the same for all curves C of genus g. In fact, by a theorem of Narasimhan and Seshadri [73] and Donaldson [92], it is essentially the parameter space for equivalence classes of representations of the fundamental group (more precisely, a central extension of the fundamental group) $\pi_1(C)$ in $SU(2)$. Consequently it depends only on the topology of C. We can therefore choose the curve to be hyperelliptic, and for this case we consider the first direct image W of a certain line bundle on $C \times \mathcal{N}_C^-$. This W is a vector bundle on $\mathcal{N}_C^-$ of rank $4g$ on which the hyperelliptic involution of C induces an involution which decomposes it as a direct sum $W^+ \oplus W^-$. By the Riemann-Roch Theorem for curves with involution, the bundles W^+ and W^- have ranks $3g + 1$ and $g - 1$, respectively, and a consequence of this is that:

$$c_g(W^-) = c_{g+1}(W^-) = c_{g+2}(W^-) = 0.$$
(12.5)

Moreover, by Grothendieck-Riemann-Roch with involution these Chern classes can be expressed explicitly as polynomials in the Newstead classes α, β, γ, and the identities (12.5) therefore give relations among these cohomology classes. These are called the *Mumford relations* (Atiyah and Bott [89] §9,

Zagier [115] §6) and are precisely analogous to the relatons $c_{n-r+1}(\mathcal{Q}) = \cdots = c_n(\mathcal{Q}) = 0$ in the cohomology ring of the Grassmannian $\mathbb{G}(r, n)$ (Section 8.1(e)).

In Section 12.4 we deduce (12.2) and (12.4) from the Mumford relations. Although this is purely an exercise in computation, it turns out not to be so easy. The ring we have to work with is

$$\mathbb{Q}[\alpha, \beta, \gamma]/(c_g(W^-), c_{g+1}(W^-), c_{g+2}(W^-)),$$

and as a limbering-up exercise we examine the subring $\gamma = 0$ (called the *secant ring* for genus g). In this case the intersection numbers that we get are none other than the natural numbers called the secant numbers (Definition 12.9).

The Verlinde formula can be generalised to the moduli spaces $\mathcal{N}_{g,n}$ of rank 2 stable quasiparabolic vector bundles on a curve C of genus g with n marked points. When $g = 0$ (that is, when the curve is $\mathbb{P}^1$) the formula is

$$\dim H^0\left(\mathcal{N}_{0,n}, \mathcal{O}(-lK)\right) = \frac{1}{2l + 1} \sum_{j=0}^{2l} \frac{(-1)^{nj}}{\left(\sin \frac{2j+1}{4l+2}\pi\right)^{n-2}}, \qquad (12.6)$$

where $\mathcal{O}(-K)$ is the anticanonical line bundle, that is, the determinant of the tangent bundle, and $\mathcal{O}(-lK) = \mathcal{O}(-K)^{\otimes l}$. (When n is odd the anticanonical line bundle is primitive and $l \in \mathbb{Z}$; when n is even it possesses a square root and $l \in \frac{1}{2}\mathbb{Z}$.) We prove this, just in the case when n is odd, in the final section of the chapter. Here again one can define standard cohomology classes α, β satisfying Mumford relations of the form $c_g(W) = c_{g+1}(W) = 0$ for some vector bundle on the moduli space. In this case, in fact, the classes α, β generate the secant ring of genus g, and so again the Verlinde formula can be proved using a Riemann-Roch theorem. Finally, we indicate an alternative proof of (12.6) using the birational geometry of the moduli space $\mathcal{N}_{0,n}$.

A warning: this chapter differs from its predecessors in being an exposition of some relatively recent research. For this reason it is less self-contained, and some difficult topics may be treated with less explanation, or fewer references to the literature, than might be desirable. I hope that the reader will bear this in mind.

12.1 Sums of inverse powers of trigonometric functions

(a) Sine sums

To begin, given natural numbers $n, k \in \mathbb{N}$ we define the following sum, which can be thought of as taken over the vertices in the upper half-plane of a regular $2k$-gon:

$$V_n(k) := \sum_{j=1}^{k-1} \frac{1}{\left(\sin \frac{j\pi}{k}\right)^{2n}}.$$

Dividing by k^{2n} and taking the limit as $k \to \infty$ one obtains

$$\lim_{k \to \infty} \frac{V_n(k)}{k^{2n}} = \frac{2\zeta(2n)}{\pi^{2n}}, \tag{12.7}$$

where ζ is the Riemann zeta function

$$\zeta(s) = 1 + \frac{1}{2^s} + \frac{1}{3^s} + \frac{1}{4^s} + \cdots.$$

Proposition 12.1.

$$\sum_{n=1}^{\infty} V_n(k) \sin^{2n} x = 1 - k \tan x \cot kx.$$

We will prove this in a moment.

Remark 12.2. Substituting x/k for x and letting $k \to \infty$ yields the identity:

$$2 \sum_{n=1}^{\infty} \zeta(2n) \left(\frac{x}{\pi}\right)^{2n} = 1 - x \cot x.$$

$\square$

Moreover, by using Cauchy's residue theorem one can deduce:

Corollary 12.3.

$$V_n(k) = \operatorname*{Res}_{x=0} \left[\frac{-k \cot kx}{\sin^{2n} x} dx \right].$$

$\square$

Proof of Proposition 12.1. Using $\cos kx + i \sin kx = (\cos x + i \sin x)^k$ we see that

$$\cot kx = \frac{\sum_i (-1)^i \binom{k}{2i} \cot^{k-2i} x}{\sum_i (-1)^i \binom{k}{2i+1} \cot^{k-2i-1} x}.$$

This implies that $y = \cot\left(x + \frac{j\pi}{k}\right)$ for $j = 0, 1, \ldots, k-1$ are the distinct roots of an equation of degree k,

$$\sum_i (-1)^i \binom{k}{2i} y^{k-2i} - \cot kx \sum_i (-1)^i \binom{k}{2i+1} y^{k-2i-1} = 0.$$

Reading off the linear coefficient gives an expression for the sum of roots,

$$k \cot kx = \sum_{j=0}^{k-1} \cot \left(x + \frac{j\pi}{k} \right),$$

and from this we deduce:

$$k \tan x \cot kx = 1 + \frac{\tan x}{2} \sum_{j=1}^{k-1} \left\{ \cot \left(x + \frac{j\pi}{k} \right) + \cot \left(x - \frac{j\pi}{k} \right) \right\}$$

$$= 1 - \sum_{j=1}^{k-1} \frac{\sin^2 x}{\sin^2 \frac{j\pi}{k} - \sin^2 x}$$

$$= 1 - \sum_{n=1}^{\infty} V_n(k) \sin^{2n} x.$$

$\square$

It follows from Corollary 12.3 that $V_n(k)$ is a polynomial in k of degree $2n$, whose coefficients can be expressed in terms of the Laurent coefficients of $\cot x$ and $\operatorname{cosec}^{2n} x$ at $x = 0$. In particular, (12.7) gives its leading term:

$$V_n(k) = \left(\frac{2\zeta(2n)}{\pi^{2n}} \right) k^{2n} + \cdots. \tag{12.8}$$

(b) Variations

The sum $V_n(k)$ corresponded to the Riemann zeta function via (12.7). Corresponding to the series

$$1 - \frac{1}{2^s} + \frac{1}{3^s} - \frac{1}{4^s} + \cdots, \qquad 1 + \frac{(-1)^n}{3^s} + \frac{1}{5^s} + \frac{(-1)^n}{7^s} + \cdots,$$

we introduce, respectively, the trigonometric sums

$$V_n^-(k) := \sum_{j=1}^{k-1} \frac{(-1)^{j-1}}{\left(\sin \frac{j\pi}{k} \right)^{2n}}, \qquad U_n(k) := \sum_{j=0}^{k-1} \frac{(-1)^{nj}}{\left(\sin \frac{(2j+1)\pi}{2k} \right)^n}.$$

Remark 12.4. Note that in terms of the numbers $V_n(k)$, $V^-(k)$ and $U_n(k)$, the Verlinde formulae (12.2), (12.1) and (12.6) can be written:

$$\dim H^0(\mathcal{N}_C^+, \mathcal{L}^{\otimes k}) = \left(\frac{k+2}{2} \right)^{g-1} V_{g-1}(k+2),$$

$$\dim H^0(\mathcal{N}_C^-, \mathcal{L}^{\otimes k}) = (k+1)^{g-1} V_{g-1}^-(2k+2),$$

$$\dim H^0(\mathcal{N}_{0,n}, \mathcal{O}(-lK)) = \frac{1}{2l+1} U_{n-2}(2l+1).$$

$\square$

Clearly

$$V_n^-(k) = V_n(k) - 2V_n(k/2) \quad \text{(when } k \text{ is even)}$$

$$U_{2n}(k) = V_n(2k) - V_n(k).$$

So from Proposition 12.1 and the relations

$$\frac{1}{\sin x} = \cot \frac{x}{2} - \cot x, \qquad \tan x = \cot x - 2 \cot 2x,$$

we deduce:

Proposition 12.5.

$$\sum_{n=0}^{\infty} V_n^-(k) \sin^{2n} x = \frac{k \tan x}{\sin kx} \qquad \text{when } k \text{ is even, and}$$

$$\sum_{n=1}^{\infty} U_{2n}(k) \sin^{2n} x = k \tan x \tan kx.$$

$\square$

Corollary 12.6. *If k is even, then*

$$V_n^-(k) = \operatorname*{Res}_{x=0} \left[\frac{k}{\sin kx \sin^{2n} x} dx \right].$$

$\square$

Next, observe that $U_n(k)$ has an alternative expression using cosines:

$$U_n(k) = (-1)^{(k-1)n/2} \sum_{j=0}^{k-1} \frac{1}{\left(\cos \frac{2j\pi}{k} \right)^n}.$$

Proposition 12.7. *Suppose k is odd. Then*

$$\sum_{n=1}^{\infty} U_n(k) \sin^{n-1} x = \frac{k(1 + \sin kx)}{\cos x \cos kx}.$$

Proof. Using the trigonometric identity

$$\cos kx = \sum_i (-1)^i \binom{k}{2i} \sin^{2i} x \cos^{k-2i} x$$

we see that $y = \cos\left(x + \frac{2j\pi}{k}\right)$ for $j = 0, 1, \ldots, k-1$ are the roots of an equation of degree k,

$$\sum_i (-1)^i \binom{k}{2i}(1 - y^2)^i y^{k-2i} = \cos kx.$$

Since k is odd, the linear term in this equation is $(-1)^{(k-1)/2}ky$. Hence, by the relation between the coefficients of an equation and sum of reciprocals of its roots we obtain

$$(-1)^{(k-1)/2}\frac{k}{\cos kx} = \sum_{j=0}^{k-1} \frac{1}{\cos\left(x + \frac{2j\pi}{k}\right)},$$

and from this it follows that

$$\frac{k}{\cos kx} = (-1)^{(k-1)/2}\frac{1}{2}\sum_{j=0}^{k-1}\left\{\frac{1}{\cos\left(x + \frac{2j\pi}{k}\right)} + \frac{1}{\cos\left(x - \frac{2j\pi}{k}\right)}\right\}$$

$$= (-1)^{(k-1)/2}\sum_{j=0}^{k-1}\frac{\cos x \cos\frac{2j\pi}{k}}{\cos^2\frac{2j\pi}{k} - \sin^2 x}$$

$$= \cos x \sum_{m=0}^{\infty} U_{2m+1}(k)\sin^{2m} x,$$

where, in the last line, we have expanded $(1 - u)^{-1} = 1 + u + u^2 + \cdots$ with $u = \sin x/\cos(2j\pi/k)$. This gives an equality between the even part of the series in the proposition and the even part of the function of x on the right-hand side. The corresponding statement for the odd part of the series follows immediately from Proposition 12.5. $\square$

Corollary 12.8. *If k is odd, then*

$$\frac{U_n(k)}{k} = \operatorname*{Res}_{x=0}\left[\frac{1 + \sin kx}{\cos kx \sin^n x}dx\right].$$

In particular, $U_n(k)/k$ is a polynomial in k of degree n. $\square$

(c) Tangent numbers and secant numbers

Definition 12.9. In the Taylor expansion

$$\frac{1 + \sin x}{\cos x} = E_0 + E_1 x + E_2 \frac{x^2}{2!} + \cdots = \sum_{n=0}^{\infty} E_n \frac{x^n}{n!}$$

the coefficient E_n is called the n-th *secant number*, or *Euler number*, if n is even (these are the coefficients of $\sec x$, in other words), and is called a *tangent number* if n is odd (the coefficients of $\tan x$). ☐

For low values of n these numbers look like:

n even	0	2	4	6	8	10	12	14
E_n	1	1	5	61	1385	50521	2702765	199360981
n odd	1	3	5	7	9	11	13	15
E_n	1	2	16	272	7936	353792	22368256	1903757312

They can be expressed as:

$$E_n = \frac{2^{n+2} n!}{\pi^{n+1}} \left(1 + \frac{1}{(-3)^{n+1}} + \frac{1}{5^{n+1}} + \frac{1}{(-7)^{n+1}} + \frac{1}{9^{n+1}} + \cdots \right). \quad (12.9)$$

To see this, multiply both sides of Proposition 12.7 by $\sin x$, to give (for k odd)

$$\sum_{n=1}^{\infty} U_n(k) \sin^n x = \frac{k \tan x (1 + \sin kx)}{\cos kx}.$$

In this identity replace x by x/k and let $k \to \infty$. Noting, from the definition of $U_n(k)$, that asymptotically

$$\frac{U_n(k)}{k^n} \sim \sum_{j=0}^{k-1} \left(\frac{(-1)^j 2}{(2j+1)\pi} \right)^n$$

for large k, the identity (12.9) follows.

There is a third way to present the numbers E_n, which we will describe next. This is analogous to the Pascal triangle construction of the Catalan numbers (Section 8.1(e)). First we need:

Lemma 12.10. *For each integer $n \geq 0$ there exists a polynomial $P_n(y) \in \mathbb{Z}[y]$ with the property that*

$$\cos x \left(\frac{d}{dx} \right)^n \left(\frac{1}{\cos x} \right) = P_n(\tan x).$$

Proof. We use induction on n, starting with the trivial case $P_0(y) = 1$. For the inductive step, differentiate both sides of the equation with respect to x to give

$$\cos x \left(\frac{d}{dx}\right)^{n+1} \left(\frac{1}{\cos x}\right) - \sin x \left(\frac{d}{dx}\right)^{n} \left(\frac{1}{\cos x}\right) = \frac{1}{\cos^2 x} P_n'(\tan x).$$

This shows that

$$P_{n+1}(y) = y P_n(y) + (1 + y^2) P_n'(y). \tag{12.10}$$

$\square$

As we see from this proof, the sequence of polynomials $\{P_n(y)\}_{n \geq 0}$ is determined by the initial condition $P_0(y) = 1$ and the recurrence relation (12.10). In particular, one sees that for even and odd values of n, $P_n(y)$ is an even or odd function, respectively, and can therefore be written as

$$P_n(y) = E_{n,n} y^n + E_{n,n-2} y^{n-2} + E_{n,n-4} y^{n-4} + \cdots.$$

If n is *even*, then the constant coefficient is

$$E_{n,0} = P_n(0) = \left(\frac{d}{dx}\right)^n \left(\frac{1}{\cos x}\right)\bigg|_{x=0} = \text{secant number } E_n.$$

The recurrence relation (12.10) implies, for the coefficients $E_{n,k}$,

$$E_{n+1,k} = k E_{n,k-1} + (k+1) E_{n,k+1}. \tag{12.11}$$

In other words, the numbers $E_{n,k}$ occupy a 'twisted' Pascal triangle, in which each entry is the sum of the adjacent entries in the preceding row, multiplied by k whenever the diagonal is between columns $k - 1$ and k:

Secant numbers and $E_{n,k}$ for $n \equiv k \mod 2$

n\k	0	1	2	3	4	...
0	1					
		1				
2	1		2			
		5		6		
4	5		28		24	
		61		180		...
6	61		662		...	
		1385		...		
8	1385	...				

This table generates the secant numbers recursively. Namely, the diagonal is just the sequence $E_{n,n} = n!$, as follows from (12.11). We then twist and add down to the left.

The tangent numbers can be described in a similar manner. For $n \geq 0$ there exist polynomials $Q_n(y) \in \mathbb{Z}[y]$ such that

$$\left(\frac{d}{dx}\right)^n \tan x = Q_n(\tan x),$$

and these satisfy a recurrence relation

$$Q_{n+1}(y) = (1 + y^2)Q'_n(y).$$

Together with the initial condition $Q_0(y) = y$, this determines a sequence of functions $\{Q_n(y)\}_{n \geq 0}$ which are even or odd as n is odd or even, respectively. In this case, for $n \geq 1$,

$$Q_n(y) = E_{n,n+1}y^{n+1} + E_{n,n-1}y^{n-1} + E_{n,n-3}y^{n-3} + \cdots,$$

where the $E_{n,k}$ satisfy the same recurrence relation (12.11) as above and occupy a twisted Pascal triangle:

Tangent numbers and $E_{n-1,k}$ for $n \equiv k \bmod 2$

$n \backslash k$	1	2	3	4	5	$\cdots$
1	1					
		1				
3	2		2			
		8		6		
5	16		40		24	
		136		240		$\cdots$
7	272		1232		$\cdots$	
		1385		$\cdots$		
9	7936		$\cdots$			

As in the previous case, the numbers in the first column are precisely the tangent numbers:

$$E_{n-1,1} = \left.\frac{dQ_{n-1}}{dy}\right|_{y=0} = \left.\left(\frac{d}{dx}\right)^n \tan x\right|_{x=0} = \text{tangent number } E_n.$$

The tangent and secant numbers E_n are the 'zigzag numbers' in the book [53] of Conway and Guy.

12.2 Riemann-Roch theorems

Given a holomorphic vector bundle E on a compact complex manifold X, one often needs to know the dimension of the vector space $H^0(X, E)$ of global sections. In general this is difficult to measure, and one considers instead the easier quantity

$$\chi(X, E) := \dim H^0(E) - \dim H^1(E) + \dim H^2(E) - \dim H^3(E) + \cdots,$$

where $H^i(E) = H^i(X, E)$ is the i-th cohomology group with coefficents in the sheaf of sections of E. This is called the *Euler-Poincaré characteristic* of E. For any exact sequence

$$0 \to E_1 \to E_2 \to E_3 \to 0$$

it satisfies

$$\chi(X, E_1) - \chi(X, E_2) + \chi(X, E_3) = 0.$$

What we will see is that $\chi(X, E)$ can be computed just from the topological invariants of X and E. For more details we refer the reader to Hirzebruch [98] or Fulton [94].

Let Ω_X be the cotangent vector bundle on X, and let ω_X be its determinant line bundle, called the *canonical line bundle* on X. (For the case when X is an algebraic curve see Section 9.5.)

Kodaira Vanishing Theorem 12.11. *If L is a holomorphic line bundle on X for which $L \otimes \omega_X^{-1}$ is ample, then $H^i(X, L) = 0$ for all $i > 0$.* □

In this situation, the dimension of the space $H^0(X, L)$ of global sections reduces to the (simpler) Euler characteristic $\chi(X, L)$.

Remarks 12.12.

(i) Suppose that the set of all global sections of a line bundle M on X has no common zeros. These sections then define a morphism $X \to \mathbb{P}^N$, where $N + 1$ is the dimension of the space of sections. If this morphism is an embedding, then the line bundle M is said to be *very ample*. If some tensor power of M is very ample, then M is *ample*.

(ii) If X is a curve, then a line bundle is ample if and only if its degree is positive. In this case, therefore, the Kodaira Vanishing Theorem reduces to Theorem 9.20. In the general case it is known that ampleness of a line bundle M depends only on its Chern class $c_1(M)$. (See, for example, Griffiths and Harris [54] chapter 1.) □

(a) Some preliminaries

Let X be a compact complex manifold and E a rank r complex vector bundle on X. This has a *(total) Chern class*

$$c(E) = 1 + c_1(E) + c_2(E) + \cdots + c_r(E) \in H^*(X, \mathbb{Z}), \qquad c_i(E) \in H^{2i}(X, \mathbb{Z}).$$

On exact sequences

$$0 \to E_1 \to E_2 \to E_3 \to 0$$

this satisfies

$$c(E_1)c(E_3) = c(E_2); \tag{12.12}$$

and the Chern class of the dual vector bundle $E^\vee$ is given by

$$c_i(E^\vee) = (-1)^i c_i(E).$$

In particular, the *Chern class of X* is defined to be the Chern class of its tangent bundle T_X:

$$c(X) := c(T_X) = 1 + c_1(T_X) + c_2(T_X) + \cdots + c_n(T_X), \quad \text{where } n = \dim X.$$

Next we consider the formal power series expansion, in the ring of power series in infinitely many variables $\mathbb{Q}[[x_1, x_2, \ldots]]$, of the infinite product:

$$\frac{x_1}{1 - e^{-x_1}} \times \frac{x_2}{1 - e^{-x_2}} \times \cdots$$

The term of degree m in this expansion is a symmetric homogeneous polynomial in the variables $x_1, x_2, \ldots$ and can therefore be expressed as a polynomial in the elementary symmetric polynomials $\sigma_1, \ldots, \sigma_m$. (See Macdonald [101].) If we denote this polynomial by $\mathrm{Td}_m(\sigma_1, \ldots, \sigma_m)$, then:

$$\prod_{i=1}^{\infty} \frac{x_i}{1 - e^{-x_i}} = \sum_{m=0}^{\infty} \mathrm{Td}_m(\sigma_1, \ldots, \sigma_m)$$
$$= 1 + \frac{1}{2}\sigma_1 + \frac{1}{12}(\sigma_1^2 + \sigma_2) + \frac{1}{24}\sigma_1\sigma_2 + \cdots.$$

Definition 12.13.

(i) The *Todd class* of a vector bundle E of rank r is

$$\mathrm{td}\, E := \sum_{m=0}^{r} \mathrm{Td}_m(c_1(E), \ldots, c_r(E)) \in H^*(X, \mathbb{Q}).$$

(ii) The Todd class of a complex manifold X is defined to be that of its tangent bundle,

$$\operatorname{td} X := \operatorname{td} T_X.$$

The component in top degree $\operatorname{td}_n(X) \in H^{2n}(X, \mathbb{Q}) = \mathbb{Q}$, where $n = \dim X$, is called the *Todd characteristic* of X. □

Like the Chern class, the Todd class is multiplicative on exact sequences. That is, if

$$0 \to E_1 \to E_2 \to E_3 \to 0,$$

then it follows from (12.12) that

$$\operatorname{td}(E_1)\operatorname{td}(E_3) = \operatorname{td}(E_2). \tag{12.13}$$

The Todd class can also be computed as follows. We can write

$$\frac{x}{1 - e^{-x}} = e^{x/2} \times \frac{x/2}{\sinh x/2}.$$

The last factor here is a power series in x^2. Letting $\pi_m(x_i) = \sigma(x_i^2)$ be the elementary symmetric polynomial in the squares $x_1^2, x_2^2, \ldots$, the homogeneous term of degree $2m$ in the expansion of the infinite product

$$\prod_{i=1}^{\infty} \frac{x_i/2}{\sinh x_i/2}$$

is a polynomial in $\pi_1, \ldots, \pi_m$, which we denote by $\widehat{A}_m(\pi_1, \ldots, \pi_m)$. Then we have

$$\prod_{i=1}^{\infty} \frac{x_i}{1 - e^{-x_i}} = e^{\sigma_1/2} \sum_{i=1}^{\infty} \widehat{A}_m(\pi_1, \pi_2, \ldots, \pi_m).$$

Given a vector bundle E of rank r, set

$$p_i(E) = (-1)^i c_{2i}(E \oplus E^\vee), \qquad i = 1, \ldots, r,$$

called the *Pontryagin classes* of E. In this language, the Todd class is given by

$$\operatorname{td} E = e^{c_1(E)/2} \sum_{i=1}^{\infty} \widehat{A}_m(p_1(E), p_2(E), \ldots, p_m(E)). \tag{12.14}$$

(b) Hirzebruch-Riemann-Roch

If X is a complex manifold of dimension n with its natural orientation, the top component of the rational cohomology ring $H^{2n}(X, \mathbb{Q})$ is canonically isomorphic to $\mathbb{Q}$ by evaluation on the fundamental class of X, and we denote the composition of this isomorphism with projection of cohomology to $H^{2n}(X, \mathbb{Q})$ by

$$\int_X : H^*(X, \mathbb{Q}) \to \mathbb{Q}.$$

For $\alpha \in H^{2n}(X, \mathbb{Q})$ we shall also use the symbol (α) to denote $\int_X \alpha \in \mathbb{Q}$.

Theorem 12.14. Hirzebruch-Riemann-Roch for the structure sheaf. *The Euler-Poincaré characteristic of the structure sheaf of a complex manifold X is equal to the Todd characteristic of X,*

$$\chi(X, \mathcal{O}_X) = \int_X \operatorname{td} X.$$

$\square$

In the curve case $n = 1$, this says $\chi(X, \mathcal{O}_X) = \frac{1}{2} \deg c_1(X)$, which we have already shown in Chapter 9 (Propositions 9.25 and 9.89). In the surface case $n = 2$, it says

$$\chi(X, \mathcal{O}_X) = \tfrac{1}{12} \left(c_1(X)^2 + c_2(X) \right).$$

In the theory of algebraic surfaces this is called *Noether's formula*.

Example 12.15. Let X be projective space $\mathbb{P}^n$. Here there exists an exact sequence, called the *Euler sequence*,

$$0 \to \mathcal{O}_{\mathbb{P}^n} \to \mathcal{O}_{\mathbb{P}^n}(1)^{\oplus(n+1)} \to T_{\mathbb{P}^n} \to 0.$$

Letting $h = c_1(\mathcal{O}_{\mathbb{P}^n}(1))$, it follows from (12.13) that

$$\operatorname{td} \mathbb{P}^n = \operatorname{td} \mathcal{O}_{\mathbb{P}^n}(1)^{\oplus(n+1)} = \left(\frac{h}{1 - e^{-h}} \right)^{n+1}.$$

The class h is Poincaré dual to a hyperplane in $\mathbb{P}^n$ and has self-intersection number $(h^n) = 1$. So, applying the Hirzebruch-Riemann-Roch Theorem, we obtain

$$\chi(\mathcal{O}_{\mathbb{P}^n}) = \int_{\mathbb{P}^n} \left(\frac{h}{1 - e^{-h}} \right)^{n+1} = \operatorname*{Res}_{x=0} \frac{1}{(1 - e^{-x})^{n+1}} = 1. \qquad (12.15)$$

The last identity can be proved by induction on n using the fact that $\frac{d}{dx}(1 - e^{-x})^{-n} = n(1 - e^{-x})^{-n} - n(1 - e^{-x})^{-n-1}$. $\qquad\square$

Remark 12.16. More precisely, one can say that $\dim H^0(\mathcal{O}_{\mathbb{P}^n}) = 1$ while $H^i(\mathcal{O}_{\mathbb{P}^n}) = 0$ for all $i > 0$. The dimensions of these cohomology spaces are birational invariants, and in particular this shows that $\chi(\mathcal{O}_X) = 1$ whenever X is a rational variety. $\qquad\square$

Next we consider a line bundle L on X and its Chern class $c_1(L) \in H^2(X, \mathbb{Z})$. The exponential function of $c_1(L)$ is a finite sum and determines a rational cohomology class

$$e^{c_1(L)} = \sum_{m \geq 0} \frac{c_1(L)^m}{m!} \in H^*(X, \mathbb{Q}).$$

Theorem 12.17. Hirzebruch-Riemann-Roch for a line bundle. *The Euler-Poincaré characteristic of a (holomorphic) line bundle L on a complex manifold X is given by:*

$$\chi(X, L) = \int_X e^{c_1(L)} \operatorname{td} X.$$

In particular, $\chi(X, L^k)$ is a polynomial in k of degree $n = \dim X$. $\qquad\square$

Example 12.18. Let X be projective space $\mathbb{P}^n$ and $L = \mathcal{O}_{\mathbb{P}^n}(1)$ be its hyperplane line bundle. In this case Hirzebruch-Riemann-Roch says

$$\chi(\mathcal{O}_{\mathbb{P}^n}(k)) = \int_{\mathbb{P}^n} \left(e^{kh} \left(\frac{h}{1 - e^{-h}} \right)^{n+1} \right)$$

$$= \operatorname*{Res}_{x=0} \frac{e^{kx}}{(1 - e^{-x})^{n+1}}$$

$$= \binom{n + k}{n}.$$

Of course, this is the number of linearly independent forms of degree k in $n + 1$ variables. (See the remark following Example 1.8.) $\qquad\square$

Example 12.19. If X is a complex torus, then the tangent bundle T_X is trivial and so $\operatorname{td}(X) = 1$. Hence in this case

$$\chi(X, L) = \frac{(c_1(L)^n)}{n!}.$$

In other words, the Euler characteristic is equal to the self-intersection of the line bundle divided by $n!$. In particular, this shows that the self-intersection of a line bundle on an n-dimensional complex torus is always divisible by $n!$.

The (analytic) Jacobian Jac C of a curve C of genus g is a g-dimensional complex torus, and its cohomology ring is therefore equal to the exterior algebra generated by $H^1(C, \mathbb{Z})$:

$$H^*(\text{Jac } C, \mathbb{Z}) \cong \bigwedge H^1(C, \mathbb{Z}).$$

Letting $\alpha_1, \ldots, \alpha_g, \beta_1, \ldots, \beta_g \in H^1(C, \mathbb{Z})$ be a symplectic basis (the dual basis of (9.42)), it is known that the theta line bundle $\mathcal{O}_J(\Theta)$ on the Jacobian has Chern class

$$c_1(\theta) = \alpha_1 \wedge \beta_1 \wedge \cdots \wedge \alpha_g \wedge \beta_g \in H^2(\text{Jac } C, \mathbb{Z}).$$

Thus

$$(\alpha_1 \wedge \beta_1 \wedge \cdots \wedge \alpha_g \wedge \beta_g)^g = g! \prod_{i=1}^{g} \alpha_i \wedge \beta_i,$$

and hence $\mathcal{O}_J(\Theta)$ has self-intersection $g!$. It follows from Hirzebruch-Riemann-Roch above that

$$\dim H^0(\text{Jac } C, \mathcal{O}_J(k\Theta)) = k^g. \qquad (12.16)$$

The Verlinde formulae (12.1) and (12.2) can be viewed as nonabelian versions of this formula.   □

In order to generalise from line bundles to vector bundles we need to replace the class $e^{c_1(L)}$ by the Chern character:

Definition 12.20. The sum of powers $\sum_i x_i^m$ is for each $m \geq 1$ a symmetric polynomial in the variables x_i and is therefore a polynomial in the elementary symmetric polynomials $\sigma_1, \ldots, \sigma_m$, which we denote by $s_m(\sigma_1, \ldots, \sigma_m)$. If E is a vector bundle, then we define

$$\text{ch}_0(E) = \text{rank } E, \qquad \text{ch}_m(E) = \frac{1}{m!} s_m(c_1(E), \ldots, c_m(E)) \quad \text{for } m \geq 1.$$

The sum

$$\text{ch } E = \sum_{m \geq 0} \text{ch}_m(E) \in H^*(X, \mathbb{Q})$$

is called the *Chern character* of E.   □

Note that the Chern character of a line bundle L is ch $L = e^{c_1(L)}$.

Example 12.21. In general, if E has Chern classes $c_1, c_2, c_3, \ldots$, then

$$\mathrm{ch}_1(E) = c_1,$$

$$\mathrm{ch}_2(E) = \tfrac{1}{2}c_1^2 - c_2,$$

$$\mathrm{ch}_3(E) = \tfrac{1}{6}c_1^3 - \tfrac{1}{2}c_1 c_2 + \tfrac{1}{2}c_3.$$

□

The Chern character, unlike the Chern class, is additive on exact sequences, that is,

$$0 \to E_1 \to E_2 \to E_3 \to 0 \quad \Longrightarrow \quad \mathrm{ch}\, E_1 - \mathrm{ch}\, E_2 + \mathrm{ch}\, E_3 = 0. \quad (12.17)$$

Theorem 12.22. Hirzebruch-Riemann-Roch for a vector bundle. *The Euler-Poincaré characteristic of a (holomorphic) vector bundle E on a complex manifold X is given by:*

$$\chi(X, E) = \int_X \mathrm{ch}(E)\,\mathrm{td}(X).$$

□

(c) Grothendieck-Riemann-Roch for curves

Given a proper morphism $f : X \to Y$ between algebraic varieties and an algebraic vector bundle $\mathcal{E}$ on X, one can define its direct image sheaf $f_*\mathcal{E}$ and its higher direct images $R^q f_*\mathcal{E}$, for $q > 0$, on Y (Definition 11.10). In general, these sheaves are not vector bundles, but let us pretend that they are. (This assumption is justified by the fact that the direct images are coherent sheaves. This means that they have resolutions by locally free sheaves, so that the Chern character is defined using the additivity property on exact sequences.) Then it is possible to express the alternating sum of Chern characters

$$\sum_q (-1)^q \mathrm{ch}\, R^q f_*\mathcal{E} \in H^*(Y, \mathbb{Q})$$

in terms of the Chern character ch $\mathcal{E}$ and the 'Todd character' of f. This generalises Hirzebruch-Riemann-Roch (where Y is a point) and is called the *Grothendieck-Riemann-Roch Theorem*. Here we shall consider only the special case $X = C \times Y$, where C is a curve and where $f : C \times Y \to Y$ is the projection to the second factor. (For a more general statement the reader may refer to Fulton [94], for example.)

First, take an affine open cover $\bigcup_i$ Spm A_i of Y and consider the restriction of $\mathcal{E}$ to each open set $C \times$ Spm A_i. This defines a vector bundle on the curve C_{A_i} in the sense of Definition 11.9 and so determines A_i modules $H^0(\mathcal{E}_i)$ and $H^1(\mathcal{E}_i)$. If these are locally free modules, then they define, by gluing over the open cover, a pair of vector bundles $f_*\mathcal{E}$ and $R^1 f_*\mathcal{E}$, which are the direct image bundles.

Künneth's Theorem says that the cohomology ring of the product $C \times Y$ is the tensor product of the cohomology rings of C and Y. In other words, there is an isomorphism of graded rings

$$H^*(C \times Y) \cong H^*(C) \otimes H^*(Y).$$

The cohomology of the curve C has three components $H^0(C)$, $H^1(C)$ and $H^2(C)$, and so the cohomology of $C \times Y$ is the direct sum of three pieces:

$$H^0(C) \otimes H^*(Y), \quad H^1(C) \otimes H^*(Y), \quad H^2(C) \otimes H^*(Y).$$

The Chern character of a vector bundle $\mathcal{E}$ on the product $C \times Y$ can therefore be decomposed as

$$\text{ch}\,\mathcal{E} = \text{ch}^{(0)}\mathcal{E} + \text{ch}^{(1/2)}\mathcal{E} + \text{ch}^{(1)}\mathcal{E}, \quad \text{ch}^{(i)}\mathcal{E} \in H^{2i}(C) \otimes H^*(Y).$$

Now $\text{ch}^{(0)}\mathcal{E}$ can be viewed as an element of $H^*(Y)$. On the other hand, the fundamental class of a point $\eta \in H^2(C, \mathbb{Z})$ determines a natural isomorphism $\int_C : H^2(C, \mathbb{Q}) \xrightarrow{\sim} \mathbb{Q}$, and using this isomorphism we can view $\text{ch}^{(1)}\mathcal{E}$ as an element of $H^*(Y)$, which we will denote by $\text{ch}^{(1)}(\mathcal{E})/\eta$.

Theorem 12.23. Grothendieck-Riemann-Roch for $f : C \times Y \to Y$.

$$\text{ch}\,f_*\mathcal{E} - \text{ch}\,R^1 f_*\mathcal{E} = \text{ch}^{(1)}(\mathcal{E})/\eta - (g-1)\text{ch}^{(0)}(\mathcal{E}).$$

When Y is a point this is nothing but the Riemann-Roch formula 10.10.   □

From this formula one reads off the Chern character (of $f_*\mathcal{E}$ or $R^1 f_*\mathcal{E}$). The following remark is useful for recovering the Chern classes from the Chern character in general. First, for any vector bundle F consider the derived Chern class

$$c'(F) = \sum_{i \geq 1} i\, c_i(F). \tag{12.18}$$

From the relations between elementary symmetric polynomials and sums of powers, this derived class satisfies

$$c'(F) = \left(\sum_{i \geq 1} (-1)^{i-1} i! \operatorname{ch}_i(F) \right) c(F). \tag{12.19}$$

Since the logarithmic derivative $c'(F)/c(F)$ of the Chern class is additive on exact sequences, it follows that the Chern class can be written

$$c(F) = \exp \int \frac{c'(F)}{c(F)},$$

where $\int$ denotes the formal inverse of the derivative (12.18). Hence

$$c(F) = \exp \int \sum_{i \geq 1} (-1)^{i-1} i! \operatorname{ch}_i(F).$$

(d) Riemann-Roch with involution

Suppose that the curve C has an involution $\sigma : C \to C$ (that is, an automorphism of order 2), and suppose that σ lifts to a vector bundle E (still with order 2). In this case σ acts also on the vector spaces $H^0(E)$ and $H^1(E)$. We will denote the invariant and anti-invariant subspaces (that is, the eigenspaces of ± 1) by $H^i(E)^+$ and $H^i(E)^-$, and write

$$\chi(E)^{\pm} = \dim H^0(E)^{\pm} - \dim H^1(E)^{\pm}.$$

Moreover, if $p \in C$ is a fixed point of σ, then the involution acts in the fibre E_p and we denote by $E_p^+, E_p^- \subset E_p$ the invariant and antiinvariant subspaces. The following is proved in Desale and Ramaman [91].

Proposition 12.24. *Given an involution σ acting on a vector bundle E as above, we have*

$$\chi(E)^+ - \chi(E)^- = \frac{1}{2} \sum_{p \in \operatorname{Fix}(\sigma)} \left(\dim E_p^+ - \dim E_p^- \right),$$

where $\operatorname{Fix}(\sigma) \subset C$ denotes the set of fixed points. □

Example 12.25. Let F be any vector bundle on the curve C.
(i) If $E := F \oplus \sigma^* F$, then $\chi(E)^+ - \chi(E)^- = 0$.

(ii) If $E := F \otimes \sigma^* F$, then

$$\chi(E)^+ - \chi(E)^- = \frac{1}{2} \sum_{p \in \mathrm{Fix}(\sigma)} \left(\dim S^2 F_p - \dim \bigwedge^2 F_p \right) = \frac{1}{2} \mathrm{rank}\, F \times |\mathrm{Fix}(\sigma)|.$$

$\square$

Let $f : F \to F'$ be a homomorphism of rank 2 vector bundles, and suppose that the two bundles have an isomorphic determinant line bundle. Then, by tensoring the dual map $f^\vee : F'^\vee \to F^\vee$ with $\det F = \det F'$ we obtain a bundle map

$$f^{\mathrm{adj}} := f^\vee \otimes 1_{\det} : F' \to F,$$

called the *adjoint* of f. (See also Exercise 10.2.)

Example 12.26. Suppose that F is a rank 2 vector bundle on C whose determinant is σ-invariant, that is, $\det F \cong \sigma^* \det F$. Given any (local) homomorphism $f : F \to \sigma^* F$, we can pull back the adjoint map $f^{\mathrm{adj}} : \sigma^* F \to F$ to obtain a homomorphism $\sigma^* f^{\mathrm{adj}} : F \to \sigma^* F$. Moreover, the mapping

$$f \mapsto \sigma^* f^{\mathrm{adj}},$$

composed with itself, recovers f. It therefore defines a lift of σ to the vector bundle

$$E := \mathcal{H}om(F, \sigma^* F) = F^\vee \otimes \sigma^* F.$$

At a fixed point $p \in C$, the invariant subspace E_p^+ is 1-dimensional (spanned by the identity endomorphism in F_p) while the antiinvariant subspace E_p^- is the 3-dimensional space $\mathfrak{sl}(F_p)$ of tracefree endomorphisms. It follows that

$$\chi(E)^+ - \chi(E)^- = -|\mathrm{Fix}(\sigma)|.$$

$\square$

Proposition 12.24 can be globalised in much the same way as the Riemann-Roch Theorem globalises to Grothendieck-Riemann-Roch. Let Y be a complete nonsingular variety, and suppose that the involution $\sigma \times 1_Y$ of $C \times Y$ lifts to a vector bundle E on the product. Then the involution acts in the direct images $\pi_* E$ and $R^1 \pi_* E$, where $\pi : C \times Y \to Y$ is the projection, and we set

$$\chi_\pi(E)^\pm := \mathrm{ch}(\pi_* E)^\pm - \mathrm{ch}(R^1 \pi_* E)^\pm.$$

For each fixed point $p \in C$ the vector bundle $E|_{p \times Y}$ decomposes into invariant and antiinvariant subbundles $E|_{p \times Y}^\pm$, and we have:

Proposition 12.27.

$$\chi_\pi(E)^+ - \chi_\pi(E)^- = \frac{1}{2} \sum_{p \in \mathrm{Fix}(\sigma)} \left(\mathrm{ch}(E|^+_{p \times Y}) - \mathrm{ch}(E|^-_{p \times Y}) \right).$$

$$\square$$

12.3 The standard line bundle and the Mumford relations

Let C be a (complete nonsingular algebraic) curve of genus g, let L be a line bundle on C of odd degree, and let $\mathcal{U}$ be a universal vector bundle on the product $C \times SU_C(2, L)$. For every $E \in SU_C(2, L)$ the restriction to $C \times [E]$ of the determinant line bundle $\det \mathcal{U}$ is isomorphic to L, and so $\det \mathcal{U}$ can be expressed as a tensor product,

$$\det \mathcal{U} \cong L \boxtimes \Phi, \tag{12.20}$$

where Φ is some line bundle pulled back from $SU_C(2, L)$. (It is enough to take Φ to be the direct image of $\det \mathcal{U} \otimes L^{-1}$. That this is a line bundle follows from the base change theorem in the form of Remark 11.19.) In particular, the first Chern class of $\mathcal{U}$ can be written:

$$c_1(\mathcal{U}) = c_1(L) \otimes 1 + 1 \otimes \phi, \qquad \phi = c_1(\Phi) \in H^2(SU_C(2, L)). \tag{12.21}$$

(a) The standard line bundle

We are now going to define a natural line bundle on the moduli variety $SU_C(2, L)$: this is the line bundle $\mathcal{L}$ which appears in the left-hand side of (12.2). First of all, we observe that by using the universal bundle $\mathcal{U}$ we can construct, given a point $p \in C$, two vector bundles on $SU_C(2, L)$, namely the direct image and the restriction:

$$\pi_* \mathcal{U}, \qquad \mathcal{U}|_{p \times SU_C(2,L)},$$

where $\pi : C \times SU_C(2, L) \to SU_C(2, L)$ is projection on the second factor. In order for $\pi_* \mathcal{U}$ to be a vector bundle we assume that $d := \deg L$ is sufficiently large, so that

$$H^1(E) = 0 \quad \text{for all } E \in SU_C(2, L), \tag{12.22}$$

and the direct image is locally free by Lemma 11.16 and Theorem 11.18. As in Chapters 10 and 11, we set $N := d + 2 - 2g$: this is the dimension of $H^0(E)$ for each $E \in SU_C(2, L)$ and hence the rank of $\pi_* \mathcal{U}$.

Definition 12.28.

(i) If $L \in \text{Pic } C$ satisfies the condition (12.22), then the *standard line bundle* (not to be confused with the canonical line bundle!) on the moduli variety $SU_C(2, L)$ is defined to be

$$\mathcal{L} := \left(\det \mathcal{U}|_{p \times SU_C(2,L)} \right)^N \otimes (\det \pi_* \mathcal{U})^{-2}.$$

This is also called the *determinant line bundle*.

(ii) If $L \in \text{Pic } C$ does not satisfy the condition (12.22), then we choose a line bundle ξ on C of sufficiently high degree that $L \otimes \xi^2$ satisfies (12.22). The standard line bundle on $SU_C(2, L)$ is then defined to be the pullback of $\mathcal{L}$ under the isomorphism

$$SU_C(2, L) \xrightarrow{\otimes \xi} SU_C(2, L \otimes \xi^2).$$

$\square$

At first sight it appears that this definition depends on various choices, but in fact this is not the case. First of all, $\det \left(\mathcal{U}|_{p \times SU_C(2,L)} \right)$ is isomorphic to the line bundle Φ in (12.20), and this shows that $\mathcal{L}$ does not depend on the choice of $p \in C$. How about its dependence on $\mathcal{U}$? Any other universal bundle is of the form $\mathcal{U} \otimes \pi^* \Psi$, where Ψ is some line bundle on $SU_C(2, L)$, and replacing $\mathcal{U}$ with $\mathcal{U} \otimes \pi^* \Psi$ has the effect of tensoring $\det \left(\mathcal{U}|_{p \times SU_C(2,L)} \right)$ by Ψ^2. On the other hand, it follows from the projection formula

$$\pi_*(\mathcal{U} \otimes \pi^* \Psi) = \Psi \otimes \pi_* \mathcal{U}$$

that $\det \pi_* \mathcal{U}$ gets tensored by Ψ^N. Hence the tensor product in the definition remains unchanged, and so $\mathcal{L}$ as defined in 12.28(i) is independent of the choice of universal bundle.

Proposition 12.29. *The standard line bundle defined in Definition 12.28(ii) is independent of the choice of ξ.* $\square$

Proof. It is sufficient to prove this in the case $\xi = \mathcal{O}_C(q)$ for some point $q \in C$. The pull-back of $\mathcal{L}$ under $\otimes \xi$ is the line bundle defined as in (i) but using the (pulled back) universal bundle $\mathcal{U}' := \mathcal{O}_C(q) \otimes \mathcal{U}$. In part (i) of the definition we note that $\mathcal{U}_{p \times SU_C(2,L)}$ remains unchanged when we replace $\mathcal{U}$ by $\mathcal{U}'$. On the other hand, $\pi_* \mathcal{U}'$ fits into an exact sequence

$$0 \to \pi_* \mathcal{U} \to \pi_* \mathcal{U}' \to \mathcal{U}_{q \times SU_C(2,L)} \to 0,$$

so that N becomes $N + 2$, while $\det \pi_* \mathcal{U}$ gets tensored by $\det \mathcal{U}_{q \times SU_C(2,L)}$. Altogether, then, replacing $\mathcal{U}$ by $\mathcal{U}'$ has the effect of tensoring $\mathcal{L}$ by

$$\left(\det \mathcal{U}|_{p \times SU_C(2,L)}\right)^2 \otimes \left(\det \mathcal{U}|_{q \times SU_C(2,L)}\right)^{-2}.$$

But we have already noted that each determinant is isomorphic to Φ, and so the product is trivial.   □

We are now going to construct global sections of the standard line bundle. We consider a vector bundle obtained by 'pinching' the trivial rank 2 bundle $\mathcal{O}_C \oplus \mathcal{O}_C$ at points $p_1, \ldots, p_N \in C$. This means the subsheaf $F \subset \mathcal{O}_C \oplus \mathcal{O}_C$ that is constructed as follows. For each $i = 1, \ldots, N$ we choose $(a_i : b_i) \in \mathbb{P}^1$ and define

$$F = \{(s, t) \mid a_i s + b_i t \equiv 0 \bmod \mathfrak{m}_i \text{ for all } i = 1, \ldots, N\} \subset \mathcal{O}_C \oplus \mathcal{O}_C, \tag{12.23}$$

where $\mathfrak{m}_i \subset \mathcal{O}_C$ is the maximal ideal of the point p_i. By pulling F back to $C \times SU_C(2, L)$, tensoring with $\mathcal{U}$ and then taking the direct image, we obtain an exact sequence of vector bundles on $SU_C(2, L)$:

$$\pi_* (F \otimes \mathcal{U}) \longrightarrow \pi_* \left(\mathcal{U}^{\oplus 2}\right) \stackrel{f}{\longrightarrow} \bigoplus_{i=1}^{N} \mathcal{U}|_{p_i \times SU_C(2,L)} \longrightarrow R^i \pi_* (F \otimes \mathcal{U}).$$

$$\tag{12.24}$$

In this sequence, f is a bundle map between vector bundles of equal rank $2N$. Consequently, its determinant $\det f$ defines a global section of

$$\left(\det \pi_* \left(\mathcal{U}^{\oplus 2}\right)\right)^{-1} \otimes \det \left(\bigoplus_{i=1}^{N} \mathcal{U}|_{p_i \times SU_C(2,L)}\right) \cong \mathcal{L}.$$

At a point $[E]$ in the moduli space f has fibre map

$$(\mathrm{ev}_{p_1}, \ldots, \mathrm{ev}_{p_N}) : H^0(E)^{\oplus 2} \to \bigoplus_{i=1}^{N} E/\mathfrak{m}_i E,$$

with kernel $H^0(E \otimes F)$. Thus the zero set of the global section $\det f \in H^0(\mathcal{L})$ consists exactly of points $[E]$ for which $H^0(E \otimes F) \neq 0$.

Remark 12.30. While the homomorphism f depends on the choices made in (12.23), the global section of $\mathcal{L}$ depends only on the isomorphism class of F. This is because the zero-set of $\det f$ corresponds to the *Fitting ideal* of the torsion sheaf $R^1 \pi_* (F \otimes \mathcal{U})$. (For uniqueness of the Fitting ideal see

Northcott [105].) The section det $f \in H^0(\mathcal{L})$ is called the *determinantal section* defined by F.   □

Recall the construction of $SU_C(2, L)$ in Chapter 10, as Proj of the ring of semiinvariants of the group action $GL(N) \curvearrowright \mathrm{Alt}_{N,2}(H^0(L))$. It therefore carries line bundles of the form $\mathcal{O}(i)$ (see Section 8.5(b)) whose space of global sections is the space of semiinvariants of weight i. In fact, we have already observed in Remark 10.73 that (when deg L is odd) there are no semiinvariants of odd weight. In this case, therefore, there is no line bundle $\mathcal{O}(1)$, and we will refer to $\mathcal{O}(2)$ as the tautological line bundle.

Proposition 12.31. *The standard line bundle $\mathcal{L}$ on $SU_C(2, L)$ is isomorphic to the tautological line bundle $\mathcal{O}(2)$ on the Proj quotient $\mathrm{Alt}^s_{N,2}(H^0(L))/GL(N)$. Moreover, under this identification the determinantal section of F, at a Gieseker matrix $T \in \mathrm{Alt}_N(H^0(L))$, is*

$$\mathrm{Pfaff}\left(\mathrm{ev}_{p_1}\left(\widetilde{T}_{(a_1,b_1)}\right) + \cdots + \mathrm{ev}_{p_N}\left(\widetilde{T}_{(a_N,b_N)}\right)\right), \tag{12.25}$$

where

$$\widetilde{T}_{(a,b)} := \begin{pmatrix} a^2 T & abT \\ abT & b^2 T \end{pmatrix} \in \mathrm{Alt}_{2N}(H^0(L)).$$

Proof. For each $p \in C$ there is a skew-symmetric isomorphism

$$\mathcal{U}|_{p \times SU_C(2,L)} \xrightarrow{\sim} \mathcal{U}^\vee|_{p \times SU_C(2,L)} \otimes \Phi.$$

Now consider the following composition of f from (12.24) with its tranpose tensored with Φ:

$$\pi_* \mathcal{U}^{\oplus 2} \longrightarrow \bigoplus_{i=1}^N \mathcal{U}|_{p_i \times SU_C(2,L)} \xrightarrow{\sim} \bigoplus_{i=1}^N \mathcal{U}^\vee|_{p_i \times SU_C(2,L)} \otimes \Phi \longrightarrow \left(\pi_* \mathcal{U}^{\oplus 2}\right)^\vee \otimes \Phi.$$

By construction this is skew-symmetric and has rank 2 everywhere. Its pullback via the quotient map

$$\mathrm{Alt}^s_{N,2}(H^0(L)) \to SU_C(2, L)$$

is a homomorphism of rank $2N$ trivial vector bundles which is represented at $T \in \mathrm{Alt}^s_{N,2}(H^0(L))$ by the matrix

$$\mathrm{ev}_{p_1}\left(\widetilde{T}_{(a_1,b_1)}\right) + \cdots + \mathrm{ev}_{p_N}\left(\widetilde{T}_{(a_N,b_N)}\right).$$

It is clear that the function which assigns to T the Pfaffian of this matrix is a weight 2 semiinvariant, and hence so is the determinantal section of the vector bundle F. But this implies that the standard line bundle is isomorphic to $\mathcal{O}(2)$.   □

Remark 12.32. In our proof of stability of Gieseker points in Chapter 10 (see Lemma 10.74) we actually used the special case of the above proposition in which:

$$(a_i : b_i) = \begin{cases} (1 : 0) & \text{if } 1 \leq i \leq \frac{N-1}{2} \\ (0 : 1) & \text{if } \frac{N+1}{2} \leq i \leq N - 1 \\ (1 : 1) & \text{if } i = N. \end{cases}$$

In this case (12.25) takes the form (in the notation of Lemma 10.74)

$$\text{Pfaff} \begin{pmatrix} f(T) & \text{ev}_{p_N} T \\ \text{ev}_{p_N} T & f'(T) \end{pmatrix}.$$

The matrix $\text{ev}_{p_N} T$ has rank ≤ 2, and so by the formula of Example 10.55 the Pfaffian above is equal to the semiinvariant $(\text{rad } f(T))^t h(T)\text{rad } f'(T)$ used in the proof of Proposition 10.70. (In this case the vector bundle F is obtained from the direct sum of two line bundles by pinching at one point.)   □

(b) The Newstead classes

In this section we are assuming that deg is odd, so $SU_C(2, L) = \mathcal{N}_C^-$ in the notation of the introduction to this chapter, and we will often write just $\mathcal{N}$ to denote this moduli space. The second Chern class of the universal bundle has a Künneth decomposition which we will write as:

$$c_2(\mathcal{U}) = \eta \otimes \omega + \psi + 1 \otimes \chi,$$

where $\eta \in H^2(C)$ is the fundamental class of a point and

$$\omega \in H^2(\mathcal{N}), \qquad \psi \in H^3(\mathcal{N}) \otimes H^1(C), \qquad \chi \in H^4(\mathcal{N}).$$

By (12.21) and Grothendieck-Riemann-Roch, the direct image $\pi_*\mathcal{U}$ has the first Chern class

$$c_1(\pi_*\mathcal{U}) = -\omega + (d + 1 - g)\phi.$$

Hence, by Definition 12.28, the standard line bundle $\mathcal{L}$ has the Chern class

$$c_1(\mathcal{L}) = (d + 2 - 2g)\phi - 2(-\omega + (d + 1 - g)\phi) = 2\omega - d\phi =: \alpha.$$

One can see the uniqueness of the class α – that is, its independence of the choice of universal bundle $\mathcal{U}$ – again in the following way. Since $\mathcal{U}$ is unique up to tensoring with a line bundle from $\mathcal{N}$, the tensor product $\mathcal{U}^\vee \otimes \mathcal{U}$ does not depend on the choice of $\mathcal{U}$ and therefore neither does the second Chern class or its Künneth decomposition

$$c_2(\mathcal{U}^\vee \otimes \mathcal{U}) = 4c_2(\mathcal{U}) - c_1(\mathcal{U})^2 = 2\eta \otimes (2\omega - d\phi) + 4\psi - 1 \otimes (\phi^2 - 4\chi).$$

In particular, the first term $2\eta \otimes \alpha$, and hence α, is independent of $\mathcal{U}$. But the same is true also of the last term, which we will denote by

$$\beta := \phi^2 - 4\chi \in H^4(\mathcal{N}).$$

Finally, the middle term ψ is uniquely determined, but this is not itself a cohomology class on $\mathcal{N}$. Squaring, however, does determine a class on the moduli space:

$$\psi^2 =: \eta \otimes \gamma, \qquad \gamma \in H^6(\mathcal{N}).$$

These three cohomology classes α, β, γ on $\mathcal{N}$ are called the *Newstead classes* on the moduli space (Newstead [104]). In terms of these we can now take the first step towards the Verlinde formula (12.2).

Proposition 12.33.

$$\dim H^0(\mathcal{N}_C^-, \mathcal{L}^k) = \int_{\mathcal{N}} e^{(k+1)\alpha} \left(\frac{\sqrt{\beta}/2}{\sinh \sqrt{\beta}/2} \right)^{2g-2}.$$

We prove this by applying Hirzebruch-Riemann-Roch to the line bundle $\mathcal{L}^k$. This says:

$$\chi(\mathcal{N}, \mathcal{L}^k) = \int_{\mathcal{N}} e^{k\alpha} \operatorname{td}(\mathcal{N}). \tag{12.26}$$

To compute the Todd class we use:

Lemma 12.34. *The tangent bundle of $\mathcal{N}_C^-$ is isomorphic to $R^1\pi_* \mathfrak{sl}\, \mathcal{U}$.*   □

We have seen in Theorem 10.1 that the tangent space at the point $[E]$ is the vector space $H^1(\mathfrak{sl}\, E)$. The isomorphism

$$T_{\mathcal{N},[E]} \overset{\sim}{\to} H^1(\mathfrak{sl}\, E)$$

is called the *Kodaira-Spencer map*, and the proof of the lemma involves globalising this to an isomorphism of vector bundles on $\mathcal{N}$, though we omit the details here.

Proof of Proposition 12.33. We apply Grothendieck-Riemann-Roch for the projection $\pi : C \times \mathcal{N} \to \mathcal{N}$ to the bundle $\mathfrak{sl}\,\mathcal{U}$. By definition of the Newstead classes we have

$$c_2(\mathfrak{sl}\,\mathcal{U}) = c_2(\mathcal{U}^\vee \otimes \mathcal{U}) = 2\eta \otimes \alpha + 4\psi - 1 \otimes \beta.$$

The total Chern class of $\mathfrak{sl}\,\mathcal{U}$ is just $1 + c_2(\mathfrak{sl}\,\mathcal{U})$, so the odd degree components of the Chern character vanish and the even degree components are

$$\mathrm{ch}_{2r}(\mathfrak{sl}\,\mathcal{U}) = \frac{2(-1)^r c_2(\mathfrak{sl}\,\mathcal{U})^r}{(2r)!}.$$

By the Grothendieck-Riemann-Roch Theorem 12.23, therefore,

$$c_1(T_\mathcal{N}) = 2\alpha, \qquad \mathrm{ch}_{2r}(T_\mathcal{N}) = \frac{2(g-1)\beta^r}{(2r)!} \quad \text{for } r > 0. \tag{12.27}$$

In particular,

$$\mathrm{ch}(T_\mathcal{N}) + \mathrm{ch}(T_\mathcal{N}^\vee) = 6g - 6 + \sum_{r>0} \frac{4(g-1)\beta^r}{(2r)!},$$

and it follows from (12.19) that the logarithmic derivative of the Chern class of $T_\mathcal{N} \oplus T_\mathcal{N}^\vee$ is

$$\frac{c'(T_\mathcal{N} \oplus T_\mathcal{N}^\vee)}{c(T_\mathcal{N} \oplus T_\mathcal{N}^\vee)} = -4(g-1) \sum_{r>0} \beta^r = \frac{-4(g-1)\beta}{1-\beta}.$$

Hence $c(T_\mathcal{N} \oplus T_\mathcal{N}^\vee) = (1 - \beta)^{2g-2}$. So by (12.14) the Todd class is

$$\mathrm{td}(\mathcal{N}) = e^{c_1(T_\mathcal{N})/2}\widehat{A}(\mathcal{N}) = e^\alpha \left(\frac{\sqrt{\beta}/2}{\sinh \sqrt{\beta}/2} \right)^{2g-2}.$$

Hence from (12.26) we obtain

$$\chi(\mathcal{L}^k) = \int_\mathcal{N} e^{(k+1)\alpha} \left(\frac{\sqrt{\beta}/2}{\sinh \sqrt{\beta}/2} \right)^{2g-2}.$$

Since $c_1(T_\mathcal{N}) = 2\alpha$ (by (12.27)), it follows that the anticanonical line bundle $K_\mathcal{N}^{-1}$ is ample (Remark 12.12), and so the proposition follows from the Kodaira Vanishing Theorem 12.11. □

(c) The Mumford relations

We will assume now that C is a hyperelliptic curve, that is, a 2-sheeted cover of the projective line $C \xrightarrow{2:1} \mathbb{P}^1$ (see Example 9.7). We let $\sigma : C \to C$ be the covering involution. The fixed points of σ are the ramification points of C over $\mathbb{P}^1$, and there are precisely $2g + 2$ of them which we will denote by

$p_1, \ldots, p_{2g+2} \in C$. We fix one of these points $p = p_i$ and consider the moduli space $SU_C(2, \mathcal{O}_C(p))$. If E is a stable rank 2 vector bundle on C with $\det E \cong \mathcal{O}_C(p)$, then by Exercise 10.6 we have

$$\text{Hom}(E, \sigma^* E(-p)) = 0. \tag{12.28}$$

We consider the vector bundle $F := E^\vee \otimes \sigma^* E \otimes \mathcal{O}_C(-p)$. This is a bundle of rank 4 to which the involution σ lifts, and with $H^0(F) = 0$ by (12.28). By Riemann-Roch, $\chi(F) = -4g$ and therefore $H^1(F)$ has dimension $4g$. By Example 12.26, its invariant and antiinvariant subspaces satisfy

$$-\dim H^1(F)^+ + \dim H^1(F)^- = -2g - 2,$$

and hence

$$\dim H^1(F)^+ = 3g + 1, \qquad \dim H^1(F)^- = g - 1. \tag{12.29}$$

We can globalise this bundle over the moduli space. The hyperelliptic involution induces an involution, which we will also denote by σ, on $C \times SU_C(2, \mathcal{O}_C(p))$ by acting on the first factor. Let

$$W := R^1\pi_* \left(\mathcal{U}^\vee \otimes \sigma^* \mathcal{U} \otimes_{\mathcal{O}_C} \mathcal{O}_C(-p) \right).$$

Again by (12.28) this is a vector bundle on $SU_C(2, \mathcal{O}_C(p))$ of rank $4g$. Since the (0-th) direct image is zero, Grothendieck-Riemann-Roch 12.23 says:

$$-\text{ch}(W) = \text{ch}^{(1)} \left(\mathcal{U}^\vee \otimes \sigma^* \mathcal{U} \otimes \mathcal{O}_C(-p) \right) / \eta - (g - 1)\text{ch}^{(0)}$$
$$\left(\mathcal{U}^\vee \otimes \sigma^* \mathcal{U} \otimes \mathcal{O}_C(-p) \right). \tag{12.30}$$

The hyperelliptic involution σ acts on the cohomology groups $H^i(C, \mathbb{Z})$ as multiplication by $(-1)^i$, so that

$$c_1(\sigma^* \mathcal{U}) = c_1(\mathcal{U}), \qquad c_2(\sigma^* \mathcal{U}) = \eta \otimes \omega - \psi + 1 \otimes \chi,$$

and hence

$$c_1(\mathcal{U}^\vee \otimes \sigma^* \mathcal{U}) = c_3(\mathcal{U}^\vee \otimes \sigma^* \mathcal{U}) = 0,$$
$$c_2(\mathcal{U}^\vee \otimes \sigma^* \mathcal{U}) = 2\eta \otimes \alpha - 1 \otimes \beta,$$
$$c_4(\mathcal{U}^\vee \otimes \sigma^* \mathcal{U}) = 4\eta \otimes \gamma.$$

We now compute using the logarithmic derivative of the Chern class of W. Using (12.30) and the base change Theorem 11.15 we obtain

$$-\frac{c'(W)}{c(W)} = \frac{(1 - g)(-2\beta) - 2(\alpha - \beta + 2\gamma)}{1 - \beta} = -\frac{2(\alpha - g\beta + 2\gamma)}{1 - \beta}.$$

We now decompose W under the involution. By base change and (12.29) the subbundles W^+, W^- have ranks $3g + 1$ and $g - 1$, respectively; while by Proposition 12.27 their Chern characters satisfy

$$-\text{ch}(W^+) + \text{ch}(W^-) = \frac{1}{2} \sum_{i=1}^{2g+2} \left(\text{ch}(\mathcal{O}_N) - \text{ch}(\mathfrak{sl}\,\mathcal{U}|_{p_i \times N}) \right)$$

$$= -(g + 1)\left(e^{\sqrt{\beta}} - e^{-\sqrt{\beta}} \right).$$

It follows that

$$-\frac{c'(W^+)}{c(W^+)} + \frac{c'(W^-)}{c(W^-)} = -(g + 1)\left(\frac{\sqrt{\beta}}{1 + \sqrt{\beta}} - \frac{\sqrt{\beta}}{1 - \sqrt{\beta}} \right) = \frac{(2g + 2)\beta}{1 - \beta},$$

and hence

$$\frac{c'(W^+)}{c(W^+)} = \frac{\alpha - (2g + 1)\beta + 2\gamma}{1 - \beta}, \qquad \frac{c'(W^-)}{c(W^-)} = \frac{\alpha + \beta + 2\gamma}{1 - \beta}. \qquad (12.31)$$

Now, since W^- has rank $g - 1$, it follows that in the expansion of

$$c(W^-) = \exp \int \frac{\alpha + \beta + 2\gamma}{1 - \beta}$$

all terms of degree $\geq 2g$ vanish. The resulting relations among the Newstead classes $\alpha, \beta, \gamma \in H^*(N)$, generated by

$$c_g(W^-) = c_{g+1}(W^-) = c_{g+2}(W^-) = 0, \qquad (12.5)$$

are called the *Mumford relations*.

12.4 From the Mumford relations to the Verlinde formula

We consider the polynomial ring in three indeterminates $\mathbb{Q}[A, B, C]$ in which the degree of a polynomial is computed using weights

$$\deg A = 1, \quad \deg B = 2, \quad \deg C = 3.$$

A, B, C will correspond to the Newstead classes $\alpha, \beta, 2\gamma$, and we can interpret the second identity in (12.31), from which the Mumford relations are to be read off, as follows. If $c_i(W^-) = \xi_i(\alpha, \beta, 2\gamma)$, then the sequence of polynomials ξ_i can be put together in a generating function

$$F(t) := \sum_{r \geq 0} \xi_r(A, B, C)t^r \in \mathbb{Q}[A, B, C][[t]],$$

and this function satisfies a differential equation:

$$(1 - Bt^2)\frac{dF}{dt} = (A + Bt + Ct^2)F(t), \qquad F(0) = 1. \qquad (12.32)$$

This initial value problem is equivalent to the following recurrence relation for the coefficients:

$$\xi_0 = 1, \qquad (r + 1)\xi_{r+1} = A\xi_r + rB\xi_{r-1} + C\xi_{r-2}, \quad \text{for } r \geq 0, \qquad (12.33)$$

where we let $\xi_r = 0$ when $r < 0$. For example, $\xi_1 = A$, $\xi_2 = \frac{1}{2}(A^2 + B)$ and $\xi_3 = \frac{1}{6}(A^3 + 5AB + 2C)$.

Equivalently, the differential equation can also be solved to give

$$F(t) = \frac{1}{\sqrt{1-Bt^2}} \exp\left(\frac{-Ct}{B} + \frac{2(C+AB)}{B\sqrt{B}} \tanh^{-1} \sqrt{B}t \right)$$

$$\qquad (12.34)$$

$$= \frac{1}{\sqrt{1-Bt^2}} \left(\frac{1+t\sqrt{B}}{1-t\sqrt{B}} \right)^{\frac{C+AB}{2B\sqrt{B}}}.$$

By the *Mumford relations* we will mean the ideal of $\mathbb{Q}[A, B, C]$ generated by $\xi_g(A, B, C), \xi_{g+1}(A, B, C), \dots$

(a) Warming up: secant rings

As a warm-up we are going to examine the case where $C = 0$. Although it will not be needed in the proof of (12.2), it is an interesting exercise in itself and will reappear in Section 12.5 (b).

In this case we have a sequence of polynomials in two variables $\xi_r(A, B)$, determined by a differential equation

$$\frac{F'(t)}{F(t)} = \frac{A + Bt}{1 - Bt^2}, \qquad F(0) = 1,$$

for the generating function $F(t) = \sum_{r \geq 0} \xi_r(A, B)t^r$. This is equivalent to a recurrence relation:

$$(r + 1)\xi_{r+1}(A, B) = A\xi_r(A, B) + rB\xi_{r-1}(A, B),$$

$$\xi_0 = 1, \quad \xi_r = 0 \text{ for } r < 0. \qquad (12.35)$$

The polynomials $\xi_r(A, B)$ for $r \geq g$ generate a homogeneous ideal $I_g \subset \mathbb{Q}[A, B]$, and because of the recurrence relation (12.35) this is generated by just two polynomials, $\xi_g(A, B)$ and $\xi_{g+1}(A, B)$. The residue ring

$$R_g := \mathbb{Q}[A, B]/I_g$$

is a graded ring, called the *secant ring* of genus g. What we will see is that it generates the secant numbers of Section 12.1(c) in the same way that the cohomology ring of the Grassmannian $\mathbb{G}(2, n)$ generates the Catalan numbers (Section 8.1(e)).

Example 12.35. One sees $\xi_1 = A$, $2\xi_2 = A^2 + B$, $3\xi_3 = \frac{1}{2}A(A^2 + B) + 2AB$ and $4\xi_4 = \frac{1}{6}A(A^3 + 5AB) + \frac{3}{2}(A^2 + B)B$. Hence

$$R_1 = \mathbb{Q}[A, B]/(A, A^2 + B) = \mathbb{Q}[A, B]/(A, B) = \mathbb{Q},$$

$$R_2 = \mathbb{Q}[A, B]/(A^2 + B, A^3 + 5AB) = \mathbb{Q}[A]/(A^3),$$

$$R_3 = \mathbb{Q}[A, B]/(A^3 + 5AB, A^4 + 14A^2B + 9B^2).$$

$\square$

We will denote by $\alpha, \beta \in R_g$ the residue classes modulo I_g of $A, B \in \mathbb{Q}[A, B]$.

Proposition 12.36.

 (i) *The polynomials $\xi_r(A, B)B^s$, as r, s range though all nonnegative integers, are a basis for $\mathbb{Q}[A, B]$ as a vector space over $\mathbb{Q}$.*
 (ii) *The subset of $\xi_r(A, B)B^s$ with $r + s \geq g$ is a basis for I_g.*
(iii) *The secant ring has basis $\{\xi_r(\alpha, \beta)\beta^s\}_{r+s\leq g-1}$.*

Proof.

 (i) Monomials $A^r B^s$ can be ordered lexicographically with $A > B$. With this ordering, the recurrence relation (12.35) implies that the maximal monomial (with nonzero coefficient) in $\xi_r(A, B)$ is A^r. Hence for every r, s the monomial $A^r B^s$ appears as the maximal monomial in $\xi_r(A, B)B^s$. The elements $\xi_r(A, B)B^s$ are therefore expressed in terms of the basis $\{A^r B^s\}$ by a unimodular triangular matrix.
 (ii) Let $J_g \subset \mathbb{Q}[A, B]$ be the vector subspace with basis $\{\xi_r(\alpha, \beta)\beta^s\}_{r+s\geq g}$. Clearly $BJ_g \subset J_g$, and so it follows from (12.35) that $AJ_g \subset J_g$ as well. Hence J_g is an ideal. On the other hand, $\xi_g, \xi_{g+1} \in J_g$, so that $I_g \subset J_g$. So we just have to show that $\xi_r(\alpha, \beta)\beta^s$ belongs to I_g whenever $r + s \geq g$, and for this we use induction on s. For $s = 0$ it is true by definition of I_g, while if $\xi_r B^s \in I_g$ for $s > 0$, then

$$\xi_r B^{s+1} = (\xi_r B)B^s = -\frac{1}{r+1}A(\xi_{r+1}B^s) + \frac{r+2}{r+1}\xi_{r+2}B^s \in I_g.$$

(iii) This is an immediate consequence of (i) and (ii). $\square$

Corollary 12.37. *Let $R_g = \bigoplus(R_g)_{(m)}$ as a graded ring. Then $(R_g)_{(2g-2)}$ is spanned by β^{g-1}, while $(R_g)_{(m)}$ vanishes for $m \geq 2g - 1$.* $\square$

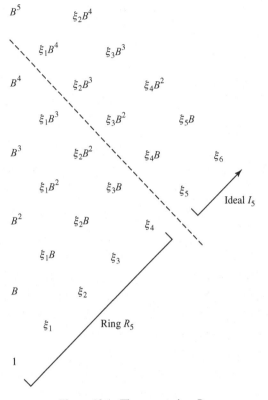

Figure 12.1: The secant ring R_5

Corollary 12.38. *The secant ring R_g has dimension $g(g+1)/2$ over $\mathbb{Q}$, and Hilbert series*

$$\sum_{m=0}^{2g-2} t^m \dim_{\mathbb{Q}}(R_g)_{(m)} = \frac{(1-t^g)(1-t^{g+1})}{(1-t)(1-t^2)}.$$

$\square$

Remark 12.39. More generally, suppose that $\mathbb{Q}[x_1, \ldots, x_n]/(F_1, \ldots, F_n)$ is finite-dimensional, where $F_1, \ldots, F_n \in \mathbb{Q}[x_1, \ldots, x_n]$ and degrees are weighted by $\deg x_i = a_i \in \mathbb{N}$. Then the quotient $\mathbb{Q}[x_1, \ldots, x_n]/(F_1, \ldots, F_n)$ has dimension $\prod_i \deg F_i / \prod_i a_i$ and Hilbert series (as a graded ring) $\prod_i (1 - t^{\deg F_i}) / \prod_i (1 - t^{a_i})$. (See Proposition 1.9 in Chapter 1.) $\square$

R_g is generated in top degree by $(-\beta)^{g-1}$, and we shall use this element to identify $(R_g)_{(2g-2)} \xrightarrow{\sim} \mathbb{Q}$. Composing with projection $R_g \to (R_g)_{(2g-2)}$ defines a linear map which we will formally denote by

$$\int : R_g \to \mathbb{Q}.$$

If we consider the 'intersection numbers' $\int \xi_i \alpha^j (-\beta)^k$ (where $i + j + 2k = 2g - 2$, and so only depends on i, k), then we find nothing other than the two index secant numbers of Section 12.1(c):

Proposition 12.40. *For i, j, k satisfying $i + j + 2k = 2g - 2$ we have*

$$\int \xi_i(\alpha, \beta) \alpha^j (-\beta)^k = E_{2g-2-2k-i,i}.$$

In particular, $\int \alpha^{2g-2}$ is equal to the secant number E_{2g-2}.

Proof. Each side is defined by the same recurrence relation, (12.35) and (12.11), respectively, and so the identity follows inductively. □

The basis of R_5 and the twisted Pascal triangle of its intersection numbers

degree			
8	β^4		
	$\xi_1\beta^3$		
6	β^3	$\xi_2\beta^2$	
	$\xi_1\beta^2$	$\xi_3\beta$	
4	β^2	$\xi_2\beta$	ξ_4
	$\xi_1\beta$	ξ_3	
2	β	ξ_2	
	ξ_1		
0	1		

$8-2k-i \setminus i$	0	1	2	3	4
0	1				
		1			
2	1		2		
		5		6	
4	5		28		24
		61		180	
6	61		662		
		1385			
8	1385				

Remark 12.41. Notice that the secant ring R_g is isomorphic to the intersection ring of the Grassmannian $\mathbb{G}(2, g + 1)$ additively, though not multiplicatively. What we see is that, just as the Pascal triangle of Section 8.1(e) tabulates the degrees of the cohomology classes $s_i B^k$ with respect to the Plücker hyperplane class A, so the Pascal triangle above records the degrees of the classes $\xi_i (-\beta)^k$ with respect to the 'hyperplane class' $\alpha = c_1(\mathcal{L})$. □

This proposition implies that

$$\int e^{k\alpha}(-\beta)^n = \frac{k^{2g-2-2n} E_{2g-2-2n}}{(2g-2-2n)!} = \operatorname*{Res}_{x=0} \left[\frac{x^{2n} dx}{x^{2g-1} \cos kx} \right],$$

and from this it follows that for any power series $f(x)$ we have

$$\int e^{k\alpha} f(-\beta) = \operatorname*{Res}_{x=0} \left[\frac{f(x^2)dx}{x^{2g-1}\cos kx} \right].$$

In particular, taking $f(x) = (\sqrt{x}/\sin\sqrt{x})^{2g-1}$ and applying Corollary 12.8 yields:

$$\int e^{k\alpha} \left(\frac{\sqrt{\beta}}{\sinh\sqrt{\beta}} \right)^{2g-1} = \operatorname*{Res}_{x=0} \left[\frac{dx}{\cos kx \, \sin^{2g-1} x} \right] \tag{12.36}$$

$$= \frac{U_{2g-1}(k)}{k} \qquad \text{when } k \text{ is odd.}$$

This illustrates our strategy for evaluating the right-hand side of Proposition 12.33 in order to prove the Verlinde formula. Of course, to do this we must work in the full ring $\mathbb{Q}[A, B, C]/I_g$ without the restriction $C = 0$.

(b) The proof of formulae (12.2) and (12.4)

We now return to consider the original polynomials in three variables defined by the recurrence relation (12.33) and to the homogeneous ideal $I_g \subset \mathbb{Q}[A, B, C]$ generated by $\xi_r(A, B, C)$ for $r \geq g$. Because of (12.33) it is generated by the three polynomials $\xi_g, \xi_{g+1}, \xi_{g+2}$. Our aim is to study the residue ring $\mathbb{Q}[A, B, C]/I_g$.

Just as for the secant rings, we can construct a basis of $\mathbb{Q}[A, B, C]/I_g$ as a rational vector space, but in this case polynomials such as $\xi_r(A, B, C)C^s$ are not enough, and we have to work a bit harder. Let t_1, t_2 be independent variables and consider the product $F(t_1)F(t_2)$ of two generating functions, each defined by the differential equation (12.32) with coefficients $\xi_r(A, B, C)$. This product is symmetric in t_1, t_2 and can therefore be expressed as a power series in the elementary symmetric polynomials, which we will write as

$$x = t_1 + t_2, \qquad y = -t_1 t_2.$$

(We change the sign of $t_1 t_2$ in order to be consistent with the literature.) This series can be written, then, as

$$H(x, y) := F(t_1)F(t_2) = \sum_{r,s\geq 0} \xi_{r,s} x^r(-y)^s \tag{12.37}$$

for some coefficients $\xi_{r,s} = \xi_{r,s}(A, B, C) \in \mathbb{Q}[A, B, C]$. Note that $\deg \xi_{r,s} = r + 2s$, where A, B, C have degrees 1, 2, 3.

Remark 12.42. Each $\xi_{r,s}$ is a polynomial in $\xi_1, \xi_2, \xi_3, \ldots$. For example, by multiplying out (12.37) we find $\xi_{1,0} = \xi_1$, $\xi_{0,1} = \xi_1^2 - 2\xi_2$, $\xi_{2,0} = \xi_2$ and

$\xi_{1,1} = \xi_1\xi_2 - 3\xi_3$. These polynomials can also be read off by observing that the left-hand side of (12.37) is nothing other than the formal resultant of $F(t) = \sum_{r\geq 0} \xi_r t^r$ and $G(t) := t^2 - xt - y$, that is:

$$H(x, y) = \begin{vmatrix} 1 & \xi_1 & \xi_2 & \xi_3 & \xi_4 & \xi_5 & \xi_6 & \xi_7 & \cdots \\ & 1 & \xi_1 & \xi_2 & \xi_3 & \xi_4 & \xi_5 & \xi_6 & \xi_7 & \cdots \\ -y & -x & 1 \\ & -y & -x & 1 \\ & & -y & -x & 1 \\ & & & -y & -x & 1 \\ & & & & -y & -x & 1 \\ & & & & & -y & -x & 1 \\ & & & & & & -y & -x & 1 \\ & & & & & & & \ddots & \ddots & \ddots \end{vmatrix}.$$

(See Section 1.3(a).) For example, by noting that the x^r term comes from the top-left $(r + 2) \times (r + 2)$ block, we see that $\xi_{r,0} = \xi_r$. $\square$

We now look for recurrence relations satisfied by the polynomials $\xi_{r,s}(A, B, C)$. First of all, by the chain rule, the partial derivatives satisfy

$$\frac{\partial H}{\partial t_1} = \frac{\partial H}{\partial x} - t_2 \frac{\partial H}{\partial y}, \qquad \frac{\partial H}{\partial t_2} = \frac{\partial H}{\partial x} - t_1 \frac{\partial H}{\partial y}.$$

Solving these equations we get:

$$\frac{\partial H}{\partial x} = \frac{1}{t_1 - t_2}\left(t_1 \frac{\partial H}{\partial t_1} - t_2 \frac{\partial H}{\partial t_2}\right)$$

$$= \frac{1}{t_1 - t_2}\left(\frac{t_1(A + Bt_1 + Ct_1^2)}{1 - Bt_1^2} - \frac{t_2(A + Bt_2 + Ct_2^2)}{1 - Bt_2^2}\right) H(x, y)$$

$$= \frac{(1 - By)(A + Cy) + x(B + Cx)}{(1 - By)^2 - Bx^2} H(x, y),$$

$$\frac{\partial H}{\partial y} = \frac{1}{t_1 - t_2}\left(\frac{\partial H}{\partial t_1} - \frac{\partial H}{\partial t_2}\right) = \frac{(1 - By)(B + Cx) + Bx(A + Cy)}{(1 - By)^2 - Bx^2} H(x, y).$$

These equations can be written in matrix form as

$$\begin{pmatrix} \partial H/\partial x \\ \partial H/\partial y \end{pmatrix} = \frac{1}{\det \mathbf{M}} \mathbf{M}\begin{pmatrix} A + Cy \\ B + Cx \end{pmatrix} H, \quad \text{where } \mathbf{M} = \begin{pmatrix} 1 - By & x \\ Bx & 1 - By \end{pmatrix}.$$

Inverting, it follows that

$$-Bx\frac{\partial H}{\partial x} + (1 - By)\frac{\partial H}{\partial y} = (B + Cx)H(x, y),$$

$$(1 - By)\frac{\partial H}{\partial x} - x\frac{\partial H}{\partial y} = (A + Cy)H(x, y).$$

From these equations we obtain two recurrence relations (in which we take $\xi_{r,s} = 0$ if r or $s < 0$):

$$s\xi_{r,s} = -(r + s)B\xi_{r,s-1} - C\xi_{r-1,s-1}, \tag{12.38}$$

$$(r + 1)\xi_{r+1,s} - (r + 1)B\xi_{r+1,s-1} - (s + 1)\xi_{r-1,s+1} = A\xi_{r,s} + C\xi_{r,s-1}. \tag{12.39}$$

The first of these determines all the polynomials $\xi_{r,s}$, given the boundary polynomials $\xi_{r,0}$ and $\xi_{0,s}$. It also determines $\xi_{0,s} = (-B)^s$ (since $\xi_{0,0} = 1$). The boundary polynomials $\xi_{r,0}$, on the other hand, are equal to ξ_r, as we have observed in Remark 12.42, or as can be seen using (12.38) (with $s = 1$) and (12.39) (with $s = 0$), which together show that $\xi_{r,0}$ satisfies the recurrence relation (12.33). The polynomials $\xi_{r,s} \in \mathbb{Q}[A, B, C]$ are therefore completely determined.

Proposition 12.43.

(i) *As a vector space over $\mathbb{Q}$ the polynomial ring $\mathbb{Q}[A, B, C]$ has basis $\xi_{r,s}(A, B, C)C^t$, where r, s, t range through all nonnegative integers.*

(ii) *The subset consisting of $\xi_{r,s}(A, B, C)C^t$ for $r + s + t \geq g$ is a basis for the ideal $I_g \subset \mathbb{Q}[A, B, C]$.*

Proof. For both parts the idea is exactly the same as for Proposition 12.36.

(i) We order monomials lexicographically with $A > B > C$ and note that, by (12.33), the maximal monomial appearing (with nonzero coefficient) in $\xi_r(A, B, C)$ is A^r. Hence by the recurrence relation (12.38) and the fact that $\xi_{r,0} = \xi_r$ (Remark 12.42), $A^r B^s$ appears as the maximal monomial in $\xi_{r,s}(A, B, C)$, and this implies that the set of $\xi_{r,s}(A, B, C)C^t$ is a basis of the polyomial ring.

(ii) Denote by $J_g \subset \mathbb{Q}[A, B, C]$ the vector subspace spanned by $\xi_{r,s}(A, B, C)C^t$ with $r + s + t \geq g$. This is an ideal: it is clear that $CJ_g \subset J_g$, and $AJ_g \subset J_g$, $BJ_g \subset J_g$ follow from (12.33) and (12.38).

Since $\xi_r \in J_g$ for all $r \geq g$, it follows that $I_g \subset J_g$. So it just remains to see that $\xi_{r,s}(A, B, C)C^t \in I_g$ for all $r + s + t \geq g$. This can be shown by a double induction using Remark 12.42. $\qquad\square$

It follows from this proposition that the residue ring $\mathbb{Q}[A, B, C]/I_g$ has a basis

$$\xi_{r,s}(A, B, C)C^t \quad \mod I_g, \qquad r+s+t \leq g-1. \qquad (12.40)$$

The basis element with (strictly) highest degree is C^{g-1} with degree $3g-3$ (noting that $\deg \xi_{r,s}(A, B, C)C^t = r + 2s + 3t$), and we see that C^{g-1} spans all monomials of degree $3g-3$ modulo I_g.

Proposition 12.44. *Let $m, n \geq 0$ be nonnegative integers satisfying $m + 2n = 3g - 3$. Then*

$$A^m B^n \equiv (-1)^n b_{g-1-n} \frac{m!}{g!} C^{g-1} \quad \mod I_g,$$

where $b_k \in \mathbb{Q}$ is a rational number defined by the Taylor expansion $x/\sin x = \sum_k b_k x^{2k}$ when $k \geq 0$ and is zero when $k < 0$.

Proof. We introduce a polynomial ring in one variable $\mathbb{Q}[T]$ and define a linear map

$$E : \mathbb{Q}[A, B, C] \rightarrow \mathbb{Q}[T], \qquad f \mapsto \sum_{h=0}^{\infty} E_h(f)T^h, \qquad (12.41)$$

where E_h is a linear map $\mathbb{Q}[A, B, C] \rightarrow \mathbb{Q}$ defined on monomials by

$$E_h : A^m B^n C^p \mapsto \begin{cases} (-1)^n m! p! \binom{h+1}{p} b_{h-n-p} & \text{if } m + 2n + 3p = 3h, \\ 0 & \text{otherwise.} \end{cases} \qquad (12.42)$$

(Note that (12.41) is a finite sum since, for any polynomial f, the number $E_h(f)$ is nonzero for only finitely many h.)

Claim: Under the linear map $\mathbb{Q}[A, B, C][[x, y]] \rightarrow \mathbb{Q}[T][[x, y]]$ induced by E, the generating function $H(x, y)$ of (12.37) transforms to the constant $1 \in \mathbb{Q}[T][[x, y]]$.

The claim shows that E kills all $\xi_{r,s}(A, B, C)$ with r, s not both zero. By Proposition 12.43 and (12.40) it follows that E descends to the residue ring

$$\mathbb{Q}[A, B, C]/I_g \rightarrow \mathbb{Q}[T],$$

with the 1-dimensional image coming from the component in the top degree $\langle C^{g-1}\rangle$. The proof of the proposition is then completed by applying E to each of C^{g-1} and $A^m B^n$ (that is, applying E_{g-1} to each of these):

$$E : C^{g-1} \mapsto g! T^{g-1}, \qquad A^m B^n \mapsto (-1)^n m! b_{g-1-n} T^{g-1}.$$

Since both monomials belong, modulo I_g, to the span $\langle C^{g-1} \rangle$, on which E is injective, the proposition follows.

Before proving the claim, we will write down the map E in terms of A, B, C^*, where $C^* := C + AB$. This is given by

$$E_h(A^m B^n (C^*)^p) = \sum_{j=0}^{p} \binom{p}{j} E_h(A^{m+j} B^{n+j} C^{p-j})$$

$$= (-1)^{n+p} p! m! b_{h-n-p} \sum_{j=0}^{p} (-1)^{p-j} \binom{m+j}{j} \binom{h+1}{p-j}.$$

We now use the binomial coefficient identity

$$\sum_{j=0}^{p} (-1)^{p-j} \binom{a+j}{j} \binom{b+p}{p-j} = \binom{a-b}{p} \tag{12.43}$$

with $a = m, b = h + 1 - p$ to deduce that (12.42) is equivalent to the following. We write $k := h - n - p$.

$$E_h : A^m B^n (C^*)^p \mapsto \begin{cases} \dfrac{(-1)^k m! (2k-1)!}{(2k-1-p)!} b_k & \text{if } m + 2n + 3p = 3h, \\ 0 & \text{otherwise.} \end{cases} \tag{12.44}$$

We now prove the claim. First, substituting (12.34) into the defining formulae (12.37) gives

$$H(x, y) = \frac{1}{\sqrt{(1 - By)^2 - Bx^2}} \left(\frac{1 + x\sqrt{B} - By}{1 - x\sqrt{B} - By} \right)^{C^*/2B\sqrt{B}}$$

$$= \frac{1}{(1 - By)\sqrt{1 - \frac{Bx^2}{(1-By)^2}}} \exp\left\{ -\frac{Cx}{B} + \frac{C^*}{B\sqrt{B}} \tanh^{-1} \frac{x\sqrt{B}}{1 - By} \right\}$$

$$= \frac{\sinh \theta}{x\sqrt{B}} \exp\left\{ Ax + \left(\frac{\theta}{B\sqrt{B}} - \frac{x}{B} \right) C^* \right\},$$

where we have set

$$\theta = \theta(B) := \tanh^{-1} \frac{x\sqrt{B}}{1 - By}.$$

From (12.44) we see that

$$E\left(B^n e^{Ax+C^*z}\right) = x^n T^n \frac{\sqrt{x^3 T}}{\sinh\left(\sqrt{x^3 T}(1+zT)\right)}.$$

So for general power series $f(B)$, $w(B)$ we have

$$E\left(f(B)e^{Ax+C^*w(B)}\right) = f(xT)\frac{\sqrt{x^3 T}}{\sinh\left(\sqrt{x^3 T}\,(1+w(xT)T)\right)}.$$

Applying this to

$$f(B) = \frac{\sinh\theta(B)}{x\sqrt{B}}, \qquad w(B) = \frac{\theta(B)}{B\sqrt{B}} - \frac{x}{B},$$

we obtain $E\left(H(x, y)\right) = 1$, as claimed. □

We can now prove the intersection formula (12.4) and the Verlinde formula (12.2). We will compute the right-hand side of Proposition 12.33 by the same methods that we used for (12.36).

First of all, since the Newstead classes α, β, γ satisfy the Mumford relations, it follows from Proposition 12.44 that they satisfy

$$(\alpha^m \beta^n) = (-1)^n b_{g-1-n} \frac{m!}{g!}(2\gamma)^{g-1}.$$

This identity implies that

$$\int_{\mathcal{N}} e^{k\alpha}(-\beta)^n = ck^{3g-3-2n}b_{g-1-n} = ck^{g-1}\operatorname*{Res}_{x=0}\left[\frac{x^{2n}k\,dx}{x^{2g-2}\sin kx}\right],$$

where $c := (2\gamma)^{g-1}/g!$. An arbitrary power series $f(x)$ therefore satisfies

$$\int_{\mathcal{N}} e^{k\alpha} f(-\beta) = ck^{g-1}\operatorname*{Res}_{x=0}\left[\frac{f(x^2)k\,dx}{x^{2g-2}\sin kx}\right].$$

In particular, if we take

$$f(x) = \left(\frac{\sqrt{x}/2}{\sin\sqrt{x}/2}\right)^{2g-2}$$

and apply Corollary 12.6, then we obtain:

$$\int_{\mathcal{N}} e^{k\alpha} \left(\frac{\sqrt{\beta}/2}{\sinh \sqrt{\beta}/2} \right)^{2g-2} = \frac{ck^{g-1}}{4^{g-1}} \operatorname*{Res}_{x=0} \left[\frac{k\,dx}{\sin kx \, \sin^{2g-2}(x/2)} \right]$$

$$= \frac{ck^{g-1}}{4^{g-1}} V_{g-1}^{-}(2k).$$

On the other hand, irreducibility of $\mathcal{N}$ (see Section 11.2(a)) implies that when $k = 1$ the left-hand side of Proposition 12.33 is 1, and hence $c = 4^{g-1}$. From this and Proposition 12.33, the formulae (12.2), (12.4) and also

$$(\gamma^{g-1}) = g!2^{g-1}$$

all follow.

12.5 An excursion: the Verlinde formula for quasiparabolic bundles

The Verlinde formula has various generalisations. One can replace vector bundles of rank 2 by higher rank bundles, or one can replace the structure group $SL(2)$ by more general algebraic groups. Here, however, we shall generalise the curve C to a curve with marked points and prove the formula (12.6).

(a) Quasiparabolic vector bundles

On the projective line $\mathbb{P}^1$ there do not exist stable vector bundles of rank greater than 1 (Theorem 10.31). However, there do exist stable quasiparabolic vector bundles, and one can construct a moduli space for these.

Definition 12.45.

(i) A *quasiparabolic vector bundle* of rank r on a curve C is a pair consisting of a vector bundle E of rank r and a subsheaf $E' \subset E$ which is a vector bundle of the same rank r.

(ii) The *degree* of a quasiparabolic vector bundle is

$$\deg(E' \subset E) := \tfrac{1}{2}(\deg E' + \deg E).$$

(iii) A rank 2 quasiparabolic vector bundle $E' \subset E$ is *stable* if every line subbundle $\xi \subset E$ satisfies

$$\deg(\xi' \subset \xi) < \tfrac{1}{2} \deg(E' \subset E), \quad \text{where } \xi' := \xi \cap E'.$$

(As usual, if $<$ is replaced by $\leq$, then $E' \subset E$ is *semistable*.) $\square$

A quasiparabolic vector bundle $(E' \subset E)$ has a determinant (quasiparabolic) line bundle

$$\det(E' \subset E) := (\det E' \subset \det E).$$

We necessarily have

$$\det E' = \det E \otimes \mathcal{O}_C(-D)$$

for some positive divisor D, and this divisor is uniquely determined by the inclusion homomorphism $E' \subset E$. In what follows we will consider the case where $D = p_1 + \cdots + p_n$ for distinct points $p_1, \ldots, p_n \in C$. The quasiparabolic bundle $E' \subset E$ is then equivalent to data consisting of the vector bundle E together with a codimension 1 vector subspace $U_i \subset E_{p_i}$ in the fibre of E at each point $p_1, \ldots, p_n$ (Figure 12.2).

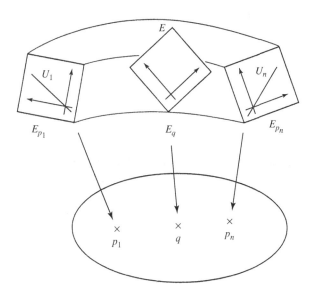

Figure 12.2: A rank 2 quasiparabolic vector bundle

For a fixed quasiparabolic line bundle $L' \subset L$ we denote:

$$SU_C^{\mathrm{par}}(2, L' \subset L) := \left\{ \begin{array}{l} \text{stable quasiparabolic rank 2} \\ \text{vector bundles } E' \subset E \\ \text{with } \det(E' \subset E) \cong (L' \subset \\ L) \end{array} \right\} \Big/ \text{isomorphism}.$$

If $(L' \subset L) = (L(-p_1 - \cdots - p_n) \subset L)$ (with the standard inclusion – that is, as the sheaf of local sections of L vanishing at the points p_i), then $E' \subset E$ can be viewed as a quasiparabolic bundle on the marked curve $(C; p_1, \ldots, p_n)$. In this case we shall also use the notation $SU_C^{\mathrm{par}}(2, L; p_1, \ldots, p_n)$.

Theorem 12.46. (Mehta-Seshadri [102] for $g \geq 2$; **Bauer [90]** when $g = 0$.) *Let $p_1, \ldots, p_n \in C$ be distinct points on a curve C of genus g. Then there exists a nonsingular quasiprojective variety $\mathcal{N}_{g,n}$ of dimension $3g - 3 + n$ with $SU_C^{\mathrm{par}}(2, L; p_1, \ldots, p_n)$ as its set of points and with the following properties.*

   (i) *If n is odd, then $\mathcal{N}_{g,n}$ represents the moduli functor for a stable, quasi-parabolic of rank 2 and determinant $L(-p_1 - \cdots - p_n) \subset L$. This means that there exists a universal quasiparabolic bundle $\mathcal{U}' \subset \mathcal{U}$ on $C \times \mathcal{N}_{g,n}$, from which every family of stable quasiparabolic bundles is obtained (up to equivalence) as a pull-back. Moreover, $\mathcal{N}_{g,n}$ is projective in this case.*

  (ii) *If n is even, then $\mathcal{N}_{g,n}$ is a best approximation to the functor for stable quasiparabolic bundles.*

 (iii) *At the point $[E' \subset E] \in \mathcal{N}_{g,n}$ corresponding to a stable quasiparabolic bundle $(E' \subset E)$ the tangent space is isomorphic to*

$$T_{[E' \subset E]} \mathcal{N}_{g,n} = H^1(C, \mathfrak{sl}(E' \subset E)).$$

□

Remark 12.47. Although we will not carry this out here, we will indicate briefly how one can prove this theorem in the same spirit as the construction of $SU_C(2, L)$ as a quotient variety in Chapter 10. There, we represented $SU_C(2, L)$ as the underlying points of $\mathrm{Alt}_{N,2}(H^0(L)) /\!/ GL(N)$, where $\mathrm{Alt}_{N,2}(H^0(L))$ consists of $N \times N$ skew-symmetric matrices with entries in $H^0(L)$ and rank ≤ 2 over the function field $k(C)$. We will modify this by considering the diagonal action of $GL(N)$ on

$$\mathrm{Alt}_{N,2}(H^0(L)) \times \underbrace{\mathbb{A}^N \times \cdots \times \mathbb{A}^N}_{n \text{ times}},$$

where each $\mathbb{A}^N$ has the natural linear action of $GL(N)$. In this product consider the subset

$$\mathrm{Alt}_{N,2}^{p_1, \ldots, p_n}(H^0(L)) = \{(T, u_1, \ldots, u_n) \mid \mathrm{ev}_{p_i}(T) \wedge u_i = 0 \text{ for all } i\}.$$

It is easy to see that this is an affine subvariety which is preserved by the $GL(N)$ action. Moreover, this action extends to the larger group $G = GL(N) \times \mathbb{G}_m \times \cdots \times \mathbb{G}_m$, where $\mathbb{G}_m \times \cdots \times \mathbb{G}_m$ acts diagonally on $\mathbb{A}^N \times \cdots \times \mathbb{A}^N$ by homotheties. With these definitions one can show that $\mathcal{N}_{g,n}$ is the quotient variety $\mathrm{Alt}_{N,2}^{p_1, \ldots, p_n}(H^0(L)) /\!/ G$.

□

The last statement in Theorem 12.46 has the following meaning. Let $\mathcal{E}nd(E' \subset E)$ be the vector bundle of (local) endomorphisms of E which map the subsheaf E' to itself. The trace of an element of $\mathcal{E}nd(E' \subset E)$ takes values in the line bundle $\mathcal{O}_C(-p_1 - \cdots - p_n)$, and there is a direct sum decomposition

$$\mathcal{E}nd(E' \subset E) \cong \mathcal{O}_C(-p_1 - \cdots - p_n) \oplus \mathfrak{sl}(E' \subset E),$$

where $\mathfrak{sl}(E' \subset E)$ denotes the subbundle of tracefree endomorphisms.

Part (iii) of the theorem globalises as follows.

Proposition 12.48. *When n is odd the tangent bundle of $\mathcal{N}_{g,n}$ is isomorphic to the 1st direct image:*

$$T_{\mathcal{N}_{g,n}} = R^1 \pi_* \mathfrak{sl}(\mathcal{U}' \subset \mathcal{U}), \qquad \pi : C \times \mathcal{N}_{g,n} \to \mathcal{N}_{g,n},$$

where $(\mathcal{U}' \subset \mathcal{U})$ is the universal quasiparabolic bundle. $\square$

Let $E' \subset E$ and $F' \subset F$ be two rank 2 quasiparabolic vector bundles.

Definition 12.49. A *quasiparabolic homomorphism* from $E' \subset E$ to $F' \subset F$ is a bundle homomorphism $E \to F$ which maps $E' \to F'$; we denote the space of quasiparabolic homomorphisms by $\mathrm{Hom}\big((E' \subset E), (F' \subset F)\big)$. These are the global sections of a subsheaf

$$\mathcal{H}om\big((E' \subset E), (F' \subset F)\big) \subset \mathcal{H}om(E, F)$$

consisting of local homomorphisms which take $E' \to F'$. This is a vector bundle of rank 4. $\square$

The following is a quasiparabolic version of Exercise 10.6.

Lemma 12.50. *Let $E' \subset E$ and $F' \subset F$ be semistable rank 2 quasiparabolic vector bundles with $\deg(E' \subset E) > \deg(F' \subset F)$. Then $\mathrm{Hom}\big((E' \subset E), (F' \subset F)\big) = 0$.* $\square$

Suppose that $E' \subset E$ is a rank 2 quasiparabolic bundle with $\det E' = \det E \otimes \mathcal{O}_C(-p_1 - \cdots - p_n)$. Then $E'' := E \otimes \mathcal{O}_C(-p_1 - \cdots - p_n)$ is a subsheaf of E' and so we obtain a new quasiparabolic bundle $E'' \subset E'$. It is then easy to show the following.

Lemma 12.51. *$E' \subset E$ is (semi)stable if and only if $E'' \subset E'$ is (semi)stable.* $\square$

Since $\deg(E'' \subset E') = \deg(E' \subset E) - n$, we can put these two lemmas together to obtain:

Proposition 12.52. *Suppose that a rank 2 quasiparabolic bundle $E' \subset E$ is semistable and that ξ is a line bundle with $\deg \xi > n/2$, where $\det E' = \det E \otimes \mathcal{O}_C(-p_1 - \cdots - p_n)$. Then*

$$\mathrm{Hom}\left((E'' \subset E') \otimes \xi, (E' \subset E)\right) = 0.$$

$\square$

Now consider the 4-dimensional vector space $\mathrm{End}(E_{\mathrm{gen}})$ over $k(C)$. This is self-dual via the trace inner product

$$\mathrm{End}(E_{\mathrm{gen}}) \times \mathrm{End}(E_{\mathrm{gen}}) \to \mathcal{O}_C, \qquad (f, g) \mapsto \mathrm{tr}\, fg.$$

The two vector bundles $\mathcal{E}nd(E' \subset E)$ and $\mathcal{H}om\left((E'' \subset E'), (E' \subset E)\right)$ both have $\mathrm{End}(E_{\mathrm{gen}})$ as their total set and, in fact, they are mutually dual with respect to this inner product. Combining this observation with the duality Theorem 10.11 we deduce:

Corollary 12.53. *If $E' \subset E$ is a semistable rank 2 quasiparabolic bundle and ξ is any line bundle with $\deg \xi > n/2$, where $\det E' = \det E \otimes \mathcal{O}_C(-p_1 - \cdots - p_n)$, then*

$$H^1\left(C, \mathcal{E}nd(E' \subset E) \otimes \Omega_C \otimes \xi\right) = 0.$$

$\square$

(b) A proof of (12.6) using Riemann-Roch and the Mumford relations

From now on we take $C = \mathbb{P}^1$ and fix an odd number $n \geq 3$, and n points $p_1, \ldots, p_n \in \mathbb{P}^1$. By Theorem 12.46 we have a moduli space $\mathcal{N}_{0,n}$ for rank 2 stable quasiparabolic vector bundles on $\mathbb{P}^1$, and $\mathcal{N}_{0,n}$ is a nonsingular projective variety of dimension $n - 3$. Moreover, there exists a universal quasiparabolic bundle $\mathcal{U}' \subset \mathcal{U}$ on $\mathbb{P}^1 \times \mathcal{N}_{0,n}$, and this determines cohomology classes $\alpha_0 \in H^2(\mathcal{N}_{0,n})$ and $\beta \in H^4(\mathcal{N}_{0,n})$ by

$$c_2(\mathcal{E}nd\,\mathcal{U}) = 2h \otimes \alpha_0 - 1 \otimes \beta, \qquad (12.45)$$

where $h \in H^2(\mathbb{P}^1)$ is the class of a point in $\mathbb{P}^1$ (that is, $h = c_1(\mathcal{O}_{\mathbb{P}^1}(1))$).

For each of the points $p_1, \ldots, p_n \in \mathbb{P}^1$ we can consider the restriction $\mathcal{U}|_{p_i \times \mathcal{N}_{0,n}}$. In this restriction the subsheaf $\mathcal{U}'$ defines a line subbundle and its quotient:

$$0 \to \mathcal{M}_i^{\mathrm{sub}} \to \mathcal{U}|_{p_i \times \mathcal{N}_{0,n}} \to \mathcal{M}_i^{\mathrm{quo}} \to 0,$$

and for each $i = 1, \ldots, n$ these line bundles determine a class

$$\delta_i := c_1(\mathcal{M}_i^{\mathrm{quo}}) - c_1(\mathcal{M}_i^{\mathrm{sub}}) \in H^2(\mathcal{N}_{0,n}).$$

By definition of β (12.45) it follows that

$$\delta_i^2 = \beta. \tag{12.46}$$

We will also put

$$\alpha := 2\alpha_0 + \sum_{i=1}^n \delta_i. \tag{12.47}$$

We now apply Grothendieck-Riemann-Roch to the rank 3 vector bundle $\mathfrak{sl}(\mathcal{U}' \subset \mathcal{U})$ on the product $\mathbb{P}^1 \times \mathcal{N}_{0,n}$. Using the exact sequence (alternatively, one can just compute directly)

$$0 \to \pi_* \mathfrak{sl}\, \mathcal{U}' \to \bigoplus_{i=1}^n (\mathcal{M}_i^{\mathrm{sub}})^{-1} \otimes \mathcal{M}_i^{\mathrm{quo}} \to R^1 \pi_* \mathfrak{sl}(\mathcal{U}' \subset \mathcal{U}) \to 0 \tag{12.48}$$

we obtain

$$\mathrm{td}\left(R^1 \pi_* \mathfrak{sl}(\mathcal{U}' \subset \mathcal{U})\right) = \mathrm{td}(\pi_* \mathfrak{sl}\, \mathcal{U}')^{-1} \prod_{i=1}^n \mathrm{td}\left((\mathcal{M}_i^{\mathrm{sub}})^{-1} \otimes \mathcal{M}_i^{\mathrm{quo}}\right)$$

$$= e^{\alpha_0} \left(\frac{\sqrt{\beta}/2}{\sinh \sqrt{\beta}/2}\right)^2 e^{\sum \delta_i/2} \left(\frac{\sqrt{\beta}/2}{\sinh \sqrt{\beta}/2}\right)^n$$

$$= e^{\alpha/2} \left(\frac{\sqrt{\beta}/2}{\sinh \sqrt{\beta}/2}\right)^{n-2}.$$

By Proposition 12.48 we note that this is equal to the Todd class of $\mathcal{N}_{0,n}$, and in particular $\alpha = c_1(T_{\mathcal{N}_{0,n}})$. Thus, by Hirzebruch-Riemann-Roch:

$$\chi\left(\mathcal{N}_{0,n}, \mathcal{O}(-lK)\right) = \int_{\mathcal{N}} e^{l\alpha} \mathrm{td}\, \mathcal{N}_{0,n} = \int_{\mathcal{N}} e^{(2l+1)\alpha/2} \left(\frac{\sqrt{\beta}/2}{\sinh \sqrt{\beta}/2}\right)^{n-2}. \tag{12.49}$$

We will now write $n = 2g + 1$ (where $g \geq 1$) and consider the direct image sheaf

$$W := \pi_*\left(\mathfrak{sl}(\mathcal{U}' \subset \mathcal{U}) \otimes \mathcal{O}_{\mathbb{P}^1}(g-1)\right).$$

This is a vector bundle by Corollary 12.53, and by Grothendieck-Riemann-Roch (we leave the computation to the reader)

$$\frac{c'(W)}{c(W)} = \frac{\alpha + \beta}{1 - \beta}.$$

The classes α, β therefore satisfy the relations of the secant ring R_g (Section 12.4(a)). In other words, since by Riemann-Roch the vector bundle W has rank $g - 1$, we see that

$$\xi_r(\alpha, \beta) = 0 \quad \text{for all } r \geq g.$$

One can call these the *quasiparabolic Mumford relations* (and, by the usual recurrence relations, they are generated by the two relations for $r = g, g + 1$). As we saw in Proposition 12.40 (taking $i = 0$), these relations imply that

$$(\alpha^j \beta^k) = E_{2g-2-k}(\beta^{g-1}) \quad \text{for } j + 2k = 2g - 2.$$

Hence, using (12.49) and (12.36) we obtain

$$\chi\left(\mathcal{N}_{0,2g+1}, \mathcal{O}(-lK)\right) = \int_{\mathcal{N}} e^{(2l+1)\alpha/2} \left(\frac{\sqrt{\beta}/2}{\sinh \sqrt{\beta}/2}\right)^{2g-1}$$

$$= \frac{1}{2l+1} U_{2g-1}(2l+1)(\beta/4)^{g-1}.$$

We shall see shortly (Theorem 12.56) that the varieties $\mathcal{N}_{0,n}$ are rational. This implies that $\chi\left(\mathcal{N}_{0,2g+1}, \mathcal{O}\right) = 1$ (Remark 12.16), and so the case $l = 0$ of the above formula says that $(\beta)^{g-1} = 4^{g-1}$. From this we obtain the quasiparabolic Verlinde formula (12.6)

$$\chi\left(\mathcal{N}_{0,2g+1}, \mathcal{O}(-lK)\right) = \frac{1}{2l+1} \sum_{j=0}^{2l} \frac{(-1)^j}{\left(\sin \frac{2j+1}{4l+2}\pi\right)^{2g-1}}.$$

On the other hand, the line bundle $\mathcal{O}(-K)$ is ample, and so by Kodaira Vanishing 12.11 this formula actually computes, for $l \geq 0$, the dimension of the space $H^0(\mathcal{N}_{0,2g+1}, \mathcal{O}(-lK))$.

Remark 12.54. Here is a table of low values of $h^0\left(\mathcal{N}_{0,n}, \mathcal{O}(-lD)\right)$ coming from the Verlinde formula (12.6), where D generates the Picard group of

$\mathcal{N}_{0,n}$ – that is, $D = K$ when n is odd, and $D = K/2$ when n is even:

n \ l	0	1	2	3	4	5	6	7	8
3	1	1	1	1	1	1	1	1	1
4	1	2	3	4	5	6	7	8	9
5	1	5	13	25	41	61	85	113	145
6	1	4	11	24	45	76	119	$\cdots$	
7	1	21	141	521	1401	3101	$\cdots$		
8	1	8	43	160	461	$\cdots$			
9	1	85	1485	10569	$\cdots$				
10	1	16	171	$\cdots$					
11	1	341	15565	$\cdots$					

Notice that at level 1:

$$h^0\left(\mathcal{N}_{0,n}, \mathcal{O}(-D)\right) = \begin{cases} 2^8 & \text{when } n = 2g + 2, \\ 1 + 4 + 4^2 + \cdots + 4^{g-1} & \text{when } n = 2g + 1. \end{cases}$$

Similarly, at level 2:

$$h^0\left(\mathcal{N}_{0,n}, \mathcal{O}(-2D)\right) = \begin{cases} 1 + \sum_{j=1}^{g} 2^{2j-1} & \text{when } n = 2g + 2, \\ 1 + 4\sum_{j=1}^{g} \binom{2g+1}{2j} 5^{j-1} & \text{when } n = 2g + 1. \end{cases}$$

(c) Birational geometry

One can express Definition 12.45(iii) by saying that 'stability of a quasiparabolic vector bundle $E' \subset E$ is the average of stability of the vector bundles E and E''. Generalising this idea, one can consider a 'weighted average' of the stabilities of E and E'. In other words, one can define a degree

$$\deg_a(E' \subset E) = (1 - a)\deg E + a \deg E', \qquad a \in \mathbb{R},$$

and then define stability accordingly for each value of the parameter $a \in \mathbb{R}$. If we note that $\deg_a(E' \subset E) = \deg E - an$, where $n = \deg E - \deg E'$, then we see that there is a further generalisation. Namely, suppose that $\det E' = \mathcal{O}_C(-p_1 - \cdots - p_n) \otimes \det E$, and attach a number $a_i \in \mathbb{R}$ (called a *weight*) to each point $p_i \in C$. Then we can define a degree

$$\deg_\mathbf{a}(E' \subset E) := \deg E - \sum_{i=1}^{n} a_i \mathrm{length}_{p_i}(E/E'), \qquad \text{where } \mathbf{a} = (a_1, \ldots, a_n).$$

We then define stability for the pair $E' \subset E$ using this notion of degree.

Definition 12.55.

(i) A quasiparabolic vector bundle $E' \subset E$ together with assigned weights $a_1, \ldots, a_n \in \mathbb{R}$ at the points $p_1, \ldots, p_n \in C$, where $\det E' = \mathcal{O}_C(-p_1 - \cdots - p_n) \otimes \det E$, is called a *parabolic vector bundle*.

(ii) A parabolic vector bundle $(E' \subset E; a_1, \ldots, a_n)$ of rank 2 is *stable* if, for every line subbundle $\xi \subset E$,

$$\deg_{\mathbf{a}}(\xi' \subset \xi) < \tfrac{1}{2} \deg_{\mathbf{a}}(E' \subset E), \qquad \text{where } \xi' = \xi \cap E'.$$

□

Note that if the weights a_i are all 0, then this coincides with the stability of the vector bundle E; if all $a_i = 1$, then it coincides with the stability of E'; and if all $a_i = 1/2$, then it coincides with the stability of $E' \subset E$ as a quasiparabolic bundle (Definition 12.45).

Of course, the key point here is that there should exist a moduli space for parabolic vector bundles with fixed determinant $\det(E' \subset E)$ and weights $\mathbf{a} = (a_1, \ldots, a_n)$, and this is indeed the case, by Mehta and Seshadri [102].

Now restrict, as in the last section, to the projective line $C = \mathbb{P}^1$, with an odd number $n = 2g + 1$ of points $p_1, \ldots, p_{2g+1} \in \mathbb{P}^1$ fixed. We let $\overline{p}_1, \ldots, \overline{p}_{2g+1} \in \mathbb{P}^{2g-2}$ be the images of $p_1, \ldots, p_{2g+1}$ under the Veronese embedding $v : \mathbb{P}^1 \hookrightarrow \mathbb{P}^{2g-2}$ of degree $2g - 2$. For $\mathbf{a} \in \mathbb{R}^{2g+1}$, let $\mathcal{N}_{0,\mathbf{a}}$ be the moduli space of rank 2 stable parabolic bundles on $(\mathbb{P}^1; p_1, \ldots, p_{2g+1})$ with weights $\mathbf{a}$.

Theorem 12.56 (Bauer [90]). *For g suitably chosen weights $\mathbf{a}^{(0)}, \mathbf{a}^{(1)}, \ldots,$ $\mathbf{a}^{(g-1)} \in \mathbb{R}^{2g+1}$ the moduli spaces*

$$\mathcal{N}_{0,\mathbf{a}^{(i)}} =: \widetilde{\mathbb{P}}^{2g-2}_{(i)}, \qquad 0 \le i \le g - 1,$$

are nonsingular projective varieties of dimension $2g - 2$ and have the following structure.

(i) *$\widetilde{\mathbb{P}}^{2g-2}_{(0)}$ is isomorphic to $\mathbb{P}^{2g-2}$.*

(ii) *$\widetilde{\mathbb{P}}^{2g-2}_{(1)}$ is the blow-up of $\mathbb{P}^{2g-2}$ at the points $\overline{p}_1, \ldots, \overline{p}_{2g+1}$.*

(iii) *$\widetilde{\mathbb{P}}^{2g-2}_{(i+1)}$ is obtained from $\widetilde{\mathbb{P}}^{2g-2}_{(i)}$ by the following flop (or more precisely, a reverse flip). (See Section 6.3(a).) $\widetilde{\mathbb{P}}^{2g-2}_{(i)}$ is blown up along a finite number*

$$\binom{2g + 1}{i + 1} + \binom{2g + 1}{i - 1} + \binom{2g + 1}{i - 3} + \cdots$$

of subvarieties $\cong \mathbb{P}^i$, and then the exceptional divisors are contracted down in a different direction to subvarieties $\cong \mathbb{P}^{2g-3-i}$ in $\widetilde{\mathbb{P}}^{2g-2}_{(i+1)}$.

(iv) $\widetilde{\mathbb{P}}^{2g-2}_{(g-1)}$ is isomorphic to the moduli space of quasiparabolic bundles $\mathcal{N}_{0,2g+1}$.

□

$$\widetilde{\mathbb{P}}^{2g-2}_{(1)} \longleftrightarrow \widetilde{\mathbb{P}}^{2g-2}_{(2)} \longleftrightarrow \cdots \longleftrightarrow \widetilde{\mathbb{P}}^{2g-2}_{(g-2)} \longleftrightarrow \widetilde{\mathbb{P}}^{2g-2}_{(g-1)} = \mathcal{N}_{0,2g+1}$$

$\downarrow$ Big Bang

$\mathbb{P}^{2g-2}$

Examples 12.57. $g = 2$. In this case $\mathcal{N}_{0,5}$ is the blow-up of $\mathbb{P}^2$ at 5 points. This is called the *quartic del Pezzo surface* and embeds in $\mathbb{P}^4$ as a complete intersection of two quadrics.

$g = 3$. In this case $\mathcal{N}_{0,7}$ is obtained from the blow-up $\widetilde{\mathbb{P}}^4_{(1)}$ of $\mathbb{P}^4$ at 7 points by flopping along 22 lines, each line transforming to a $\mathbb{P}^2$ in the moduli space. These lines are

(a) the proper transforms of the $\binom{7}{2} = 21$ lines in $\mathbb{P}^4$ joining pairs of the points $p_1, \ldots, p_7$; and

(b) the proper transform of the rational normal quartic $v(\mathbb{P}^1) \subset \mathbb{P}^4$ through the 7 points.

$g = 4$. The moduli space $\mathcal{N}_{0,9}$ is obtained from $\mathbb{P}^6$ as follows. First blow up at 9 points. Then flop along the following 37 lines, transforming them into $\mathbb{P}^4$s:

(a) the proper transforms of the $\binom{9}{2} = 36$ lines in $\mathbb{P}^6$ joining pairs of the 9 points; and

(b) the proper transform of the rational normal sextic $v(\mathbb{P}^1) \subset \mathbb{P}^6$ through the 9 points.

Finally, flop along the following 93 $\mathbb{P}^2$s, transforming them into $\mathbb{P}^3$s:

(c) the proper transforms of the $\binom{9}{3} = 84$ planes in $\mathbb{P}^6$ containing 3 of the 9 points; and

(d) the proper transforms of the 9 cones over $v(\mathbb{P}^1) \subset \mathbb{P}^6$ with vertices at each of the 9 points.

□

An alternative proof of the Verlinde formula for quasiparabolic bundles might be given by analysing how the Euler-Poincaré characteristic $\chi(\mathcal{N}_{0,\mathbf{a}^{(i)}}, \mathcal{O}(-lK))$ changes under the flops described by Theorem 12.56. (This is a parabolic version of Thaddeus's proof [112] of the Verlinde formula (12.2).)

We will not do this here, but just remark on the following interesting consequence. The sequence of flops determines a birational isomorphism between $\widetilde{\mathbb{P}}^{2g-2}_{(1)}$ and $\mathcal{N}_{0,2g+1}$, which is an isomorphism in codimension 1. This gives us an isomorphism between spaces of global sections of corresponding line bundles, and in particular it gives us an equality:

$$\dim H^0(\widetilde{\mathbb{P}}^{2g-2}_{(1)}, \mathcal{O}(-lK)) = \frac{1}{2l+1} \sum_{i=0}^{2l} \frac{(-1)^i}{\left(\sin \frac{2i+1}{4l+2}\pi\right)^{2g-1}} = \frac{U_{2g-1}(2l+1)}{2l+1}.$$

(12.50)

This gives the dimension of the vector space of forms of degree $l(2g-1)$ in the homogeneous coordinates $x_0, x_1, \ldots, x_{2g-2}$ which vanish at the points $\overline{p}_1, \ldots, \overline{p}_{2g+1} \in \mathbb{P}^{2g-2}$.

Now, in the case of the plane (that is, $g = 2$) such linear systems blowing up finite sets of points have been thoroughly studied classically; for ≤ 8 points in a general position, for example, the resulting surfaces are the *del Pezzo surfaces*, which play an important role in various different geometrical contexts. For $\mathbb{P}^3$, too, the blow-ups along finite point sets are also fairly well understood. (See, for example, Semple and Roth [107] chapter 8 §2.) However, for $\mathbb{P}^4$ and higher, such an extraordinary formula as (12.50) is completely unexpected and is, I think, a truly remarkable discovery. Yet all that is needed is to pass to the nice model $\mathcal{N}_{0,2g+1}$, and the dimension formula simply reduces, using Kodaira vanishing, to a Riemann-Roch calculation.

Bibliography

We list here some useful references that came to mind, but the author has made no attempt at completeness.

Chapter 1

[1] J.W.S. Cassels: *Lectures on Elliptic Curves*, LMS Student Texts **24**, Cambridge University Press 1991.
[2] J.A. Dieudonné, J.B. Carrel: *Invariant Theory Old and New*, Academic Press 1970.
[3] J. Fogarty: *Invariant Theory*, W. A. Benjamin, New York-Amsterdam 1969.
[4] F. Klein: *The Icosahedron and the Solution of Equations of the Fifth Degree* (1884) Dover edition 1956.
[5] F. Klein: The development of mathematics in the 19th century (1928) translated by M. Ackerman, in *Lie Groups: History, Frontiers and Applications, Volume IX*, R. Hermann, Math. Sci. Press 1979.
[6] V.L. Popov, E.B. Vinberg: Invariant theory, in *Algebraic Geometry IV*, ed. A.N. Parshin, I.R. Shafarevich, *Encyclopaedia of Mathematical Sciences* **55**, Springer-Verlag 1994.
[7] J.-P. Serre: *A Course in Arithmetic*, Graduate Texts in Mathematics **7**, Springer-Verlag 1973.
[8] T.A. Springer: *Invariant Theory*, Lecture Notes in Mathematics **585**, Springer-Verlag 1977.

Chapter 2

A standard text on rings and modules is:

[9] M.F. Atiyah, I.G. Macdonald: *Introduction to Commutative Algebra*, Addison-Wesley, 1969.
[10] S. Mukai: Counterexample to Hilbert's Fourteenth Problem for the 3-dimensional additive group, *RIMS preprint* **1343**, Kyoto (2001).
[11] M. Nagata: On the fourteenth problem of Hilbert, *International Congress of Mathematicians*, Edinburgh 1958.
[12] M. Nagata: On the 14th problem of Hilbert, *Amer. J. Math.* **81** (1959) 766–722.
[13] M. Nagata: Note on orbit spaces, *Osaka Math. J.* **14** (1962) 21–31.
[14] D. Rees: On a problem of Zariski, *Illinois J. Math.* **2** (1958) 145–149.

[15] C.S. Seshadri: On a theorem of Weitzenböck in invariant theory, *J. Math. Kyoto Univ.* **1** (1962) 403–409.

[16] R. Weitzenböck: Über die Invarianten von Linearen Gruppen, *Acta Math.* **58** (1932) 230–250.

Chapter 3

A good introduction to algebraic varieties is

[17] G. Kempf: *Algebraic Varieties*, London Mathematical Society Lecture Note Series, **172**, Cambridge University Press 1993.

Chapter 4

A useful collection of Hilbert's papers can be found in:

[18] D. Hilbert: Hilbert's invariant theory papers, translated by M. Ackerman, comments by R. Hermann, in *Lie Groups: History, Frontiers and Applications Volume VIII*, Math Sci Press 1978.

[19] D. Hilbert: Über die Theorie der algebraischen Formen, *Math. Annalen* **36** (1890) 473–534.

[20] D. Hilbert: Über die vollen Invariantensysteme, *Math. Annalen* **42** (1893) 313–373.

[21] S. Aronhold: Zur Theorie der homogenen Functionen dritten Grades von drei Variablen, *Crelle J.* **39** (1850) 140–159.

[22] D.A. Buchsbaum, D. Eisenbud: Algebra structures for finite free resolutions, and some structure theorems for ideals of codimension 3, *Amer. J. Math.*, **99** (1977) 447–485.

[23] A. Cayley: A second memoir upon quantics, *Collected Mathematical Papers II*, Cambridge (1889) 250–275.

[24] J. Dixmier, D. Lazard: Minimum number of fundamental invariants for the binary form of degree 7, *J. Symbolic Computation* **6** (1988) 113–115.

[25] M. Nagata: Complete reducibility of rational representations of a matrix group, *J. Math. Kyoto Univ.* **1** (1961) 87–99.

[26] I. Schur: *Vorlesungen über Invarianttheorie*, Grundlehren 143, Springer-Verlag 1968.

[27] T. Shioda: On the graded ring of invariants of binary octavics, *Amer. J. Math.* **89** (1967) 1022–1046.

Cayley's Ω process is explained in chapter 4 of:

[28] B. Sturmfells: *Algorithms in Invariant Theory*, Springer-Verlag 1993.

[29] B. Sturmfells: *Groebner Bases and Convex Polytopes*, University Lecture Series **8**, American Mathematical Society 1996.

Chapter 5

The standard text on quotient varieties is:

[30] D. Mumford, J. Fogarty, F. Kirwan: *Geometric Invariant Theory*, Springer-Verlag (1965) 3rd edition 1994.

[31] D. Mumford: Projective invariants of projective structures and applications, in *Proceedings of the International Congress of Mathematicians, Stockholm 1962*, 526–530.

[32] H. Matsumura. P. Monsky: On the automorphisms of hypersurfaces, *J. Math. Kyoto Univ.* **3** (1964) 347–361.

[33] E.B. Elliot: *An Introduction to the Algebra of Quantics*, Chelsea, 1895, 2nd edition reprinted 1964.

[34] C. Jordan: Mémoire sur l'equivalence des formes, *J. École Polytechnique* **48** (1880) 112–150.

[35] Salmon: *Higher Plane Curves*, Cambridge University Press 1873.

Chapter 6

Toric varieties as quotient varieties:

[36] D. Cox: The homogeneous coordinate ring of a toric variety, *J. Alg. Geom.*, **8** (1995) 17–50.

For the general theory of toric varieties, see:

[37] W. Fulton: *An Introduction to Toric Varieties*, Princeton University Press 1993.

[38] G. Gelfand, M.M. Kapranov, A.V. Zelevinsky: *Discriminants, Resultants and Multidimensional Determinants*, Birkhäuser, 1994.

For flips, flops and minimal models:

[39] J. Kollár, S. Mori: *Birational Geometry of Algebraic Varieties*, Cambridge Tracts in Mathematics **134**, Cambridge University Press 1998.

[40] McDuff, D. Salamon: *Introduction to Symplectic Topology*, Oxford Mathematical Monographs, Oxford University Press, 1998.

[41] T. Oda: *Convex bodies and Algebraic Geometry: an introduction to the theory of toric varieties*, Springer-Verlag 1988.

Chapter 7

[42] I. Dolgachev, D. Ortland: *Point sets in projective spaces and theta functions*, *Astérisque* **165** (1988).

[43] D. Allcock: The moduli space of cubic threefolds, *Astérisque* **165** (1988).

[44] D. Gieseker: Geometric invariant theory and applications to moduli spaces, preprint (2001).

[45] J. Harris, I. Morrison: *Moduli of Curves*, Springer-Verlag 1998.

Moduli can also be constructed without the use of invariant theory. See:

[46] J. Kollár: Projectivity of complete moduli, *J. Diff. Geom.* **32** (1990) 235–268.

Numerical semistability was first introduced by Hilbert [20], and later generalised by Mumford [30]. For its application to the moduli of curves, see:

[47] D. Mumford: Stability of projective varieties, *l'Enseign. Math.*, **23** (1977) 39–110.

[48] M. Schlessinger: Functors of Artin rings, *Trans. Am. Math. Soc.*, **130** (1968) 208–222.

Related to the topic of Section 7.2(b) is

[49] M. Yokoyama: Stability of cubic 3-folds, *Tokyo J. Math.*, to appear.

Chapter 8

[50] Z.I. Borevich, I.R. Shafarevich: *Number Theory*, Academic Press, 1966.

[51] R. Bott, L. Tu: *Differential Forms in Algebraic Topology*, Springer Graduate Texts in Mathematics **82**, Springer-Verlag, 2nd edition 1995.

[52] C. Chevalley: Sur les décompositions cellulaire des éspàces G/B, (with a preface by A. Borel) *Proc. Symp. Pure Math.*, **56** (1994) Part 1, 1–23.
A delightful reference for Catalan (and many other) numbers is:

[53] J.H. Conway, R.K. Guy: *The Book of Numbers*, Copernicus, 1996.
For Grassmannians, see Chapter 1, §5 of:

[54] P.A. Griffiths, J. Harris: *Principles of Algebraic Geometry*, Wiley, 1978.
Or chapter 14 of:

[55] W.V.D. Hodge, D. Pedoe: *Methods of Algebraic Geometry*, Cambridge University Press 1952.

[56] H. Hiller: Combinatorics and intersections of Schubert varieties, *Comment. Math. Helv.*, **57** (1982) 41–59.

[57] J. Igusa: On the arithmetic normality of the Grassmann variety, *Proc. Nat. Acad. Sci. USA*, **40** (1954) 309–313.

[58] R. Stanley: *Enumerative Combinatorics*, vol 2, Cambridge University Press 1999.

[59] O. Taussky: Introduction into connections between algebraic number theory and integral matrices, Appendix in H. Cohn, *A Classical Invitation to Algebraic Numbers and Class Fields*, Universitex, Springer-Verlag 1978.

[60] H. Weyl: *The Classical Groups*, Princeton University Press 1939.

Chapter 9

[61] D. Eisenbud: *Commutative Algebra with a view toward Algebrtaic Geometry*, Graduate Texts in Mathematics **150**, Springer-Verlag 1995.

[62] A. Grothendieck: *Fondaments de la Géometrie Algébrique*, collected Bourbaki talks, Paris 1962.

[63] K. Iwasawa: *Algebraic Function Theory*, Iwanami 1952 (Japanese).

[64] D. Mumford: *Lectures on Curves on an Algebriac Surface*, Princeton University Press 1960.
For curves, Jacobians and theta functions, see:

[65] D. Mumford: *Tata lectures on Theta II*, Birkhäuser 1984.

[66] S. Bosch, W. Lütkenbohmert, M. Raynaud: *Neron Models*, Springer-Verlag 1990.

[67] R.J. Walker: *Algebraic Curves*, Princeton University Press 1950.

[68] A. Weil: *Zur algebraischen Theorie der algebraischen Functionen, Collected Papers I* [1938b] 227–231, Springer-Verlag 1979.

[69] T. Kawada: *Introduction to the Theory of Algebraic Curves*, Shibundo 1968 (Japanese).
An alternative proof of the Riemann-Roch Theorem can be found in chapter 8 of Kempf [17].

Chapter 10

[70] M.F. Atiyah: On the Krull-Schmidt theorem with application to sheaves, *Bull. Soc. Math. France*, **84** (1956) 307–317.

[71] M.F. Atiyah: Vector bundles over an elliptic curve, *Proc. London Math. Soc.*, 7 (1957) 414–452.

[72] D. Gieseker: On the moduli of vector bundles on an algebraic surface, *Ann. Math.* **106** (1977) 45–60.

For stable vector bundles, unitary representations of the fundamental group, and construction of moduli:

[73] M.S. Narasimhan, C.S. Seshadri: Stable and unitary vector bundles on a compact Riemann surface, *Ann. Math.* **82** (1965) 540–567.

[74] S. Ramanan: The moduli space of vector bundles over an algebraic curve, *Math. Annalen* **200** (1973) 69–84.

The generalised theta divisor first appeared in:

[75] M. Raynaud: Sections des fibrés vectoriels sur une courbe, *Bull. Soc. Math. France* **110** (1982) 103–125.

[76] C. Seshadri: Fibrés vectoriels sur les courbes algébriques, *Astérisque*, **96**, 1982.

[77] C. Seshadri: Vector bundles on curves, *Contemporary Math.* **153** (1993) 163–200.

Chapter 11

For spectral curves, see:

[78] A. Beauville: Jacobiennes des courbes spectrales et systéme hamiltonian complétement intégrables, *Acta Math.* **164** (1990) 211–235.

[79] A. Beauville, M.S. Narasimhan, S. Ramanan: Spectral curves and the generalised theta divisor, *J. für Reine Angew. Math.* **398** (1989) 169–179.

[80] R.W.T.H. Hudson: *Kummer's Quartic Surface*, with a foreword by W. Barth, revised reprint of the 1905 original, Cambridge Mathematical Library, Cambridge University Press 1990.

[81] D. Mumford: *Abelian Varieties*, Tata Institute of Fundamental Research Studies in Mathematics **5**, Oxford University Press, 1970.

[82] M.S. Narasimhan, S. Ramanan: Geometry of Hecke cycles I, in *Ramanujan – A Tribute*, 291–345, Springer 1978.

[83] P.E. Newstead: Stable bundles of rank 2 and odd degree over a curve of genus 2 *Topology* **7** (1968) 205–215.

[84] P.E. Newstead: *Introduction to Moduli Problems and Orbit Spaces*, Tata Institute of Fundamental Research Lectures on Mathematics and Physics, 51, Narosa Publishing House 1978.

[85] S. Seshadri: Theory of Moduli, *Proc. Symp. Pure Math.* **29** (1975) 225–304.

[86] S. Seshadri: Desingularisation of the moduli varieties of vector bundles on curves, *Algebraic Geometry* 155–184, Kinokuniya 1977.

[87] Y. Teranishi: The ring of invariants of matrices, *Nagoya Math. J.* **104** (1986) 149–161.

[88] A. Weil: Remarques sur un mémoire d'Hermite, *Arch. d. Math.* 5 (1954) 197–202.

Chapter 12

[89] M.F. Atiyah, R. Bott: The Yang-Mills equations over Riemann surfaces, *Phil. Trans. Roy. Soc. London* **A 308** (1982) 337–345.

[90] S. Bauer: Parabolic bundles, elliptic surfaces and $SU(2)$ representation spaces of genus zero Fuchsian groups, *Math. Annalen* **290** (1991) 509–526.

[91] U.V. Desale, S. Ramanan: Classification of vector bundles of rank 2 on hyperelliptic curves, *Inventiones* **38** (1976) 161–185.

[92] S.K. Donaldson: A new proof of a theorem of Narasimhan and Seshadri, *J. Diff. Geom.* **18** (1983) no. 2, 269–277.

[93] G. Faltings: Stable G-bundles and projective connections, *J. Algebraic Geom.* **2** (1993) no. 3, 507–568.

[94] W. Fulton: *Intersection Theory*, 2nd edition. Ergebnisse der Mathematik und ihrer Grenzgebiete **2**, Springer-Verlag 1998.

[95] M. Furuta, B. Steer: Seifert fibred homology 3-spheres and the Yang-Mills equations on Riemann surfaces with marked points, *Adv. Math.* **96** (1992) 38–102.

[96] G. Harder: Eine Bemerkung zu einer Arbeit von P.E. Newstead, *J. für Math.* **242** (1970) 16–25.

[97] G. Harder, M.S. Narasimhan: On the cohomology groups of moduli spaces of vector bundles on curves, *Math. Annalen* **212** (1975) 215–248.

[98] F. Hirzebruch: *Topological Methods in Algebraic Geometry*, 3rd edition, Springer-Verlag 1966.

[99] Y.-H. Kiem: The equivariant cohomology ring of the moduli space of vector bundles over a Riemann surface, *Geometry and Topology: Aarhus (1998)*, *Contemp. Math.* **258** (2000) 249–261.

[100] D.E. Knuth, J. Buckholtz: Computation of tangent, Euler and Bernoulli numbers, *Math. Comp.* **21** (1967) 663–688.

[101] I.G. Macdonald: *Symmetric functions and Hall polynomials*, 2nd edition, with contributions by A. Zelevinsky, Oxford Mathematical Monographs, Oxford University Press 1995.

On the moduli space of parabolic vector bundles:

[102] V. Mehta, C.S. Seshadri: Moduli of bundles on curves with parabolic structure, *Math. Ann.* **248** (1980) 205–239.

For moduli, topology and the Weil conjectures, pre-Verlinde:

[103] P.E. Newstead: Topological properties of some spaces of stable bundles, *Topology* **6** (1967) 241–262.

[104] P.E. Newstead: Characteristic classes of stable bundles of rank 2 on an algebraic curve, *Trans. AMS* **169** (1968) 337–345.

[105] D.G. Northcott: *Finite Free Resolutions*, Cambridge University Press 1976.

[106] C. Pauly: Espaces de modules de fibrés paraboliques et blocs conformes, *Duke Math. J.* **84** (1996), no. 1, 217–235.

A classical reference for the birational geometry of rational threefolds is:

[107] J.G. Semple, L. Roth: *Introduction to Algebraic Geometry*, Oxford University Press 1949.

[108] B. Siebert, G. Tian: Recursive relations for the cohomology ring of moduli spaces of stable bundles, *Turkish J. Math.* **19** (1995) no. 2, 131–144.

[109] A. Szenes: Hilbert polynomials of moduli spaces of rank 2 vector bundles I, *Topology* **32** (1993) 587–597.

[110] A. Bertram, A. Szenes: Hilbert polynomials of moduli spaces of rank 2 vector bundles II, *Topology* **32** (1993) 599–609.

[111] M. Thaddeus: Conformal field theory and the cohomology ring of the moduli spaces of stable bundles, *J. Differential Geometry* **35** (1992) 131–149.

An algebraic geometric proof of the Verlinde formula is given in:

[112] M. Thaddeus: Stable pairs, linear systems and the Verlinde formula, *Invent. Math.* **117** (1994) 317–353.

[113] K. Ueno: *Advances in Moduli Theory*, AMS Translations of Mathematical Monographs **206** (2002).

The Verlinde formulae appeared originally in:

[114] E. Verlinde: Fusion rules and modular transformations in $2d$-conformal field theory, *Nuclear Physics B* **306** (1988) 360–376.

[115] D. Zagier: On the cohomology of moduli spaces of rank two vector bundles over curves, in *The Moduli Space of Curves*, R. Dijgraaf *et al.* (eds), Birkhäuser 1995, 533–563.

For a spectacular account of the Verlinde formula, inverse trigonometric series, and their relationship with Euler and Bernoulli numbers, see:

[116] D. Zagier: Elementary aspects of the Verlinde formula and of the Narasimhan-Harder-Atiyah-Bott formula, *Israel Math. Conf. Proc.* **9** (1996) 445–462.

Index

1-parameter subgroup (1-PS), 110, 212
  normalised, 224

A_5, 17
affine quotient map, 159
Abel, xix
Abel's Theorem, 288, 334, 345, 347
Abel-Jacobi map, 287, 334
abelian differential, 337
abelian varieties, xv, 399
action of an affine algebraic group, 103
action of ray type, 322
adjoint of a bundle map, 456
adjoint functors, 286
adjoint representation, 127
adjugate matrix, 58
affine algebraic group, 100
affine algebraic variety, 85
affine charts, 95
affine curves, 288
affine hyperelliptic curve, 234, 271
affine hypersurface, 81
affine Jacobian
  of a plane curve, 424
  of a spectral curve, 424
affine line, 172
  is not complete, 104
affine quotient map, 164, 176, 182, 184, 190,
  315, 402
affine space, 77, 78, 158, 163, 181
affine variety, 81, 123
algebraic curve, xvii, xviii
algebraic de Rham cohomology, 327, 332
algebraic family of vector bundles on C
  parametrised by Spm A, 408
algebraic function field, xvii
  of Spm R, 85
algebraic group, xvii, 77, 100, 399
  is always nonsingular, 124
algebraic Jacobian, 399

algebraic number field, 270, 397
algebraic representation, 117
algebraic torus, 122
  is linearly reductive, 131
algebraic variety, xvi, 77, 91, 398
alternating group, 10
ample, 447
analytic Jacobian, 341, 452
anharmonic, 32
Apollonius, 7
approximation of the quotient functor, 402
Approximation Theorem, 112
arithmetic genus, 287, 293, 337
arithmetic Jacobian, 426
Artin ring, 312
ascending chain condition, 54
automorphic form, 43
averaging map, 14
axis of a double point, 224

Base Change Theorem, 400, 405
base point free, 308
basic open set, 79
best approximation (by an algebraic variety),
  399, 402
Betti numbers, 336
Big Bang, 485
bigraded ring, 73
  support of, 70
bilinear relations for differentials of divisor
  type, 344–345
bilinear relations for differentials of the second
  kind, 339–341
binary dihedral group, 17
binary forms, 139, 177, 196, 219
binary icosahedral group, 17
binary octics, 177
binary polyhedral groups, 18
binary quartic, 1, 26, 29, 172, 176, 220
  classical invariants of, 27

Printed in the United States
By Bookmasters